Methods in Enzymology

Volume 189
RETINOIDS
Part A
Molecular and Metabolic Aspects

METHODS IN ENZYMOLOGY

EDITORS-IN-CHIEF

John N. Abelson   Melvin I. Simon

DIVISION OF BIOLOGY
CALIFORNIA INSTITUTE OF TECHNOLOGY
PASADENA, CALIFORNIA

FOUNDING EDITORS

Sidney P. Colowick and Nathan O. Kaplan

Methods in Enzymology

Volume 189

Retinoids

Part A

Molecular and Metabolic Aspects

EDITED BY

Lester Packer

DEPARTMENT OF MOLECULAR AND CELL BIOLOGY
UNIVERSITY OF CALIFORNIA, BERKELEY
BERKELEY, CALIFORNIA

Editorial Advisory Board

ACADEMIC PRESS, INC.
Harcourt Brace Jovanovich, Publishers
San Diego   New York   Boston
London   Sydney   Tokyo   Toronto

Academic Press, Inc.
San Diego, California 92101

United Kingdom Edition published by
Academic Press Limited
24-28 Oval Road, London NW1 7DX

Library of Congress Catalog Card Number: 54-9110

ISBN   0-12-182090-4   (alk. paper)

Printed in the United States of America
90 91 92 93   9 8 7 6 5 4 3 2 1

Table of Contents

Section I. Structure and Analysis

A. Structure

B. Analysis

Section II. Receptors, Transport, and Binding Proteins

A. Extracellular

B. Intracellular

C. Specific Methods

Contributors to Volume 189

Article numbers are in parentheses following the names of contributors.
Affiliations listed are current.

PATRICIO ABARZÚA (35), *Department of Oncology and Virology, Roche Research Center, Nutley, New Jersey 07110*

ALICE J. ADLER (21), *Eye Research Institute, Boston, Massachusetts 02114*

N. R. AL-MALLAH (16), *Laboratoire de Pharmacocinétique et Toxicocinétique, Faculté de Pharmacie, 13385 Marseille, Cedex 5, France*

C. AUBERT (16), *Laboratoire de Pharmacocinétique et Toxicocinétique, Faculté de Pharmacie, 13385 Marseille, Cedex 5, France*

JOHN STUART BAILEY (37), *Banting and Best Department of Medical Research, University of Toronto, Toronto, Ontario M5G 1L6, Canada*

MARK D. BALL (50), *Department of Chemistry, Rose–Hulman Institute of Technology, Terre Haute, Indiana 47803*

FAAN WEN BANGERTER (44), *Department of Biological Chemistry and Molecular Pharmacology, Harvard Medical School, Boston, Massachusetts 02115*

ARUN B. BARUA (13), *Department of Biochemistry and Biophysics, Iowa State University, Ames, Iowa 50011*

DAVID A. BERNLOHR (38), *Department of Biochemistry, University of Minnesota, St. Paul, Minnesota 55108*

PAUL S. BERNSTEIN (56), *Jules Stein Eye Institute, School of Medicine, University of California at Los Angeles, Los Angeles, California 90024*

HANS K. BIESALSKI (18), *Department of Physiological Chemistry and Pathobiochemistry, University of Mainz, D-6500 Mainz, Federal Republic of Germany*

WILLIAM S. BLANER (19, 27), *Institute of Human Nutrition, Columbia University, New York, New York 10032*

THEODORE R. BREITMAN (23), *Laboratory of Biological Chemistry, Division of Cancer Treatment, National Cancer Institute, National Institutes of Health, Bethesda, Maryland 20892*

C. D. B. BRIDGES (5, 20, 33), *Department of Biological Sciences, Purdue University, West Lafayette, Indiana 47907*

MELISSA K. BUELT (38), *Department of Biochemistry, University of Minnesota, St. Paul, Minnesota 55108*

H. BUN (16), *Laboratoire de Pharmacocinétique et Toxicocinétique, Faculté de Pharmacie, 13385 Marseille, Cedex 5, France*

CHRISTER BUSCH (32), *Department of Pathology, Uppsala University, S-751 85 Uppsala, Sweden*

LOUISE M. CANFIELD (45), *Department of Biochemistry, University of Arizona, Tucson, Arizona 85721*

J. P. CANO (16), *Sanofi Recherche, 34082 Montpellier, France*

GERALD J. CHADER (21), *Laboratory of Retinal Cell and Molecular Biology, National Eye Institute, National Institutes of Health, Bethesda, Maryland 20892*

RONALD F. CHILDS (11), *Department of Chemistry, McMaster University, Hamilton, Ontario L8S 4L8, Canada*

LAURIE L. CHINANDER (38), *Department of Biochemistry, University of Minnesota, St. Paul, Minnesota 55108*

ANDREW J. CLIFFORD (9), *Department of Nutrition, University of California at Davis, Davis, California 95616*

DALE A. COOPER (55), *Edible Oil Products Division, Procter & Gamble Company, Cincinnati, Ohio 45224*

FRANS J. M. DAEMEN (43), *Department of Biochemistry, University of Nijmegen, 6500 HB Nijmegen, The Netherlands*

ANN K. DALY (24, 31), *Department of Pharmacological Sciences, University of Newcastle upon Tyne, Medical School, Newcastle upon Tyne NE2 4HH, England*

M. I. DAWSON (2), *Life Sciences Division, SRI International, Menlo Park, California 94025*

WILLEM J. DE GRIP (43), *Department of Biochemistry, University of Nijmegen, 6500 HB Nijmegen, The Netherlands*

ANDREW P. DE LEENHEER (4, 10), *Laboratoria voor Medische Biochemie en voor Klinische Analyse, Rijksuniversiteit Gent, B-9000 Gent, Belgium*

ULF ERIKSSON (32), *Ludwig Institute for Cancer Research, Stockholm Branch, S-104 01 Stockholm, Sweden*

RONALD M. EVANS (22), *Howard Hughes Medical Institute, The Salk Institute for Biological Studies, La Jolla, California 92138*

GÖRAN FEX (42), *Department of Clinical Chemistry, University Hospital of Lund, S-221 85 Lund, Sweden*

SHAO-LING FONG (20, 33), *Department of Ophthalmology, Indiana University, Indianapolis, Indiana 46202*

THOMAS A. FRITZ (45), *Department of Biochemistry, University of Arizona, Tucson, Arizona 85721*

KEIKO FUNA (32), *Ludwig Institute for Cancer Research, University Hospital, S-751 85 Uppsala, Sweden*

HAROLD C. FURR (8, 9) *Department of Biochemistry and Biophysics, Iowa State University, Ames, Iowa 50011*

L. E. GERBER (47), *Department of Food Science and Nutrition, University of Rhode Island, Kingston, Rhode Island 02881*

VINCENT GIGUÈRE (22), *Division of Endocrinology, Research Institute, The Hospital for Sick Children, Toronto, Ontario M5G 1X8, Canada*

DEWITT S. GOODMAN (19, 29), *Department of Medicine, Columbia University, College of Physicians and Surgeons, New York, New York 10032*

JEFFERY I. GORDON (58), *Departments of Medicine, Biochemistry and Molecular Biophysics, Washington University School of Medicine, St. Louis, Missouri 63110*

GERHART GRAUPNER (26), *La Jolla Cancer Research Foundation, La Jolla, California 92126*

JOSEPH F. GRIPPO (25), *Department of Toxicology and Pathology, Hoffman-La Roche Ltd., Nutley, New Jersey 07110*

MARY LOU GUBLER (60), *Department of Oncology and Virology, Roche Research Center, Nutley, New Jersey 07110*

EARL H. HARRISON (52), *Department of Physiology and Biochemistry, The Medical College of Pennsylvania, Philadelphia, Pennsylvania 19129*

THOMAS HERMANN (26), *La Jolla Cancer Research Foundation, La Jolla, California 92126*

DONALD L. HILL (3), *Biochemistry Department, Southern Research Institute, Birmingham, Alabama 35255*

P. D. HOBBS (2), *Life Sciences Division, SRI International, Menlo Park, California 94025*

ANTON M. JETTEN (25), *Cell Biology Group, Laboratory of Pulmonary Pathobiology, National Institute of Environmental Health Sciences, Research Triangle Park, North Carolina 27709*

GUNVOR JOHANNESSON (42), *Department of Clinical Chemistry, University Hospital of Lund, S-221 85 Lund, Sweden*

A. DANIEL JONES (9), *Facility for Advanced Instrumentation, University of California at Davis, Davis, California 95616*

T. ALWYN JONES (28), *Institute for Molecular Biology, Biomedical Center, Uppsala University, S-751 24 Uppsala, Sweden*

PERE JULIÀ (48), *Departament de Bioquímica i Biologia Molecular, Facultat de Ciències, Universitat Autònoma de Barcelona, 08193 Barcelona, Spain*

LAWRENCE A. KAPLAN (15), *Department of Pathology, Bellevue Hospital Medical Center, New York, New York 10016*

MICHIMASA KATO (34), *First Department of Internal Medicine, Gifu University School of Medicine, Tsukasa-machi 40, Gifu 500, Japan*

MICHAEL KLAUS (1), *Department of Pharmaceutical Research, Hoffmann-La Roche Ltd., CH-4002 Basel, Switzerland*

M. RAJ LAKSHMAN (46), *Lipid Research Laboratory, Veterans Administration Medical Center, Washington, D.C. 20422*

WILLY E. LAMBERT (10), *Laboratoria voor Medische Biochemie en voor Klinische Analyse, Rijksuniversiteit Gent, B-9000 Gent, Belgium*

GARY M. LANDERS (6), *Jules Stein Eye Institute, UCLA School of Medicine, Los Angeles, California 90024*

JUERGEN M. LEHMANN (26), *La Jolla Cancer Research Foundation, La Jolla, California 92126*

M. A. LEO (54, 59), *Department of Medicine, Mount Sinai School of Medicine, City University of New York, New York, New York 10021*

MARC S. LEVIN (58), *Department of Medicine, Washington University School of Medicine, St. Louis, Missouri 63110*

ELLEN LI (58), *Departments of Medicine, Biochemistry and Molecular Biophysics, Washington University School of Medicine, St. Louis, Missouri 63110*

C. S. LIEBER (54, 59), *Department of Medicine and Pathology, Mount Sinai School of Medicine, City University of New York, New York, New York 10021*

ZHENG-SHI LIN (33), *Department of Neurobiology and Behavior, State University of New York at Stony Brook, Stony Brook, New York 11794*

MARIA A. LIVREA (39, 57), *Istituto di Chimica Biologica, Università di Palermo Policlinico, 90127 Palermo, Italy*

WILLIAM A. MACCREHAN (17), *Organic Analytical Research Division, Center for Analytical Chemistry, National Institute of Standards and Technology, Gaithersburg, Maryland 20899*

PAUL N. MACDONALD (51), *Department of Biochemistry, University of Arizona, Tucson, Arizona 85724*

BERNARD MARTIN (24), *Centre International De Recherches Dermatologiques, Sophia Antipolis, 06565 Valbomme, Cedex, France*

VALERIE MATARESE (38), *Department of Biochemistry, University of Minnesota, St. Paul, Minnesota 55108*

JUDITH A. MILLER (15), *Medical Research Laboratories, Cincinnati, Ohio 45219*

YASUTOSHI MUTO (34), *First Department of Internal Medicine, Gifu University School of Medicine, Tsukasa-machi 40, Gifu 500, Japan*

JOSEPH L. NAPOLI (52, 53), *Department of Biochemistry, School of Medicine and Biomedical Sciences, The State University of New York at Buffalo, Buffalo, New York 14214*

H. J. NELIS (4), *Laboratoria voor Medische Biochemie en voor Klinische Analyse, Rijksuniversiteit Gent, B-9000 Gent, Belgium*

CLARA NERVI (25), *Cell Biology Group, Laboratory of Pulmonary Pathobiology, National Institute of Environmental Health Sciences, Research Triangle Park, North Carolina 27709*

MARCIA NEWCOMER (28), *Department of Biochemistry, Vanderbilt University School of Medicine, Nashville, Tennessee 37232*

HANS NORDLINDER (32), *Department of Pathology, Uppsala University, S-751 85 Uppsala, Sweden*

MASATAKA OKUNO (34), *First Department of Internal Medicine, Gifu University School of Medicine, Tsukasa-machi 40, Gifu 500, Japan*

JAMES ALLEN OLSON (46), *Department of Biochemistry and Biophysics, Iowa State University, Ames, Iowa 50011*

DAVID E. ONG (51), *Department of Biochemistry, Vanderbilt University, Nashville, Tennessee 37232*

XAVIER PARÉS (48), *Departament de Bioquímica i Biologia Molecular, Facultat de Ciències, Universitat Autònoma de Barcelona, 08193 Barcelona, Spain*

MAGNUS PFAHL (26), *La Jolla Cancer Research Foundation, La Jolla, California 92126*

ROBERT R. RANDO (44, 56), *Department of Biological Chemistry and Molecular Pharmacology, Harvard Medical School, Boston, Massachusetts 02115*

CHRISTOPHER P. F. REDFERN (24, 31), *Department of Dermatology, University of Newcastle upon Tyne, Newcastle upon Tyne NE1 4LP, England*

A. CATHARINE ROSS (7, 49), *Department of Physiology and Biochemistry, Medical College of Pennsylvania, Philadelphia, Pennsylvania 19129*

PUSPHA SAKENA (32), *Department of Pathology, Uppsala University, Uppsala, Sweden*

BRAHMA P. SANI (3, 36), *Biochemistry Department, Southern Research Institute, Birmingham, Alabama 35255*

GARY S. SHAW (11), *Department of Biochemistry, University of Alberta, Edmonton, Alberta T6G 2H7, Canada*

MICHAEL I. SHERMAN (35, 60), *Department of Oncology and Virology, Roche Research Center, Nutley, New Jersey 07110*

FRIEDRICH SIEBERT (12), *Max-Planck-Institut für Biophysik, D-6000 Frankfurt am Main, Federal Republic of Germany*

GEORGES SIEGENTHALER (30, 61), *Clinique Dermatologie, Hôpital Cantonal Universitaire, CH-1211 Geneva, Switzerland*

K. L. SIMPSON (47), *Department of Food Science and Nutrition, University of Rhode Island, Kingston, Rhode Island 02881*

CHI-HUNG SIU (37), *Banting and Best Department of Medical Research, University of Toronto, Toronto, Ontario M5G 1L6, Canada*

DIANNE ROBERT SOPRANO (29), *Department of Biochemistry, Temple University School of Medicine, Philadelphia, Pennsylvania 19140*

MEIR J. STAMPFER (15), *Department of Epidemiology, Harvard Medical School of Public Health, and The Channing Laboratory, Harvard Medical School and Brigham and Woman's Hospital, Boston, Massachusetts 02115*

EVAN A. STEIN (15), *Medical Research Laboratories, Cincinnati, Ohio 45219*

WILLIAM STILLWELL (40, 41), *Department of Biology, Indiana University—Purdue University at Indianapolis, Indianapolis, Indiana 46205*

NORIKO TAKAHASHI (23), *Laboratory of Biological Chemistry, Division of Cancer Treatment, National Cancer Institute, National Institutes of Health, Bethesda, Maryland 20892*

THOMAS E. TARARA (45), *Department of Biochemistry, University of Arizona, Tucson, Arizona 85721*

LUISA TESORIERE (39, 57), *Istituto di Chimica Biologica, Universitá di Palermo Policlinico, 90127 Palermo, Italy*

MATY TZUKERMAN (26), *La Jolla Cancer Research Foundation, La Jolla, California 92126*

STEPHEN R. WASSALL (40, 41), *Department of Physics, Indiana University—Purdue University at Indianapolis, Indianapolis, Indiana 46205*

BARBARA WIGGERT (21), *Laboratory of Retinal Cell and Molecular Biology, National Eye Institute, National Institutes of Health, Bethesda, Maryland 20892*

KEN N. WILLS (26), *La Jolla Cancer Research Foundation, La Jolla, California 92126*

RONALD WYSS (14), *Department of Drug Metabolism, Pharmaceutical Research, Hoffmann-La Roche Ltd., CH-4002 Basel, Switzerland*

XIAO-KUN ZHANG (26), *La Jolla Cancer Research Foundation, La Jolla, California 92126*

Preface

Spectacular progress and unprecedented interest in the field of retinoids prompted us to consider this topic for two volumes in the *Methods in Enzymology* series: Volume 189, Retinoids, Part A: Molecular and Metabolic Aspects and Volume 190, Retinoids, Part B: Cell Differentiation and Clinical Applications.

From a historical perspective we know that studies in the 1930s showed that vitamin A (retinol) and retinal had a role in the visual process. It was also recognized that some link between vitamin A and cancer incidence existed. Several decades ago it was discovered that retinoic acid had a dramatic effect on the chemically induced DMBA mouse skin carcinogenesis model in which enormous reductions in the tumor burden were observed. This led to the realization that retinoids had important effects on cell differentiation. This resulted almost immediately in the synthesis and evaluation of new retinoids. Indeed, the effects of retinoids on cell differentiation appear to be more universal and of greater importance than their light-dependent role in vision and microbial energy transduction.

Progress has been rapid, and the importance of accurate methodology for this field is imperative to its further development. The importance of methodology applies to the use of retinoids in basic research in molecular, cellular, and developmental biology, and in clinical medicine. In medicine, applications have been mainly to cancer and in dermatology to the treatment of skin diseases and skin aging. As new retinoids are being tested in biological models and in clinical medicine, interest in the nutrition and pharmacology of retinoids has arisen. Moreover, the beneficial effects of retinoids in pharmacological treatment have led to a recognition of the "double-edged sword" of toxicity (teratogenicity).

In Section I of this volume, Structure and Analysis, retinoid structure, stability, photosensitivity characteristics, and analytical procedures are included. Section II, Receptors, Transport, and Binding Proteins, focuses on protein isolation and the current status of methods for characterization, assay, and distribution. As the role of binding proteins is not completely clear and new ones are still being identified, the methods for investigating their presence are of importance. In particular, different types of nuclear receptors for retinoic acid having sequence homologies with hormones have increased interest in this area, and the role of nuclear receptors in retinoid-modulated gene expression is an area that is under rapid development. In Section III, Enzymology and Metabolism, retinoid enzymology and analogies with retinoproteins in nature, such as in bacterial systems, are included. The behavior of retinoids in membranes is a subject of importance, particularly for pharmacological uses when retinoids may not be bound to a protein.

Volume 190 will cover the effects of retinoids in various cell differentiation systems and nutritional and pharmacological methods.

I am very grateful to the Advisory Board—Frank Chytil, DeWitt Goodman, Maria A. Livrea, Leonard Milstone, Concetta Nicotra, James A. Olson, and Stanley S. Shapiro—for their unique input, advice, counsel, and encouragement in the planning and organization of this volume. In most instances, I met with every member of the board on one or more occasions to discuss the topics and to identify the most important contributors. Indeed, we found almost universal acceptance, and virtually no one turned down our invitation to contribute to this volume. In fact it was somewhat autocatalytic in that many contributors, realizing the timeliness and significance of having all of the methods dealing with retinoids included, made suggestions for additional contributions which were evaluated by the board. In a few instances we may have been somewhat over-zealous, and more than one article on a method has been included. We do apologize for this slight redundancy for the sake of completeness.

LESTER PACKER

METHODS IN ENZYMOLOGY

Section I

Structure and Analysis

A. Structure
Articles 1 through 3

B. Analysis
Articles 4 through 18

[1] Structure Characteristics of Natural and Synthetic Retinoids

By Michael Klaus

Introduction

The term retinoids comprises the natural occurring vitamin A derivatives such as retinol, retinal, and retinoic acid as well as the large number of synthetic analogs prepared since the late 1960s, regardless of whether they have biological activity.[1] The goal of this intensive effort at synthesis in academia as well as in the pharmaceutical industry was to improve on the ratio of toxic effects (e.g., hypervitaminosis A) to therapeutic activity compared to retinoic acid. In addition, compounds with better selectivity toward the various therapeutic indications (dermatology, oncology, rheumatology, immunology) have been sought.

Structure of Retinoids

Formally the molecule of retinoic acid consists of three main sections: a lipophilic part at one end, connected via a polyunsaturated chain as a spacer to a hydrophilic group at the other end of the molecule (Fig. 1). All three parts have been modified extensively. In the following a selection of compounds with typical variations is given.

Changes in Hydrophilic Part

Different Oxidation Level of Polar End Group. Compounds in which the oxidation level of the polar end group is different include the natural vitamin A derivatives retinol (1), retinal (2), retinoic acid (3), and close analogs such as esters (4) and amides (5). Because the free acid is generally considered to be the active principle, for *in vitro* experiments only compounds with a carboxylic acid end group should be used. Esters and

1   *2*

---

[1] M. B. Sporn and A. B. Roberts, *Ciba Found. Symp.* **113**, 1 (1985).

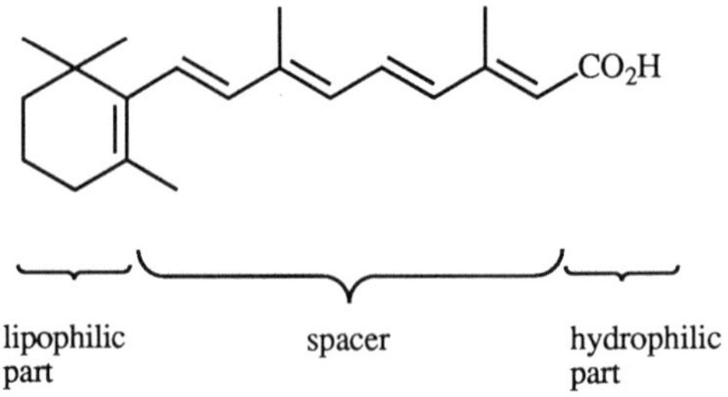

FIG. 1. Structural subunits of the retinoic acid molecule.

amides may be inactive owing to their stability toward hydrolysis under the experimental conditions.

Other Polar Groups. The carboxylic acid end group of retinoic acid (**3**) has been replaced by other acidic or polar moieties. Examples are sulfonic and phosphonic acid derivatives **6** and **7**,[2] sulfone **8**,[2] diketo derivative **9**,[3] and the phenol **10**.[4]

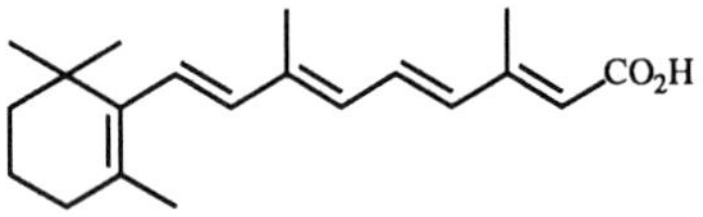

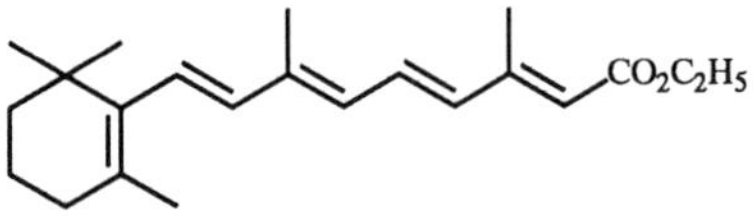

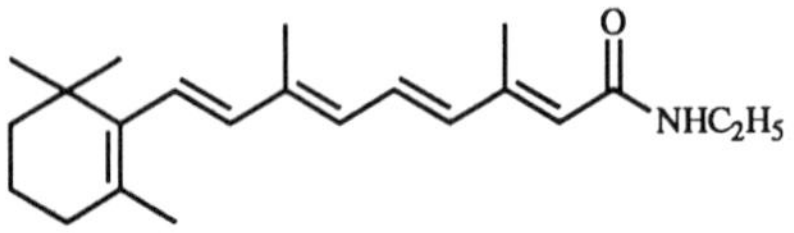

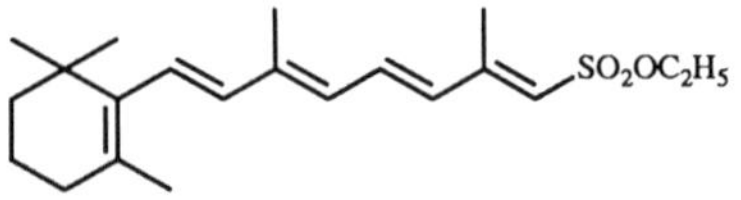

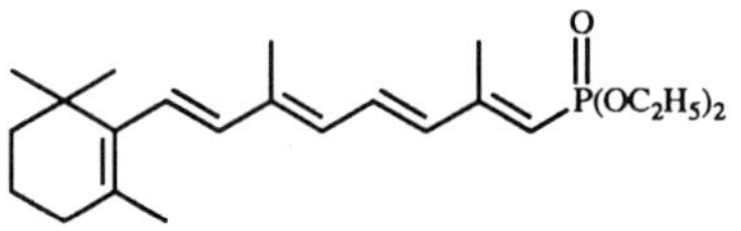

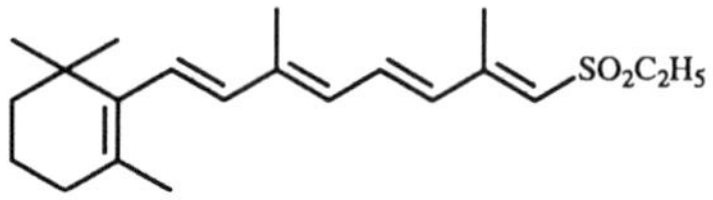

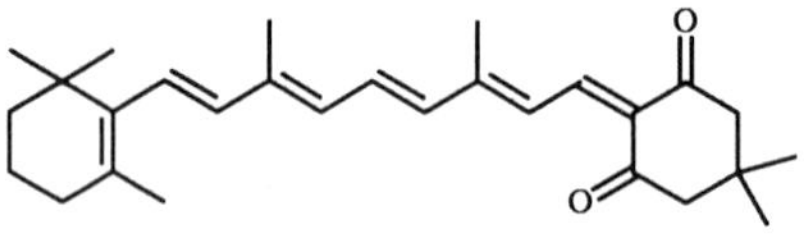

[2] M. Klaus, unpublished (1982 and 1984).
[3] N. Acton, A. Brossi, D. L. Newton, and M. B. Sporn, *J. Med. Chem.* **23**, 805 (1980).
[4] B. Loev and W.-K. Chan, U.S. Patent 4,605,675 (1986).

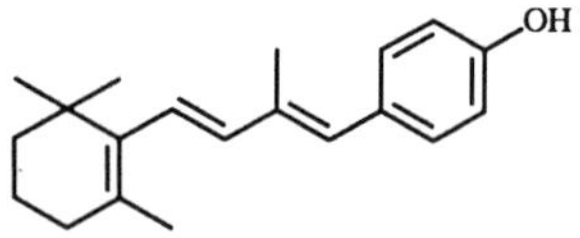

Changes in Lipophilic Part

Substitution Pattern of Cyclohexenyl Ring. The group of analogs with an altered substitution pattern of the cyclohexenyl ring consists mainly of oxygenated metabolites of retinoic acid, for example, 4-hydroxyretinoic acid **(11)**,[5] 18-hydroxyretinoic acid **(12)**,[6] and 5,6-epoxyretinoic acid **(13)**,[7] but also comprises compounds such as **14** and **15**.

Aromatization of Cyclohexenyl Ring. Replacement of the trimethylcyclohexenyl group with a suitably substituted benzene ring leads to compounds such as etretinate **(16)** and motretinid **(17)**[8] having an increased therapeutic index. A certain lipophilicity in this part of the molecule seems to be necessary for biological activity. The unsubstituted analog **18** is virtually inactive in most of the relevant test systems.

[5] M. Rosenberger, *J. Org. Chem.* **47**, 1698 (1982).
[6] M. Rosenberger and C. Neukom, *J. Org. Chem.* **47**, 1782 (1982).
[7] K. V. John, M. R. Lakshmanan, and H. R. Cama, *Biochem. J.* **103**, 539 (1967).
[8] H. Mayer, W. Bollag, R. Hänni, and R. Rüegg, *Experientia* **34**, 1105 (1978).

Other Cyclic or Acyclic Structures. Further examples of ring-modified retinoids are the norbornene analog **19**[9] and the acyclic compounds **20**[10] and **21**.[11]

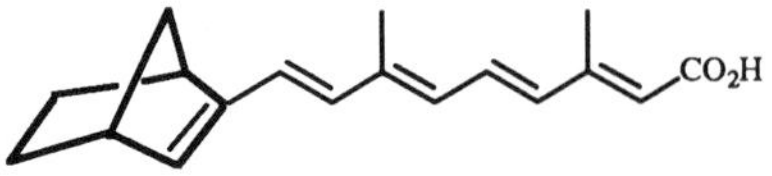

19 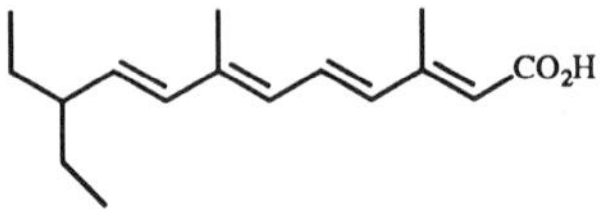   20

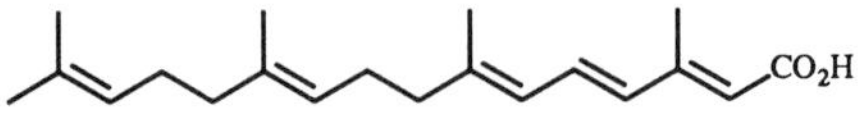

21

Changes in Nature of Spacer

Cis/Trans Isomerization of Double Bonds. With respect to the cis/trans isomers of retinoic acid (16 are possible), as well as the other retinoids, only the 13-*cis*-retinoic acid **(22)** (isotretinoin) proved to be superior to the all-trans isomer.

Hydrogenation or Dehydrogenation of Double Bonds. Many analogs have been synthesized with partially hydrogenated or dehydrogenated side chains, such as **23**,[12] **24**,[13] and **25**.[2]

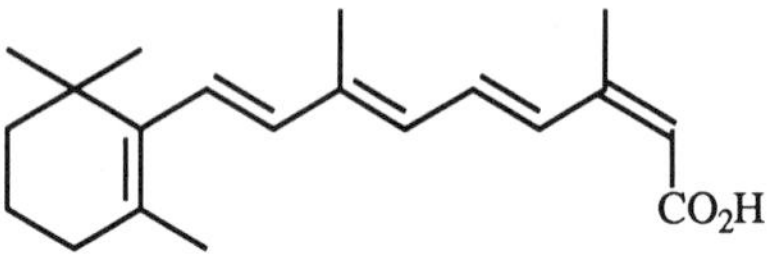

22 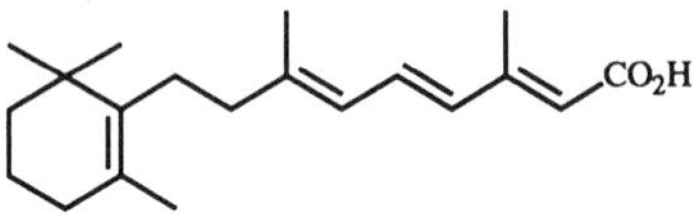   23

[9] M. I. Dawson, P. D. Hobbs, K. Kuhlmann, V. A. Fung, C. T. Helmes, and W. R. Chao, *J. Med. Chem.* **23**, 1013 (1980).

[10] R. K. Crouch, *J. Am. Chem. Soc.* **104**, 4946 (1982).

[11] Y. Muto, H. Moriwaki, and M. Omori, *Jpn. J. Cancer Res. (Gann)* **72**, 974 (1981).

[12] B. A. Pawson, H.-C. Cheung, R.-J. Han, P. W. Trown, M. Buck, R. Hansen, W. Bollag, U. Ineichen, H. Pleil, R. Rüegg, N. M. Dunlop, D. L. Newton, and M. B. Sporn, *J. Med. Chem.* **20**, 918 (1977).

[13] J. Attenburrow, A. F. Cameron, J. H. Chapman, R. M. Evans, B. A. Hems, A. B. Jansen, and T. Walker, *J. Chem. Soc.*, 1094 (1952); J. Paust, W. Hoffmann, and M. Baumann, Ger. Offen. 2,843,901 (1980).

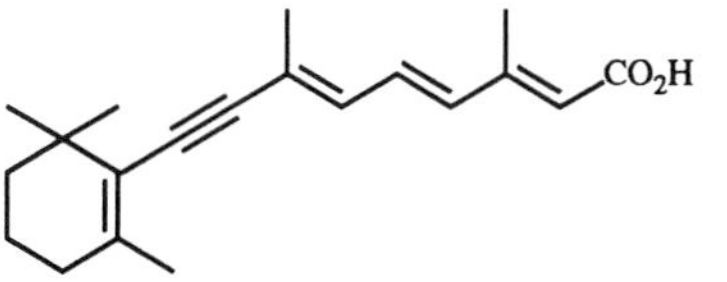

24 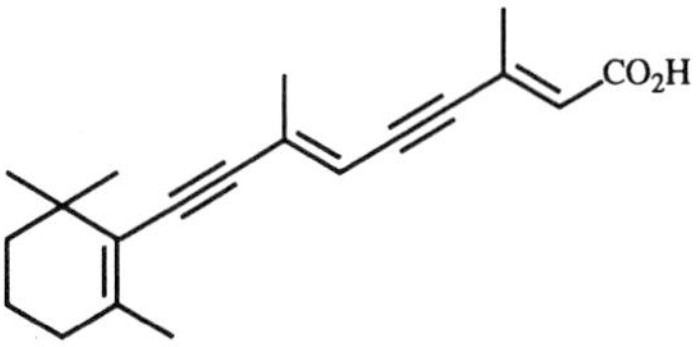   25

Compounds with Sterically Restricted Conformation of Side Chain.
The high flexibility of the side chain of retinoic acid (e.g., rotation around
the different single bonds) can be restricted to certain conformations by
incorporation of aromatic ring systems at various positions. In Fig. 2 this
concept is illustrated, with some examples of possible compounds (struc-

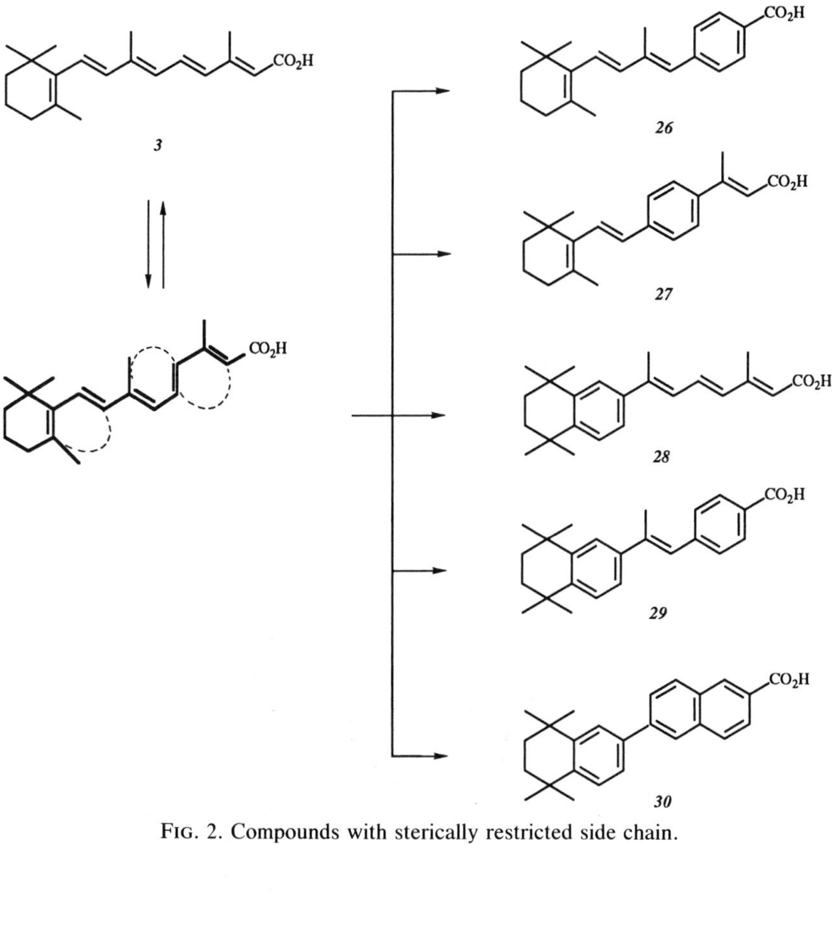

FIG. 2. Compounds with sterically restricted side chain.

TABLE I
Characteristics of Important Retinoids

Nomenclature	Structure	Molecular weight
Retinol, vitamin A alcohol, (all-*E*)-3,7-dimethyl-9-(2,6,6-trimethyl-1-cyclohexen-1-yl)-2,4,6,8-nonatetraen-1-ol		286.46
all-*trans*-Retinyl acetate, vitamin A acetate, (all-*E*)-3,7-dimethyl-9-(2,6,6-trimethyl-1-cyclohexen-1-yl)-2,4,6,8-nonatetraenyl acetate		328.50
3,4-Didehydroretinol, vitamin A₂, (all-*E*)-3,7-dimethyl-9-(2,6,6-trimethyl-1,3-cyclohexadien-1-yl)-2,4,6,8-nonatetraen-1-ol		284.44
all-*trans*-Retinal, vitamin A aldehyde (all-*E*)-3,7-dimethyl-9-(2,6,6-trimethyl-1-cyclohexen-1-yl)-2,4,6,8-nonatetraenyl		284.44
11-*cis*-Retinal		284.44

8

all-*trans*-Retinoic acid, all-*trans*-vitamin A acid, tretinoin, (all-*E*)-3,7-dimethyl-9-(2,6,6-trimethyl-1-cyclohexen-1-yl)-2,4,6,8-nonatetraenoic acid

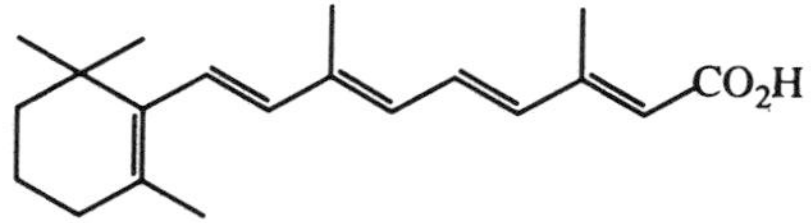

300.42

CO_2H

all-*trans*-4-Hydroxyretinoic acid

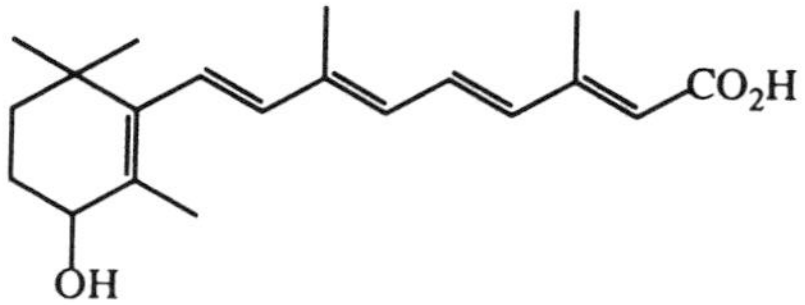

316.44

CO_2H

OH

all-*trans*-4-Oxoretinoic acid

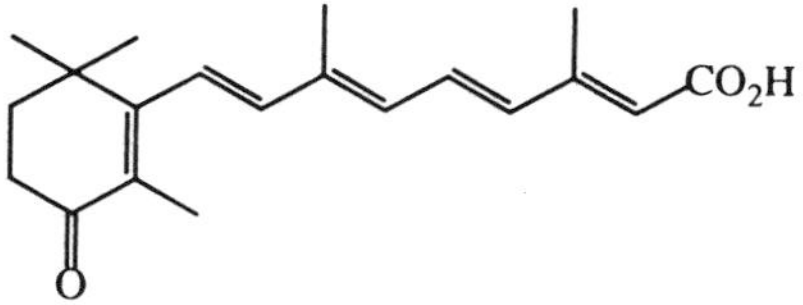

314.43

CO_2H

5,6-Epoxyretinoic acid

316.44

CO_2H

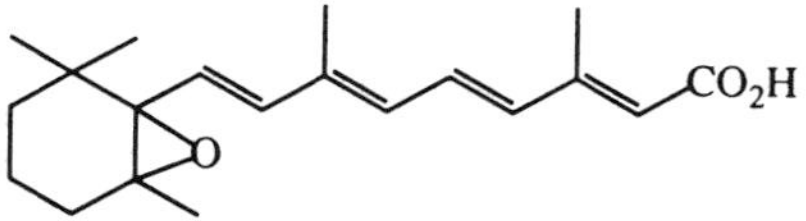

N-Ethylretinamide

327.51

NHC_2H_5

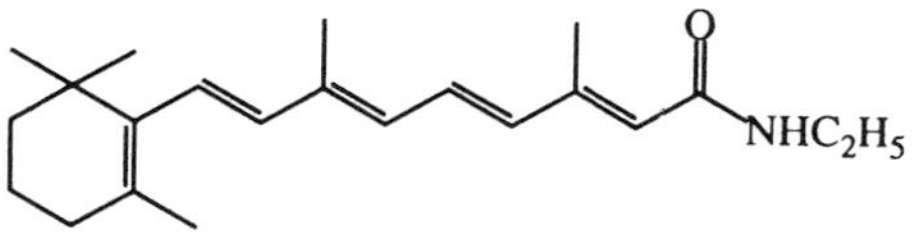

N-(4-Hydroxyphenyl)retinamide (4-HPR)

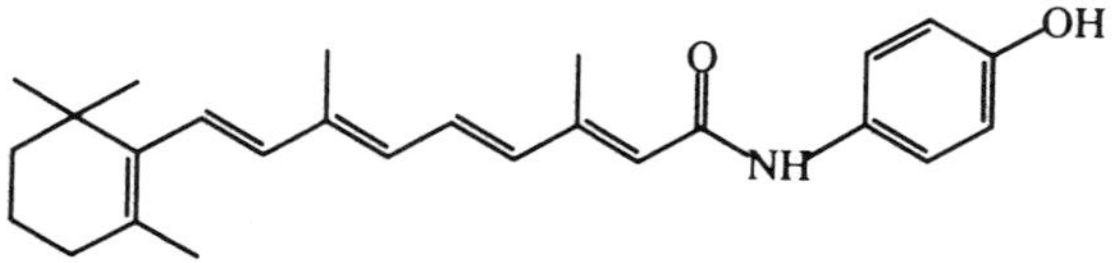

391.56

NH

OH

9

(continued)

TABLE I (continued)

Nomenclature	Structure	Molecular weight
3,4-Didehydroretinoic acid, all-*trans*-vitamin A₂ acid, (all-E)-3,7-dimethyl-9-(2,6,6-trimethyl-1,3-cyclohexadien-1-yl)-2,4,6,8-nonatetraenoic acid		298.43
13-*cis*-Retinoic acid, isotretinoin		300.42
13-*cis*-4-Hydroxyretinoic acid		316.44
E5166, (all-E)-3,7,11,15-tetramethyl-2,4,6,10,14-hexadecapentaenoic acid		302.46
Acitretin, (all-E)-9-(4-methoxy-2,3,6-trimethylphenyl)-3,7-dimethyl-2,4,6,8-nonatetraenoic acid		326.44
Etretinate, ethyl (all-E)-9-(4-methoxy-2,3,6-trimethylphenyl)-3,7-di-methyl-2,4,6,8-nonatetraenoate		354.49

Motretinid, (all-E)-N-ethyl-9-(4-methoxy-2,3,6-trimethylphenyl)-3,7-dimethyl-2,4,6,8-nonatetraenamide	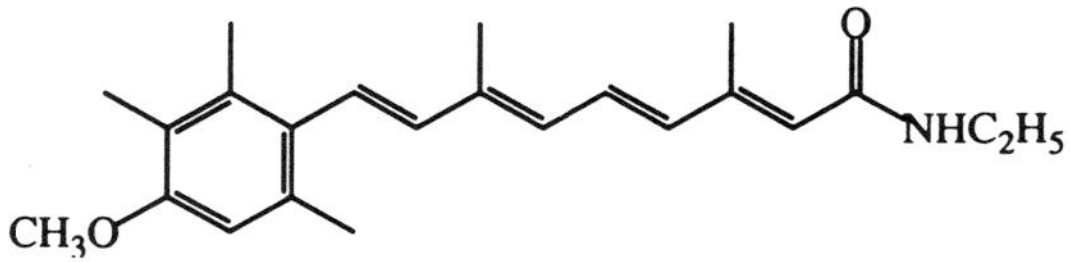	353.51
(all-E)-9-(4-Hydroxy-2,3,6-trimethylphenyl)-3,7-dimethyl-2,4,6,8-nonatetraenoic acid	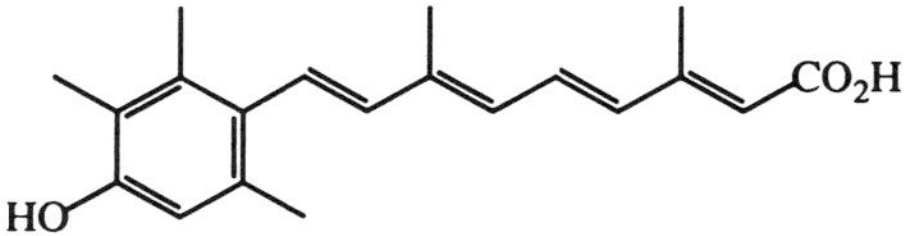	312.41
(all-E)-9-(4-Methoxy-2,3,6-trimethylphenyl)-3,7-dimethyl-2,4,6,8-nonatetraen-1-ol	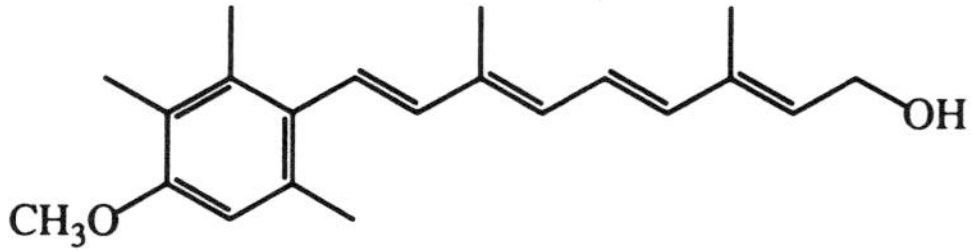	312.45
p-[(all-E)-2-Methyl-4-(2,6,6-trimethyl-1-cyclohexen-1-yl)-1,3-butadienyl]benzoic acid	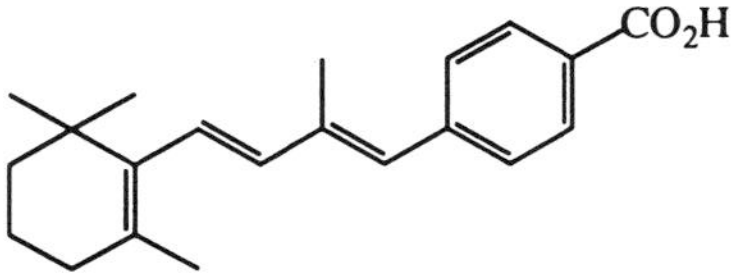	310.44
(all-E)-3-Methyl-7-(5,6,7,8-tetrahydro-5,5,8,8-tetramethyl-2-naphthyl)-2,4,6-octatrienoic acid	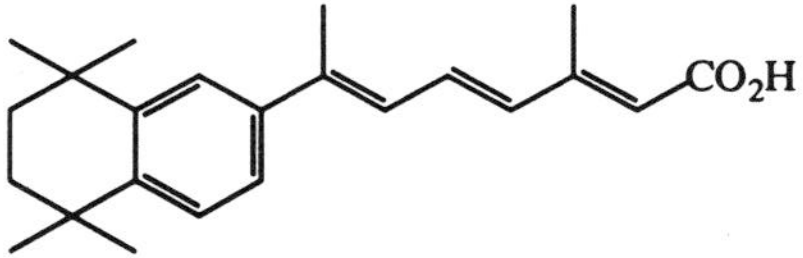	338.49
Arotinoid, TTNPB, Ro 13-7410, p-[(E)-2-(5,6,7,8-tetrahydro-5,5,8,8-tetramethyl-2-naphthyl)propenyl]benzoic acid	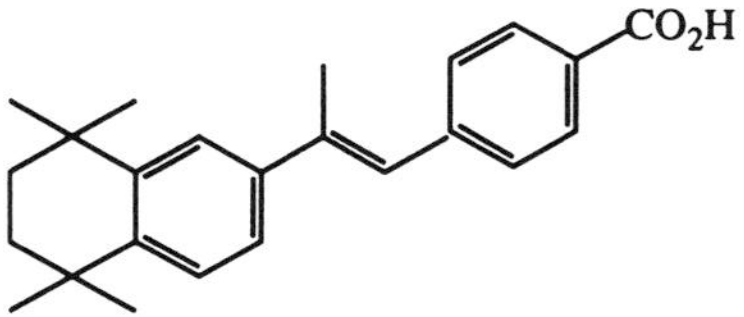	348.49

11

(continued)

TABLE I (continued)

Nomenclature	Structure	Molecular weight
Arotinoid ethyl ester, Ro 13-6298, ethyl p-[(E)-2-(5,6,7,8-tetrahydro-5,5,8,8-tetramethyl-2-naphthyl)propenyl]benzoate	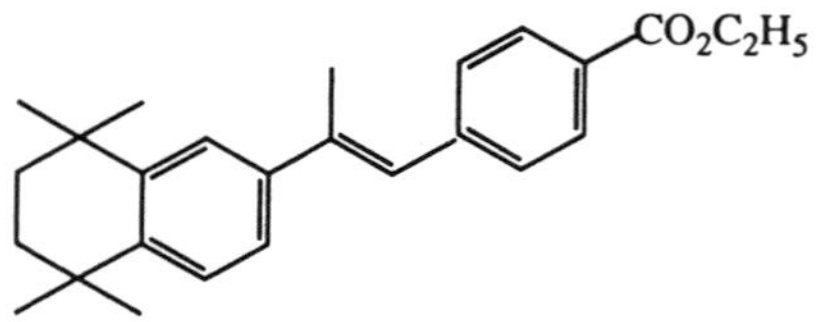	376.54
Ethyl p-[(E)-2-(5,6,7,8-tetrahydro-5,5,8,8-tetramethyl-2-naphthyl)propenyl]phenyl sulfone	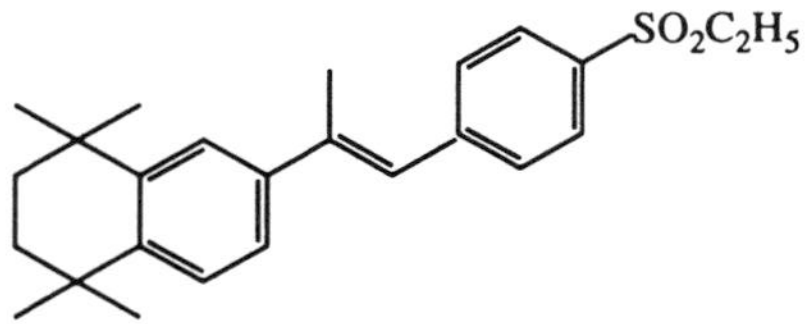	396.59
Temarotene, 1,2,3,4-tetrahydro-1,1,4,4-tetramethyl-6-[(E)-α-methyl-styryl]naphthalene	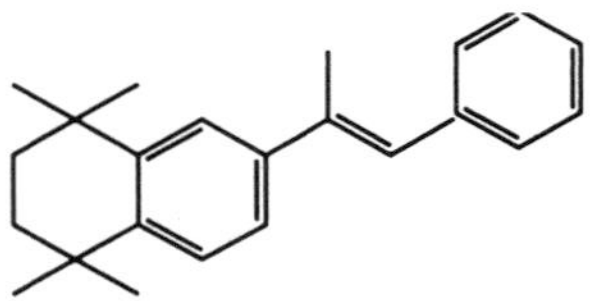	304.48

12

TTNN,
5',6',7',8'-tetrahydro-5',5',8',8'-tetramethyl-[2,2'-bi-
naphthalene]-6-carboxylic acid

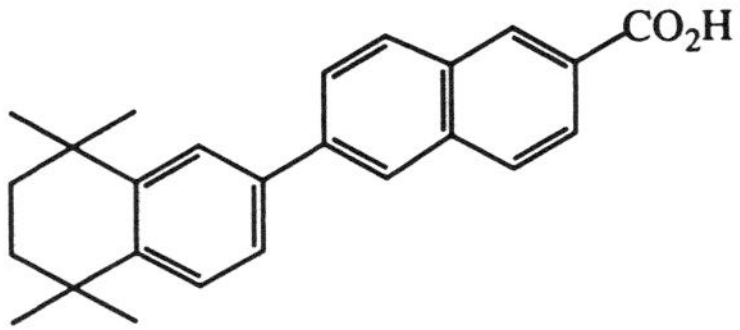

CO_2H

358.48

AM 580,
p-(5,6,7,8-tetrahydro-5,5,8,8-tetramethyl-2-naphthalene-
carboxamido)benzoic acid

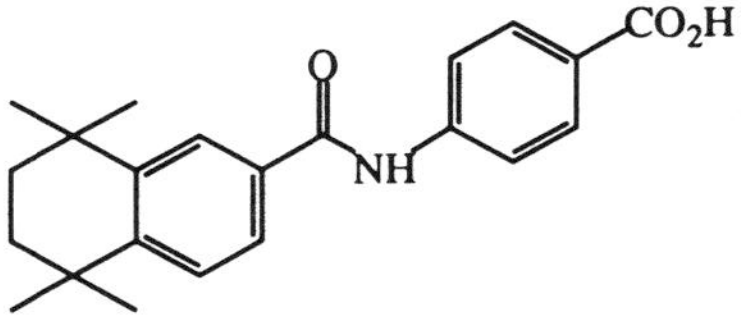

CO_2H

351.45

CD 271,
6-[3-(1-adamantyl)-4-methoxyphenyl]-2-naphthalenecar-
boxylic acid

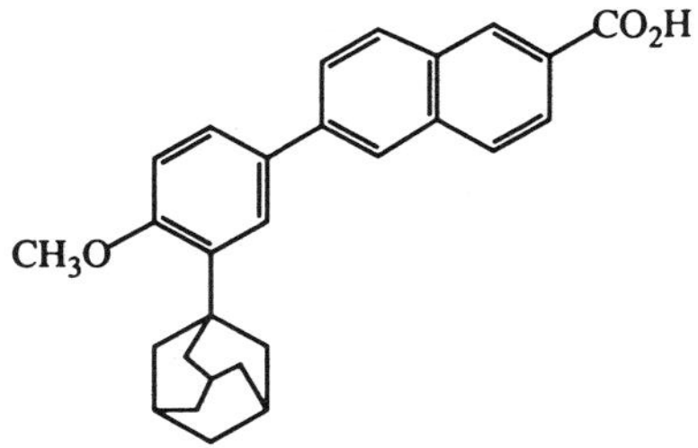

CO_2H

CH_3O

412.53

13

tures **26–30**) being shown. Thus, in compound **26**[14] the C-12–C-13 single bond of retinoic acid is locked in a cisoid conformation; in compound **27**[15] it is the C-10–C-11 single bond. Work along these lines has led to the synthesis of retinoids such as **29**, the so-called arotinoids,[16] with biological activities about one thousand times greater than retinoic acid.

Replacement of Selected Double Bonds with Amide, Sulfonamide, or Other Groups. Utilizing an interesting new concept, Shudo and co-workers[17] have synthesized compounds like **31–34**. Although the structural similarity to retinoic acid is not at all obvious, these molecules should still be considered as retinoids because of their biological properties.

Compilation of Important Retinoids

Table I lists important and often used retinoids. The structural formula, molecular weight, systematic nomenclature, and trivial, as well as generic, name where appropriate are provided for each compound.

[14] M. I. Dawson, P. D. Hobbs, R. L. Chan, W.-R. Chao, and V. A. Fung, *J. Med. Chem.* **24,** 583 (1981).

[15] F. Frickel, *in* "The Retinoids" (M. B. Sporn, A. B. Roberts, and D. S. Goodman, ed.), Vol. 1, p. 121. Academic Press, Orlando, Florida, 1984; M. Klaus, *Actual. Chim. Ther.* **12,** 63 (1985).

[16] P. Loeliger, W. Bollag, and H. Mayer, *Eur. J. Med. Chem.* **15,** 9 (1980); M. Klaus, W. Bollag, P. Huber, and W. Küng, *Eur. J. Med. Chem.* **18,** 425 (1983).

[17] H. Kagechika, E. Kawachi, Y. Hashimoto, T. Himi, and K. Shudo, *J. Med. Chem.* **31,** 2182 (1988); H. Kagechika, E. Kawachi, Y. Hashimoto, and K. Shudo, *J. Med. Chem.* **32,** 834 (1989); H. Kagechika, E. Kawachi, Y. Hashimoto, and K. Shudo, *Recent Adv. Chemother. Proc. Int. Congr. Chemother. 14th 1985,* p. 227.

[2] Synthetic Retinoic Acid Analogs: Handling and Characterization

By M. I. DAWSON and P. D. HOBBS

Introduction

The natural retinoids (Fig. 1) present some difficulty in manipulation because of the conjugated double-bond system that makes them sensitive to light, oxygen, and heat, and, therefore, susceptible to bond isomerization, oxidation, and other degradative processes. Of the oxidative states of these natural retinoids, retinoic acid is by far the easiest compound to handle, followed by retinal and then retinol, because of the latter's acid lability. Of the double bond isomers, the all-trans or *E* isomer is generally the most stable; however, light can readily cause isomerization to a mixture of isomers.

Retinoids have been defined by Sporn[1] as those compounds having retinoidlike activity, namely, the ability to control and regulate cell differentiation or regulate the reproductive or visual process in the same manner as the natural retinoids do. The synthetic retinoids have structural similarity to the natural compounds in that they possess the characteristic lipophilic head group and polar terminus separated by a spacer group.[2] Generally the trend in the synthesis of analogs of (*E*)-retinoic acid has been the design of more stable, readily manipulatable species. Therefore, the care with which they are treated under experimental conditions is generally dependent on their structural similarity to the natural retinoid skeleton. Generally, the greater the number of conjugated olefinic bonds present in the structure, the more caution should be exercised in its use in the laboratory. The procedures presented below are ones that have been employed in the preparation, characterization, storage, and handling of (*E*)-retinoic acid and its synthetic analogs.

Synthetic Manipulation

Retinoids having extended conjugated olefinic bond systems (three or more) should not be handled routinely under fluorescent light. In some

---

[1] M. B. Sporn, *in* "The Retinoids" (M. B. Sporn, A. B. Roberts, and D. S. Goodman, eds.), Vol. 1, p. 1. Academic Press, Orlando, Florida, 1984.

[2] K. Shudo and H. Kagechika, *in* "Chemistry and Biology of Synthetic Retinoids" (M. I. Dawson and W. H. Okamura, eds.), p. 275. CRC Press, Boca Raton, Florida, 1990.

METHODS IN ENZYMOLOGY, VOL. 189

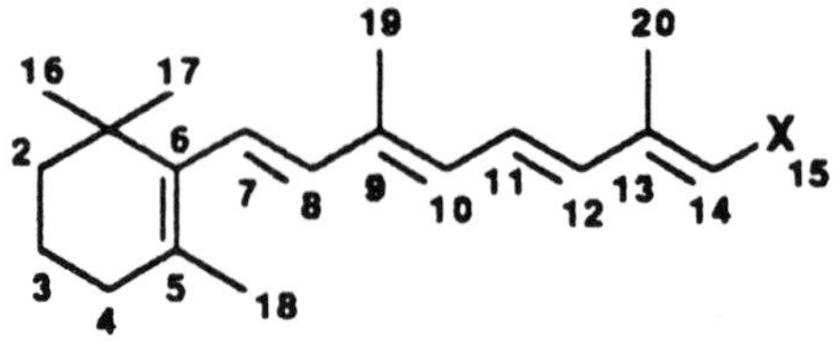

FIG. 1. Structure of the naturally occurring retinoids, (E)-retinoic acid $(X = CO_2H)$, (E)-retinal $(X = CHO)$, and (E)-retinol $(X = CH_2OH)$, indicating the retinoid numbering system.

cases it may be necessary to use photographic red light in the laboratory; however, yellow light (either fluorescent or incandescent) is usually satisfactory. Because of the sensitivity of the retinoid double-bond system to oxidation, it is necessary to conduct synthetic steps under inert gas, especially when more than one double bond is present in the structure. Antioxidants may be added to prevent oxidation. Two sample procedures are presented below.

Preparation of Ethyl (E)-Retinoate-4,4,18,18,18-d₅

The reaction is conducted in a 50-ml, 14/20 three-necked, round-bottomed flask containing a magnetic stirring bar and fitted with a gas-inlet tube that is connected to a vacuum system (50 mm) and an argon line. The remaining joints are capped by a low-temperature thermometer and a wired-on serum cap. To a suspension (degassed under argon) of 7.9 g (16.3 mmol) of [6,6-dimethyl-2-(methyl-d_3)cyclohexen-3,3-d_2-yl]methyltriphenylphosphonium bromide[3] in 15 ml of tetrahydrofuran (distilled under argon from lithium aluminum hydride) at $-30°$ (cooled in a dry ice/acetone bath) under argon is added by syringe 13.5 ml of 1.19 M n-butyllithium (16 mmol) in hexane (Aldrich, Milwaukee, WI). The deep-red suspension is stirred magnetically and allowed to warm to $0°$ over an 1-hr period before a solution of 3.75 g (18 mmol) of freshly crystallized (from cyclohexane) (E)-7-carbethoxy-2,6-dimethyl-2,4,6-heptatrienal[4] in 6 ml of degassed (argon) tetrahydrofuran is introduced by syringe. The reaction mixture is stirred at room temperature for 17 hr and then heated at $60°$ for 1 hr to complete the reaction. The dark suspension is cooled to room temperature, poured into a 250-ml separatory funnel containing 120 ml of water, 1 ml of glacial acetic acid, and 0.1 g of *tert*-butylhydroquinone, and extracted with 100 ml of 10% ethyl acetate/hexane (HPLC grade). Solvents are flushed with argon before use in the

---

[3] M. I. Dawson, P. D. Hobbs, R. L. Chan, and W. Chao, *J. Med. Chem.* **24**, 1214 (1981).
[4] W. Bollag, R. Rüegg, and G. Ryser (Hoffmann-La Roche, Inc.), U.S. Patent 4,163,103 (July 31, 1979).

extraction procedure. The organic extract is washed twice with saturated brine, transferred to a 500-ml Erlenmeyer flask, dried over anhydrous Na_2SO_4, and concentrated at reduced pressure using a rotary evaporator (Büchi, Flawil, Switzerland) under argon. The rotary evaporator is vented to the argon line.

The residue is chromatographed on a 4 × 40 cm column of silica gel 60 (EM Sciences, Cherry Hill, NJ) using 1.5% ethyl acetate/hexane (HPLC grade). The ester is eluted in fractions 5–8 (200 ml). These fractions are pooled and concentrated under argon to give 4.17 g of crude ester as a yellow gum. Analytical HPLC (Waters Associates, Milford, MA, ALC 210 equipped with a Radial-PAK B cartridge, detection at 260 nm at a flow rate of 2.0 ml/min of 1% ethyl acetate/hexane) indicated the presence of a minor, less-polar impurity. Therefore, the ester is purified by preparative HPLC (Waters Associates Prep LC/System 500 instrument equipped with two Prep Pak 500/silica gel cartridges in series, detection by refractive index, 0.2 liter/min of 1% ethyl acetate/hexane) to give, after removal of the solvent at reduced pressure (50 followed by 0.05 mmHg), 3.5 g (66% yield) of the ester as a viscous yellow oil.

Subsequent experiments indicated that the antioxidant could be eliminated if all extraction and purification steps were performed using solvents that had been degassed under argon and all manipulations were performed in an argon atmosphere. Nitrogen may also be used in place of argon.

(E)-Retinoic Acid-4,4,18,18,18-d_5

The reaction is conducted in a 25-ml, 14/20 three-necked, round-bottomed flask that is equipped with a magnetic stirring bar, reflux condenser connected to an argon/vacuum line, thermometer, and wired-on serum cap. A degassed (argon) solution of 1.48 g (4.4 mmol) of ethyl (E)-retinoate-4,4,18,18,18-d_5, prepared as described above, in 4.0 ml of ethanol is added by syringe under argon to a degassed (4 times under argon) solution of 0.7 g (12 mmol) of 85% KOH in 2.4 ml of water and 4.0 ml of ethanol. The stirred suspension is heated to an internal temperature of 80° in an oil bath over a period of 20 min and then maintained at this temperature for an additional 12 min. The oil dissolves at 80°.

The reaction work-up is performed under argon. The cooled reaction mixture is acidified with 10 ml of 50% aqueous acetic acid and then transferred to a 100-ml separatory funnel, where it is diluted with 50 ml of water. The precipitated carboxylic acid is extracted into two 30-ml portions of ether (peroxide-free). The organic extract is washed with two 15-ml portions of saturated brine, dried over anhydrous Na_2SO_4, and concentrated. The crude acid is extracted with 20 ml of hot methanol under argon. The hot extract is filtered through a 3-cm glass frit (medium porosity) into a filtration tube and cooled under argon to give 0.75 g (56% yield) of the carboxylic acid as bright yellow needles, mp 175.5–177.5°.[3]

Retinoid Purification

In cases where reaction procedures give more than one isomer a chromatographic separation is often necessary for purification rather than using crystallization. This is especially true in the case of the polyolefinic retinoids. Although preparative thin-layer chromatography on silica gel plates can often be used to separate olefinic bond isomers, oxidation on the plate can be a problem even when the attempt is made to run the plate in an inert atmosphere. High-performance liquid chromatography (HPLC) is particularly effective for separating E and Z isomers, especially when the recycle technique is employed. This technique has the added advantage that the chromatography is performed under an inert carrier gas such as nitrogen. The separations are best performed on the esters of analogs of (E)-retinoic acid, followed by hydrolysis to the acid under nonisomerizing conditions. Analogs of ethyl (E)-retinoate and ethyl (E)-4-[2-methyl-4-(2,6,6-trimethylcyclohexenyl)buta-1,3-dienyl]benzoate have been purified using 0.5–2% ether or 0.25–2% ethyl acetate/hexane solvent mixtures.[3,5,6] Unfortunately, these separations require a large volume of solvent. Solvent removal must be done under inert gas and the contents of the flask vented to the inert gas rather than atmosphere. For example, sodium (E)-(4-retinamidophenyl β-D-glucopyranoside)uronate is purified by reversed phase HPLC (20% water/methanol) and the eluant concentrated at reduced pressure under argon to remove the methanol.[7] The water is removed by lyophilization to give a fluffy bright-yellow powder. Venting of the lyophilization flask to the atmosphere results in almost complete oxidation of the product. However, venting to argon affords 97% pure material.

Crystallization should also be conducted under inert gas. Crude retinoids may contain insoluble matter, including polymeric oxidation products and silica derived from reaction of glassware with alkali or from chromatography. Before crystallization is undertaken for purification, the

[5] M. I. Dawson, R. Chan, P. D. Hobbs, W. Chao, and L. J. Schiff, *J. Med. Chem.* **26**, 1282 (1983).
[6] M. I. Dawson, P. D. Hobbs, K. Derdzinski, R. L. Chan, J. Gruber, W. Chao, S. Smith, R. W. Thies, and L. J. Schiff, *J. Med. Chem.* **27**, 1516 (1984).
[7] M. I. Dawson and P. D. Hobbs, *Carbohydr. Res.* **85**, 121 (1980).

crude products should first be extracted into a solvent at room temperature. For example, tetrahydrofuran (distilled from lithium aluminum hydride under argon) is effective for the highly insoluble aromatic retinoids. The solutions should then be filtered through a medium porosity glass frit or glass-fiber filter disk to remove insoluble material, and the filtrate should be concentrated under reduced pressure. The residue is then crystallized under argon in a Schlenk tube. Polyolefinic retinoid carboxylic acids have been crystallized from methanol, whereas the aromatic carboxylic acids are crystallized from ethanol. Hexane, hexane/ethyl acetate, or ethanol has been used as the solvent for crystallizing esters. Degassed solvent is introduced by syringe through the serum-capped side arm. The hot solution is then cooled under argon until crystallization is complete. The crystalline product may then be filtered on a sintered glass frit without protection from air and dried at reduced pressure. If the solution must be filtered while hot or if the product is highly oxygen sensitive, crystallization should be performed using special apparatus (e.g., the double-ended filter available from Kontes Scientific Glassware, Vineland, NJ).

In contrast, the aromatic retinoids, for example, compounds in the (E)-4-[2-(5,6,7,8-tetrahydro-5,5,8,8-tetramethyl-2-naphthalenyl)propen-1-yl]benzoic acid,[8] 6-(5,6,7,8-tetrahydro-5,5,8,8-tetramethyl-2-napthalenyl)-2-naphthalenecarboxylic acid,[9] and 4-(5,6,7,8-tetrahydro-5,5,8,8-tetramethyl-2-anthracenyl)benzoic acid[10] series, are generally stable, and the above precautions need not be exercised, except operations on the first series should be conducted under reduced light to minimize photoisomerization. However, the oxidation of retinoids in these series can occur in cases in which a methyl group is absent from the tetrahydronaphthalene ring or a 4,4-dimethyl-3,4-dihydro-2H-1-benzothiopyran ring replaces that ring.[6] The latter oxidation is far more rapid than the former.

Handling and Storage

Retinoids having conjugated polyolefinic chains should be stored under inert gas at freezer temperatures ($-20°$ or below). Storage is most easily accomplished by aliquoting portions of the sample into small glass ampoules (Wheaton, Millville, NJ) that can be opened separately when

[8] P. Loeliger, W. Bollag, and H. Mayer, *Eur. J. Med. Chem.* **15,** 9 (1980).
[9] M. I. Dawson, R. L. Chan, K. Derdzinski, P. D. Hobbs, W. Chao, and L. J. Schiff, *J. Med. Chem.* **26,** 1653 (1983).
[10] M. I. Dawson, P. D. Hobbs, K. A. Derdzinski, W. Chao, G. Frenking, G. H. Loew, H. M. Vu, A. M. Jetten, J. L. Napoli, J. B. Williams, B. P. Sani, J. J. Wille, Jr., and L. J. Schiff, *J. Med. Chem.* **32,** 1504 (1989).

needed. Amberized glass (Pierce, Rockford, IL) can be used for the more light-sensitive retinoids. Both crystalline solids and mobile liquids are readily transferred to the ampoules. The ampoule is evacuated (0.05 mm) and filled with argon 3 times and then sealed by torch under a partial pressure of argon (~20 mmHg below atmospheric pressure). Gums and viscous oils are best dissolved in an inert volatile solvent that can be removed at reduced pressure. The solution is then transferred by pipette or syringe into the ampoule. Alternately, solution aliquots can be frozen and the ampoule sealed. It may be necessary to transfer fine powders into ampoules in an argon-filled glove bag to minimize oxidation. Because sealing can be imperfect or cracks can occur on storage, labile retinoids should be analyzed by regular- and reversed-phase HPLC prior to use. Regular-phase chromatography can be used to detect the presence of bond isomers, whereas reversed-phase can be used to determine if oxidation to more polar species has occurred.

Characterization

Stereoselective methods[11] are frequently used to prepare retinoid esters. The esters are purified by crystallization or chromatography to give stereochemically pure intermediates. Analogs of retinoic acid, retinal, or retinol are then obtained by chemical modification of the purified esters. Analytical HPLC is employed to follow the purification of the intermediates and to determine the purity of the final products. Spectral comparison of the E and Z double bond isomers, often on these ester intermediates, is useful for the assignment of the configuration.

Analytical High-Performance Liquid Chromatography

Regular-phase HPLC is used to determine isomeric purity, whereas reversed-phase HPLC can be used both to determine isomeric purity and to detect impurities derived from the atmospheric oxidation of double bonds or thia groups. Retinoid esters have been analyzed using silica gel regular-phase HPLC (Table I, structures 1–7) or reversed-phase HPLC (Table II, structures 8–25). All analytical operations with polyolefinic retinoids should be performed under yellow or photographic red light. Dim ambient light should be used for the stilbene retinoids.

The following solvent systems have been used for regular-phase HPLC on esters: 0.5–2% ethyl acetate/hexane,[9] 0.5–5% diethyl ether/

---

[11] F. Frickel, in "The Retinoids" (M. B. Sporn, A. B. Roberts, and D. S. Goodman, eds.), Vol. 1, p. 121. Academic Press, Orlando, Florida, 1984.

hexane,[3] 1.5% diisopropyl ether/hexane,[12] 2% *tert*-butyl methyl ether/ hexane, 45% toluene/hexane,[13] 35% dichloromehane/hexane,[13] 0.1% dioxane/hexane.[14] Retinal analogs have been resolved using 12% diethyl ether/hexane[15] and 2.5% dioxane/hexane.[14] Labile retinol isomers have been separated using 5% acetone/hexane,[13] 15% methyl ethyl ketone/ hexane,[16] 5% dioxane/hexane,[14] 35% dichloromethane, and 18.5% diisopropyl ether/hexanes.[17] Napoli has resolved retinoic carboxylic acids using 0.2% acetic acid/5% dichloromethane/hexane.[13] Ether solvents must be free of peroxides, which can react with olefinic bonds.

Reversed-phase HPLC is invaluable for analyzing for the oxidation products of both nonpolar and polar retinoids and for detecting and isolating very polar derivatives, especially metabolites.[18–20] The reversed-phase HPLC analysis of retinoids and their metabolites was described by Ross in a previous volume in this series.[21] The solvent system used depends on the choice of column. Gradient elution using 1% aqueous ammonium acetate/acetonitrile or 0.5–1.0% acetic acid/acetonitrile is frequently used for analysis. Curley reported superior resolution using these mobile phases on an irregular particle, nonend-capped C_{18} column (Ultrasil ODS, Beckman, San Ramon, CA), whereas a spherical particle, fully end-capped C_{18} column (Ultrasphere ODS, Beckman) gave improved resolution when aqueous methanol was used.[22]

A Novapak 4-μm ODS radial compression cartridge (8 × 100 mm) in a RCM 8 × 10 Radial-PAK cartridge holder (Waters Associates) was effective in the analysis of both polyolefinic and aromatic retinoidal carboxylic acid esters using 0–5% water/acetonitrile as the isocratic solvent phase (Table II). The polyolefinic retinoids were detected by their ultraviolet absorbance at 325 or 350 nm, whereas the other retinoids were detected at 260 or 280 nm. Detection at 260 nm increases the peak area for the 9_R-Z (retinoid numbering; see Table I and Fig. 1) double-bond isomer relative to that of the 9_R-E isomer and can be used to analyze for the presence of

[12] G. Englert, S. Weber, and M. Klaus, *Helv. Chim. Acta.* **61,** 2697 (1978).

[13] J. L. Napoli, this series, Vol. 123, p. 112.

[14] J. E. Paanakker and G. W. T. Groenendijk, *J. Chromatogr.* **168,** 125 (1979).

[15] K. Tsukida, A. Kodama, and M. Ito, *J. Chromatogr.* **134,** 331 (1977).

[16] B. Stancher and F. Zonta, *J. Chromatogr.* **234,** 244 (1982).

[17] G. M. Landers and J. A. Olson, *J. Chromatogr.* **291,** 51 (1984).

[18] P. V. Bhat, L. M. De Luca, and M. L. Wind, *Anal. Biochem.* **102,** 243 (1980).

[19] C. A. Frolik, B. N. Swanson, L. L. Dart, and M. B. Sporn, *Arch. Biochem. Biophys.* **208,** 344 (1981).

[20] K. L. Skare, H. K. Schnoes, and H. F. De Luca, *Biochemistry* **21,** 3308 (1982).

[21] A. C. Ross, this series, Vol. 123, p. 68.

[22] R. W. Curley, Jr., D. L. Carson, and C. N. Ryzewski, *J. Chromatogr.* **370,** 188 (1986).

TABLE I
REGULAR-PHASE HPLC OF RETINOIDAL ESTERS[a]

Retinoid structure	Structure number	Mobile phase[b]	Retention volume (ml)
	E-1	A	39.0
	9_RZ-1	A	33.4
	13_RZ-1	A	26.8
	E-2	B	12.0
	9_RZ-2	B	10.6
	3	C	12.0
	4	C	11.2
	13_RE-5	C	13.8
	13_RZ-5	C	26.8

TABLE I (continued)

Retinoid structure	Structure number	Mobile phase[b]	Retention volume (ml)
	6	C	14.4
	7	C	12.0

[a] Instrumentation used was a Waters Associates Model 6000A solvent delivery system equipped with a 8 × 100 mm μ-Porasil 10-μm Radial-PAK radial compression cartridge in a RCM 100 module. Absorbance was monitored at 260 nm with a Schoeffel Instrument Corp. GM 770 monochromator and a SF 770 Spectroflow monitor. Peaks were recorded using a linear chart recorder.

[b] (A) 0.5% ether/hexane at 1.0 ml/min, (B) 2.0% ether/hexane at 2.0 ml/min, (C) 1.0% ethyl acetate/hexane at 2.0 ml/min.

9_R-Z double-bond impurities in the E isomer of the stilbene class of retinoids (e.g., E-13 and E-15). Unfortunately, because the E and Z double-bond isomers of the heteroatom-substituted (oxygen and sulfur) retinoids, especially those in the stilbene series, such as compound E-14, did not resolve satisfactorily in this system, the elutant was changed to 5–10% aqueous acetic acid (1%)/methanol. Retinoid carboxylic acids (E-8, E-10, E-15, 17, 18, 19, 21, 22, and 25) have been analyzed using 10% aqueous acetic acid (1%)/methanol. This same solvent system may be employed for analyzing mixtures containing both esters and the parent carboxylic acids (e.g., 6 and 19). Trifluoroacetic acid cannot be used in place of acetic acid for analysis of polyolefinic retinoids and other acid-sensitive compounds. No improvement in resolution of the stable aromatic retinoidal carboxylic acids was found using 10% water/methanol containing 0.1% trifluoroacetic acid on the Novapak column. Far inferior resolution for the carboxylic acids of both the polyolefinic and aromatic retinoid classes was observed when 30–45% aqueous ammonium acetate (1%)/acetonitrile was employed on this column.

Ionic retinoids have been purified by ion-exchange chromatography. De Luca and co-workers used a Zorbax SAX column (Du Pont, Inc., Wilmington, DE) with gradient elution of aqueous ammonium acetate in methanol to isolate a taurine-containing metabolite of (E)-retinoic acid.[20]

TABLE II
REVERSED-PHASE HPLC OF RETINOIDS[a]

Retinoid structure	Structure number	Mobile phase[b]	Retention volume[c] (ml)
	E-8	A D	11.6 13.8
	13_RZ-8	A D	11.0 11.8
	E-9, R = Et	B	17.2
	E-10, R = H	A D	16.8 16.0
	13_RZ-9	B	14.8
	E-11	B	13.4
	E-12	B	7.8
	E-13	B	11.0
	9_RZ-13	B	9.8
	E-14	C D	6.8 14.3
	9_RZ-14	C D	7.2 12.4
	E-15	D	9.2

TABLE II (continued)

Retinoid structure	Structure number	Mobile phase[b]	Retention volume[c] (ml)
	9_RZ-15	D	7.4
	16, R = Et	B	12.4
	17, R = H	D	17.0
	18	D	21.6
	6, R = Et	B D	13.0 41.6
	19, R = H	D	15.6
	20, R = Me	B	11.0
	21	D	18.8
	22	B	18.2
	23	D	22.2
	24, R = Me	B	14.4
	25, R = H	D	22.2

[a] Instrumentation was as listed in Table I except that a Waters Associates 8 × 100 mm Nova-Pak C_{18} 4-μm radial compression cartridge in a RCM 8 × 10 module was used. The absorbance for E-8, 13Z-8, E-9, and E-10 was monitored at 325 nm; the others were monitored at 260 nm.

[b] Mobile phase at a flow rate of 2.0 ml/min: 30% aqueous ammonium acetate (1%)/acetonitrile (A), 2% water/acetonitrile (B), 5% water/acetonitrile (C), 10% aqueous acetic acid (1%)/methanol (D).

Infrared Spectroscopy

Infrared (IR) spectroscopy of retinoids has been described in detail by Rockley *et al.* in an earlier volume in this series.[23] Infrared spectra, particularly the characteristic absorption bands for functional groups, are used routinely in the structure determination of retinoids. Although double-bond configurations can be determined from resonance-Raman spectra, IR spectroscopy is usually employed for this purpose by using spectral comparison with known compounds. Solution spectra are obtained in chloroform or carbon tetrachloride when retinoid solubility permits. Many aromatic retinoid carboxylic acids are insufficiently soluble in these solvents, and therefore spectra are obtained on KBr disks or Nujol mulls. Fourier-transform IR spectra have been run on very small retinoid samples (25 μg).[20,23]

Proton Nuclear Magnetic Resonance Spectroscopy

Proton nuclear magnetic resonance (^{1}H NMR) spectroscopy has proved indispensible for the accurate structure determination of novel retinoids. Spectra are usually obtained in deuteriochloroform, with tetramethylsilane as the internal standard, whenever solubility permits. Dimethyl-d_6 sulfoxide is used for the more insoluble polycyclic aromatic retinoid carboxylic acids. Acetone-d_6 and methanol-d_4 have also been used. Sample concentrations are usually 0.5–2.0 mg/ml. High-field NMR (300–400 MHz) must be used to completely resolve signals, although spectrum-matching techniques were employed[24] for analysis before high-field instruments were available. These techniques, together with homonuclear decoupling and nuclear Overhauser effect (NOE) experiments, were used to assign proton chemical shifts for many of the geometrical isomers and analogs of retinoic acid, retinal, and retinol.[25] These reference spectra are invaluable for assigning the configurations of synthetic polyolefinic retinoid analogs (Table III, structures **26–28**).[26,27] For example, signals of specific protons in the vicinity of a *Z* double bond are shifted relative to the corresponding signals of the *E* double-bond isomer because of shielding differences caused by the differences in bond stereochemistry (e.g., isomers of **1, 2, 5, 14, 19,** and **26** in Table III). Double-

[23] N. L. Rockley, M. G. Rockley, B. A. Halley, and E. C. Nelson, this series, Vol. 123, p. 92.

[24] B. A. Halley and E. C. Nelson, *J. Chromatogr.* **175,** 113 (1979).

[25] W. Vetter, G. Englert, N. Rigassi, and U. Schwieter, *in* "Carotenoids" (O. Isler, ed.), p. 189. Birkhaeuser, Basel, 1971.

[26] K. D. Hope, D. D. Muccio, P. D. Hobbs, and M. I. Dawson, unpublished (1988).

[27] M. I. Dawson and P. D. Hobbs, unpublished (1989).

TABLE III
CHEMICAL SHIFT POSITIONS FOR ^{1}H NMR SIGNALS OF RETINOIDS[a]

Structure	Structure number	Chemical shift (ppm)			
		8_R-H	12_R-H	14_R-H	20_R-Me
Analogs of ethyl retinoate					
(structure)	E-1	6.08[b]	6.27	5.76	2.34
(structure)	$9_R Z$-1	6.58[b]	6.20	5.76	2.36
(structure)	$13_R Z$-1	6.08[b]	7.77	5.63	2.05
(structure)	E-9	6.17[c]	6.28	5.77	2.35
(structure)	$13_R Z$-9	6.18[c]	7.78	5.64	2.07

(continued)

TABLE III (continued)

Structure	Structure number	Chemical shift (ppm)			
		9_R-H	10_R-H	14_R-H	20_R-Me
	E-**26**	6.25[d]	6.32	5.78	2.37
	9_RZ-**26**	6.79[d]	6.24	5.77	2.29
	13_RZ-**26**	6.25[d]	7.82	5.66	2.08
Aromatic benzoates					
	E-**14**		6.78[e]		
	9_RZ-**14**		6.45[e]		

	1-H	3-H	4-H	5-H	7-H	8-H	1'-H	3'-H	4'-H
E-2			6.83[e]						
9$_R$Z-2			6.48[e]						
E-5					6.18[c]	2.58			
13$_R$Z-5					5.92[c]	2.21			
Naphthalene carboxylates									
17, R = H	8.61[f]	7.99	8.07	8.24	7.90	8.17	7.73	7.55	7.44
27, R = Me	8.62[f]	8.07	8.21		87.49	8.02	7.36	7.19	7.46

(continued)

TABLE III (continued)

Structure	Structure number	Chemical shift (ppm)						
		2,6-H	3,5-H	1'-H	3'-H	4'-H	9'-H	10'-H
Anthracenyl benzoates								
	6	8.15[g]	7.79	8.02	7.66	7.84	7.83	7.87
	28	8.13[g]	7.78	8.68	7.68	7.77		7.82

[a] Spectra were obtained using a Varian XL 400-MHz, GE NT 300-MHz, a Nicolet 300-MHz, or a Bruker HX 270-MHz spectrometer on deuteriochloroform solutions except that (dimethyl sulfoxide)-d_6 was used for compounds 17 and 27. Tetramethylsilane was the internal reference.
[b] Ref. 3.
[c] Ref. 27.
[d] Ref. 12.
[e] Ref. 6.
[f] Ref. 26.
[g] Ref. 10.

bond configurations of polyolefinic retinoids may be assigned by comparison with the published ^{1}H NMR chemical shift data for isomers of the natural retinoids[24] and ethyl 3,7-dimethyl-9-(4-methoxy-2,3,6-trimethylphenyl)-2,4,6,8-nonatrienoate (Etretinate®, *E*-**26**).[12] The 9_RZ and *E* double-bond isomers of the 9_R- and 10_R-methyl analaogs of the α-methylstilbene retinoids are readily distinguished by the approximately 0.4 ppm upfield shift of the olefinic proton in the *Z* double-bond isomer relative to the equivalent signal in the *E* isomer, as well as smaller shifts of similar signals for the *gem*-dimethyl and aromatic protons caused by the greater shielding of these protons by the aromatic groups in the *Z* isomer.

The biaryl retinoid carboxylic acids and other nonolefinic retinoids are important tools in retinoid structure–activity studies. The complete assignment of the chemical shifts for the protons in this series of retinoids requires additional techniques, which are well illustrated by the analysis of the solution conformation of **17** and **27,** in which all the ^{1}H and ^{13}C NMR chemical shifts were unequivocally assigned using two-dimensional (2D) NMR.[25] Homonuclear selective decoupling was used to establish the through-bond connectivity of signals arising from the aromatic protons. Two-dimensional heteronuclear correlations were made both for the carbons bearing hydrogen and for the long-range coupling between fully substituted carbons and the protons two carbon atoms distant. The [1]H–[13]C connectivities permitted the unambiguous assignment of all the proton and carbon chemical shifts for the aromatic protons and carbons. Two-dimensional phase-sensitive NOE enhancement was used to confirm the chemical shift assignments and to assign the nonaromatic proton chemical shifts. The NOE signal enhancements observed between protons in different aromatic rings were used to calculate the corresponding interproton distances in order to determine the dihedral angle between the planes of the biaryl aromatic systems. Of course, proton chemical shifts are rarely assigned as rigorously as in the above example. Two-dimensional heteronuclear experiments require high sample concentrations (e.g., 13 mg/ml in dimethyl-d_6 sulfoxide in this case). NOE enhancements are generally useful for the proton chemical shift assignments of the aromatic retinoids. For example, the ^{1}H NMR signals for the *gem*-dimethyl groups and the aromatic proton signals of **3** and **4** (Table I) were assigned by measuring the NOE [1]H–[1]H interactions in a 2D NOE experiment, which permitted the unequivocal identification of these isomers.

Carbon-13 Nuclear Magnetic Resonance Spectroscopy

^{13}C NMR spectra of retinoids are usually obtained on samples dissolved in deuteriochloroform if the solubility is adequate. Dimethyl-d_6 sulfoxide) must frequently be used as the solvent for the polycyclic aro-

matic retinoids. Retinoid concentrations of 10 mg/ml on a 100.6-MHz instrument (Varian, Palo Alto, CA, Model XL 400) or 25–50 mg/ml on a 22.5-MHz spectrometer (Jeol, Tokyo, Japan, Model FX90Q) have been used. Tetramethylsilane has been used as an internal standard. Chemical shift assignments for the aromatic retinoids have been made by 2D heteronuclear (^{1}H–^{13}C) correlation spectroscopy.[26-28] Prior to the introduction of this technique, Englert published extensive tables of ^{13}C NMR assignments for the polyolefinic retinoids by utilizing ^{13}C chemical shift differences on complexation of the retinoid polar terminus with the lanthanide shift reagent ytterbium tris(dipivaloylmethane), together with proton-coupled spectra.[12,29] These tables were used to assign configurations to synthetic polyolefinic retinoids, including isomers of ethyl 3,7-dimethyl-9-(4-methoxy-2,3,6-trimethylphenyl)-2,4,6,8-nonatetraenoate. Several of the observed chemical shift differences between isomers clearly indicate the Z double-bond positions (Table IV, structures **29–31**). For example, in isomers *E*-**9** and 13$_R$*Z*-**9** the 7 ppm downfield shift of the signal due to the 20$_R$-methyl carbon in the Z double-bond isomer relative to that in the E isomer demonstrates the presence of the 13Z double bond.

Ultraviolet Spectroscopy

Ultraviolet (UV) spectra are normally run on ethanolic solutions of retinoids, except for the carboxylic acids, which are frequently obtained in 95% ethanol to prevent molecular association. Absorption data and concentrations are illustrated in Table V (structures **32–35**). Geometrical isomers about a single double bond are readily identified by comparison of the absorption maxima of the two isomers because differences in the longest-wavelength absorption maxima are not large compared to shifts caused by differences in substitution. Important differences between isomeric pairs include (1) a consistent increase in the absorption maximum in the double-bond series of 7*E*,9*Z*,11*E*,13*E*; 7*E*,9*E*,11*E*,13*E*-; and 7*E*,9*E*,11*E*,13*Z*-polyolefinic retinoids (compare isomers of **9** and **32**, Table V), (2) a longer wavelength maximum for the 9$_R$*E*-stilbene retinoids compared with their 9$_R$*Z* isomers (isomers of **14** and **29**), and (3) a shorter wavelength maximum for the ortho-substituted biaryl retinoids compared with unsubstituted analogs (**17** and **27**). In **21** the absorption maximum at 285–305 nm, typically found for compounds in this series, is masked because of overlap with the principal maximum at 258 nm. Large shifts of the absorption maximum to shorter wavelength indicate considerably re-

[28] K. M. Waugh, K. D. Berlin, W. T. Ford, E. M. Holt, J. P. Carrol, P. R. Schomber, M. D. Thompson, and L. J. Schiff, *J. Med. Chem.* **28**, 116 (1985).

[29] G. Englert, *Helv. Chim. Acta* **58**, 2367 (1975).

TABLE IV
CHEMICAL SHIFTS OF ^{13}C NMR SIGNALS OF RETINOIDS[a]

Structure	Structure number	Chemical shift (ppm)		
		12_R-C	19_R-C	20_R-C
Analogs of ethyl retinoate				
	E-1	135.0[b]	13.1	13.7
	9_RZ-1	134.3[b]	21.3	13.9
	13_RZ-1	129.5[b]	13.1	20.7
	E-9	135.2[c]	12.9	13.8
	13_RZ-9	129.4[c]	12.8	20.9

(continued)

TABLE IV (continued)

Structure	Structure number	Chemical shift (ppm)		
	E-**26**	135.9[d]	12.9	13.8
	9_RZ-**26**	135.1[d]	20.8	13.9
	13_RZ-**26**	130.2[d]	12.9	20.8
Aromatic benzoates				
	E-**13**		17.4[e]	
	9_RZ-**13**		26.8[e]	

(continued)

E-14	18.0^f	
E-29	17.8^f	
9$_R$Z-29	27.5^f	
E-30	14.1^b	
9$_R$Z-30	21.3^b	
E-31	13.8^b	

TABLE IV (continued)

Naphthalene carboxylates

Chemical shift (ppm)

Structure number	1-C	2-C	3-C	4-C	4a-C	5-C	6-C	7-C	8-C	8a-C
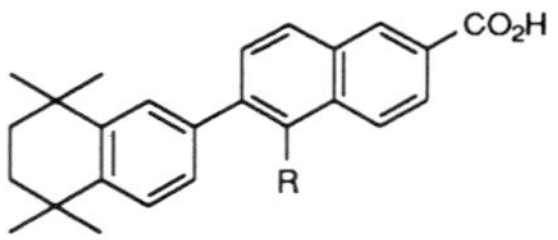 17, R = H	130.3[g]	127.8	125.5	128.4	135.4	124.7	140.1	126.1	129.8	131.2
27, R = Me	130.9[g]	127.5	125.6	125.1	134.6	130.4	140.9	129.1	127.3	131.4

Structure number	1'-C	2'-C	3'-C	4'-C	4a'-C	8a'-C	Me	CO
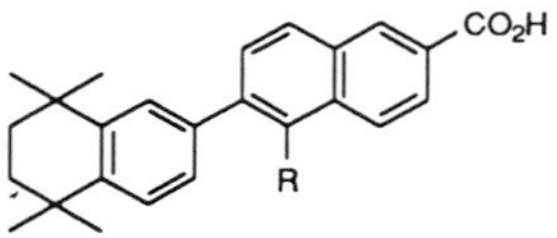 17, R = H	125.0[g]	136.7	124.4	127.2	145.1	144.3		167.4
27, R = Me	127.8[g]	138.4	126.8	126.4	144.2	143.2	16.2	167.5

[a] Spectra were run on a Varian XL 400-MHz spectrometer at 100.6 MHz, a GE NT 300-MHz or a Nicolet 300-MHz spectrometer at 75.5 MHz, or a Bruker HX-270 instrument at 68 MHz, using deuteriochloroform as the solvent except for 17 and 27, which were run in dimethyl-d_6 sulfoxide.
[b] Ref. 3.
[c] M. I. Dawson, P. D. Hobbs, R. L. Chan, W. Chao, and V. A. Fung, J. Med. Chem. 24, 583 (1981).
[d] Ref. 27.
[e] Ref. 12.
[f] Ref. 6.
[g] Ref. 26.

duced conjugation. In the Z isomer of $\mathbf{5}$[27] the terminal double bond is not coplanar[30–32] with the aromatic ring; therefore, the conjugation of the system is reduced and the absorption maximum is shifted to 280 nm compared with that of 300 nm in the E isomer. The absorption maximum of a retinoid metabolite appeared at considerably shorter wavelength than that of retinoic acid because of partial saturation and deconjugation of the polyene chain.[20]

UV absorption measurements have been used in conformational studies of retinals, other polyolefinic retinoids[33,34] and aromatic retinoids.[26] Suzuki derived a linear relationship between $\delta_{\lambda max}$ and the angle between the planes of the aromatic systems in biphenyl compounds. This relationship has been found to hold true for the biaryl retinoids.[26]

Melting Point Temperature

Melting points of polyolefinic retinoids must be determined on samples that are sealed under inert gas in capillary tubes because the rate of atmospheric decomposition for these compounds is enhanced on heating.

Elemental Analysis

Crystalline retinoidal carboxylic acids have occasionally been obtained as solvates and so the proton NMR spectra of these compounds show solvent signals. A solvent-free compound may be obtained by heating under reduced pressure, usually below 100°. For retinoids that are not thermally stable, elemental analyses must be corrected for the solvent present, which can be determined by ^{1}H NMR peak integration. High-resolution mass spectral determination of the molecular formula may also be employed as an analytical method, especially for valuable samples or for those retinoids that are gums or oils from which it is extremely difficult to remove residual traces of solvent.

Conformational Analysis

Deducing the conformation that a retinoid adopts is important for understanding the interaction and binding of the retinoid with its biological receptor sites. NMR and UV spectroscopy, X-ray crystallographic measurements, and molecular mechanics calculations are useful techniques for determining or predicting the minimum-energy conformations

[30] Y. Urushibara and M. Hirota, *Nippon Kagaku Zasshi* **82,** 351 (1961).
[31] Y. Urushibara and M Hirota, *Nippon Kagaku Zasshi* **82,** 354 (1961).
[32] Y. Urushibara and M. Hirota, *Nippon Kagaku Zasshi* **82,** 358 (1961).
[33] R. R. Birge and B. M. Pierce, *J. Chem. Phys.* **70,** 165 (1979).
[34] B. Honig, A. Warshel, and M. Karplus, *Acc. Chem. Res.* **8,** 92 (1975).

TABLE V
Ultraviolet Absorption Maximum of Retinoids[a]

Retinoid	Structure number	λ_{max}(nm)	$\varepsilon \times 10^{-4}$
	E-9	357[b]	4.5
	13_RZ-9	360[b]	3.9
	E-32	345[c]	4.8
	9_RZ-32	341[c]	4.1
	E-33, R = Et	314[d]	2.8
	E-31, R = H	309[d]	2.9
	E-34	272[e]	2.7
	E-35	287[e]	2.1
	E-29	308[e]	2.7
	9_RZ-29	300[e]	1.7

TABLE V (*continued*)

Retinoid	Structure number	$\lambda_{max}(nm)$	$\varepsilon \times 10^{-4}$
	E-14	316[e]	2.4
	9_RZ-14	306[e]	1.2
	17, R = H	303[f]	3.0
	27, R = Me	294[f]	1.2
	19, R = H	267[g]	5.1
		303	1.8
	21, R = Me[h]	258[b]	3.1
	E-5	300[b]	2.6
	13_RZ-5	280[b]	1.8

[a] Spectra were obtained at retinoid concentrations of $1.0–4.0 \times 10^{-5}$ M; ethanol as solvent. Only the longest wavelength maxima are noted because these are used for isomer comparison (with the exception of the spectrum for 21).

[b] Ref. 27.

[c] Ref. 3.

[d] M. I. Dawson, P. D. Hobbs, R. L. Chan, W. Chao, and V. A. Fung, *J. Med. Chem.* 24, 583 (1981).

[e] Ref. 6.

[f] Ref. 26.

[g] Ref. 10.

[h] 95% ethanol.

of retinoids. These conformations are considered to be similar to the conformation that a ligand adopts on binding to its receptor. For example, knowledge of the preferred conformers of E- and 11Z-retinal (**36**) (Fig. 2, structures **36–44**), is important for studies on the binding of 11Z-retinal to opsin in the visual process.[34] The low-energy conformations of these isomers have been predicted using molecular mechanics calculations[35] and determined by X-ray[36,37] and NMR[38] techniques to give the conformation of the retinoid in the crystal and in solution, respectively.

The calculated potential of the cyclohexenyl ring of E-**36** for rotation about the 6–7 bond showed a broad energy minimum for a torsion angle of 40–120° with a minimum at 50° for the s-cis conformation, and a second higher energy minimum very close to the s-*trans* geometry. Two X-ray crystallographic determinations gave ring–chain torsion angles of 58° and 62°.[36,37] The polyolefinic chain was slightly bent normal to the plane, but curved considerably in the plane because of the 19- and 20-methyl group interactions. Packing interactions may have distorted the structure compared to the solution conformation. Both the ring[38] and chain[35] conformation of E-**36** in solution have been studied by measurement of NMR long-range coupling constants and NOE enhancements. The NOE interactions between the 16-, 17-, and 18-methyl protons and the 7- and 8-olefinic protons indicated s-*cis* geometry between the ring and the side chain with an increase of the torsion angle with temperature. At ambient temperature the range of rotation was approximately 25–90°. The side-chain region from C-7 to C-15 was essentially planar.

The other isomers of retinal, as well as (E)-retinoic acid (E-**8**)[39] and (E)-retinyl acetate (E-**37**),[40] have been investigated. The minimum-energy conformations of several aromatic retinoids were determined theoretically[10] and by crystallographic,[8,29] NMR,[26] and UV[26] spectroscopic techniques in order to relate the tertiary structure of these compounds to their biological activity. The orientation of the ring and the side chain of the other isomers of retinal[41,42] is very similar to that of E-**36**. For example, calculations give a torsion angle of 41° for 11Z-**36**, which is confirmed by

[35] R. Rowan, A. Warshel, B. D. Sykes, and M. Karplus, *Biochemistry* **13**, 970 (1974).
[36] R. D. Gilardi, I. L. Karle, and J. Karle, *Acta Crystallogr., Sect. B* **28**, 2605 (1972).
[37] T. Hamanaka, T. Mitsui, T. Ashida, and M. Kakudo, *Acta Crystallogr., Sect. B* **28**, 214 (1972).
[38] B. Honig, B. Hudson, B. D. Sykes, and M. Karplus, *Proc. Natl. Acad. Sci. U.S.A.* **68**, 1289 (1971).
[39] C. H. Stam, *Acta Crystallogr., Sect. B* **28**, 2936 (1972).
[40] W. E. Oberhaensli, H. P. Wagner, and O. Isler, *Acta Crystallogr., Sect. B* **30**, 161 (1974).
[41] R. Rowan, III, and B. D. Sykes, *J. Am. Chem. Soc.* **97**, 1023 (1975).
[42] B. H. S. Lienard and A. J. Thomson, *J. Chem. Soc. Perkin Trans. 2*, p. 1400 (1977).

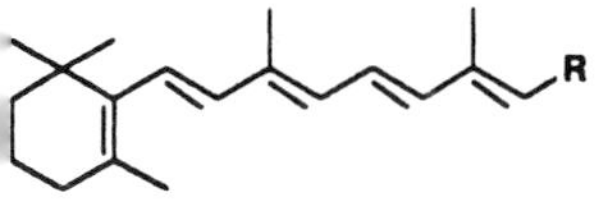

E-8   R = CO₂H
E-36   R = CHO
E-37   R = CH₂OAc

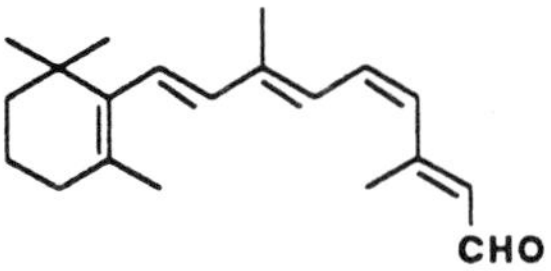

11Z-36

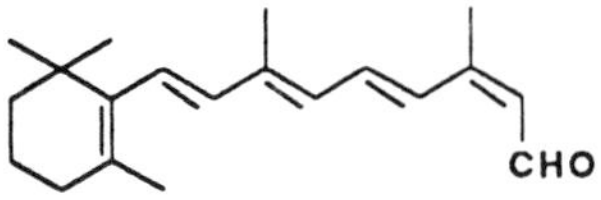

13Z-36

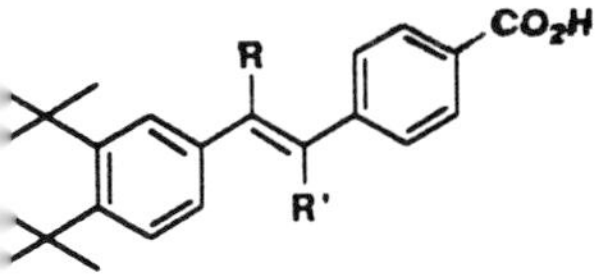

E-38   R = Me, R' = H
E-39   R = H, R' = Me
E-40   R = Me, R' = Me

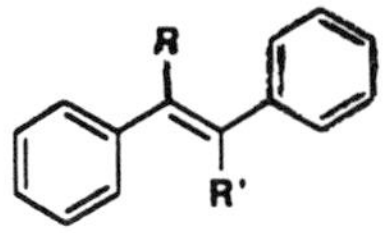

E-41   R = Me, R' = H
E-42   R = Me, R' = Me

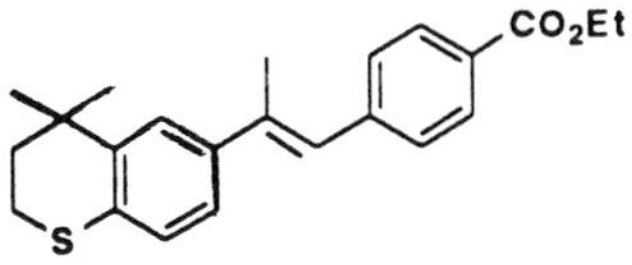

E-43

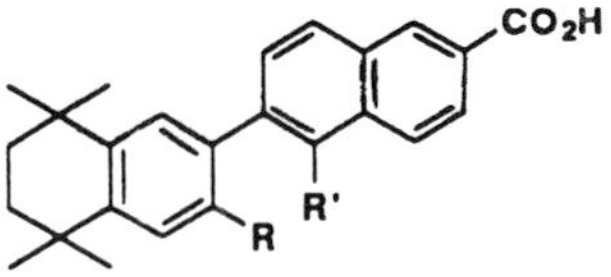

17 R = H, R' = H
27 R = H, R' = Me
44 R = Me, R' = H

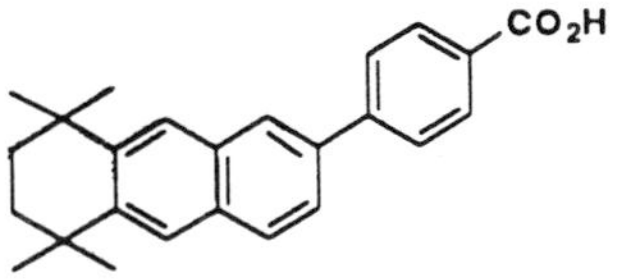

19

FIG. 2. Structures of retinoids on which conformational analyses have been performed.

the 44° angle found by X-ray crystallography.[36] In addition, X-ray, UV, and ^{1}H NMR studies all indicate that the side-chain geometry is predominantly 12-s-*cis*. According to ^{1}H NMR measurements, the chain is planar from C-7 to C-10 and from C-13 to C-15[35] but slightly twisted from planarity about the 10–11 bond. The 12–13 bond is rotated approximately 40°. ^{1}H NMR studies also revealed the contribution of a distorted s-*trans* conformer for this bond.[35] ^{1}H and ^{13}C NMR relaxation measurements and chemical shifts demonstrated that the side chains of *E*-36, 9*Z*-36, and 13*Z*-36 were planar.[41,43] The X-ray structure of 13*Z*-36[44] revealed the presence

[43] R. S. Becker, S. Berger, D. K. Dalling, D. M. Grant, and R. J. Pugmire, *J. Am. Chem. Soc.* **96**, 7008 (1974).

[44] C. J. Simmons, R. S. H. Liu, M. Denny, and K. Seff, *Acta Crystallogr., Sect. B* **37**, 2197 (1981).

of both 6-s-*cis* and almost planar 6-s-*trans* conformers in the triclinic crystal.

There were two crystalline forms for (*E*)-retinoic acid (E-**8**). The triclinic crystal contained the 6-s-*cis* conformer, and the monoclinic form had an almost planar 6-s-*trans* geometry.[39] The X-ray structure of (*E*)-retinyl acetate (*E*-**37**) showed a 6-s-*cis* configuration with a torsion angle of 58° about the 6–7 bond.[40] As in the case of *E*-**36**, the polyolefinic side chain showed high in-plane bending, which is caused by nonbonded interaction of the 19- and 20-methyl groups.

The conformations of several synthetic retinoid carboxylic acids have been determined either experimentally or by calculations. The stilbene series of retinoids are capable of conformational flexibility about the 8_R–9_R (torsion angle α_1) and 10_R–11_R (torsion angle α_2) single bonds. Molecular mechanics calculations on *E*-**38**[10] yield four energy minima for rotation about the first bond and, for each of these rotamers, two minima for rotation about the second bond. The conformational minima were within 2 kcal/mol and the barriers between minima were less than 4 kcal/mol. The interplanar angle of the aromatic rings of *E*-**38** was 77° for the conformers having perpendicular rings, which is in good agreement with the 71° angle for **6** found by X-ray crystallography[8] and the 69° angle (assuming α_1 and α_2 are equal) for (*E*)-α-methylstilbene (*E*-**41**) calculated by Suzuki[45] from the UV absorption spectrum.

Studies were also reported for analogs of *E*-**38**. The torsion angles for *E*-**39**, the 10_R-methyl analog of *E*-**38**, were calculated to be almost the inverse of those of *E*-**38**, whereas both were found to be 58° for the 9_R,10_R-dimethyl analog *E*-**40**.[10] This result is in complete agreement with the torsion angles Suzuki calculated from the UV spectrum of α,α'-dimethylstilbene (*E*-**42**).[45] X-ray studies indicated that the torsion angle between the planes of the aromatic rings of *E*-**43** was 84–86°.[28]

The biaryl retinoids **17**, **19**, **27**, and **44** possess conformational flexibility about one single bond (8_R–9_R or 10_R–11_R) of the 7_R–11_R side chain. Molecular mechanics calculations predicted a torsion angle of 32° for **19** and 39° for **17**.[10] UV absorption data suggested a torsion angle of 40–43° for biphenyl. Molecular mechanics calculations predicted an angle of 56° for **27** and one of 54° for **44**, the *o*-methyl analogs of **17**.[10] The torsion angle of *o*-methylbiphenyl was calculated to be 58° from the UV spectrum.[46] Two-dimensional NOE enhancement calculations gave a torsion angle of 34° for **17**[26] and one of 45° for both **27** and **44**, which are smaller than that determined by the other two methods.

[45] H. Suzuki, *Bull. Chem. Soc. Jpn.* **33**, 396 (1960).
[46] H. Suzuki, *Bull. Chem. Soc. Jpn.* **32**, 1350 (1959).

Acknowledgements

Support of this work by U.S. Public Health Service Grants CA30512 and CA32428 is gratefully acknowledged.

[3] Structural Characteristics of Synthetic Retinoids

By BRAHMA P. SANI and DONALD L. HILL

Introduction

The word retinoid is a general term that includes both the naturally occurring compounds with vitamin A activity and the synthetic analogs of retinol, with or without biological activity. The IUPAC–IUB Joint Commission on Biochemical Nomenclature[1] states that "retinoids are a class of compounds consisting of four isoprenoid units joined in a head-to-tail manner. All retinoids may be formally derived from a monocyclic parent compound containing five carbon–carbon double bonds and a functional group at the terminus of the acyclic portion. To avoid confusion with previously used names in this field no parent hydrocarbon is named."

The first total synthesis of a retinoid was reported by Kuhn and Morris in 1937.[2] Arens and Van Dorp reported the first synthesis of retinoic acid.[3] Isler *et al.*[4] at Hoffmann-La Roche developed the first synthesis of retinol on an industrial scale. In this field, there have been major advances in synthetic organic chemistry, resulting in the synthesis of about 2000 retinoids. Some have little resemblance to retinol and retinoic acid but still retain the biological activities associated with them.

Structures of Retinoids

The classic retinoid structure is generally subdivided into three segments, namely, the polar terminal end, the conjugated side chain, and the cyclohexenyl ring. The basic structures of the most common natural retinoids are called retinol, retinaldehyde, and retinoic acid (Fig. 1). Structures of several derivatives of the naturally occurring retinoids and examples of typical synthetic retinoids are listed below. For many of these,

---

[1] IUPAC–IUB Joint Commission on Biochemical Nomenclature, *Eur. J. Biochem.* **129**, 1 (1982).
[2] R. Kuhn and C. J. O. R. Morris, *Chem. Ber.* **70**, 853 (1937).
[3] J. F. Arens and D. A. Van Dorp, *Nature (London)* **175**, 190 (1946).
[4] O. Isler, W. Huber, A. Ronco, and M. Kofter, *Helv. Chim. Acta* **30**, 1911 (1947).

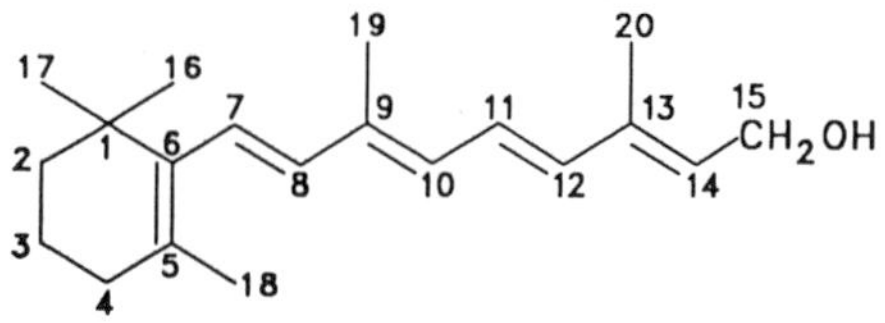

all-*trans*-Retinol; (all-*E*)-3,7-dimethyl-9-(2,6,6-trimethyl-1-yl)-2,4,6,8-nonatetraen-1-ol

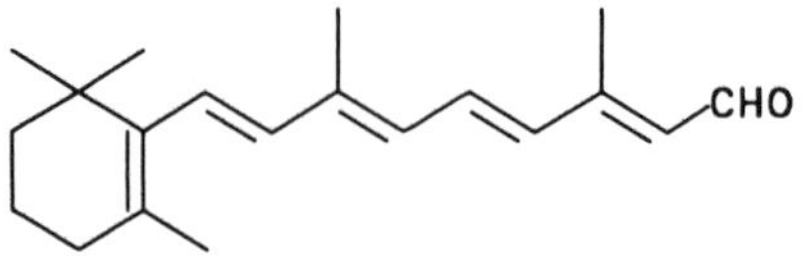

all-*trans*-Retinaldehyde; all-*trans*-retinal; (all-*E*)-3,7-dimethyl-9-(2,6,6-trimethyl-1-cyclo-hexen-1-yl)-2,4,6,8-nonatetraenal

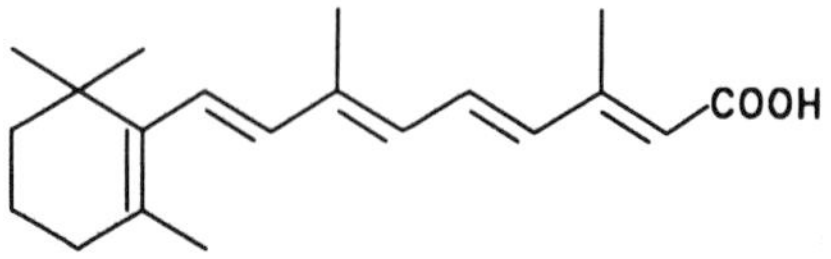

all-*trans*-Retinoic acid; retinoic acid; (all-*E*)-3,7-dimethyl-9-(2,6,6-trimethyl-1-cyclohexen-1-yl)-2,4,6,8-nonatetraenoic acid

FIG. 1. Structure of retinol, retinaldehyde, and retinoic acid.

structure–activity relationships and chemical and physical properties have been reviewed at length.[5,6]

1. all-*trans*-Retinyl ethers

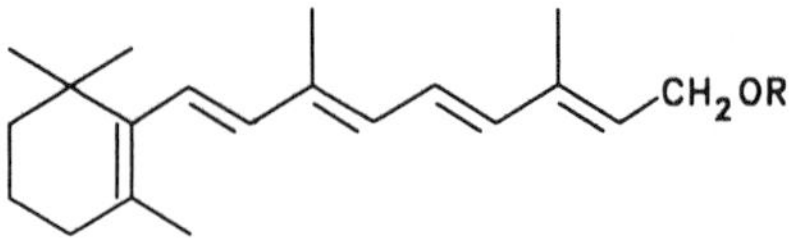

R = H, retinol
= CH$_3$, retinyl methyl ether
= C$_4$H$_9$, retinyl butyl ether
= C$_6$H$_5$, retinyl phenyl ether
= C$_8$H$_{17}$, retinyl octyl ether
= glucuronic acid, retinyl β-glucuronide

[5] D. L. Newton, W. R. Henderson, and M. B. Sporn, *Cancer Res.* **40**, 3413 (1980).
[6] F. Frickel, *in* "The Retinoids" (M. B. Sporn, A. B. Roberts, and D. S. Goodman, eds.), Vol. 1, p. 7. Academic Press, Orlando, Florida, 1984.

2. all-*trans*-Retinyl esters

R = COCH$_3$, retinyl acetate
  = COC$_{15}$H$_{31}$, retinyl palmitate

3. all-*trans*-Retinylamine derivatives

R$_1$, R$_2$ = H, COCH$_3$, *N*-acetylretinylamine
     = H, COC$_6$H$_5$, *N*-benzoylretinylamine
     = CH$_3$, COCH$_3$, *N*-methyl-*N*-acetylretinylamine
     = CH$_3$, COC$_6$H$_5$, *N*-methyl-*N*-benzoylretinylamine

4. all-*trans*-Retinal derivatives

R = O, retinal
  = NOH, retinal oxime
  = NNHCOCH$_3$, retinal acetylhydrazone
  = C(COCH$_2$CH$_2$CH$_3$)$_2$, 5-retinylidene-4,6-nonanedione
  = C(COCH$_2$)$_2$, 2-retinylidene-1,3-cyclopentanedione
  = C(COCH$_2$)$_2$CH$_2$, 2-retinylidene-1,3-cyclohexanedione
  = C(COCH$_2$CH$_2$)$_2$CH$_2$, 2-retinylidene-1,3-cyclooctanedione

5. all-*trans*-Retinoic acid esters

R = H, retinoic acid
  = CH$_3$, retinoic acid methyl ester
  = C$_2$H$_5$, retinoic acid ethyl ester
  = glucuronic acid, retinoyl β-glucuronide.

6. all-*trans*-Retinoylamino acids

R = glycine, retinoylglycine
  = leucine, retinoylleucine
  = phenylalanine, retinoylphenylalanine
  = tyrosine, retinoyltyrosine

7. all-*trans*-retinamides

R = C_2H_5, *N*-ethylretinamide
  = C_3H_7, *N*-propylretinamide
  = 2-C_2H_4OH, *N*-(2-hydroxyethyl)retinamide
  = 2-C_3H_6OH, *N*-(2-hydroxypropyl)retinamide
  = 3-C_3H_6OH, *N*-(3-hydroxypropyl)retinamide
  = C_6H_5, *N*-phenylretinamide
  = 2-C_6H_4OH, *N*-(2-hydroxyphenyl)retinamide
  = 4-C_6H_4OH, *N*-(4-hydroxyphenyl)retinamide
  = 2-C_6H_4COOH, *N*-(2-carboxyphenyl)retinamide
  = 4-C_6H_4-$COOH$, *N*-(4-carboxyphenyl)retinamide

8. 5,6-all-*trans*-Epoxyretinoids

R = COOH, 5,6-epoxyretinoic acid
  = $COOCH_3$, 5,6-epoxyretinoic acid methyl ester
  = CH_2OCOCH_3, 5,6-epoxyretinyl acetate

9. 13-*cis*-Retinoic acid derivatives

R = OH, 13-*cis*-retinoic acid
  = NHC$_2$H$_5$, *N*-ethyl-13-*cis*-retinamide
  = NH(2-C$_2$H$_4$OH), *N*-(2-hydroxyethyl)-13-*cis*-retinamide
  = NH(4-C$_6$H$_4$OH), *N*-(4-hydroxyphenyl)-13-*cis*-retinamide
  = leucine, *N*-(13-*cis*-retinoyl)leucine
  = phenylalanine, *N*-(13-*cis*-retinoyl)phenylalanine

10. Bifunctional retinoic acid analogs

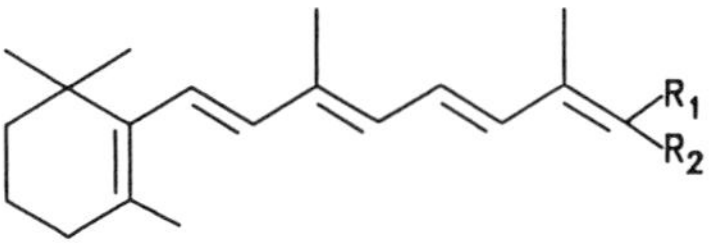

R$_1$, R$_2$ = COOH, COOH, 14-carboxyretinoic acid
   = COOC$_2$H$_5$, COOC$_2$H$_5$, ethyl 14-(ethoxycarbonyl)retinoate
   = CONHC$_2$H$_5$, COOH, 14-[(ethylamino)carbonyl]-13-*cis*-retinoic acid

11. Ring-modified all-*trans*-retinoic acid analogs

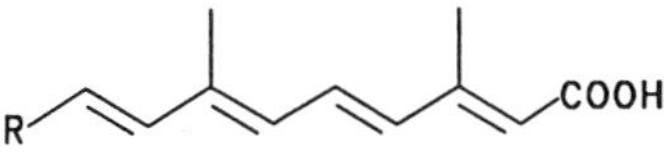

R = 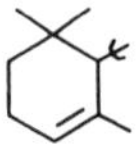 α-retinoic acid   R = , 3-pyridyl analog of retinoic acid

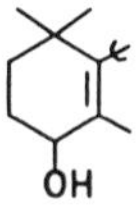

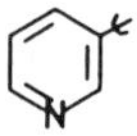

= , 4-hydroxyretinoic acid

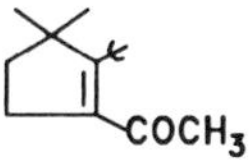

= , dimethylacetylcyclopentenyl analog of retinoic acid

= 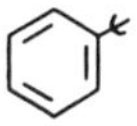 , phenyl analog of retinoic acid

= 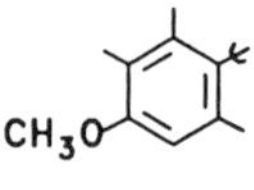 , 4-methoxy-2,3,6-trimethylphenyl analog of retinoic acid

= , 2-furyl analog of retinoic acid

=  , 3-thienyl analog of retinoic acid

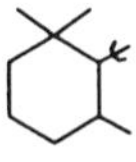

= , 5,6-dihydroretinoic acid

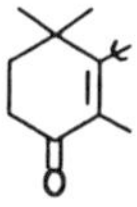

= , 4-oxoretinoic acid

12. Side-chain-modified all-*trans*-retinoic acid analogs

$R =$ COOH   , C_{15} analog of retinoic acid

$=$ COOH   , C_{17} analog of retinoic acid

$=$ COOH , C_{22} analog of retinoic acid

$=$ COOH   , aryltriene analog of retinoic acid

   COOH   , 7,8-dihydroretinoic acid

$=$ COOH   , 9,10-dihydroretinoic acid

$=$ COOH   , 11,12-dihydroretinoic acid

13. Ring- and side-chain-modified analogs of all-*trans*-retinoic acid

(*E*)-4-[2-(5,6,7,8-Tetrahydro-5,5,8,8-tetramethyl-2-naphthalenyl)-1-propenyl]benzoic acid

(*E*)-4-[2-(5,6,7,8-Tetrahydro-8,8-dimethyl-2-naphthalenyl)-1-propenyl]benzoic acid

(*E*)-4-[(5,6,7,8-Tetrahydro-5,5,8,8-tetramethyl-2-naphthalenyl)carbamolyl]benzoic acid

(E)-4-[(5,6,7,8-Tetrahydro-5,5,8,8-tetramethyl-2-naphthalenyl)car-
boxamido]benzoic acid

(E)-4-[2-(2,3-Dihydro-1,1,2,3,3-pentamethyl-1H-inden-5-yl)-1-pro-
penyl]benzoic acid

6-(5,6,7,8-Tetrahydro-5,5,8,8-tetramethyl-2-naphthalenyl)-2-naphtha-
lenecarboxylic acid

6-(5,6,7,8-Tetrahydro-5,5,8,8-tetramethyl-2-naphthalenyl)-5-methyl-
2-naphthalenecarboxylic acid

2-(5,6,7,8-Tetrahydro-5,5,8,8-tetramethyl-2-naphthalenyl)-6-benzo[b]
thiophenecarboxylic acid

4-(5,6,7,8-Tetrahydro-5,5,8,8-tetramethyl-2-anthracenyl)benzoic acid

(E)-4-[3-(3,5-Di-*tert*-butylphenyl)-3-oxo-1-propenyl]benzoic acid

[4] High-Performance Liquid Chromatography of Retinoids in Blood

By A. P. DE LEENHEER and H. J. NELIS

Introduction

The principal form of vitamin A in blood under fasting conditions is retinol.[1] This compound is mobilized from retinyl ester stores in the liver and circulates in association with its specific transport protein [retinol-

---

[1] G. A. J. Pitt, *in* "Fat-Soluble Vitamins—Their Biochemistry and Applications" (A. T. Diplock, ed.), p. 1. Heinemann, London, 1985.

binding protein (RBP) + transthyretin].[1] Plasma levels of retinol are of the order of 200–800 μg/liter.[2] Retinoic acid, an oxidative metabolite of retinol[1] occurs in much lower concentration, namely, 3–5 μg/liter.[3] A number of polar metabolites of retinoic acid have been isolated,[4] including 5,6-epoxyretinoic acid, 4-ketoretinoic acid, and retinoyl-β-glucuronide, some of which are reportedly endogenous compounds of plasma as well.[5] Appreciable amounts of retinyl esters are found in the blood only after administration of vitamin A (vitamin A tolerance test) and in conditions of hypervitaminosis A.[1,6]

This chapter is concerned with the determination of the two plasma retinoids of major interest, retinol and retinoic acid. Analytical procedures for retinoic acid metabolites and retinyl esters are discussed elsewhere in this volume. Three liquid chromatographic (HPLC) methods developed in the authors' laboratory are described in detail. References to other literature highlight recent progress that has been made in the area.

Principles

The choice between normal- or reversed-phase chromatography for retinol or retinoic acid depends on several factors, including chromatographic convenience, the availability of an internal standard, sensitivity aspects, and the desired extent of sample pretreatment. For example, normal-phase chromatography on silica is more subject to variability than its reversed-phase counterpart. The selection of a suitable structural analog for internal standardization is less straightforward in normal- than in reversed-phase chromatography. The fluorescence quantum yield of retinol is highest in nonpolar (normal-phase) eluents. Sample pretreatment in reversed-phase chromatography of retinol is usually more complex because of the need for an evaporation (concentration) step. Two of the methods below are based on normal-phase, one on reversed-phase chromatography. All three employ absorption rather than fluorescence detection.

[2] A. P. De Leenheer, H. J. Nelis, W. E. Lambert, and R. M. Bauwens, *J. Chromatogr.* **429**, 3 (1988).

[3] A. P. De Leenheer, W. E. Lambert, and I. Claeys, *J. Lipid Res.* **23**, 1362 (1982).

[4] C. A. Frolik, *in* "The Retinoids" (M. B. Sporn, A. B. Roberts, and D. S. Goodman, eds.), Vol. 2, p. 177. Academic Press, Orlando, Florida, 1984.

[5] A. B. Barua and J. A. Olson, *Am. J. Clin. Nutr.* **43**, 481 (1986).

[6] M. G. M. De Ruyter and A. P. De Leenheer, *Clin. Chem.* (*Winston-Salem, N.C.*) **24**, 1920 (1978).

Chromatography Procedures

Standard Compounds

all-*trans*-Retinol and all-*trans*-retinoic acid were purchased from Fluka (Buchs, Switzerland). The internal standard for retinol in normal phase chromatography (I.S. 1), all-*trans*-9-(4-methoxy-2,3,6-trimethylphenyl)-3,7-dimethyl-2,4,6,8-tetraenol, is synthesized by $LiAlH_4$ reduction of ethyl all-*trans*-9-(4-methoxy-2,3,6-trimethylphenyl)-3,7-dimethyl-2,4,6,8-tetraenoate, donated by Hoffmann-La Roche (Basel, Switzerland). The final solution of this compound in methanol has an absorbance of 0.895 at 327 nm in a 1.00 cm cell. Retinyl propionate (I.S. 2, used in reversed-phase chromatography of retinol), was obtained from AEC (Commentry, France). all-*trans*-13-Demethylretinoic acid (I.S. 3) was a gift from Hoffmann-La Roche (Nutley, NJ). Stock solutions of retinol, retinoic acid, and the internal standards, prepared in ethanol or methanol, are stored in amberized tubes at $-20°$.

Liquid Chromatographic Instrumentation

Any isocratic HPLC system equipped with a variable wavelength detector will suffice for either procedure. In our laboratory different pumps, namely, Varian 4100 or Varian 5020 (Varian Associates, Palo Alto, CA) or Pye Unicam LC3-XP (Pye Unicam, Cambridge, UK) and detectors (Varian Varichrom UV-10 or Pye Unicam LC3-UV) are used. Injections are made using a Valco N60 valve (Valco, Houston, TX) fitted with a 50-μl loop, or using a Varian stop-flow injector. The detection wavelength is 328–330 nm (retinol) or 350 nm (retinoic acid).

Liquid Chromatographic Conditions

The conditions of the different liquid chromatographic systems used are as follows.

1. Retinol/normal-phase chromatography[7]:
   Column, 10 μm MicroPak Si-10, 15 × 0.2 cm; mobile phase, petroleum ether : dichloromethane : 2-propanol, 80 : 19.3 : 0.7 (v/v/v); flow rate, 0.5 ml/min; temperature, ambient; injection volume, 100 μl.
2. Retinol/reversed-phase chromatography[8]:

---

[7] M. G. M. De Ruyter and A. P. De Leenheer, *Clin. Chem.* (*Winston-Salem, N.C.*) **22,** 1593 (1976).

[8] H. J. C. F. Nelis, J. De Roose, H. Vandenbavière, and A. P. De Leenheer, *Clin. Chem.* (*Winston-Salem, N.C.*) **29,** 1431 (1983).

Column, 7 μm Zorbax ODS, 25 × 0.46 cm; mobile phase, acetonitrile : dichloromethane : methanol, 70 : 15 : 15 (v/v/v); flow rate, 1 ml/min; temperature, ambient; injection volume, 50 μl.

3. Retinoic acid/normal-phase chromatography[3]:
   Column, 5 μm RSIL, 15 × 0.32 cm; mobile phase, petroleum ether : acetonitrile : acetic acid, 99.5 : 0.2 : 0.3 (v/v/v); flow rate, 0.75 ml/min; temperature, ambient; injection volume, 50 μl.

Preparation of Samples for Retinol Determination

Because both retinol and retinoic acid are susceptible to isomerization and oxidation, precautions must be taken to minimize these degradation reactions. All manipulations are carried out in subdued light, in amberized glassware, and, wherever possible, at reduced temperature. For the evaporation of extracts vacuum or nitrogen is used but no antioxidants are added before or during the extraction.

Normal-Phase Chromatography.[7] To 100 μl of serum in a conical centrifuge tube (7.6 × 0.9 cm) is added, under vigorous mixing (Super mixer 1291, Lab-line Instruments, Melrose Park, IL), 15 μl of internal standard solution (I.S. 1), 100 μl of methanol, and 200 μl of the extraction solvent, that is, the chromatographic solvent. After 60 sec of mixing and centrifugation (3000 rpm, 2 min), a 100-μl aliquot of the supernatant is injected. Standardization is carried out by adding known amounts of retinol (100–1500 μg/liter) to water and analyzing these aqueous samples by the above procedure.

Reversed-Phase Chromatography.[8] To 100 μl of serum in a brown conical-tip tube are added, under vortex mixing (see above) 15 μl of internal standard solution (I.S. 2, 4.5 mg/liter in ethanol), 200 μl of acetonitrile, and 3 ml of hexane. Extraction is carried out on a rotary mixer (Cenco Instruments, Breda, The Netherlands) for 5 min. After centrifugation (3000 rpm, 5 min), the supernatant is isolated and evaporated to dryness at room temperature under reduced pressure (Evapo-Mix, Büchler Instruments, Fort Lee, NJ) or using a gentle stream of nitrogen. The residue is reconstituted with 80 μl of the chromatographic solvent and a 50-μl aliquot is injected. For standardization, 100-μl samples of a serum pool supplemented with known amounts of retinol (100–800 μg/liter) are analyzed.

Preparation of Samples for Retinoic Acid Determination[3]

Before extraction, the serum sample (3.5 ml) is supplemented with 25 μl of an ethanolic solution of the internal standard (all-*trans*-13-demethyl-retinoic acid, 1.78 mg/liter). Ethanol (3.5 ml), 2 M NaOH (1.5 ml), and hexane (7 ml) are added, and the neutral and basic lipophilic compounds are extracted on a rotary mixer (see above) (10 min, 4°). After centrifugation the organic layer is discarded. The aqueous layer is acidified with 3 ml of 2 M HCl and reextracted for 10 min at 4° with 7 ml of hexane. The hexane layer is evaporated to dryness under reduced pressure (Evapo-Mix) or by using a gentle stream of nitrogen, the residue is redissolved in 100 μl of the chromatographic solvent, and a 50-μl aliquot is injected. Standardization is carried out by analyzing 3.5-ml aliquots of a serum pool supplemented with retinoic acid (1–9 μg/liter).

Discussion

Retinol by Normal-Phase Chromatography

A typical chromatogram of a serum extract analyzed by the normal-phase approach is depicted in Fig. 1. The analytical recovery of retinol from serum was $96.5 \pm 3.7\%$ (concentration range 975–1201 μg/liter) ($n =$ 6). Replicate analysis of a serum pool yielded a coefficient of variation (CV) of 2.5% ($\bar{x} = 589$ μg/liter, $n = 8$). A major advantage of this procedure is its speed and simplicity, because the extract can be directly injected, without the need for evaporation and reconstitution of the residue with the chromatographic solvent. Although the lack of a concentration step obviously sacrifices some sensitivity, a similar approach underlies the development of a micromethod for retinol, using as little as 5 μl of serum.[9] This ultimate reduction in sample size was achieved owing to the use of a short silica column (8 × 0.46 cm), in conjunction with fluorescence detection in a nonpolar eluent, yielding a detection limit of 50 nmol/liter (14.3 μg/liter). Several other workers have used normal-phase chromatography on silica[10–13] or polar bonded phases[14,15] and fluorescence[10,11,14,15] or absorption[10,12,13] detection as part of "macro" methods (200 μl of serum) for the determination of retinol alone or together

[9] A. J. Speek, C. Wongkham, N. Limratana, S. Saowakontha, and W. H. P. Schreurs, *J. Chromatogr.* **382,** 284 (1986).
[10] G. A. Woollard and D. C. Woollard, *HRC & CC, J. High Resol. Chromatogr. Chromatogr. Commun.* **7,** 466 (1984).
[11] A. T. Rhys Williams, *J. Chromatogr.* **341,** 198 (1985).
[12] D. D. Bankson, R. M. Russell, and J. A. Sadowski, *Clin. Chem. (Winston-Salem, N.C.)* **32,** 35 (1986).
[13] R.-K. Aaran and T. Nikkari, *J. Pharm. Biomed. Anal.* **6,** 853 (1988).
[14] H. K. Biesalski, W. Ehrenthal, M. Gross, G. Hafner, and O. Harth, *Int. J. Vitam. Nutr. Res.* **53,** 130 (1983).
[15] H. Biesalski, H. Greiff, K. Brodda, G. Hafner, and K. H. Bässler, *Int. J. Vitam. Nutr. Res.* **56,** 319 (1986).

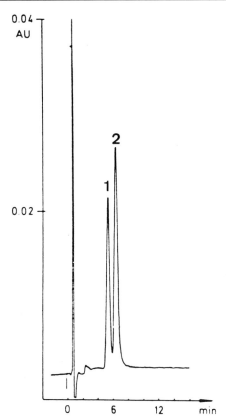

FIG. 1. HPLC chromatogram of a sample of human serum obtained by normal-phase chromatography. Column, 10 μm MicroPak Si-10, 15 × 0.2 cm; mobile phase, petroleum ether : dichloromethane : 2-propanol, 80 : 19.3 : 0.7 (v/v/v); flow rate, 0.5 ml/min; detection, 328 nm (0.04 AUFS on recorder); temperature, ambient; injection volume, 100 μl. Peak 1, retinol; Peak 2, I.S. 1.

with other isoprenoid compounds, for example, β-carotene,[13] α-tocopherol,[11,13,15] and retinyl esters.[12]

However, all recent methods either do not involve internal standardization[9,11,14,15] or include compounds structurally unrelated to retinol, such as α-naphthol[10] or tocol.[13] The internal standard used above as well as another compound suggested by Napoli[16] are structural analogs of retinol, but differ from it in the degree of saturation of the nucleus (aromatic versus cyclohexene ring). Several other analogs have been experi-

[16] J. L. Napoli, this series, Vol. 123, p. 112.

mentally tested as possible candidates for internal standardization, including 3-dehydroretinol, 13-demethylretinol, α-retinol, retroretinol, 13-cis-retinol, and 15-methylretinol.[17] However, none of them concurrently meets the basic requirements of stability, chromatographic resolution from retinol, and absence in serum. One paper describes the use of retinyl acetate as an internal standard,[12] a compound which is more appropriate for reversed-phase chromatography.

Retinol by Reversed-Phase Chromatography

Figure 2 shows a representative chromatogram of a serum extract obtained on the Zorbax ODS reversed-phase column using the nonaqueous elution conditions (nonaqueous reversed-phase chromatography, NARP). The standard curve was linear to at least 2 mg/liter, whereas the detection limit for retinol was estimated at 15 μg/liter. Within-run and between-run precision of the method was 2.6% (CV) ($\bar{x}$ = 800 μg/liter, n = 14) and 2.7% (CV) ($\bar{x}$ = 860 μg/liter, n = 10), respectively. A particular advantage of the nonaqueous chromatographic solvent is its capability to thoroughly solubilize the lipid residue obtained after evaporation of the extract. Many recent reversed-phase methods for retinol are still based on semiaqueous eluents,[18-21] which may fail to completely dissolve lipid-rich residues.[6] However, reversed-phase cannot compete with normal-phase methods in terms of simplicity of sample pretreatment: the incompatibility of the former with injection of extracts in hexane makes an evaporation step mandatory. Some investigators have tried to combine the advantages of reversed-phase chromatography with a direct injection of extracts in butanol–acetonitrile[18] or butanol–ethyl acetate,[20] but this can only be done at the expense of a substantial loss in sensitivity (sample volumes of 500–1000 μl, as opposed to 50–250 μl in alternative methods that do include an evaporation/concentration step[8,22-24]).

Lower retinyl esters are useful internal standards in reversed-phase

[17] H. J. Nelis and A. P. De Leenheer, paper presented at the 6th Fat-Soluble Vitamin Group Meeting, Leeds, UK, 6–7 April 1984, Abstr. 6.

[18] S. W. McClean, M. E. Ruddel, E. G. Gross, J. J. DeGiovanna, and G. L. Peck, *Clin. Chem. (Winston-Salem, N.C.)* **28**, 693 (1982).

[19] W. J. Driskell, J. W. Neese, C. C. Bryant, and M. M. Bashor, *J. Chromatogr.* **231**, 439 (1982).

[20] D. W. Nierenberg, *J. Chromatogr.* **311**, 239 (1984).

[21] D. Cuesta Sanz and M. Castro Santa-Cruz, *J. Chromatogr.* **380**, 140 (1986).

[22] M-L. Huang, G. J. Burckart, and R. Venkataramanan, *J. Chromatogr.* **380**, 331 (1986).

[23] W. A. MacCrehan and E. Schönberger, *Clin. Chem. (Winston-Salem, N.C.)* **33**, 1585 (1987).

[24] D. I. Thurnham, E. Smith, and P. Singh Flora, *Clin. Chem. (Winston-Salem, N.C.)* **34**, 377 (1988).

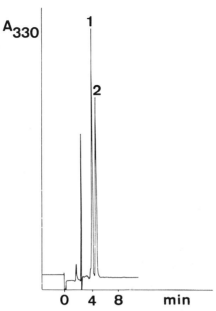

FIG. 2. HPLC chromatogram of a sample of human serum obtained by nonaqueous reversed-phase chromatography. Column, 7 μm Zorbax ODS, 25 × 0.46 cm; mobile phase, acetonitrile : dichloromethane : methanol, 70 : 15 : 15 (v/v/v); flow rate, 1 ml/min; detection, 330 nm (0.02 AUFS on recorder); temperature, ambient; injection volume, 50 μl. Peak 1, retinol; Peak 2, retinyl propionate (I.S. 2).

systems for retinol, although *in vitro* hydrolysis has been suspected.[16] Retinyl acetate has gained wide popularity,[18–21] but in some systems it is incompletely separated from retinol.[24] Retinyl propionate, as used in the above procedure, is a valuable alternative. Both compounds have also found application as internal standards in methods for the simultaneous determination of retinol, tocopherols, and carotenoids,[25] or retinol and retinyl esters,[6] respectively. However, for retinol itself structural analogs with a free terminal hydroxyl group (e.g., 13-demethylretinol, 15-methylretinol) may be more appropriate internal standards than retinyl esters. Unlike in normal-phase systems, both analogs can be readily separated from retinol by reversed-phase chromatography.

Retinoic Acid by Normal-Phase Chromatography

The occurrence of retinoic acid in blood under physiological conditions has been demonstrated by two independent techniques, namely,

[25] D. B. Milne and J. Botnen, *Clin. Chem. (Winston-Salem, N.C.)* **32**, 874 (1986).

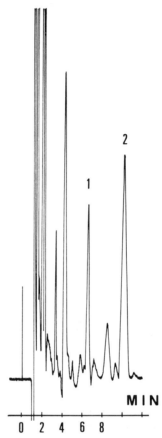

FIG. 3. HPLC profile of a serum extract obtained by normal-phase chromatography using an acidic eluent. Column, 10 μm RSIL, 15 × 0.32 cm; mobile phase, petroleum ether : acetonitrile : acetic acid, 99.5 : 0.2 : 0.3 (v/v/v); flow rate, 0.75 ml/min; detection, 350 nm (0.005 AUFS on recorder); temperature, ambient; injection volume, 50 μl. Peak 1, all-*trans*-retinoic acid; Peak 2, all-*trans*-13-demethylretinoic acid (I.S. 3).

HPLC and gas chromatography–mass spectrometry (GC–MS).[3,26] Both methods, individually[3,16] or in combination,[16] can also be considered for the quantitation of physiological levels of retinoic acid. In our HPLC assay the compound could be quantitated at concentrations down to 1 μg/liter serum, whereas the detection limit was estimated at 0.3 μg/liter. The recovery of retinoic acid at four different levels (2.4, 3.6, 4.7, and 5.8 μg/liter) was 69.0 ± 5.2% (n = 8). The within-run and between-run repro-

[26] M. G. De Ruyter, W. E. Lambert, and A. P. De Leenheer, *Anal. Biochem.* **98**, 402 (1979).

ducibility (CV) averaged 6.9% ($\bar{x}$ = 2.7 μg/liter, n = 9) and 10.2% ($\bar{x}$ = 2.9 μg/liter, n = 10), respectively. A chromatogram of a serum extract showing the peaks of retinoic acid and 13-demethylretinoic acid, the internal standard, is presented in Fig. 3. The occurrence of interfering peaks is minimized by including a preextraction of the alkalinized sample (addition of 2 M NaOH) at 4° with hexane. However, other workers recommend weaker bases (25 mM NaOH) to prevent possible degradation of retinoic acid at room temperature.[16]

Conclusions: Present Status of Retinoid Analysis in Blood

The basic principles of the normal- and reversed-phase method for the determination of retinol in human serum/plasma, outlined above, have remained valid up to date. Extraction procedures have, in general, undergone no major changes. Normal-phase chromatography permits a simple sample pretreatment, but it suffers from the nonavailability of a suitable internal standard and from lower intrinsic sensitivity. The latter drawback can be largely overcome by using fluorescence instead of absorbance detection. Further recent improvements mainly pertain to column selection (bonded phases versus silica) and to the simultaneous determination of retinol and other isoprenoid compounds.

Reversed-phase chromatography continues to be the most popular technique because of column stability and reproducibility. Again, a major trend consists of combining the determination of retinol with that of carotenoids and tocopherols. In these cases, the detection wavelength is necessarily a compromise (e.g., 292 nm for retinol plus tocopherols). Alternatively, a sophisticated detector permitting wavelength programming can be used. A major challenge for the future remains the development of micromethods (<50 μl of serum) capable of accommodating not only the analysis of "normal" but also that of hypovitaminotic samples, with sufficient accuracy and precision.

Except for the work of Napoli[16] and methods dealing with pharmacological concentrations of its 13-cis isomer, the HPLC determination of retinoic acid has not become "routine" practice. Nonetheless, despite operating at the limits of absorbance detection, HPLC gives results comparable to those obtained by GC–MS. Of both approaches the former is obviously more accessible and convenient,[16] but less sensitive and specific.

Acknowledgments

H.J.N. acknowledges his position of Research Associate from the Belgian Foundation for Scientific Research (N.F.W.O.).

[5] High-Performance Liquid Chromatography of Retinoid Isomers: An Overview

By C. D. B. Bridges

Introduction

The separation and quantitation of different retinoid isomers is essential in studies of retinoids in the visual systems of vertebrates and invertebrates, in investigations of retinoid isomer interconversions in other tissues (e.g., retinoic acid in rat liver), and in determining the biopotency of vitamin A in foodstuffs and food supplements such as fish liver oils. In nature, vitamin A compounds can exist as the aldehyde, ester, or acid forms of retinol, 3-hydroxyretinol, or 3,4-didehydroretinol. Additionally, the aldehydes are often derivatized with hydroxylamine in the tissue and extracted in the form of the corresponding oximes. In practice, the separation of the various mono-cis isomers is comparatively easy provided only a single class of retinoid derivatives is under study. More usually, however, several retinoid classes of widely differing polarities are present in the mixture being investigated. If these classes include retinyl esters and retinols, the researcher may attempt the separation in a single injection, usually employing a step gradient. The disadvantage of this approach is the long time it takes the column to reequilibrate to low-polarity eluting conditions after the final separating step that utilizes a high-polarity eluent mixture. An alternative approach is to carry out all the retinyl ester separations in a series of injections under low-polarity, isocratic conditions, then change the mobile phase to an appropriately more polar mixture of solvents and separate the retinol isomers in a second series of injections.

Extraction Methods

Methods of retinoid extraction are fairly straightforward in most tissues, and have been described previously.[1,2] However, special procedures must be used to extract retinoids in their native isomeric configuration from visual pigments. This is because the simple use of a denaturing

---

[1] G. W. T. Groenendijk, P. A. A., Jansen, S. L. Bonting, and F. J. M. Daemen, this series, Vol. 67, p. 203.

[2] C. D. B. Bridges and R. A. Alvarez, this series, Vol. 81, p. 463.

METHODS IN ENZYMOLOGY, VOL. 189

solvent such as methanol (even under red or gold light) causes the 11-*cis*-retinal prosthetic group to isomerize to other mono-cis (notably the 13-cis) and all-trans isomers. The method devised by Groenendijk *et al.*[3] for photoreceptor membranes or detergent-solubilized material overcame this problem by carrying out the methanol denaturation in the presence of a 1000-fold molar excess of hydroxylamine. At these concentrations (10–200 mM), oxime formation was so fast that it prevented isomerization of the liberated retinal. This method, or variations of it, has been applied to visual pigment systems based on retinal, 3,4-didehydroretinal, and 3-hydroxyretinal.[4–7]

The procedure[3] is to add a 1000-fold molar excess of 1 M hydroxylamine (pH 6.5) to 10–50 μM of visual pigment in a membrane suspension. Methanol is added to a concentration of 70% (v/v), then water and dichloromethane are added until the three constituents are present in equal volumes. After vortexing, the mixture is centrifuged at 10,000 g for 1 min and the lower layer collected with a syringe. The procedure is repeated 3 times, giving a recovery of 93 ± 7%.

Suzuki *et al.*[8] used a conceptually similar method that depended on using formaldehyde as a protective agent, perhaps by blocking amino groups on protein and lipid. Pepe and Schwemer[9] described a method based on that of Pilkiewicz *et al.*[10] Retinals are extracted from rod outer segment preparations in digitonin by adding sodium dodecyl sulfate to 1% final concentration, incubating for 10 min at 22°, then stirring with dichloromethane–methanol (1 : 1, v/v) at 4°. The resulting emulsion is centrifuged, and the lower dichloromethane layer containing the retinal isomers is withdrawn. The retinals are transferred to hexane and analyzed on a Spherisorb ODS 5 μm column used in normal phase with 10% ether in hexane. This column has the advantage that contamination of the sample with detergent and water does not influence resolution. The authors note that there is variability in performance between columns, probably owing to incomplete coverage of the packing material by chloroalkylsilanes.

[3] G. W. T. Groenendijk, W. J. de Grip, and F. J. M. Daemen, *Biochem. Biophys. Acta* **617**, 430 (1980).
[4] T. H. Goldsmith, B. C. Marks, and G. D. Bernard, *Vision Res.* **26**, 1763 (1986).
[5] T. Suzuki and M. Makino-Tasaka, *Anal. Biochem.* **129**, 111 (1983).
[6] T. Seki, S. Fujishita, M. Ito, N. Matsuoka, C. Kobayashi, and K. Tsukida, *Vision Res.* **26**, 255 (1986).
[7] T. Seki, S. Fujishita, M. Ito, N. Matsuoka, and K. Tsukida, *Exp. Biol.* **47**, 95 (1987).
[8] T. Suzuki, Y. Fujita, Y. Noda, and S. Miyata, *Vision Res.* **26**, 425 (1986).
[9] I. M. Pepe and J. Schwemer, *Photochem. Photobiol.* **45**, 679 (1987).
[10] F. G. Pilkiewicz, M. J. Pettei, A. P. Yudd, and K. Nakanishi, *Exp. Eye Res.* **24**, 421 (1977).

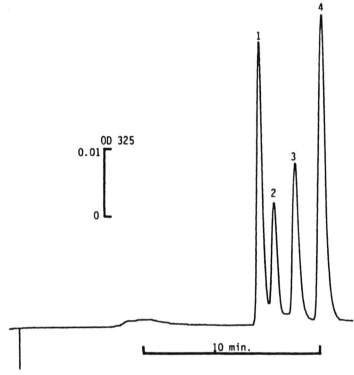

Fig. 1. Separation of isomers of retinyl stearate. Peak 1, 13-*cis*-; Peak 2, 11-*cis*-; Peak 3, 9-*cis*-; Peak 4, all-*trans*-retinyl stearate. Column, Ultrasphere 5 μm (250 × 4.6 mm, 15,000 minimum plate count); mobile phase, diethyl ether (0.4%)/*n*-hexane at 0.7 ml/min; detection at 325 nm. [From R. A. Alvarez, S. L. Fong, and C. D. B. Bridges, *Invest. Ophthalmol. Visual Sci.* **20,** 304 (1981).]

High-Performance Liquid Chromatographic Techniques

It is usually easy to develop a highly efficient method of separating the isomers of a particular class of retinoids. For most retinoid classes, with the exception of retinoic acid, normal-phase systems have been used almost exclusively. Figure 1 illustrates the separation of 13-*cis*-, 11-*cis*-, 9-*cis*-, and all-*trans*-retinyl stearates.[11] Figure 2 shows the separation of a mixture of retinal isomers obtained by irradiation of all-*trans*-retinal in ethanol: the mixture contains about 22% 13-*cis*-, 29% 11-*cis*-, 11% 9-*cis*-,

[11] R. A. Alvarez, S.-L. Fong, and C. D. B. Bridges, *Invest. Ophthalmol. Visual Sci.* **20,** 304 (1981).

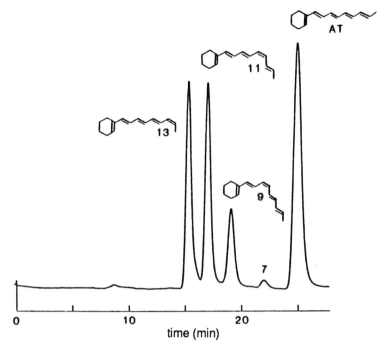

FIG. 2. Separation of isomers of retinal in a mixture generated by irradiation of a solution of all-*trans*-retinal in ethanol. Peak 13, 13-*cis*-; Peak 11, 11-*cis*-; Peak 9, 9-*cis*-; Peak 7, 7-*cis*-; Peak AT, all-*trans*-retinal. Column, μPorasil 10 μm (3026 minimum plate count); mobile phase, diethyl ether (12.5%)/*n*-hexane at 0.6 ml/min; detection at 310 nm. [From C. D. B. Bridges, S.-L. Fong, G. I. Liou, R. A. Alvarez, and R. A. Landers, *in* "Progress in Retinal Research" (N. N. Osborne and G. J. Chader, eds.), p. 137. Pergamon, Oxford, 1983.]

1% 7-*cis*-, and 37% all-*trans*-retinal (minor amounts of di-cis isomers may be present, but they are not resolved).[12]

Normal-phase systems have also been most useful in separating the isomers present in a mixture of different classes of retinoids. Figure 3 shows an example of the separation of a variety of retinoid isomers ranging in polarity from 11-*cis*-retinyl palmitate to all-*trans*-3-dehydroretinol. In this instance, a step-gradient elution mode was used, starting with 0.2% dioxane in *n*-hexane and finishing with 12% dioxane in *n*-hexane.[13] In Fig. 4, a quaternary mixture of solvents was used in the isocratic mode to

[12] C. D. B. Bridges, S.-L. Fong, G. I. Liou, R. A. Alvarez, and R. A. Landers, *in* "Progress in Retinal Research" (N. N. Osborne and G. J. Chader, eds.), p. 137. Pergamon, Oxford, 1983.

[13] C. D. B. Bridges, S.-L. Fong, and R. A. Alvarez, *Vision Res.* **20,** 355 (1980).

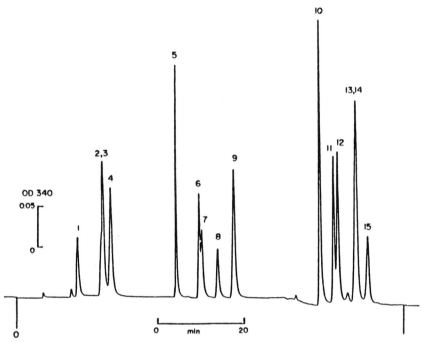

FIG. 3. Step-gradient separation of retinyl esters, retinals, retinal oxime, retinols, and 3-dehydroretinol. Peak 1, 11-*cis*-retinyl palmitate; Peak 2, all-*trans*-retinyl stearate; Peak 3, all-*trans*-retinyl palmitate; Peak 4, all-*trans*-retinyl oleate; Peak 5, all-*trans*-retinyl acetate; Peak 6, 13-*cis*-retinal; Peak 7, 11-*cis*-retinal; Peak 8, 9-*cis*-retinal; Peak 9, all-*trans*-retinal; Peak 10, all-*trans*-retinal oxime (syn conformer); Peak 11, 13-*cis*-retinol; Peak 12, 11-*cis*-retinol; Peak 13, 9-*cis*-retinol; Peak 14, all-*trans*-retinol; Peak 15, all-*trans*-3-dehydroretinol. Column, Lichrosorb 10 μm; mobile phase, dioxane/n-hexane run as follows: 0.2% dioxane for 21 min, 2% for 27 min, 12% for 25 min; flow rate, 1 ml/min; detection at 340 nm. [From C. D. B. Bridges, S. L. Fong, and R. A. Alvarez, *Vision Res.* **20**, 355 (1980).]

obtain excellent separation of retinal, retinal oxime (syn and anti conformers), and retinol isomers.[14]

A summary of normal-phase systems for separating retinoid isomers is given in Table I. The major component of the mobile phase is nearly always n-hexane, although other solvents have been sporadically used (e.g., benzene, pentane, heptane, and octane have been used occasionally by the authors). When retinyl ester isomers are being separated, a small

[14] G. M. Landers and J. A. Olson, *J. Chromatogr.* **438**, 383 (1988).

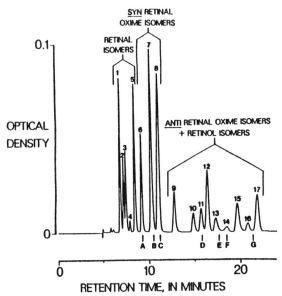

FIG. 4. Separation of retinol, retinal, and retinal oxime isomers using a quaternary mixture as the mobile phase. Peak 1, 13-*cis*-retinal; Peak 2, 11-*cis*-retinal; Peak 3, 9-*cis*-retinal; Peak 4, 7-*cis*-retinal; Peak 5, all-*trans*-retinal; Peak 6, *syn*-11-*cis*-retinal oxime; Peak 7, *syn*-all-*trans*-retinal oxime; Peak 8, *syn*-9-*cis*-retinal oxime + *syn*-13-*cis*-retinal oxime; Peak 9, *anti*-13-*cis*-retinal oxime; Peak 10, *anti*-11-*cis*-retinal oxime; Peak 11, 11-*cis*-retinol; Peak 12, 13-*cis*-retinol; Peak 13, *anti*-9-*cis*-retinal oxime; Peak 14, 9,11-*cis*-retinol; Peak 15, *anti*-all-*trans*-retinal oxime; Peak 16, 9-*cis*-retinol; Peak 17, all-*trans*-retinol. The letters refer to the retention times for the following compounds: A, *syn*-9,11-*cis*-retinal oxime; B, *syn*-7-*cis*-retinal oxime; C, *syn*-9,13-di-*cis*-retinal oxime; D, *anti*-9,13-*cis*-retinal oxime; E, *anti*-9,11-*cis*-retinal oxime; F, *anti*-7-*cis*-retinal oxime; G, 7-*cis*-retinol. Column, two Lichrosorb Si 60 5 μm (250 × 4 mm) in series; mobile phase, ethyl acetate (11.2%)/dioxane (2%)/1-octanol (1.4%)/*n*-hexane; flow rate, 1 ml/min; detection at 325 nm. [From G. M. Landers and J. A. Olson, *J. Chromatogr.* **438,** 383 (1988).]

proportion of diethyl ether or dioxane is added as a polar modifier to the mobile phase, whereas for the more polar retinoids such as the retinals and retinols larger proportions are employed. The use of ternary or even quaternary mixtures has become more common in recent work, particularly where the isomers of different retinoid classes are being separated from the same mixture.

In general, reversed-phase columns have not been used to separate the isomers of any retinoid class except retinoic acid (Table II). A notable recent exception has been the successful separation of mono- and di-*cis*-

TABLE I
NORMAL-PHASE SYSTEMS USED FOR SEPARATING RETINOID ISOMERS

Retinoid[a]	(1)[b]	(2)	(3)	(4)	Column	Ref.
		Mobile phase (%)				
B, D, F	h	Dioxane (14–18)	—	—	Lithosorb 5 μm	c
B, G	h	1-Octanol (3.8)	2-Propanol (0.2)	—	Si-60 (Merck) 5 μm	d
B, C, G	h	Dioxane (1.0)	2-Propanol (0.1)	—	Si-60 (Merck) 5 μm	e
B–E	h	Ethyl acetate (11.2)	Dioxane (2.0)	1-Octanol (1.4)	Lichrosorb Si-60 (two in series)	f
A–C, G	h	Diethyl ether (1–7)	—	—	Zorbax SIL 7 μm	g
C	h	Diethyl ether (10)	—	—	Spherisorb 5 μm ODS	h
B, D, F	b	tert-Butyl methyl ether (1–6)	Ethanol (0.12–1.1)	—	YMC-PACK 3 μm	i
B, D, F	h	Diethyl ether (8)	Ethanol (0.08)	—	YMC-PACK 3 μm	i
B, D, G	h	Diethyl ether (7)	Ethanol (0.075)	—	Sorbax SIL 7 μm	j
C, D	h	Diethyl ether (5)	—	—	Lichrosorb 5 μm	k
B	h	2-Octanol (1)	—	—	Partisil 10-ODS and Zorbax CN 5 μm	l
B	h	Dioxane (5)	—	—	Partisil 10-ODS and Zorbax CN 5 μm	l
C, F	h	Diethyl ether (15–50)	Ethanol (0.15–0.05)	—	Lichrosorb Si-60 5 μm	m
C, D, G	h	Diethyl ether (6–7)	Ethanol (0.05–0.075)	—	Zorbax SIL	n
A–D, G	h	Diethyl ether (0.4–20)	—	—	Lichrosorb 10 μm or μPorasil	o
A–D, G	h	Dioxane (0.2–12)	—	—	Lichrosorb 10 μm or μPorasil	o
D	h	Dioxane (6)	—	—	MicroPak Si-5 or Lichrosorb Si-60-5	p
A	h	tert-Butyl methyl ether (0.2–0.5)	—	—	Supelcosil LC-Si 3 μm or Ultrasphere Si-60 5 μm	q, r
B	h	Dioxane (4–5)	—	—		
A	h	Diethyl ether (0.4)	—	—	Ultrasphere Si-60 5 μm and Spherisorb CN 5 μm	s

TABLE I (*continued*)

Retinoid[a]	(1)[b]	Mobile phase (%)			Column	Ref.
		(2)	(3)	(4)		
B, C	h	Dioxane (9)	Diethyl ether (0.4)	—		
C, G	h	Diethyl ether (6–12)	—	—	μPorasil	t

[a] A, Retinyl esters; B, retinol; C, retinal; D, retinal oxime; E, retinoic acid; F, hydroxyretinol and derivatives; G, 3-dehydroretinol and derivatives.

[b] The major component (1) of the eluent is nearly always *n*-hexane (h); benzene (b) was used in one instance. One of the authors has also used *n*-pentane, *n*-heptane, *n*-octane, and isooctane.

[c] T. H. Goldsmith, B. C. Marks, and G. D. Bernard, *Vision Res.* **26,** 1763 (1986).

[d] B. Stancher and F. Zonta, *J. Chromatogr.* **312,** 423 (1984).

[e] F. Zonta and B. Stancher, *J. Chromatogr.* **301,** 65 (1984).

[f] G. M. Landers and J. A. Olson, *J. Chromatogr.* **438,** 383 (1988).

[g] T. Suzuki, Y. Maeda, Y. Toh, and E. Eguchi, *Vision Res.* **28,** 1061 (1988).

[h] I. M. Pepe and J. Schwemer, *Photochem. Photobiol.* **45,** 679 (1987).

[i] T. Seki, S. Fujishita, M. Ito, N. Matsuoka, and K. Tsukida, *Exp. Biol.* **47** (1987).

[j] T. Suzuki and M. Makino-Tasaka, *Anal. Biochem.* **129,** 111 (1983).

[k] Y. Shichida, K. Nakamura, T. Yoshizawa, A. Trehan, M. Denny, and R. S. H. Liu, *Biochemistry* **27,** 6495 (1988).

[l] P. V. Bhat and A. Lacroix, this series, Vol. 123, p. 75.

[m] T. Seki, S. Fujishita, M. Ito, N. Matsuoka, C. Kobayashi, ad K. Tsukida, *Vision Res.* **26,** 255 (1986).

[n] T. Suzuki, Y. Fujita, Y. Noda, and S. Miyata, *Vision Res.* **26,** 425 (1986).

[o] C. D. B. Bridges, S.-L. Fong, and R. A. Alvarez, *Vision Res.* **20,** 355 (1980).

[p] G. W. T. Groenendijk, W. J. de Grip, and F. J. M. Daemen, *Biochim. Biophys. Acta* **617** (1980).

[q] C. D. Bridges and R. A. Alvarez, *Science* **236,** 1678 (1987).

[r] C. D. B. Bridges, R. A. Alvarez, S.-L. Fong, G. I. Liou, and R. J. Ulshafer, *Invest. Ophthalmol. Visual Sci.* **28,** 613 (1987).

[s] C. D. B. Bridges, R. A. Alvarez, S.-L. Fong, F. Gonzalez-Fernandez, D. M. K. Lam, and G. I. Liou, *Vision Res.* **24,** 1581 (1984).

[t] K. Tsukida, R. Masahara, and M. Ito, *J. Chromatogr.* **192,** 395 (1980).

retinols with a column packed with wide-pore (300 Å) silica bonded to a polymeric C_{18} stationary phase (Fig. 5).[15]

High-Performance Liquid Chromatograms of Radiolabeled Retinoid Isomers

In certain metabolic studies it is often necessary to obtain an accurate radioactive elution profile. This can be a problem if the isomer of interest

[15] W. A. MacCrehan and E. Schonberger, *J. Chromatogr.* **417,** 65 (1987).

TABLE II
REVERSED-PHASE SYSTEMS FOR SEPARATING RETINOID ISOMERS

Retinoid	Mobile phase	Column	Ref.
Retinol	Methanol–n-butanol–water (65 : 10 : 25, v/v/v) 10 mM ammonium acetate (pH 3.2)	Vydac 201 TP, C_{18}, 5 μm	a
Retinoic acid	Methanol–water (75 : 25, v/v), 10 mM ammonium acetate	Ultrasphere XL ODS, 3 μm or Zorbax ODS, 5 μm	b
Retinoic acid	Acetonitrile–dichloromethane (90 : 10) 10 mM acetic acid	Zorbax NH$_2$, 5 μm	c

^{a} W. A. MacCrehan and E. Schonberger, *J. Chromatogr.* **417**, 65 (1987).
^{b} R. W. Curley, Jr., and J. W. Fowble, *Photochem. Photobiol.* **47** (1988).
^{c} P. V. Bhat and A. Lacroix, this series, Vol. 123, p. 75.

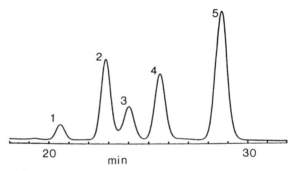

FIG. 5. Separation of retinol isomers by reversed-phase high-performance liquid chromatography. Peak 1, di-*cis*-; Peak 2, 11-*cis*-; Peak 3, 9-*cis*-; Peak 4, 13-*cis*-; Peak 5, all-*trans*-retinol. Column, Vydac 201 TP 5 μm C_{18} (250 × 4.6 mm); mobile phase, methanol/n-butanol/water (65 : 10 : 25) containing 10 mM ammonium acetate (pH 3.2) at 1 ml/min; detection at 325 nm. [From W. A. MacCrehan and E. Schonberger, *J. Chromatogr.* **417**, 65 (1987).]

elutes within a short time of another. An example is 11-*cis*-retinol, which is generated enzymatically when all-*trans*-retinol is incubated with homogenates of ocular pigment epithelium.[16,17] However, its peak is closely followed by that of 13-*cis*-retinol, an isomer that is formed nonspecifically, even in homogenates from tissues lacking the isomerase (e.g., brain, liver) or when the enzyme itself has been inactivated by heat,

[16] C. D. Bridges and R. A. Alvarez, *Science* **236**, 1678 (1987).
[17] P. S. Bernstein, W. C. Law, and R. R. Rando, *Proc. Natl. Acad. Sci. U.S.A.* **84**, 1849 (1987).

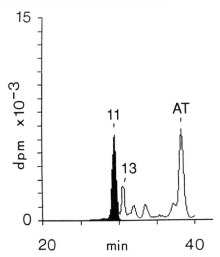

FIG. 6. Separation of radiolabeled retinol isomers extracted from a homogenate of pigment epithelium–choroid incubated with all-*trans*-[11,12-^{3}H]retinol [conditions as described by C. D. B. Bridges and R. A. Alvarez, *Science* **236**, 1678 (1987)]. Peak 11 (shaded), 11-*cis*-retinol; Peak 13, 13-*cis*-retinol; Peak AT, all-*trans*-retinol; the small peak on the rising phase of the AT peak is probably 9-*cis*-retinol, but the other peaks have not been identified. Column, Supelcosil 3 μm (150 × 4.6 mm); mobile phase, dioxane (4%)/*n*-hexane at 0.6 ml/min; fractions were manually collected and counted on a Packard Tricarb liquid scintillation counter.

trypsin, or phenymethylsulfonyl fluoride (PMSF).[16] In order to resolve unequivocally and clearly the radioactivities associated with these two isomers when they are generated from tritiated all-*trans*-retinol, it has been found necessary to count fractions that have been collected at 0.1-min intervals. This procedure is most reliably carried out manually.[16] A typical radioactive profile derived by this approach is illustrated in Fig. 6. More recently, comparatively acceptable results have been obtained in the authors' laboratory with a carefully adjusted "Foxy" fraction collector (Isco, Lincoln, NE).

[6] High-Performance Liquid Chromatography of Retinoid Isomers

By Gary M. Landers

Introduction

Isomers of retinol, retinal, and retinyl esters play a crucial role in the visual cycle in the vertebrate neural retina and retinal pigment epithelium, and efficient methods for the extraction and analysis of retinoid isomers are essential for the study of visual aspects of vitamin A metabolism. Retinol isomers and retinyl ester isomers are readily extracted by dichloromethane, but retinal is difficult to recover in high yield, presumably because of its propensity for forming Schiff bases with primary amines, which are abundant in most biological samples. If the retinal is derivatized with hydroxylamine before extraction virtually all the retinal oxime can be recovered,[1] and because retinal oxime is more resistant to isomerization than free retinal,[2] artifactual isomerization is also reduced.

However, because the oximes exist as both syn and anti isomers, the total number of retinal isomers which must be separated and quantitated is doubled. An additional difficulty arises because the *anti*-retinal oxime isomers generally elute in the same range as the cis isomers of retinol, and both retinol and retinal are generally present in ocular tissues. One solution to these difficulties is to utilize a chromatographic method, such as the one described here, to simultaneously resolve both the *anti*-retinal oxime isomers and the cis isomers of retinol.[3] This isocratic HPLC (high-performance liquid chromatography) method also resolves the isomers of retinal and allows estimation of total retinyl esters. A separate HPLC method is outlined for the determination of retinyl ester isomer composition.

A different solution to the difficulties of analyzing retinal oxime isomers in the presence of retinol isomers is to extract free retinal in the presence of excess formaldehyde.[4] However, the recovery of retinal in the presence of formaldehyde is slightly less complete than that achieved by extraction as the oxime in the presence of hydroxylamine, and free retinal is more susceptible to artifactual isomerization during extraction

[1] G. W. T. Groenendijk, P. A. A. Jansen, S. L. Bonting, and F. J. M. Daemen, this series, Vol. 67, p. 203.
[2] R. Hubbard, *J. Am. Chem. Soc.* 78, 4662 (1956).
[3] G. M. Landers and J. A. Olson, *J. Chromatogr.* 438, 383 (1988).
[4] T. Suzuki, Y. Fujita, Y. Noda, and S. Miyata, *Vision Res.* 26, 425 (1986).

than is retinal oxime. For these reasons, use of this extraction technique is not covered in this chapter, although it is nonetheless a good method.

General Precautions in Handling Retinoids

Retinoids are very susceptible to oxidation and isomerization during extraction and analysis. Oxidation of the retinoids present in the sample can be controlled by avoiding unnecessary introduction of oxygen into the sample during mixing and handling and by adding 0.1 mg/ml BHT (butylated hydroxytoluene) to all extraction solvents. The added BHT does not interfere with subsequent analysis of retinal oxime or retinol isomers, because BHT elutes much earlier than the cis isomers of retinal; however, some error in estimation of unresolved retinyl esters in the retinol/retinal oxime method might result from excessive use of BHT. It is convenient to add 100 μl of a concentrated 10 mg/ml solution of BHT in hexane to each 100 ml of methanol, dichloromethane, or hexane used in the extraction process.

Artifactual isomerization may be minimized by avoiding exposure of the sample to photoisomerizing light and by keeping the sample as cold as possible during extraction and handling to avoid thermal isomerization. All sources of white light ($\lambda < 500$ nm) in the laboratory, such as windows, doorways, and standard fluorescent and incandescent lamps, should be covered, removed, or turned off while the retinoid samples are being manipulated. It is convenient to replace most of the standard white fluorescent lamps in the laboratory with F40GO "gold" fluorescent lamps, which are available in standard sizes to fit existing fixtures (Sylvania, Warren, PA, or Westinghouse, Minneapolis, MN) and which do not photoisomerize retinoids.[5] A number of white fluorescent lamps on a separate circuit or circuits may be used to increase the light intensity and enhance color perception in the laboratory when no retinoid samples are present. (Of course, when rhodopsins or other photopigment samples are being extracted, even the yellow light sources are not suitable, and red safelights should be employed until after the photopigments have been denatured.) In addition, keeping the samples in a covered black ice bucket with ice or dry ice when they are not actually being manipulated is a good practice.

Care must also be taken in storing retinoid isomers. Thermal isomerization will be minimized if the samples are stored in a freezer at -20 or $-70°$. Amber glassware and bottles should be used when possible, and if the freezer will be opened in white light, it is best to enclose the samples

---

[5] G. M. Landers and J. A. Olson, *J. Assoc. Off. Anal. Chem.* **69**, 50 (1986).

in opaque containers (such as black plastic index card boxes or black 35-mm film containers) during storage. Wrapping the samples in aluminum foil is a common practice, but opaque plastic containers are more effective and more convenient.

Preparation of Extracts

The extraction procedure described here is basically that of Groenendijk et al.,[1,6,7] and the volumes suggested are appropriate for extraction of a single bovine retina.

To the tissue sample is added 250 μl of Hanks' buffer, pH 6.5, as well as 700 μl of methanol (containing 0.1 mg/ml BHT). After addition of 50 μl aqueous 1 M hydroxylamine (buffered to pH 6.5 with bicarbonate) to provide at least a 1000-fold excess of hydroxylamine over retinal and a final methanol concentration of 70%, the sample is homogenized. Heat generated during homogenization is a potent cause of thermal isomerization, especially when ultrasonic homogenizers are used.[8] Care should be taken not to homogenize the samples too rapidly or too extensively, and the samples should be thoroughly cooled before, during, and after homogenization. The homogenizer probe or pestle is rinsed with 1 ml of a 14 : 5 : 1 (v/v/v) mixture of methanol, buffer, and hydroxylamine solution, and the rinsings are combined with the homogenate.

The homogenate is allowed to stand 10 min on ice, after which 1 ml of dichloromethane (containing 0.1 mg/ml BHT) is added, followed by 1 ml of water or buffer. The sample is mixed by swirling and then centrifuged 10 min at about 1000 g (preferably in a refrigerated centrifuge) to separate the organic and aqueous phases. A precipitate of denatured homogenate usually accumulates at the interface between the phases, and a Pasteur pipette is used to carefully pierce through the precipitate to collect the lower dichloromethane layer. The aqueous residue is extracted twice more with dichloromethane, and the three dichloromethane extracts are pooled. A clean Pasteur pipette is then used to remove any residual water droplets which may be floating on top of the extract.

After drying over sodium sulfate in the dark for about 1 hr, the dichloromethane extract is transferred with a Pasteur pipette to the barrel of a 5-ml syringe fitted with a 0.45-μm nylon or Teflon filter (such as Gelman Sciences, Ann Arbor, MI, Accro LC13), and the extract is filtered into a

[6] G. W. T. Groenendijk, W. J. de Grip, and F. J. M. Daemen, *Biochim. Biophys. Acta* **617**, 430 (1980).
[7] G. W. T. Groenendijk, W. J. de Grip, and F. J. M. Daemen, *Anal. Biochem.* **99**, 304 (1979).
[8] M. Tsuda, this series, Vol. 88, p. 552.

clean glass tube. The sodium sulfate is rinsed with one or two small volumes of dichloromethane, and these washes are transferred to the syringe, filtered, and pooled with the extract. If analysis of retinyl ester isomers is desired, the filtered extract should be divided into two equal portions, one to be used for analysis of the oxime and retinol isomers and one for ester analysis. The dichloromethane is evaporated under a stream of nitrogen or argon, and the residue is redissolved in 200 μl of the HPLC mobile phase. Condensation of atmospheric moisture in the sample tube as it is cooled by evaporation of the dichloromethane may be prevented by fitting the tube with a stopper with only small holes for the entrance and exit of the nitrogen.[8]

HPLC of Retinoid Isomers

Equipment

Minimal equipment for this procedure consists of an HPLC pump, a sample injection valve, a UV detector capable of operating at 325 nm and/ or 360 nm, and a chart recorder or (preferably) a digital integrator. A second detector and integrator are desirable, but not essential. Provided that adequate sensitivity is obtained, a photodiode-array detector might be used to great advantage instead of one or both of the conventional UV detectors.

General Precautions for HPLC

After the mobile-phase solvent is mixed, it should be filtered under reduced pressure, both to remove particulates which might otherwise damage the equipment and to degas the mobile phase to prevent pump cavitation or formation of bubbles in the detector cell. In addition, a small guard column packed with pellicular silica (such as Pellosil from Whatman, Clifton, NJ) is a prudent precaution to protect the analytical column from particulates and from nonretinoid components of sample extracts (such as phospholipids). Finally, all tubing connecting the precolumn to the injector, the precolumn and columns to each other, the column to the detector, and the first detector to the optional second detector should be as short as possible and of small inner diameter (0.010 inch or less) to minimize loss of resolution owing to system dead volume.

HPLC Analysis of Isomers of Retinol and Retinal Oxime

A small guard column filled with pellicular silica is used with two Merck 4 × 250 mm 5-μm Si-60 Lichrosorb columns in series. The mobile phase is 11.2% ethyl acetate, 2% dioxane, 1.4% 1-octanol in hexane (v/v), at a rate of 1 ml/min. One detector may be used to monitor absorption at 325 or at 360 nm, or two detectors may be used in series to monitor both wavelengths, thus providing some capability of distinguishing between retinol isomers and the *anti*-oxime isomers on the basis of differential absorption. Less than one-tenth of the extract of endogenous retinoids from a single bovine retina is necessary to achieve excellent sensitivity, so an injection volume of 40–50 μl would be appropriate from a total extract volume of 200 μl from one-half of a bovine retina. Provided that detectability is not limiting, injection of a relatively small fraction of the total extract provides the capability of repeated analyses of a sample for statistical purposes, and results in less loading of the column with extracted phospholipids.

The chromatographic method outlined here provides separation of retinal oxime isomers from any retinal which might remain underivatized, and it also separates all the 11-cis and all-trans isomers of retinol and retinal oxime from each other and from the 9-cis and 13-cis isomers (Fig. 1). Although the *syn-13-cis*-retinal oxime and *syn-9-cis*-retinal oxime isomers are not resolved, the 9-*cis*- and 13-*cis*-retinal oximes may be estimated from the corresponding anti isomers, which are well separated.[3]

It is essential that the samples always be dried with sodium sulfate and that the mobile phase not be exposed to overly humid air, because the balance of selectivity factors in this chromatographic method is easily perturbed by moisture. It is also important that the columns be well equilibrated with the mobile phase before samples are analyzed. In practice, it is useful to pump 500–1000 ml of mobile phase through the columns during the night before analyses are to begin. Samples should not be analyzed until the retention times of successive injections of standard isomer mixtures have stabilized, and it is recommended that the columns be dedicated to this mobile phase to minimize problems with solvent equilibration. Should the resolution of retinol isomers from *anti*-retinal oxime isomers become unsatisfactory, it may in some cases be corrected by simply preparing fresh mobile phase and reequilibrating the columns.

HPLC Analysis of Isomers of Retinyl Esters

A small guard column filled with pellicular silica is used with a 4.6 × 250 5-μm silica column, such as Spherisorb. The sample is eluted with 0.3% ethyl acetate in hexane, at a flow rate of 1 ml/min. Under these conditions, retinol and retinal oxime isomers in the extract can be ignored because they either remain on the column or elute as extremely broad peaks which are not generally noticeable during analysis of subsequent

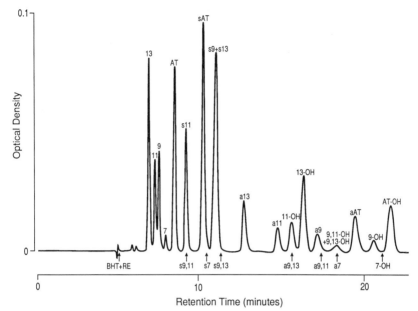

Fig. 1. Resolution of a mixture of retinal, retinal oxime, and retinol isomers. The retinol isomers and retinal oxime isomers were prepared from photoisomerized retinal and then mixed with photoisomerized retinal. Detection was by absorbance at 325 nm. Peak identities are as follows: 13, 13-*cis*-retinal; 11, 11-*cis*-retinal; 9, 9-*cis*-retinal; 7, 7-*cis*-retinal; AT, all-*trans*-retinal; s11, *syn*-11-*cis*-retinal oxime; sAT, *syn*-all-*trans*-retinal oxime; s9, *syn*-9-*cis*-retinal oxime; s13, *syn*-13-*cis*-retinal oxime; a13, *anti*-13-*cis*-retinal oxime; a11, *anti*-11-*cis*-retinal oxime; 11-OH, 11-*cis*-retinol; 13-OH, 13-*cis*-retinol; a9, *anti*-9-*cis*-retinal oxime; 9,11-OH, 9,11-*cis*-retinol; 9,13-OH, 9,13-*cis*-retinol; aAT, *anti*-all-*trans*-retinal oxime; 9-OH, 9-*cis*-retinol; and AT-OH, all-*trans*-retinol. Additional elution positions are marked for compounds not represented by peaks in this chromatogram: BHT, butylated hydroxytoluene; RE, retinyl esters; s9,11, *syn*-9,11-*cis*-retinal oxime; s7, *syn*-7-*cis*-retinal oxime; s9,13, *syn*-9,13-*cis*-retinal oxime; a9,13, *anti*-9,13-*cis*-retinal oxime; a9,11, *anti*-9,11-*cis*-retinal oxime; a7, *anti*-7-*cis*-retinal oxime; and 7-OH, 7-*cis*-retinol.

samples. The resolution of a mixture of retinyl palmitate standards using this system is shown in Fig. 2.

Normal-phase chromatography tends to separate retinyl esters on the basis of both isomeric configuration and fatty acyl length and degree of unsaturation.[9] Thus, all-*trans*-retinyl palmitate is likely to overlap with 11-*cis*-retinyl linoleate, and 13-*cis*-retinyl palmitate elutes close to 11-*cis*-retinyl stearate. This is not often a major problem, because retinyl palmi-

[9] R. A. Alvarez, C. D. B. Bridges, and S.-L. Fong, *Invest. Ophthalmol. Visual Sci.* **20,** 304 (1981).

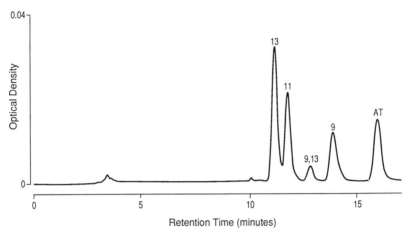

Fig. 2. Resolution of retinyl palmitate isomers prepared by acylation of isomers of retinol. Detection was by absorbance at 325 nm. Peak identities are 13, 13-*cis*-; 11, 11-*cis*-; 9,13, 9,13-*cis*-; 9, 9-*cis*-; and AT, all-*trans*-retinyl palmitate.

tate tends to predominate in vertebrate retinal pigment epithelium, accompanied by only a small amount of retinyl stearate.[9] However, when several retinyl esters are present, and the isomeric state of the esters is the parameter of interest, rather than the fatty acyl composition, it is probably best to reduce the retinyl ester fraction with lithium aluminum hydride[10] and then analyze the resulting mixture of retinol isomers.

Preparation of Standards

Purified isomer standards are necessary to confirm the identity of peaks in extracts of tissue samples, and for the preparation of standard curves.

Preparation and Purification of Retinal Isomers

Twenty-five milligrams of all-*trans*-retinal (Sigma, St. Louis, MO) is dissolved in 50 ml acetone (containing 0.1 mg/ml BHT) in a 250-ml Erlenmeyer flask. The flask is flushed with nitrogen or argon to displace oxygen, stoppered, and then placed in a beaker of ice which is supported on a ring stand. A 100-W incandescent bulb is positioned about 15 cm below the beaker, and the retinal solution is exposed to the incandescent lamp

---

[10] C. D. B. Bridges and R. A. Alvarez, this series, Vol. 81, p. 463.

for 1–2 hr. The acetone solution of retinal isomers may be stored in a freezer until needed for the preparation of pure isomers of retinal, or of mixtures of retinal oxime, retinol, or retinyl esters.

The acetone is evaporated under a stream of nitrogen, or under reduced pressure in a rotary evaporator, and the retinal is redissolved in 4% ethyl acetate (v/v) in hexane. Isomers are purified on a silica HPLC column, eluted with 4% ethyl acetate in hexane, at a rate of 1 ml/min for a 4.6-mm i.d. column, or 9 ml/min for a 9-mm ID column. Retinal isomers elute in the order 13-cis, 11-cis, 9-cis, 7-cis, all-trans under these conditions. Peaks are best collected by hand, and only the central portion of each peak should be collected if maximal purity is to be obtained. The fractions are concentrated under reduced pressure with a rotary evaporator, or under a stream of nitrogen, and stored protected from light below $-20°$.

Confirmation of Retinal Isomer Identities

Ideally, the isomeric identity of each purified retinal isomer should be confirmed by high-field (>200 MHz) proton NMR and comparison to published spectroscopic data.[11,12] However, relatively large amounts of material are required for NMR analysis, and such definitive characterization may not be necessary in all cases. An alternative is to deduce the peak identities by comparison of retinal isomer mixtures prepared by different methods. Thus, retinal photoisomerized in acetone will contain 11-cis- and 7-cis-retinal, whereas retinal photoisomerized in hexane or retinal catalytically isomerized by iodine in hexane will instead contain predominantly the 13-cis and 9-cis isomers, in addition to the parent all-trans isomer.[5,13] Similarly, photoisomerization of retinol or retinyl palmitate in chloroform or dichloromethane will result in mixtures which contain primarily the 9-cis and all-trans isomers.[5] Such a mixture of retinol isomers can be oxidized to the corresponding retinal mixture with manganese dioxide,[10] without alteration of configuration. Peak identities may be further confirmed by comparison of UV–VIS absorption maxima (determined either by scanning collected fractions or by stopped-flow scanning with a scanning or diode-array detector) to published values.[12] It should be noted that UV absorption maxima in hexane may be shifted by the presence of polar modifiers such as ethyl acetate.[14] Retinol, retinyl palmitate, and retinal oxime isomers prepared from retinal isomers by the

[11] D. J. Patel, *Nature (London)* **221**, 825 (1969).
[12] R. S. H. Liu and A. E. Asato, this series, Vol. 88, p. 506.
[13] R. R. Rando and A. Chang, *J. Am. Chem. Soc.* **105**, 2879 (1983).
[14] F. Zonta and B. Stancher, *J. Chromatogr.* **301**, 65 (1984).

methods outlined below will retain the geometric configuration of the parent retinal isomer, although some repurification may be necessary to remove minor products of thermal isomerization.

Preparation of Retinal Oxime, Retinol, and Retinyl Esters from Retinal

The methods described here are modifications of those of Bridges and Alvarez,[10] and they are ·suitable for use with mixed or pure retinal isomers. The reactions may be monitored by spotting aliquots of the reaction mixtures on silica thin-layer chromatography (TLC) plates, developed with a solvent such as 50 : 30 : 20 (v/v/v) hexane/toluene/ethyl acetate. When viewed with a 366-nm UV lamp, retinal on the plates appears dark (no fluorescence), whereas retinal oxime fluoresces orange, and retinol and retinyl esters appear as white spots, with the retinyl esters near the solvent front.

Reduction of Retinal to Retinol. The acetone or HPLC solvent should be removed from 0.1 to 0.2 mg of retinal by evaporation under reduced pressure or under a stream of nitrogen or argon, and the residue redissolved in 2 ml of methanol. (Retinal should not be stored for long periods as a methanol solution, because nonretinal derivatives will form, presumably by aldol condensation with the methanol.) One to two milligrams sodium borohydride should be added and allowed to react in the cold for a few minutes. After the reaction is complete (as shown by TLC), 2 ml hexane should be added to the methanol, followed by 2 ml water, and the hexane layer collected with a Pasteur pipette. The hexane extraction is then repeated twice more, and the hexane extracts are combined. The extract should be washed once or twice with water, and then dried over sodium sulfate. The retinol isomers may be repurified by HPLC using the same column as for the purification of retinal isomers, except with 20% ethyl acetate (v/v) in hexane as the mobile phase.

Formation of Retinal Oximes. A methanol solution of retinal is prepared, as for reduction to retinol, and reacted with sufficient aqueous bicarbonate-buffered 1 M hydroxylamine (pH 6.5) to make a 70% methanol reaction mixture. After the mixture has been allowed to react for 10 min or more, the retinal oximes may be extracted into hexane, as after reduction with sodium borohydride. Purification of the syn and anti isomers of each retinal isomer is performed with the same HPLC column and solvent used for repurification of retinol isomers.

Esterification. Approximately 10 μmol of retinol (prepared by reduction of retinal) is evaporated to dryness under a stream of nitrogen, and 100 μl of dichloromethane (containing 45 μl/ml freshly distilled triethylamine) is added. (Note that all dichloromethane used in acylation should be freshly dried over excess sodium sulfate.) One hundred microliters of a

200 mM solution of palmitoyl chloride or other fatty acyl chloride in dichloromethane is then added and allowed to react 10 min at room temperature. The reaction mixture should then be chilled on ice, and the course of the reaction monitored by TLC. The retinyl esters are extracted into hexane as described for the reduction of retinol and purified by HPLC as described for analysis of retinyl ester isomers.

Preparation of Standard Curves

Ideally, standard curves should be prepared for all of the isomers of retinal oxime, retinol, and retinyl palmitate, and probably retinyl stearate, but this is not practical. In practice, it is quite adequate to prepare standard curves for *syn*-11-*cis*-retinal oxime, *syn*-all-*trans*-oxime, *anti*-11-*cis*-retinal oxime, *anti*-all-*trans*-retinal oxime, 11-*cis*-retinol, and all-*trans*-retinol (and all-*trans*-retinyl palmitate for estimation of retinyl esters) for analysis of retinal and retinol isomers by the coupled-column method. For retinyl ester isomer analysis, preparation of standard curves for 11-*cis*- and all-*trans*-retinyl palmitate, and possibly for 11-*cis*- and all-*trans*-retinyl stearate, is sufficient. Although this may seem daunting, all the purified quantitative standards for the retinol/retinal oxime method may be mixed (after their concentrations have been determined by UV spectroscopy) and the standard curves run simultaneously, because all of the isomers are resolved. In the author's hands such standard curves are essentially linear from less than 1 to above 1500 ng per injection for each of these retinoids. A list of extinction coefficients for the 11-cis and all-trans isomers is provided in Table I. The slopes of standard curves for

TABLE I

ABSORPTION PROPERTIES OF 11-cis AND all-trans ISOMERS OF RETINAL OXIME
AND RETINOL IN HEXANE

Compound	λ_{max}[a]	ε_{max}[a]	ε_{325}	ε_{360}
syn-all-*trans*-Retinal oxime	357	52,500	32,900[b]	54,900[a]
anti-all-*trans*-Retinal oxime	361	51,700	28,700[b]	51,600[a]
syn-11-*cis*-Retinal oxime	347	35,900	25,700[b]	35,000[a]
anti-11-*cis*-Retinal oxime	351	30,000	19,600[b]	29,600[a]
all-*trans*-Retinol or retinyl esters	325	51,800	52,100[c]	10,300[b]
11-*cis*-Retinol or retinyl esters	318	34,300	34,100[c]	5,930[b]

[a] G. W. T. Groenendijk, P. A. A. Jansen, S. L. Bonting, and F. J. M. Daemen, this series, Vol. 67, p. 203.
[b] Calculated from spectra of purified standards, based on ε_{max} (*Ibid.*).
[c] Calculated from $E_{1\,cm}^{1\%}$ data of R. Hubbard, *J. Am. Chem. Soc.* **78,** 4662 (1956).

11-cis and all-trans isomers of retinol and retinal oxime have been demonstrated to be directly related to the extinction coefficient of the isomer at the wavelength of detection in the dual-column method.[3] Thus, the slopes of the standard curves for the less interesting (physiologically) 9-cis and 13-cis isomers may be estimated from a plot of standard curve slope versus extinction coefficient for the all-trans and 11-cis isomers.

HPLC of Separate Classes of Retinoids

Although the system described for the simultaneous analysis of retinal oxime and retinol isomers is quite usable, it represents a compromise. Simpler, more efficient systems have been developed for the analysis of the isomers of each of the classes of retinoids when the others are not present. Thus, for the analysis of retinol isomers when retinal oxime isomers are not present, the method of Stancher and Zonta[15] with a 4 × 250 mm 5-μm Lichrosorb column eluted with 3.8% 1-octanol, 0.2% 2-propanol in hexane at a flow rate of 0.6 ml/min is very good. Similarly, Zonta and Stancher[14] have published an excellent method for the separation of isomers of retinal, based on a 4 × 250 mm 5-μm Lichrosorb column eluted with 1% dioxane, 0.1% 2-propanol in hexane at a flow rate of 1.0 ml/min. For the analysis of retinal oxime isomers in the absence of retinol isomers, particularly when resolution of the syn-9-cis and syn-13-cis isomers is needed, a 4 × 250 mm 5-μm Lichrosorb column eluted with 0.4% dioxane, 5.6% ethyl acetate, and 1.2% 1-octanol (v/v) in hexane at a flow rate of 1.0 ml/min should be adequate (G. M. Landers and J. A. Olson, unpublished observation).

Acknowledgments

Most of the work described was performed in the laboratory of Dr. James A. Olson and funded by a grant from the National Institutes of Health (R01 EY 03677). The author thanks Dr. Olson, Dr. Harold Furr, and Dr. Arun Barua for their support and for many helpful discussions. This chapter was written while the author was supported by a grant from the National Institutes of Health (EY 07026) in the laboratory of Dr. Dean Bok.

[15] B. Stancher and F. Zonta, *J. Chromatogr.* **312,** 423 (1984).

[7] Separation of Fatty Acid Esters of Retinol by High-Performance Liquid Chromatography

By A. CATHARINE ROSS

Introduction

Long-chain fatty acid esters of retinol comprise the major storage form of vitamin A in numerous tissues. It has long been recognized that various organs have characteristic patterns of esterified retinol[1] and that differences may also exist among animal species in the retinyl ester composition within a particular organ such as the liver.[2] Thin-layer chromatographic methods have now been replaced by HPLC procedures that permit excellent resolution, as well as quantification, of individual fatty acid esters of retinol. This chapter provides a brief review and comparison of four published procedures. All are based on similar principles and provide rather comparable separations, but they utilize different column and solvent combinations. One method that was developed in this laboratory was presented in 1986 in this series[3] and is referred to here for some specific procedural information.

Principle

Retinyl esters are among the most hydrophobic of retinoids, and thus reversed-phase HPLC columns of octadecyl- or octyl-substituted silica gel allow good adsorption of these molecules, with desorption by solvents of low dielectric constant. The differential hydrophobicity of retinyl esters is determined by their fatty acyl chains, and, like fatty acids, the hydrophobicity increases with increasing acyl chain length and decreases with extent of unsaturation.[3,4] Thus, two of the predominant esters in most tissues, retinyl palmitate and retinyl oleate, have similar adsorption characteristics and provide the greatest challenge for separation. The intrinsic, characteristic ultraviolet absorbance of retinol (and its esters) at or near 325 nm permits sensitive detection of retinol and its esters without need for pre- or postcolumn derivatization. Diode-array detectors can provide additional spectral information useful in assessing the purity of

[1] D. S. Goodman, H. S. Huang, and T. Shiratori, *J. Lipid Res.* **6**, 390 (1965).
[2] S. Futterman and J. S. Andrews, *J. Biol. Chem.* **239**, 4077 (1964).
[3] A. C. Ross, this series, Vol. 123, p. 68.
[4] A. C. Ross, *Anal. Biochem.* **115**, 324 (1981).

eluted materials. Detection using an on-line fluorimeter should also be possible and may be even more sensitive, provided interference by other sample materials can be eliminated. Sample preparation will vary considerably depending on the tissue, quantity needed for analysis, and content of other lipids; this subject has been discussed previously.[3]

Materials and Methods

Standards. Few esters of retinol to serve as standards are available commercially; however, methods for synthesis are not difficult. Two acylation methods that have proved valuable are an adaptation of the method of Lentz *et al.*[5] using fatty acid anhydride, and the acyl chloride procedure of Huang and Goodman.[6] Our use of each of these methods has been described previously.[3,4] A series of even-carbon esters is needed to identify retention times of physiological retinyl esters, whereas one or more odd-carbon number retinyl esters (e.g., retinyl pentadecanoate or heptadecanoate) are useful for introduction as an internal standard to allow quantitative analysis.[3] We prefer to keep standards for long-term storage as the neat oils or solids under nitrogen or argon and at $-20°$ and to prepare working standards in column solvent as needed for short-term storage (a few days) at -20 or $4°$.

Instrumentation. Nearly any HPLC system will suffice. The four methods described below utilize isocratic solvent systems or step gradients and can be accomplished at room temperature with a single-pump HPLC system equipped with an ultraviolet absorbance monitor, either a variable-wavelength detector or a fixed-wavelength filter detector.

Columns and Solvents. The four methods that have been described vary in the choice of HPLC column and solvent system, as summarized in Table I.[7-9] Only generic information about columns is provided, as newer columns with more finely divided silica, shorter column length, and proprietary changes in substituent preparation are now available. In any case, individual columns may require some adjustment of the described solvent systems to optimize separations. The earliest method to be described, that of de Ruyter and de Leenheer,[7] utilizes a C_{18} substituted silica HPLC column with methanol, 1 ml/min, as the mobile phase. With methanol alone, retinyl palmitate and oleate are not resolved; however, good separation is achieved by inclusion of silver nitrate as a complexing

[5] B. R. Lentz, Y. Barenholz, and T. E. Thompson, *Chem. Phys. Lipids* **15**, 216 (1975).

[6] H. S. Huang and D. S. Goodman, *J. Biol. Chem.* **240**, 2839 (1965).

[7] M. G. M. de Ruyter and A. P. de Leenheer, *Anal. Chem.* **51**, 43 (1979).

[8] P. V. Bhat and A. Lacroix, *J. Chromatogr.* **272**, 269 (1983).

[9] H. G. Furr, D. A. Cooper, and J. A. Olson, *J. Chromatogr.* **378**, 45 (1986).

TABLE I
HPLC COLUMNS AND SOLVENT SYSTEM COMBINATIONS USED
FOR SEPARATION OF RETINYL ESTERS

Column type[a]	Solvent	Ref.
RP, octadecyl silica	Methanol/silver nitrate	7
RP, octyl or phenyl silica	Acetonitrile/water	4
RP, octadecyl silica	Methanol/water	8
RP, octadecyl silica	Acetonitrile/dichloromethane	9

[a] RP, Reversed-phase HPLC column.

ion for retinyl oleate in the methanol mobile phase. Elution of esters up to retinyl stearate takes place in about 16 min. Disadvantages include the cost of silver nitrate and the eventual coating of detector window surfaces by silver ions.

The method of Ross[4] was developed to overcome these disadvantages. By using an octyl- (or phenyl-) substituted silica column, good separation of retinyl palmitate and oleate, as well as other esters, could be achieved using a mobile phase of acetonitrile/water (88 : 12, v/v), 3 ml/min, to elute all esters up to palmitate, with a step to acetonitrile/water (98 : 2, v/v) to elute more hydrophobic esters (e.g., retinyl stearate, which is the most hydrophobic retinyl ester found so far in significant amounts in tissue samples). Resolution of all esters up to retinyl stearate requires about 60 min. A possible disadvantage of this method, and perhaps others as discussed previously,[3] is the rather low solubility of lipids in water-containing solvent mixtures. Thus, especially when lipid extracts contain much larger amounts of other lipid esters relative to retinyl esters (as is the case for chylomicrons, milk, and some fatty tissues of low vitamin A concentration), a preliminary chromatographic step to separate triglycerides from retinyl esters is needed.[3,4]

The method of Bhat and Lacroix[8] utilizes a C_{18} substituted column eluted with methanol/water (98 : 2), 1.5 ml/min, for esters up to retinyl oleate with elution at 2 ml/min for the rest of the run. Elution time is about 90 min. This method did not provide complete, baseline, separation for standards of retinyl palmitate and retinyl stearate and did not appear to separate these esters at all when lipid extracts of normal rat liver were analyzed. More recently, Furr et al.[9] have described the use of a C_{18} substituted column with an isocratic, nonaqueous, mobile phase of acetonitrile/dichloromethane (80 : 20, v/v), 1.5 ml/min. This system provides adequate, although not complete, separation of retinyl palmitate and stearate standards and has the advantage that a run can be completed in about 15 min.

Total Retinyl Ester and Total Retinol Analyses

The four procedures described above were developed for the individual analysis of retinyl ester species. In some cases, the investigator wishes to separate retinyl esters as a class from unesterified retinol and to quantify each. Bankson et al.[10] have described use of a normal-phase silica HPLC column and a linear gradient of dioxane/hexane (0.5 : 100, v/v) to 100% dioxane to sequentially elute retinol, retinyl acetate, and a combination of long-chain retinyl esters. In this system, retinyl stearate, oleate, linoleate, heptadecanoate, palmitate, and myristate coelute. It would be expected that medium-chain fatty acid esters of retinol, such as are found in milk,[11] would not coelute with these long-chain esters. Separation is rapid (~10 min). The reported coefficient of variation is quite good (3–6% for retinol and 8–10% for retinyl esters), and analysis of normal adult human plasma required only 100 μl of serum or plasma.

For determination of total retinol in tissues or plasma, a number of satisfactory methods have been described (e.g., see Ref. 3). Saponification[3] is required as a preliminary step and also serves to eliminate many other lipids from the tissue extract. Since our previous discussion of this method,[3] we have adopted[12] the use of a nonhydrolyzable analog of retinol, triphenylmethoxyphenylretinol (TMMP-retinol[13]), as the internal standard for quantification. This analog has real advantage over the use of retinyl acetate for quantifying retinol in that TMMP-retinol cannot be converted to the analyte. Good separation and measurement of total retinol in the extract of 50–100 μl of serum or an appropriate aliquot of tissue homogenate can be accomplished in less than 5 min using a short (5 cm × 4.6 mm i.d.), 5-μm C_8 column and methanol/water (78 : 22, v/v), 2 ml/min, as the isocratic mobile phase.

Acknowledgments

This work was supported by National Institutes of Health Grants HD-16484, HL-22633, DK-41479, and a Research Career Development Award, HD-00691, and by general support funds from the Howard Heinz Endowment.

[10] D. D. Bankson, R. M. Russell, and J. A. Sadowski, Clin. Chem. 32, 35 (1986).
[11] A. C. Ross, M. E. Davila, and M. P. Cleary, J. Nutr. 115, 1488 (1985).
[12] A. M. G. Pasatiempo, T. A. Bowman, C. E. Taylor, and A. C. Ross, Am. J. Clin. Nutr. 49, 501 (1989).
[13] J. L. Napoli, this series, Vol. 123, p. 112.

[8] Reversed-Phase High-Performance Liquid Chromatography of Retinyl Esters

By HAROLD C. FURR

Introduction

Dietary vitamin A in excess of immediate tissue requirements is stored as the ester of retinol with long-chain fatty acids; the primary site of such storage is the liver. Futterman and Andrews[1] analyzed the retinyl ester composition of human, rat, calf, sheep, rabbit, cat, frog, and trout livers by gas chromatography of the saponified fatty acids. They found that the predominant ester in every case was retinyl palmitate, with retinyl stearate and oleate the next most prevalent; small amounts of retinyl linoleate, palmitoleate, myristate, heptadecanoate, and pentadecanoate were also found (in approximately that order of abundance). Goodman *et al.*[2] confirmed these results in rat liver by use of argentation thin-layer chromatography and reversed-phase paper chromatography of the retinyl esters.

More recently, high-performance liquid chromatography (HPLC) has been used to analyze retinyl ester composition. De Ruyter and De Leenheer[3] used methanol with an octadecylsilane column to separate several long-chain retinyl esters by reversed-phase HPLC, and they later used argentation chromatography with methanol–water eluents in an ingenious approach to the separation of vitamin A esters.[4] Ross[5] and Bhat and LaCroix[6] employed acetonitrile–water mobile phases on reversed-phase columns, but their procedures are time-consuming, requiring 1 hr or more to elute retinyl stearate. Cullum and Zile[7] used a multistep-gradient reversed-phase HPLC system to separate and quantitate a wide variety of retinoids, but the "critical pair" retinyl palmitate and retinyl oleate were not resolved on this system. May and Koo recently demonstrated the linear relationships between log(capacity factor) and mobile phase composition for a given fatty acyl chain length in methanol–water mobile

[1] S. Futterman and J. S. Andrews, *J. Biol. Chem.* **239**, 4077 (1964).

[2] D. S. Goodman, H. S. Huang, and T. Shiratori, *J. Lipid Res.* **6**, 390 (1965).

[3] M. G. M. De Ruyter and A. P. De Leenheer, *Clin. Chem.* **24**, 1920 (1978).

[4] M. G. M. De Ruyter and A. P. De Leenheer, *Anal. Chem.* **51**, 43 (1979).

[5] A. C. Ross, *Anal. Biochem.* **115**, 324 (1981).

[6] P. V. Bhat and A. LaCroix, *J. Chromatogr.* **272**, 269 (1983).

[7] M. E. Cullum and M. H. Zile, *Anal. Biochem.* **153**, 23 (1986).

phases and between log(relative retention time) and carbon chain length for a given mobile phase composition; however, resolution of retinyl oleate from palmitate in biological samples was not complete with methanol–water mobile phases, even at long analysis times.[8]

Biesalski and Weiser[9] have presented an isocratic system for adsorption HPLC of retinyl esters (separating geometric isomers of the major vitamin A esters), and Bridges et al.[10] have developed a gradient method for separation of retinyl esters by adsorption HPLC. Nonaqueous reversed-phase HPLC was used for the chromatography of retinyl palmitate[11–13] but was not used by those authors for separation of other retinyl esters. Mobile phases of methanol–tetrahydrofuran can separate retinyl esters by fatty acyl chain length but do not resolve retinyl palmitate from retinyl oleate.[14] We have found, however, that mobile phases of acetonitrile with chlorinated hydrocarbons provide rapid and satisfactory separation of the most common retinyl esters in biological tissues, including the usually difficult-to-separate retinyl palmitate and retinyl oleate.[15] This method has been used for separation and quantitation of retinyl esters in bovine retinal pigment epithelium,[16] rabbit lacrimal gland,[17] and bovine milk,[18] and for determination of the effects of dietary fatty acid composition on rat liver retinyl ester composition.[19]

Instrumentation

For the isocratic method described here, a single pump, injector, and absorbance detector monitoring at 325 nm are adequate. For convenience, a variable-wavelength detector or photodiode-array detector (or two absorbance detectors in series) facilitates identification of nonretinoid components in biological samples (carotenoids, α-tocopherol, and ubiquinones). For gradient chromatography, a system capable of pumping two mobile phases is required.

[8] H. E. May and S. I. Koo, J. Liq. Chromatogr. **12,** 1261 (1989).
[9] H. K. Biesalski and H. Weiser, J. Clin. Chem. Clin. Biochem. **27,** 65 (1989).
[10] C. D. B. Bridges, S.-L. Fong, and R. A. Alvarez, Vision Res. **20,** 355 (1980).
[11] N. A. Parris, J. Chromatogr. **157,** 161 (1978).
[12] W. O. Landen, J. Assoc. Off. Anal. Chem. **63,** 131 (1980).
[13] H. J. C. F. Nelis and A. P. De Leenheer, Anal. Chem. **55,** 270 (1983).
[14] H. C. Furr, O. Amedee-Manesme, and J. A. Olson, J. Chromatogr. **309,** 299 (1984).
[15] H. C. Furr, D. A. Cooper, and J. A. Olson, J. Chromatogr. **378,** 45 (1986).
[16] J. C. Saari and D. L. Bredberg, J. Biol. Chem. **263,** 8084 (1988).
[17] J. L. Ubels, T. B. Osgood, and K. M. Foley, Curr. Eye Res. **7,** 1009 (1988).
[18] D. C. Woolard and H. Indyk, J. Micronutr. Anal. **5,** 35 (1989).
[19] H. C. Furr, A. J. Clifford, L. M. Smith, and J. A. Olson, J. Nutr. **119,** 581 (1989).

Retinyl esters are well separated on a Waters Resolve 5-μm C_{18} column (15 cm × 3.9 mm i.d.; Waters Associates, Milford, MA); alternatively, we have used a Rainin Microsorb 5-μm C_{18} column (15 cm × 4.6 mm i.d.; Rainin, Woburn, MA) or a Whatman 10-μm Partisil 10/25 ODS-2 column (25 cm × 4.6 mm i.d.; Whatman Chemical Separations, Clifton, NJ). Guard columns (2 cm × 2 mm i.d.; Upchurch Scientific, Oak Harbor, WA) packed with pellicular octadecylsilane material (Vydac, The Separations Group, Hesperia, CA) are used routinely.

For isocratic elution, mobile phases of acetonitrile–chloroform, acetonitrile–dichloromethane, or acetonitrile–dichloroethane (80 : 20, v/v) give equivalent separations; 1,2-dichloroethane is preferable to dichloromethane because its higher boiling point (almost identical to that of acetonitrile) eliminates bubble formation in pumps and detector cells and prevents changes of mobile phase composition owing to preferential evaporation of solvent. Chloroform is to be avoided because of its toxicity; the other chlorinated hydrocarbon solvents are considered less toxic but should be used with care. Addition of cyclohexene (0.1%) may suppress deleterious effects arising from free radical formation from chlorinated hydrocarbon solvents, and has no effect on retinyl ester separation. Therefore, our routine isocratic mobile phase is acetonitrile–dichloroethane (80 : 20, v/v) containing 0.1% cyclohexene, at a flow rate 1.5 ml/min (Table I). An Ultracarb ODS (30) column (7-μm, C_{18}, 250 × 4.6 mm i.d.; Phenomenex, Rancho Palos Verdes, CA) gives equivalent resolution of retinyl esters but is much more retentive, such that a much less polar mobile phase (e.g., acetonitrile–dichloroethane, 50 : 50) must be used (H. Furr, unpublished observations).

For gradient elution, to provide appropriate capacity factors for retinol as well as for retinyl esters, a linear gradient of mobile phase from acetonitrile–water (85 : 15) to acetonitrile–dichloroethane (80 : 20, v/v, plus 0.1% cyclohexene) has proved useful with the above-mentioned columns (Table I).

Retinyl Ester Standards

Retinyl ester standards are prepared conveniently by transesterification between retinyl acetate and the appropriate fatty acid methyl ester under reduced pressure.[20] Alternatively, retinyl esters can be prepared by condensing retinol with fatty acid anhydrides[5] or fatty acyl chlorides.[21]

[20] S. Futterman and J. S. Andrews, *J. Biol. Chem.* **239**, 81 (1964).
[21] H. S. Huang and D. S. Goodman, *J. Biol. Chem.* **240**, 2839 (1965).

TABLE I
CHROMATOGRAPHIC CONDITIONS

Parameter	Conditions
Isocratic HPLC	
Column	Waters Resolve 5-μm C$_{18}$, 15 cm × 3.9 mm, or Whatman Partisil ODS-2 10-μm C$_{18}$, 25 cm × 4.6 mm
Mobile phase	Acetonitrile–dichloroethane (80:20, v/v) containing 0.1% cyclohexene; flow rate 1.5 ml/min
Detection wavelength	325 nm for retinoids
Limit of detection	Retinyl palmitate, 4 pmol (5:1 signal-to-noise ratio) for Resolve column; 8 pmol retinyl palmitate for Partisil ODS-2 column
Resolution	1.2 between retinyl oleate and retinyl palmitate
Gradient HPLC	
Column	Waters Resolve 5-μm C$_{18}$, 15 cm × 3.9 mm
Mobile phase	Linear gradient over 10 min from acetonitrile–water (85:15, v/v) to acetonitrile–dichloroethane (80:20, v/v, plus 0.1% cyclohexene) with 15-min hold; flow rate 1.5 ml/min

After open-column chromatography on short silica columns, eluting with 5% ethyl acetate in hexane (to remove residual retinol and reaction byproducts), esters can be further purified by semipreparative adsorption HPLC (e.g., Whatman M20 10/50 silica column, 50 × 1.0 cm i.d.; eluted with 4% ethyl acetate in hexane, flow rate 20 ml/min). Because retinol is unstable and commercial retinol preparations are usually impure, it is convenient to prepare retinol fresh by saponification of the more stable retinyl acetate or by reduction or retinaldehyde in methanol solution with sodium borohydride.[22] Commercial retinyl acetate and retinaldehyde preparations are usually adequately pure if stored properly (sealed under argon and stored at −20°). Solutions of retinyl ester standards in hexane, methanol, or acetonitrile may be stored sealed under argon, in a refrigerator or freezer, protected from white light. Purity of retinoids may be rapidly confirmed by examination of their ultraviolet absorption spectra and by thin-layer chromatography.

Sample Preparation

A variety of sample preparation procedures for vitamin A analysis of biological tissues has been proposed.[23] A simple procedure involves

[22] R. Hubbard, P. K. Brown, and D. Bownds, this series, Vol. 18C, p. 615.
[23] C. A. Frolik and J. A. Olson, in "The Retinoids" (M. B. Sporn, A. B. Roberts, and D. S. Goodman, eds.), Vol. 2, p. 181. Academic Press, New York, 1984.

grinding the tissue with anhydrous sodium sulfate (which dehydrates and breaks apart the tissue) and extracting repetitively with dichloromethane.[24] The dichloromethane extracts are filtered and combined, then made up to appropriate volume. For example, a 0.5-g liver sample is ground with 2 g anhydrous sodium sulfate using a mortar and pestle, then extracted with 5 portions of dichloromethane of 5 ml each, with continuous grinding. The combined extracts can be filtered and diluted to 50 ml in a volumetric flask. An aliquot of the extract is evaporated under a gentle steam of argon (preferable to nitrogen because the oxygen content of commercial argon is lower), and the residue is redissolved in a small volume of 2-propanol or 2-propanol plus dichloromethane. Aliquots of 0.1–2 nmol (30–500 ng) total retinyl esters can be analyzed; injection volumes should be limited to 10 μl or less for isocratic elution, but they can be as large as 100 μl for gradient elution. All procedures can be carried out under yellow fluorescent lighting (F40 gold fluorescent tubes), which provides good illumination without danger of retinoid isomerization.[25]

Quantitation

As shown by Ross,[5] retinol and retinyl esters have equal molar absorptivities (52,275 M^{-1} cm^{-1} at 325 nm), and so standards may be quantitated by absorbance; because the retinyl esters have equal molar absorbance, retinyl palmitate may be used as quantitative standard for the other esters if peak area (not peak height) is used. It is best to scan absorbance of standard solutions over the wavelength range 250 to 400 nm, rather than to determine absorbance only at 325 nm, so as to disclose possible decomposition of standards. Standard curves of integrator peak area versus mass retinyl ester injected are usually linear over the range 0 to 2 nmol (up to 500 ng retinol component of retinyl esters), but linearity should be confirmed with the particular HPLC detector in use. It is preferable to express results in molar units (e.g., nanomoles retinyl ester per gram tissue), but "retinol equivalents" (micrograms of the retinol component of each retinyl ester) may be used if the usage is explicit; expressing results as total weight of each vitamin A ester (mass of retinol plus mass of fatty acyl component) is confusing and should be avoided.

Ross[26] has suggested the use of retinyl heptanoate (to estimate recovery of short-chain retinyl esters) and retinyl pentadecanoate (to estimate recovery of long-chain retinyl esters) as internal standards. Because re-

[24] J. A. Olson, *Nutr. Rep. Int.* **19,** 807 (1979).
[25] G. M. Landers and J. A. Olson, *J. Assoc. Off. Anal. Chem.* **69,** 50 (1986).
[26] A. C. Ross, this series, Vol. 123, p. 68.

tinyl heptanoate has not been detected in animal tissues and retinyl penta-decanoate is usually present in only trace amounts, these are appropriate internal standards. However, because retinyl pentadecanoate elutes quite near retinyl oleate, retinyl eicosanoate (retinyl arachidate, 20 : 0) may be a better choice for internal standard. We have never observed saturated retinyl esters of more than 18 carbons (no retinoids eluting after retinyl stearate in the HPLC system described here), even in trace amounts, in livers of various species of animals fed a variety of diets; hence, retinyl eicosanoate also is an appropriate internal standard for long-chain retinyl esters.

The lower limit of detection will depend on the particular HPLC detector used. With an LDC Spectromonitor III detector (LDC, Riviera Beach, FL) set at 325 nm, we have found lower limits of detection of 4.5 pmol retinyl palmitate (1.3 ng retinyl equivalents) at a signal-to-noise ratio of 5 : 1, using a 5-μm Resolve C$_{18}$ column. The within-day coefficient of variation was ±0.25% for retinyl palmitate and ±2.21% for retinyl oleate; the between-day coefficient of variation was ±1.12% for retinyl oleate. This allowed detection and quantitation of the major vitamin A esters from rat livers deficient in vitamin A, containing 1.25 nmol total retinyl esters/g liver (0.35 μg total retinyl esters/g liver) (Fig. 1). Because the 10-μm Partisil 10/25 ODS-2 column produced later-eluting, broader peaks, the limit of detection was not quite so low with this column: 8 pmol retinyl palmitate (2.2 ng vitamin A palmitate, expressed as nanograms retinol) at a 5 : 1 signal-to-noise ratio. Over the range 10 to 1000 ng total vitamin A, this isocratic method gave values for liver extracts identical to those given by a (different) gradient reversed-phase HPLC method.[15] Although it offers a lower limit of detection, the 5-μm Resolve column may be less suitable for simultaneous isocratic analysis of retinol and retinyl esters; because retinol elutes so near the solvent front (uncorrected retention time 1.4 min; capacity factor $k' = 0.6$), other components in tissue extracts may contribute to the retinol peak absorbance. Of the columns that we have tested, the 10-μm Partisil ODS-2 column retains retinol better ($k' = 2.5$) while eluting retinyl esters reasonably quickly, allowing more confident quantitation of retinol with retinyl esters.[15] Alternatively, gradient elution, as described above, may be used to determine retinol and retinyl esters in a single chromatographic run (Fig. 2).

Identification of Retinyl Esters

A semilogarithmic plot of corrected retention time (or capacity factor, k') versus carbon number of the saturated fatty acyl component of retinyl esters gives a straight line in isocratic reversed-phase HPLC (Fig. 3). (The

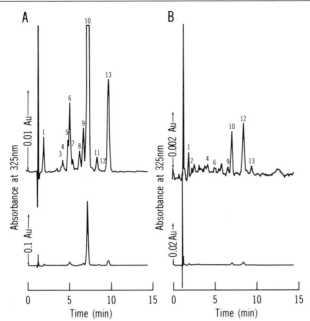

FIG. 1. Isocratic reversed-phase HPLC separations of rat liver extracts, using a 5-μm
Resolve column. Chromatographic conditions as in Table I; peak identification as in Table
II. (A) Liver extract (prepared as described in the text) from rat fed 840 nmol retinyl acetate
per day for 14 days; aliquot of extract analyzed equivalent to 1.6 mg liver, 0.98 nmol total
retinyl esters (610 nmol/g, 175 μg total retinyl esters/g liver). The upper trace is the chromatogram recorded at attenuation 5; the lower trace shows the same chromatogram recorded simultaneously at attenuation 9. (B) Liver extract from rat fed 17.5 nmol retinyl
acetate per day for 14 days; aliquot of extract analyzed equivalent to 11.1 mg liver, 0.014
nmol total retinyl esters (1.3 nmol/g, 0.4 μg total retinyl esters/g liver). The upper trace is
chromatogram recorded at attenuation 3; the lower trace shows the same chromatogram
recorded simultaneouly at attenuation 7. [From H. C. Furr, D. A. Cooper, and J. A. Olson,
J. Chromatogr. **378,** 45 (1986), with permission of the publisher.]

chromatographic dead time and hence the capacity factor can be calculated readily after chromatography of a series of saturated retinyl ester
standards.[27]) This relationship may be used to advantage in tentative identification of retinyl esters. More crudely but usefully, the uncorrected
retention time relative to retinyl palmitate (almost always the largest peak
observed) can also be used to identify tentatively the vitamin A esters
(Table II). These relative retention values should always be checked for
the particular column in use.

27 H. C. Furr, *J. Chromatogr. Sci.* **27,** 216 (1989).

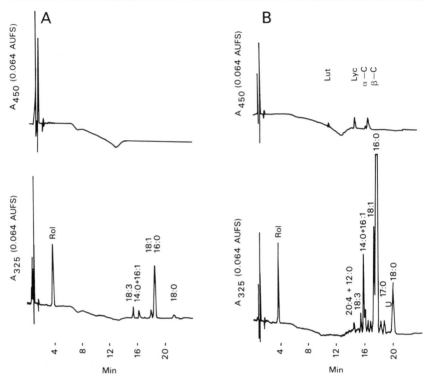

FIG. 2. Gradient reversed-phase separation of retinyl esters and carotenoids from extracts of human liver taken at liver transplantation surgery (injected amounts correspond to 4 mg liver, extracted as described in text). (A) Sample from subject with 85 nmol total vitamin A/g liver (24 μg/g). (B) Sample from subject with 746 nmol total vitamin A/g liver (215 μg/g). Upper traces: absorbance at 450 nm (carotenoids); lower traces: absorbance at 325 nm (retinoids). Rol, retinol; 14:0 + 16:1, retinyl myristate plus palmitoleate (unresolved); 15:0, retinyl pentadecanoate; 16:0, retinyl palmitate; 17:0, retinyl heptadecanoate; 18:0, retinyl stearate; 18:1, retinyl oleate, 18:2, retinyl linoleate; 18:3, retinyl linolenate; 20:4 + 12:0, retinyl arachidonate plus laurate (unresolved); U, ubiquinone 45; Lut, lutein; Lyc, lycopene; α-C, α-carotene; β-C, β-carotene. Gradient chromatographic conditions as described in Table I. (From S. A. Tanumihardjo, H. C. Furr, O. Amedee-Manesme, and J. A. Olson, unpublished results.)

The selectivity of this method, as for most HPLC methods for vitamin A, depends on the fact that few other biological compounds absorb light significantly at 325 nm. However, especially in concentrated tissue extracts, some other compounds are also detected. We have observed α-tocopherol (absorbance maximum 292 nm) eluting shortly after retinol (retention 0.36 relative to retinyl palmitate with our columns and mobile phase) as well as ubiquinone 45 (tentative identification based on ultravio-

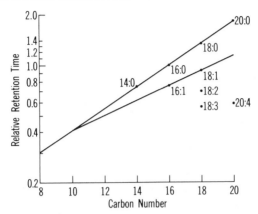

FIG. 3. Relative retention times of retinyl esters. Semilogarithmic plot of retention times of retinyl ester standards (relative to retinyl palmitate) versus carbon number of fatty acyl component, as determined on a Resolve 5-μm C$_{18}$ column. Isocratic chromatographic conditions as in Table I. [From H. C. Furr, D. A. Cooper, and J. A. Olson, *J. Chromatogr.* **378**, 45 (1986), with permission of the publisher.]

TABLE II
PEAK IDENTIFICATION OF RETINYL ESTERS

Peak[a]	Identification	Relative retention time[b]
1	Retinol	0.26
2	Nonretinoid (α-tocopherol)	0.36
3	Retinyl linolenate	0.56
4	Retinyl arachidonate plus laurate	0.59
5	Nonretinoid	0.69
6	Retinyl linoleate	0.70
7	Retinyl myristate plus palmitoleate	0.75
8	Retinyl pentadecanoate	0.87
9	Retinyl oleate	0.93
10	Retinyl palmitate	1.00
11	Retinyl heptadecanoate	1.16
12	Nonretinoid (ubiquinone 45)	1.20
13	Retinyl stearate	1.35

[a] Peak numbers as indicated in Fig. 1.
[b] Relative retention times on a Resolve 5-μm C$_{18}$ column, isocratic elution.

let absorption spectra and cochromatography with ubiquinone standards) eluting between retinyl heptadecanoate and stearate (relative retention 1.20). (Ubiquinone 50 elutes well after retinyl stearate, relative retention 1.7.) The carotenoid lycopene elutes at 0.58 relative to retinyl palmitate (between retinyl linolenate and laurate), α-carotene at 0.83, and β-carotene at 0.89 (between retinyl myristate and retinyl pentadecanoate); the presence of carotenoids can be determined by monitoring the absorbance at 450 nm. Light absorption at 325 nm by such carotenoids is typically 10% of their maximum light absorption.

Resolution of the critical pair, retinyl oleate and retinyl palmitate, has been optimized in this system (resolution 1.2), with short analysis times (capacity factors for retinyl esters from 2.8 to 8.7, all esters eluted within 10 min) and with better limits of detection than slower HPLC separations (owing to less peak broadening). Nonetheless, the quantitatively less important retinyl arachidonate and retinyl laurate are not resolved by this chromatographic system, and the resolution of retinyl myristate and palmitoleate is marginal. We have not been able to achieve separation of retinyl oleate from retinyl palmitate with methanol–tetrahydrofuran, acetonitrile–tetrahydofuran, or methanol–dichloromethane mobile phases.

Acknowledgments

Dr. Dale A. Cooper carried out some of these chromatographic studies, and Dr. Arun Barua has provided much assistance. It is a pleasure to acknowledge the support of Dr. James A. Olson for these studies. This work has been supported by the United States Department of Agriculture. (87-CRCR-1-2320), the National Institutes of Health (DK-32793, DK-39733, and CA 46406), and the Iowa Agriculture and Home Economics Experiment Station, Ames (Project No. 2534).

[9] Stable Isotope Dilution Mass Spectrometry to Assess Vitamin A Status

By Andrew J. Clifford, A. Daniel Jones, and Harold C. Furr

Introduction

Accurate assessments of total body stores of vitamin A in humans are essential for identification of individuals and populations at risk of deficiency or toxicity, and they are key elements in the evaluation of nutrition intervention programs. Current estimates are based on dietary intake, blood levels, or changes in blood levels of the vitamin (relative dose response) and stable isotope dilution techniques.

Serum vitamin A concentrations are homeostatically regulated and are inadequate indicators of status except in the rare cases when they are extremely low (10 μg/dl) or high (100 μg/dl). Most of the vitamin A reserves are stored in the liver and can be directly measured only under exceptional circumstances in humans. Total body stores of vitamin A can, however, be determined indirectly by measuring the dilution of standard doses of isotopically labeled vitamin A in blood sampled at appropriate times after dosing.[1,2]

The isotope dilution technique is a method of chemical analysis of total body vitamin A based on the administration of a known amount of labeled vitamin A followed by measurement of the ratio of labeled to unlabeled vitamin A in a quantity isolated from blood plasma or serum after appropriate equilibration. The technique entails synthesis and purification of vitamin A labeled with a stable (nonradioactive) isotope, administration of sufficient labeled vitamin to permit its detection and measurement in blood serum after equilibration with the total vitamin A pool, isolation and purification of the vitamin from blood plasma or serum, and determination of the ratio of labeled to unlabeled vitamin A using gas chromatography–mass spectrometry. Recent advances in capillary gas chromatography[3] and the widespread availability of inexpensive mass spectrometers have made stable isotope dilution experiments accessible and easy to perform.

Synthesis of Deuterated Retinoids

In the mass spectrum of retinol ($C_{20}H_{30}O$), naturally occurring heavy isotopes give ions at $[M + 1]^+$, $[M + 2]^+$, and $[M + 3]^+$ with intensities of 23, 2.7, and 0.22% relative to the molecular ion ($M^+ = 100\%$), respectively. These intensities represent background levels above which isotopic enrichment must be distinguished. To obtain reliable mass spectrometric determinations of the low levels of isotopic enrichment expected with milligram doses of labeled retinoids, deuterated retinoids should contain at least three and preferably four or more deuterium atoms. The deuterium should be incorporated into positions in the molecule that are

[1] D. R. Hughes, P. Rietz, W. Vetter, and G. A. J. Pitt, *Int. J. Vitam. Nutr. Res.* **46,** 231 (1976).
[2] H. C. Furr, O. Amedee-Manesme, A. J. Clifford, H. R. Bergen III, A. D. Jones, D. P. Anderson, and J. A. Olson, *Am. J. Clin. Nutr.* **49,** 713 (1989).
[3] M. L. Lee, F. J. Yang, and K. D. Bartle, "Open Tubular Column Gas Chromatography." Wiley, New York, 1984.

both chemically and metabolically stable, and the isotopic purity of labeled retinoids administered to subjects should be sufficiently high to ensure that administered doses do not contribute to pools of unlabeled retinoids in tissues and biological fluids.

Labeled retinoids and carotenoids are usually synthesized via the Wittig or Wittig–Horner reactions which involve the reaction of carbonyl compounds such as ketones with phosphorus ylides to form carbon–carbon double bonds. Because aliphatic hydrogen atoms at carbons alpha to carbonyls undergo base-catalyzed exchange with D_2O, β-ionone is a useful starting compound because it can be converted to trideuterated forms in high yield.[4] Subsequent steps in the synthesis of deuterated retinoids are often complicated by hydrogen–deuterium exchange that occurs during condensation reactions, but this problem is avoided by condensing deuterated ketones with deuterated triethyl phosphonoacetate (labeled by exchange with D_2O). The product of this condensation is over 98% tetradeuterated[5] and is a key precursor in the synthesis of retinol-10,19,19, 19-d_4.

Preparation of Retinol-10,19,19,19-d_4 and Retinyl Esters. Retinoids and their synthetic precursors are susceptible to photoinduced oxidation and isomerization and should be protected from exposure to light and oxidants, including air.

A scheme for the synthesis of retinol-10,19,19,19-d_4 from β-ionone-13,13,13-d_3 is adapted from that presented in detail by Bergen and co-workers[5] (Fig. 1). Deuterated starting materials are prepared via exchange with D_2O. Preparation of β-ionone-13,13,13-d_3 is accomplished by mixing β-ionone with D_2O and small amounts of NaOD and pyridine to increase the rate of hydrogen–deuterium exchange and disperse the emulsion. The exchange procedure is carried out twice, with the hexane extracts giving a 90% yield of β-ionone-13,13,13-d_3 with over 98% deuterium incorporation as determined by ^1H NMR. Deuterated triethylphosphonoacetate is prepared by stirring triethylphosphonoacetate with D_2O and a catalytic amount of NaOD for 2 hr. Again, the exchange reaction is carried out twice, with the diethyl ether extracts yielding at least 99% triethylphosphonoacetate-d_2 as determined by ^1H NMR. Because these deuteriums exchange rapidly with active hydrogens, GC–MS determinations generally underestimate deuterium incorporation into triethylphosphonoacetate and are not recommended.

The β-ionone-d_3 is condensed with the triethylphosphonoacetate-d_2

[4] J. E. Johansen and S. Liaaen-Jensen, *Acta Chem. Scand.* **28,** 349 (1974).
[5] H. R. Bergen, H. C. Furr, and J. A. Olson, *J. Labelled Compd. Radiopharm.* **25,** 11 (1988).

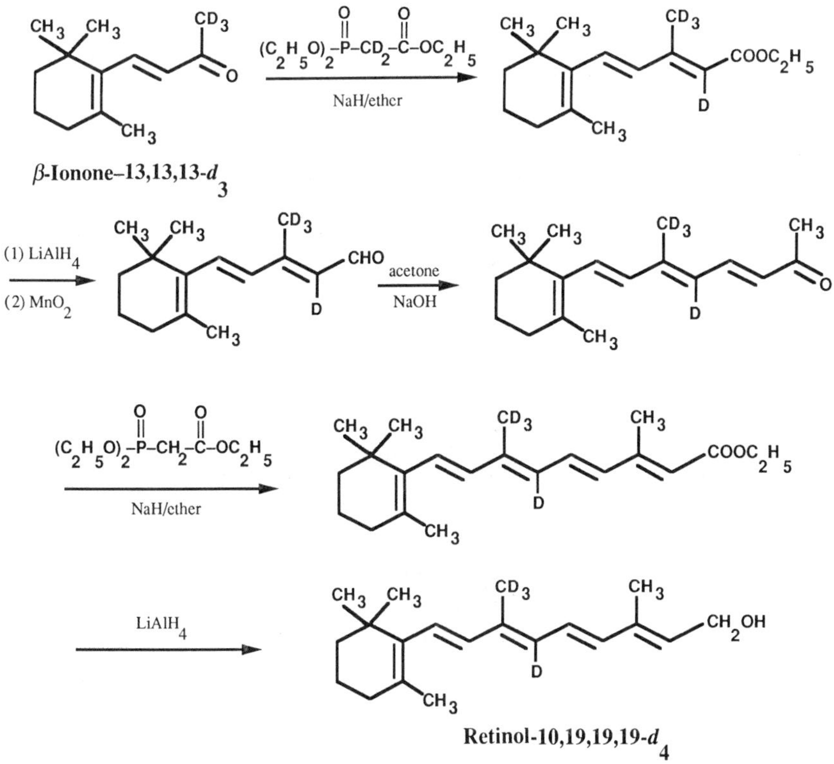

FIG. 1. A synthetic scheme for retinol-10,19,19,19-d_4, adapted from that by Bergen and co-workers.[5]

by stirring with sodium hydride in diethyl ether overnight. Because the exchangeable hydrogens in both reactants have been replaced by deuterium, replacement of deuterium by hydrogen is negligible. The product of this reaction, ethyl β-ionylideneacetate-d_4, is reduced using LiAlH$_4$ in diethyl ether at −70° to form β-ionylideneethanol-d_4. Subsequent oxidation to β-ionylideneacetaldehyde-d_4 is accomplished with MnO$_2$. Condensation of this aldehyde with acetone is conducted using base (NaOH) catalysis and yielded cis (20%) and all-trans (80%) isomers of the C$_{18}$ tetraene ketone. The all-trans isomer is separated by normal-phase HPLC (silica column; hexane/ethyl acetate, 98 : 2). Condensation of the C$_{18}$ tetraene ketone with unlabeled triethylphosphonoacetate is carried out as described above and yields retinoic acid-d_4 ethyl ester, which is reduced to retinol-d_4 using LiAlH$_4$. Acetylation of retinol-d_4 is conducted using acetic anhydride in pyridine. After isolation of the retinyl-d_4 acetate by normal phase HPLC (silica column; hexane/ethyl acetate, 98 : 2), GC–MS and ^{1}H NMR analysis showed the isotopic purity of the all-*trans*-retinyl-10,19,19,19-d_4 acetate to be in excess of 98% tetradeuterated.

Administration of Deuterated Vitamin A

Approximately 1 μmol of all-*trans*-retinyl-10,19,19,19-d_4 acetate per kilogram body weight is administered orally in corn oil. Absorption of retinyl acetate is believed to be nearly quantitative under normal conditions. An antecubital blood sample (~15 ml) is drawn 15 to 50 days later. The blood must be protected from light and is allowed to clot in a dark place, then cooled at 5° to shrink the clot, after which the serum is isolated by centrifugation, harvested, and stored at −70° (if possible) under argon, nitrogen, or other inert atmosphere in ultraclean glass tubes that have been flushed with argon. Each serum sample should be stored in several aliquots if repeated analyses are needed to avoid thawing and refreezing the sample. Serum is preferable to plasma because it minimizes the formation of precipitates during cold storage that can clog micropipettes. Use of plastic tubes and pipette tips, rubber stoppers, and paraffin film should be avoided as they often introduce plasticizers and other contaminants into the sample that can interfere with subsequent chromatography, UV detection, and mass spectrometry.

Extraction of Retinol and Retinyl Esters from Serum

The following procedure works well to extract retinol and retinyl esters from blood serum or plasma: serum is mixed with an equal volume of ethanol (which can contain either retinyl acetate or 15,15-dimethylretinol as an internal standard to quantify serum retinol) to precipitate protein, and 1–4 volumes of hexane containing 100 μg/ml of BHT (butylated hydroxytoluene) are added and thoroughly mixed to extract the retinoids. The hexane layer (upper phase) is isolated by centrifugation and carefully harvested. A stream of argon is used to evaporate the hexane, and the residue is dissolved in a solvent suitable for HPLC (2-propanol) or gas chromatography (cyclohexane). The serum extracts should be stored under an inert atmosphere to minimize oxidative degradation and should be analyzed promptly.

Separation of Retinol and Retinyl Esters

Many forms of conventional liquid chromatography can be used to separate retinol and retinyl esters in preparation for GC–MS analysis.

Reversed-phase HPLC using an analytical (μBondapak C_{18}, 30 $\times$ 3.9 mm, Waters Associates, Milford, MA) column with gradient elution (CH_3OH/H_2O, 80:20 for 5 min, followed by a linear gradient of CH_3OH/tetrahydrofuran, 50:50 from 0 to 70% over 10 min) separated retinol from retinyl esters in human serum.[6] Alternatively, retinol and retinyl esters can be separated by reversed-phase HPLC using gradient elution programming from acetonitrile/water (80:20) to acetonitrile/1,2-dichloroethane (80:20).[7] Elution of retinol and retinyl esters is monitored by UV absorbance at 325 nm, and fractions corresponding to the retinol and retinyl esters are collected for analysis by GC–MS.

Some fatty acid esters of retinol require such high temperatures for gas chromatographic separation that decomposition is likely. As a result, analysis of retinyl esters is performed by saponifying the combined retinyl ester fractions isolated by HPLC, followed by GC–MS analysis of the retinol recovered from the saponification.

Capillary Gas Chromatography of Retinol

Successful gas chromatographic separations of retinol have been infrequent owing to rapid cis–trans isomerizations and dehydration of retinol during analysis. Such decompositions proceed rapidly on hot catalytic surfaces such as heated GC injectors, surface-active sites within GC columns, and jet separators used in GC–MS interfaces. Although retinol can be analyzed by gas chromatography following pyrolytic conversion to anhydroretinol in heated injectors, anhydroretinol is believed to consist of several isomers that yield a broad chromatographic band[8] which results in poorer limits of detection. Derivatization of retinol to a more stable form may alleviate this problem, but many derivatizations subject retinol to acidic conditions that may induce decomposition and loss of material.

Successful capillary gas chromatographic analysis of intact underivatized retinol requires the use of cold on-column injection, which avoids exposing retinol to hot surface active sites.[9] It is essential that the capillary GC column be free from exposed silica sites which catalyze decomposition. Even the most hardy cross-linked and bonded-phase GC columns form active sites when subjected to repeated injections of polar solvents. As a result, GC analyses of retinol should be performed using

[6] H. C. Furr, O. Amedee-Manesme, and J. A. Olson, *J. Chromatogr.* **309**, 299 (1984).

[7] H. C. Furr, this volume [8].

[8] M. E. Cullum, J. A. Olson, and S. W. Veysey, *Int. J. Vitam. Nutr. Res.* **54**, 3 (1984).

[9] C. R. Smidt, A. D. Jones, and A. J. Clifford, *J. Chromatogr.* **434**, 21 (1988).

solutions of retinol in nonpolar hydrocarbon solvents such as cyclohexane.

The best results obtained for gas chromatographic separations of
retinol have used normal-bore (0.25 mm i.d.) capillary columns 15 m in
length, coated with a thin film (0.10 μm) of a bonded and cross-linked
nonpolar methyl silicone stationary phase (e.g., DB-1, J&W Scientific,
Folsom, CA). Helium is the preferred carrier gas, with an optimum linear
velocity of about 35 cm/sec at 150°. Retinol is introduced into the column
via cold (room temperature) on-column injections of 1.0 μl of cyclohexane solutions of extracted retinol, using commercially available on-
column injectors (either J&W Scientific or Scientific Glass Engineering,
Inc., Austin, TX) and syringes equipped with fused silica needles. The
use of a short length of deactivated fused silica capillary as a retention gap
is *not* recommended as it exposes the analyte to a reactive surface that
can catalyze decomposition. After the injection, the column is ballistically
heated to 150° and programmed from 150 to 250° at 10°/min. Retinol elutes
from the column at a column temperature of about 220° in a sharp peak
(Fig. 2), and the cis isomers are clearly resolved from all-*trans*-retinol.

Direct interfacing of capillary GC columns into mass spectrometers is
recommended for GC–MS analysis because jet separators expose retinol
to reactive surfaces that induce decomposition. The end of the column
should extend to, but not into, the ion source. The transfer line between
the gas chromatograph and the mass spectrometer should be maintained
at a temperature of about 220°.

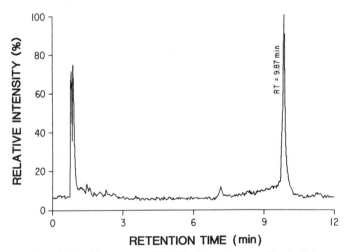

FIG. 2. Total ion current GC–MS chromatogram of retinol-d_4.

Mass Spectrometry of Retinol

Quantitative determination of ratios of labeled to unlabeled retinol are readily determined using traditional 70-eV electron ionization mass spectrometry, though several precautions should be taken to obtain optimal results. In particular, use of hydrogen as carrier gas for GC–MS analysis should be avoided as it may lead to artifactual results owing to hydrogen–deuterium exchange within the ion source of the mass spectrometer.

Electron ionization mass spectra of retinol vary a great deal among different mass spectrometers, and these variations have also been observed on the same instrument. The most striking variation is in the relative abundances of the molecular ion (M^+, m/z 286) and an ion at m/z 268 $[M - H_2O]^+$. Additional experiments have shown that the ion at m/z 268 does not arise from fragmentation of the molecular ion, but rather from dehydration of retinol that occurs within the ion source of the mass spectrometer before ionization.[10] This appears to be primarily related to the presence of adsorbed materials on the surfaces of the ion source.

Electron ionization is the ionization method of choice for analysis of retinol, in spite of the fact that the molecular ion of retinol represents less than 5% of the ions produced in electron ionization with 70-eV electrons (Fig. 3). Reducing the electron energy to 20 eV yields proportionately more molecular ion, but the sensitivity of the analysis is severely impaired. Chemical ionization of retinol is not recommended because quantitative dehydration often occurs as a result of ion–molecule collisions within the ion source. Similar results have also been obtained in attempts to perform HPLC–MS analysis of retinol using thermospray ionization. Furthermore, chemical ionization (or thermospray) conditions produce high concentrations of hydrogen-containing radicals in the mass spectrometer that may cause hydrogen–deuterium exchange and inaccurate measurements of isotope enrichment (A. J. Clifford and A. D. Jones, unpublished observations).

To circumvent analytical problems associated with variable dehydration of retinol inside the mass spectrometer ion source, isotope dilution analysis should be carried out by performing selected ion monitoring of four ions: the molecular ion of retinol-d_4 (m/z 290.2548), the molecular ion of unlabeled retinol (m/z 286.2297), $[M - H_2O]^+$ for retinol-d_4 (m/z 272.2442), and $[M - H_2O]^+$ for unlabeled retinol (m/z 268.2191). Accurate masses are provided for the benefit of those who wish to perform selected ion monitoring at high mass resolution. As a general rule, nominal mass resolution selected ion monitoring has been adequate for mea-

---

[10] A. J. Clifford, A. D. Jones, Y. Tondeur, H. C. Furr, H. R. Bergen III, and J. A. Olson, *Proc. 34th Annu. Conf. Mass Spectrom. Allied Topics*, p. 327 (1986).

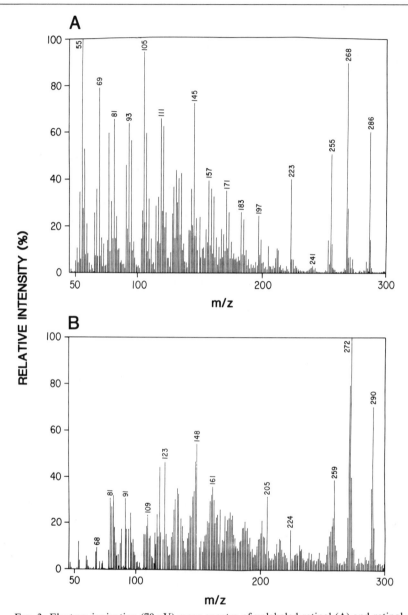

FIG. 3. Electron ionization (70 eV) mass spectra of unlabeled retinol (A) and retinol-d_4 (B).

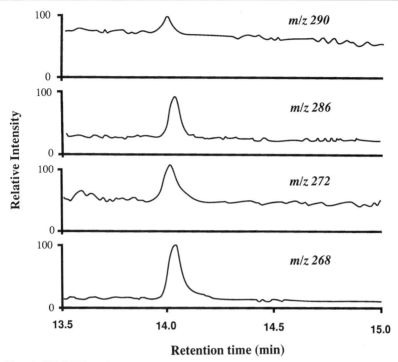

FIG. 4. GC–MS ion chromatograms of human serum extracts 14 days after administration of retinyl-d_4 acetate reconstructed for m/z 290, m/z 286, m/z 272, and m/z 268.

suring retinol-d_4 and unlabeled retinol from human serum (Fig. 4), provided that sample isolation procedures have separated retinol from other lipid contaminants.

Calibration curves should be generated using known amounts of retinol-d_4 and unlabeled retinol in various ratios. This exercise serves to verify the purity of the labeled and unlabeled standards as well as to provide an experimental determination of the limits for the ratios of labeled to unlabeled retinol that can be reliably quantified.

Calculation of Total Body Vitamin A

Calculation of total body vitamin A from the isotope dilution data has already been described in detail.[1,2] Briefly, for adult humans with daily intakes of 350 to 1500 μg of vitamin A the following formula is used to calculate total body vitamin A:

Total body pool of vitamin A (mmol) =
$$0.5 \times \text{retinyl-}d_4 \text{ acetate dose (mmol)} \times 0.65a[(H:D) - 1]$$

where 0.5 is the fraction of administered retinyl-d_4 acetate taken up by the liver,[11] 0.65 is the ratio of specific activities of retinol in plasma versus liver,[12] a adjusts the ratio of serum unlabeled to retinol-d_4 (H : D) for the half-life (140 days[13]) of liver vitamin A turnover.[2] The stable isotope dilution technique for measuring total body stores of vitamin A correlates well with direct measurement of vitamin A in liver biopsies from replete humans when the factor 0.5 (fraction of administered vitamin A-d_4 taken up by liver) is used.[2]

[11] P. Rietz, O. Wiss, and F. Weber, *Vitam. Horm. (N.Y.)* **32**, 237 (1974).
[12] V. A. Hicks, D. B. Gunning, and J. A. Olson, *J. Nutr.* **114**, 1327 (1984).
[13] H. E. Sauberlich, R. E. Hodges, D. L. Wallace, H. Kolder, J. E. Canham, J. Hood, N. Raica, and L. K. Lowry, *Vitam. Horm. (N.Y.)* **32**, 251 (1974).

[10] Mass Spectrometry of Methyl Ester of Retinoic Acid

By Andre P. De Leenheer and Willy E. Lambert

Introduction

all-*trans*-Retinoic acid (RA) is a physiological metabolite of retinol generated in the tissues where it carries out its function.[1] It retains the growth-promoting and epithelial-differentiating activity of retinol,[2,3] but it cannot maintain reproduction or vision.[4,5] Up to the late 1970s, thorough studies of the biochemical role and the physiological mechanisms of action of RA suffered from the lack of suitable analytical techniques. Later, improved analytical methodologies, such as high-performance liquid chromatography (HPLC) and gas chromatography–mass spectrometry (GC–MS),[6] were applied in this area and stimulated further research on retinoic acid metabolism and activity. However, the low physiological level together with the potential isomerization and degradation still make research on RA a real analytical challenge. In this chapter, we describe

[1] D. W. Goodman, *Fed. Proc., Fed. Am. Soc. Exp. Biol.* **39**, 2716 (1980).
[2] S. Krishnamurthy, J. G. Bieri, and F. L. Andrews, *J. Nutr.* **79**, 503 (1963).
[3] J. B. Williams and J. L. Napoli, *Proc. Natl. Acad. Sci. U.S.A.* **82**, 4658 (1985).
[4] J. N. Thompson, J. M. Howell, and G. A. Pitt, *Proc. R. Soc. London B* **159**, 510 (1964).
[5] J. E. Dowling and G. Wald, *Proc. Natl. Acad. Sci. U.S.A.* **46**, 587 (1960).
[6] W. E. Lambert, H. J. Nelis, M. G. De Ruyter, and A. P. De Leenheer, *in* "Modern Chromatographic Analysis of the Vitamins" (A. P. De Leenheer, W. E. Lambert, and M. G. De Ruyter, eds.), p. 1. Dekker, New York and Basel, 1985.

the application of GC–MS for the measurement of retinoic acid in biological matrices.

Internal Standardization

Whenever extractions and chromatographic separations are considered, the advantages of internal standardization cannot be ignored. The internal standard (IS) compensates for losses during sample pretreatment (extraction, evaporation) as well as for variations in derivatization and injection. To ensure maximal compensation, the use of a structural analog is desirable. For mass spectrometric analyses, the molecular weight of the IS has to be increased by at least 3 mass units to avoid interference of natural isotopes of the analyte on the m/z value of the labeled compound. However, a difference of more than 5 units will cause a net chromatographic separation between the IS and the analyte, thus negatively influencing the precision and lowering carrier effects.[7] Furthermore, the label must be present in the molecule at a stable position to avoid isotopic exchange. Two possibilities exist: one can use a stable isotope-labeled molecule (the most common stable isotope being ^2H), or one can use a ^{14}C- or ^3H-radiolabeled IS. The latter are less expensive and more readily available, but their use could lead to contamination of the instrument; furthermore, precautions should be taken to protect the analyst from radiation.

For the identification of endogenous retinoic acid in human serum, De Ruyter *et al.* used methyl-d_3 retinoate as an IS.[8] It was prepared using deuterated diazomethane with a deuteration yield of 91% (trideuterated), 8% (dideuterated), and 1% (monodeuterated form), respectively. Methyl-d_3 retinoate, however, does not compensate for changes in the derivatization; moreover, its behavior during extraction may differ drastically from that of retinoic acid itself. From this point of view, all-*trans*-retinoic acid-4,4,18,18,18-d_5 (Fig. 1) clearly is a better choice as an IS.[9] T.-C. Chiang used ^{14}C-labeled retinoic acid; for quantitations, however, he had to bring into account the contribution from the 62% nonlabeled retinoic acid present in the prepared IS.[10]

[7] D. Picart, F. Jacolot, F. Berthou, and H. H. Floch, *in* "Quantitative Mass Spectrometry in Life Sciences" (A. P. De Leenheer, R. R. Roncucci, and C. Van Peteghem, eds.), Vol. 2, p. 105. Elsevier, Amsterdam, 1978.

[8] M. G. De Ruyter, W. E. Lambert, and A. P. De Leenheer, *Anal. Biochem.* **98**, 402 (1979).

[9] J. L. Napoli, B. C. Pramanik, J. B. Williams, M. I. Dawson, and P. D. Hobbs, *J. Lipid Res.* **26**, 387 (1985).

[10] T.-C. Chiang, *J. Chromatogr.* **182**, 335 (1980).

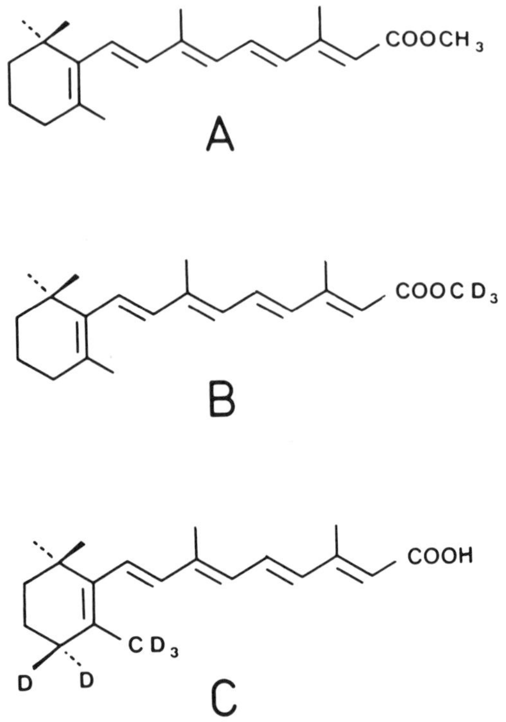

FIG. 1. Structures of all-*trans*-methyl retinoate (**A**), all-*trans*-methyl-d_3 retinoate (**B**), and all-*trans*-retinoic acid-4,4,18,18,18-d_5 (**C**).

Sample Preparation

Prior to GC–MS measurement extensive cleanup of the extracts is highly recommended. De Ruyter *et al.* denatured 2 ml of EDTA-treated plasma with 1 ml of ethanol and 100 μl of 2 N HCl. The deproteinized sample is then extracted twice with 2.5 ml of *n*-hexane.[8] The organic layers are combined, and the solvent is evaporated under reduced pressure. This residue is dissolved in 700 μl of a mixture of chloroform–*n*-hexane–methanol [65 : 35 : 1 (v/v/v)], and a 500-μl aliquot is applied to a Sephadex LH-20 column eluted with the same solvent mixture.[11] The fraction of the eluent corresponding to the elution of retinoic acid is collected, 20 ng of methyl-d_3 retinoate is added as an IS, and the mixture is evaporated under reduced pressure. The same ethanolic HCl–*n*-hexane extraction was applied by Chiang, although larger volumes were used (10

[11] Y. L. Ito, M. Zile, H. Ahrens, and H. F. De Luca, *J. Lipid Res.* **15,** 517 (1974).

ml EDTA-containing plasma, 1 ml of 2 N HCl, 10 ml ethanol, and 25 ml n-hexane). The percent recovery of RA from the spiked plasma samples was found to be between 40 and 60% (n = 7). No additional chromatographic purification is described before the GC–MS analysis.[10]

Napoli et al. developed the most sensitive assay for retinoic acid up to now. They denature 100 μl EDTA-containing plasma with an equal volume of 0.2 N HCl in methanol. After thorough mixing, the sample is extracted with three 0.5-ml portions of n-hexane. The recovery was 83 ± 2% (mean ± SD, n = 10).[9] Another isolation procedure for RA from plasma includes a preextraction in alkaline medium: 3.5 ml of plasma is denatured with an equal volume of ethanol and 1.5 ml of 2 N NaOH. The neutral and basic lipophilic constituents are extracted with 7.0 ml of n-hexane, which is discarded. After addition of 3.0 ml of 2 N HCl, the aqueous layer is extracted with 7.0 ml of n-hexane. This organic fraction is then evaporated to dryness. To prevent partial breakdown of the RA, as mentioned by Napoli et al.,[9] the extraction has to be performed at 4°, whereas the n-hexane used for the extraction must contain butylated hydroxytoluene (0.025%, w/v). The recovery as determined with radiolabeled [15-^{14}C]retinoic acid was 69 ± 5% (mean ± SD, n = 8).[12] Furthermore, it must be recognized that under the conditions described above (extreme pH values), the β-retinoylglucuronide eventually present in the sample is hydrolyzed to RA, which is then determined together with the endogenous RA.[13]

Derivatization

The most popular derivative of RA is obviously the methyl ester form. Different reagents can be proposed for methylation, such as diazomethane, anhydrous HCl–methanol, anhydrous methanol–BF_3, and dimethylformamide dimethylacetal. In view of the lability of the RA, the mildest procedure should be used. Treatment of the extract, placed in an ice bath, with a small volume (500 μl) of a freshly prepared[14] ethereal solution (0.3 M) of diazomethane results in the formation of the methyl ester in a very short time (<5 min).[9,12] However, care must be taken in the generation and handling of diazomethane as it is a highly toxic and potentially explosive gas. A less hazardous methylation reaction is the treatment of the extract with 50 μl of dimethylformamide dimethylacetal for 1 hr at 65°.[10] Using similar reagents a variety of esters can be prepared. In our

[12] A. P. De Leenheer, W. E. Lambert, and I. Claeys, J. Lipid Res. 23, 1362 (1982).
[13] A. B. Barua and J. A. Olson, Am. J. Clin. Nutr. 43, 481 (1986).
[14] Th. J. De Boer and H. J. Backer, in "Organic Synthesis" (N. Rabjohn, ed.), Vol. 4, p. 250. Wiley, New York, 1963.

laboratory, we also tried to perform trimethylsilyl (TMS) derivatization using N,O-bis(trimethylsilyl)acetamide. However, no acceptable reproducibility could be obtained.

Gas Chromatography

Among the retinoids, methyl retinoate has excellent chromatographic properties, and it is not surprising that most of the research work with GC has been done on this compound. In a study on the metabolism of retinoic acid, Dunagin et al. chromatographed cis–trans isomers of RA on Gas Chrom P coated with 3% SE-30 at 180° using an argon flow rate of 50 ml/min.[15] We carried out the GC analyses on a glass silanized column (1.80 m × 2 mm i.d.) filled with 1% FFAP (free fatty acid phase) coated on Gas Chrom Q (100–200 mesh); the carrier gas is He at a flow rate of 30 ml/min. Temperatures are 210° for the injector, 190° for the column, and 265° for the separator.[8] On a column with the same dimensions, filled with Gas Chrom Q (80–100 mesh), but coated either with 2% FFAP or 3% QF-1, we were also able to chromatograph retinoic acid as the methyl ester or TMS derivative. Separations can also be performed on a 6 ft × 2 mm i.d. column packed with 3% SE-30 on 80–100 mesh Chromosorb W HP at a column temperature of 230° and an injector temperature of 250°,[10] or on a 3 ft × 2 mm i.d. column packed with 3% SP2100-DOH on Supelcoport (100–120 mesh) at a column temperature of 230°. In the latter case, methane is used as carrier gas at a flow rate of 20 ml/min. It is noteworthy that under the latter GC conditions no separation is obtained between the cis and trans isomers of methyl retinoate,[9] whereas a marginal separation (Fig. 2) is obtained on the 3% SE-30 phase.[10]

Mass Spectrometry

The extensive possibilities of selected ion monitoring (SIM) rely on the selective monitoring of only one or a restricted number of relevant m/z values. Molecules that do not generate ions at these specific values are not detected, resulting in a very high degree of selectivity and an increased sensitivity. The electron-impact mass spectra of methyl retinoate at 70 eV show very extensive fragmentation with a large number of noncharacteristic peaks.[16] Therefore, we lowered the electron energy to 20 eV without loss of sensitivity (a diminished proportion of the sample being ionized but the molecular ion becoming more prominent). The tem-

[15] P. E. Dunagin, R. D. Zachman, and J. A. Olson, Biochim. Biophys. Acta 124, 71 (1966).
[16] H. W. Elliot and G. R. Waller, in "Biochemical Applications of Mass Spectrometry" (G. R. Waller, ed.), p. 499. Wiley, New York, 1972.

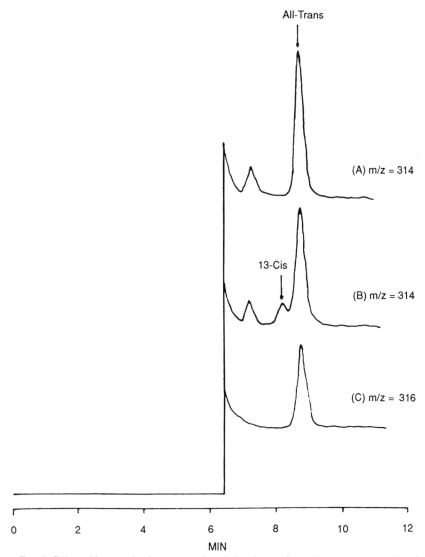

FIG. 2. Selected ion monitoring traces of methyl retinoate from plasma samples: (A) m/z 314 trace of sample with all-*trans*-retinoic acid; (B) m/z 314 trace of sample with mixture of all-*trans*- and 13-*cis*-retinoic acids; (C) m/z 316 trace of sample with ^{14}C-labeled (at the carboxylic group) all-*trans*-retinoic acid as an internal standard. Column: glass column (6 ft × 2 mm i.d.) packed with 3% SE-30 on 80–100 mesh Chromosorb W HP at 230°. (From Ref. 10, with permission.)

perature for the all-glass jet separator was 265° and 270° for the ion source. The first channel of the multiple ion detection device (MID) was focused on the ion at m/z 314 using an acceleration voltage of 3500 V (M‡, methyl retinoate), whereas for the second channel (m/z 317, M‡, methyl-d_3 retinoate) an acceleration voltage of 3466.75 V was applied. Using this technique we were the first to demonstrate the presence of retinoic acid in human plasma under normal physiological conditions.[8]

Chiang also monitored the molecular ion of methyl retinoate (m/z 314) and [^{14}C]methyl retinoate (m/z 316) in the electron-impact mode but operated at 70 eV. The separator temperature was 250°. In analyzing more than 5 ml of nonsupplemented plasma no retinoic acid could be detected. The very high injector and column temperature (250 and 230°, respectively) probably destroyed a considerable amount of the product.[10] In contrast to electron-impact mass spectrometry, chemical-ionization MS ionizes a higher portion of the sample and provides a much simpler spectrum with the quasimolecular ion as the base peak. Both factors result in an increased sensitivity. Furthermore, the sensitivity can be increased 10-fold by changing from positive chemical ionization (PCI) to negative chemical ionization (NCI). The PCI mass spectrum of methyl retinoate with methane as reagent gas shows characteristic peaks at m/z 343 [M$^+$ + C$_2$H$_5$], 315 [M$^+$ + H], 283 [M$^+$ − OCH$_3$], and 255 [M$^+$ − CO$_2$CH$_3$]. In contrast, the NCI mode is even more advantageous, as the response is concentrated in one single intense peak at m/z 314 (Fig. 3), representing the molecular ion of methyl retinoate, and at m/z 319, being the molecular ion of the pentadeuterated methyl retinoate. By applying

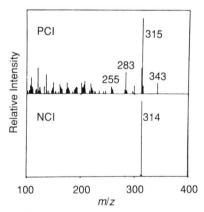

FIG. 3. Comparison of PCI and NCI mass spectra of all-*trans*-methyl retinoate. The samples (100 ng of the pure standard) were introduced into the mass spectrometer through the GLC column with methane as both GLC carrier and chemical ionization reagent gas. (From Ref. 9, with permission.)

SIM at these two m/z values (314 and 319) after NCI, retinoic acid was demonstrated and quantitated in 100-μl aliquots of plasma down to 75 pg/ml. Within-day and day-to-day coefficients of variation were 5.3 ± 0.19% (mean ± SD, $\bar{x}$ = 3.6 ng/ml, n = 10) and 6.4 ± 0.30% (mean ± SD, $\bar{x}$ = 4.7 ng/ml, n = 7).[9] The endogenous level of RA in human plasma as determined in our laboratory by HPLC (range 2.7–4.2 ng/ml, $\bar{x}$ = 3.5 ng/ml, n = 37)[12] agreed quite well with the results obtained by this GC–MS procedure[9] (range 2.8–6.6 ng/ml, $\bar{x}$ = 4.9 ng/ml, n = 12).

Two other applications of the mass spectrometer were described for the analysis of retinoic acid in biological matrices. None of the procedures uses gas chromatography, and, consequently, there is no need for derivatization of the retinoic acid molecule. The first application involves laser desorption mass spectroscopy (LDMS); this technique has the ability to provide molecular fragmentation patterns at the microprobe level. A pulsed laser is focused on an aluminum sample-mounting plate, and the desorbed species are extracted into a time-of-flight mass spectrometer, sorting the ions into the fragmentation pattern by a mass-to-charge ratio.[17] In the positive-ion spectrum, the molecular ion peak at m/z 300 is easily seen, as well as the fragments at m/z 256 (loss of CO_2) and 242 (loss of CO_2 and CH_3, followed by addition of a positive hydrogen ion). In the negative-ion spectrum, the quasimolecular ion (M + H) peak is observed at m/z 301, whereas the m/z 256 peak is comparable to the one also found in the positive-ion spectra. The most recent application is the use of direct exposure probe (DEP) mass spectrometry to confirm the identity of retinoic acid in fractions collected from an HPLC column. One microliter of the sample is placed onto a rhenium filament with the current programmed from 0 to 1.3 A at 50 mA/sec. Electron-impact analysis is then performed first to check for purity, under the following conditions: filament current, 0.25 mA; electron energy, 70 eV; multiplier voltage, 1200 V; ion source temperature, 120°. For the CI procedure ammonia at 0.40 torr is used together with a filament current of 0.15 mA, an electron energy of 140 eV, a multiplier voltage of 1600 V, and an ion source temperature of 80°. The prominent ion at m/z 300 for retinoic acid is then monitored in the selected ion monitoring mode.[18] The authors indicate a sensitivity down to 20 pg, although they do not show these data. These latest applications seem very interesting for future research on retinoic acid in biological samples.

[17] J. M. McMahon, Anal. Biochem. **147**, 535 (1985).
[18] V. M. Papa, J. Hupert, H. Friedman, P. S. Ng, E. F. Robbins, and S. Mobarhan, Biomed. Environ. Mass Spectrom. **16**, 323 (1988).

[11] Characterization of Retinylidene Iminium Salts by High-Field ^{1}H and ^{13}C Nuclear Magnetic Resonance Spectroscopy

By GARY S. SHAW and RONALD F. CHILDS

Introduction

The visual pigment rhodopsin contains an 11-*cis*-retinal chromophore linked via a protonated Schiff base to lysine-296. A similar pigment, bacteriorhodopsin, contains the all-trans isomer and is also connected to a lysine residue by a protonated Schiff base. These combinations of chromophore and protein in rhodopsin and bacteriorhodopsin give rise to absorption maxima at 506 and 568 nm, respectively. In contrast, simple retinal iminium salts, which lack the protein, have absorption maxima between 440 and 480 nm. It is now generally agreed that the large differences in absorption maxima between the natural pigments and simple retinal iminium salts are a result of interactions between charged amino acid residues of the protein and its chromophore. More recently, it has been shown that the chromophore in bacteriorhodopsin exists in a perturbed 6-s-trans conformation.[1] This conformation also gives rise to a further "red shifting" of the absorption spectrum.

One approach used in studying the conformation of the retinal chromophore and its interactions with charged groups has been to study analogs of rhodopsin and bacteriorhodopsin in solution. Typically, these analogs possess a retinal backbone and a protonated Schiff base similar to those found in rhodopsin or bacteriorhodopsin. However, in place of the protein in the natural pigments, a simple alkyl amine is usually employed in the model systems. The resulting compounds are referred to as retinylidene iminium salts (1).

Nuclear magnetic resonance (NMR) spectroscopy is a technique that is available for studying these natural pigment analogs in solution and the solid state. Pioneering studies on the retinal isomers and their conformational mobility by ^{1}H NMR spectroscopy were accomplished by Patel[2] and Rowan *et al.*[3] many years ago. Later, Shriver and co-workers were

---

[1] G. S. Harbison, S. O. Smith, J. A. Pardoen, J. M. L. Courtin, J. Lugtenburg, J. Herzfeld, R. A. Mathies, and R. G. Griffin, *Biochemistry* **24,** 6955 (1985).
[2] D. J. Patel, *Nature* (*London*) **221,** 825 (1974).
[3] R. Rowan , A. Warshel, B. D. Sykes, and M. Karplus, *Biochemistry* **13,** 970 (1974).

able to completely characterize an 11-*cis*-retinylidene iminium salt by ^{1}H and ^{13}C NMR spectroscopy at low temperatures.[4] In this chapter we describe the preparation and characterization of some retinylidene iminium salts using high-field ^{1}H and ^{13}C NMR spectroscopy.

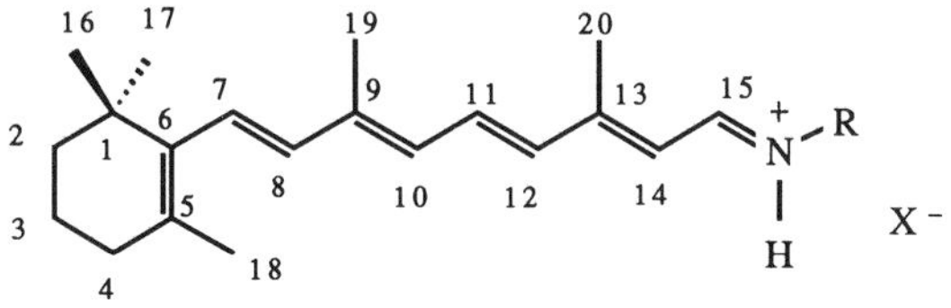

1

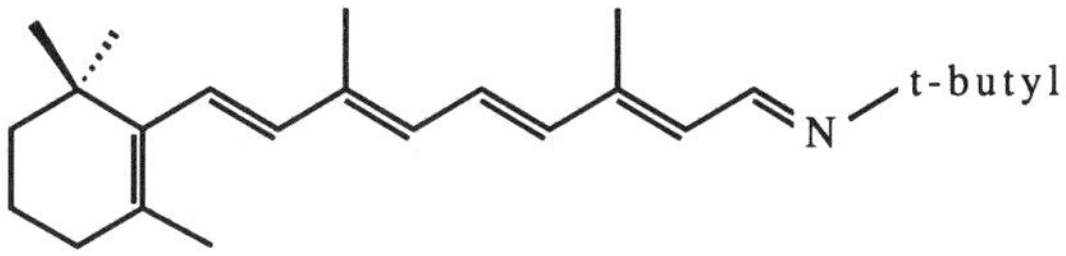

2

Preparation of Retinylidene Iminium Salts

Retinylidene iminium salts are prepared through the condensation of all-*trans*-retinal with either *n*-butyl- or *tert*-butylamine and subsequent protonation. The reactions should be carried out in the dark and in the absence of moisture. For the synthesis of the retinylidene imine **2**, *tert*-butylamine (28.2 mmol) is added to a solution of all-*trans*-retinal (2.8 mmol) in dry ether (50 ml). It is essential to use a large excess of amine to drive the reaction to completion. This solution is stirred at room temperature over 3-Å molecular sieves for 24 hr. The mixture is filtered and the ether and excess amine removed under reduced pressure. The residue is redissolved in ether and placed under reduced pressure repeatedly to ensure complete removal of unreacted amine. This yields a pale yellow residue (910 mg, 95%). The retinylidene iminium salt **3** is formed by dissolving imine **2** in ether (30 ml) and adding an ethereal solution of CF_3SO_3H dropwise until precipitation is complete. The resulting dark red precipitate is recrystallized twice from dichloromethane/ether at $-20°$ (yield 440 mg, 37%). A similar procedure is used to synthesize the retinylidene iminium salt **4**.

[4] J. W. Shriver, G. D. Mateescu, and E. W. Abrahamson, *Biochemistry* **18**, 4785 (1979).

3   R = *tert*-butyl, X⁻ = CF₃SO₃⁻

4   R = *n*-butyl, X⁻ = Cl₃CCOO⁻

¹H NMR Spectroscopy

In obtaining an ¹H NMR spectrum of a retinylidene iminium salt such as **3**, a small sample of the salt is dissolved in an inert deuterated solvent, in the absence of any water. Typically, we use a sample concentration of 4 mM (~1.0 mg in 800 μl solvent) for characterization purposes. However, with the sensitivities provided by modern high-field NMR instruments operating at over 250 MHz (ideally 500 MHz), this concentration may be decreased to approximately 40 μM if one is willing to increase the amount of time to acquire a suitable signal-to-noise ratio. A good solvent choice for these studies is deuterated dichloromethane (CD_2Cl_2) as its residual proton resonance ($CHDCl_2$) is found at 5.32 ppm and does not interfere with any of the proton resonances of the retinylidene iminium salts. At 500 MHz, a pulse width of 5 μsec (about a 30° pulse) is used with an acquisition time of about 3.3 sec (including a relaxation delay). Generally, a sweep width of 5000 Hz is employed. Data are collected at 32 K and zero-filled to 64 K prior to Fourier transformation.

The primary objectives in characterizing an iminium salt such as **3** are to ensure that protonation of the imine has occurred and to establish isomeric purity. The resonance at lowest field in **3** arises from C-15–H, the proton which is closest to the protonated Schiff base linkage (Fig. 1). In all-trans isomers such as **3**, this proton is usually found between 8.1 and 8.6 ppm. The next group of signals are clustered between 6.1 and 7.6 ppm and belong to the protons in the polyenylic portion of the molecule: C-7–H, C-8–H, C-10–H, C-11–H, C-12–H, and C-14–H. The third portion of the spectrum results from the aliphatic protons on the cyclohexenyl ring, the nitrogen alkyl group, and the methyl groups of the chromophore. The specific pattern which arises for each one of these signals has been described in an earlier volume of this series.[5] In the portion of the 500 MHz spectrum shown (Fig. 1), the characterization of the all-*trans*-iminium salt **3** is straightforward. The resonance at 8.20 ppm arises from

[5] P. Towner and W. Gärtner, this series, Vol. 88, p. 546.

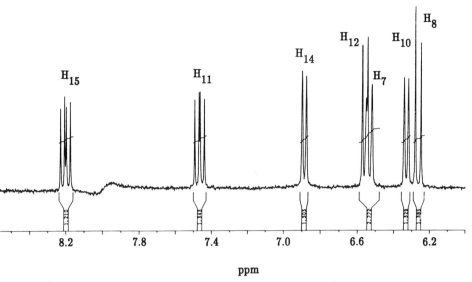

FIG. 1. Vinyl region of the 500-MHz ^{1}H NMR spectrum of **3**. ($J_{15,N}$ = 15.81, $J_{14,15}$ = 11.11, $J_{11,12}$ = 14.85, $J_{10,11}$ = 11.72, $J_{7,8}$ = 16.11 Hz.)

C-15–H, the resonance at 7.45 ppm results from C-11–H, and the other polyenylic proton resonances are located between 6.1 and 7.0 ppm as indicated in Fig. 1.

An examination of the ^{1}H NMR data in Table I shows that the odd-numbered proton resonances along the polyene skeleton of **3** experience a shift to lower field compared to those in the neutral imine **2**. Furthermore, an analysis of the coupling patterns observed for this region of the spectrum (Fig. 1), shows that C-15–H is present as a doublet of doublets in **3**. This arises from a large coupling constant ($J_{15,N}$ 15.8 Hz) across the C=N bond and a smaller scalar coupling ($J_{14,15}$ 11.1 Hz) across the C-14–C-15 single bond. This coupling pattern and coupling constants are also indicative of an *anti*-C=N arrangement as shown in **3**.

It should also be noted that since scalar coupling is observed between C-15–H and N–H, exchange of the nitrogen proton with the counteranion must be slow relative to the NMR time scale. Thus, the equilibrium for proton exchange [Eq. (1)] lies far to the right at room temperature for **3** which has a triflate counterion. In other retinylidene iminium salts, such as **4**, which use trichloroacetic acid as the protonating agent,[6] this coupling between C-15–H and N–H is generally only observed at low temper-

[6] C. Pattaroni and J. Lauterwein, *Helv. Chim. Acta* **64**, 1969 (1981).

TABLE I

^{1}H NMR Chemical Shift^{a-c} and Coupling Constant^d Data for Retinylidene Imines and Iminium Salts

Position	Compound				
	2	3	4	5	6^e
C-2–H	1.50t	1.61t	1.47t	1.61t	
C-3–H	1.62m	1.61m	1.62m	1.61m	
C-4–H	2.05t	2.06t	2.05t	2.06t	
C-7–H	6.28d	6.53d	6.51d	6.53d	
C-8–H	6.16d	6.25d	6.24d	6.29d	
C-10–H	6.19d	6.32d	6.29d	6.73d	
C-11–H	6.91dd	7.45dd	7.40dd	7.00t	7.51dd
C-12–H	6.41d	6.55d	6.56d	6.19d	
C-14–H	6.33d	6.86d	6.74d	6.86d	
C-15–H	8.32d	8.20dd	8.20d	8.23d	
C-16–H / C-17–H	1.05s	1.05s	1.04s	1.05s	
C-18–H	1.73s	1.74s	1.73s	1.74s	2.76s
C-19–H	2.01s	2.10s	2.09s	2.10s	2.09s
C-20–H	2.14s	2.31s	2.29s	2.45s	2.30s
C-1′–H	—	—	3.70t	1.50s	
C-2′–H	1.28s	1.50m	1.78m		
C-3′–H	—	—	1.41m		
C-4′–H	—	—	0.96t		
N–H	11.60bs		11.60bs		
$J_{7,8}$	16.28	16.11	16.18		
$J_{10,11}$	11.21	11.72	11.74		
$J_{11,12}$	15.13	14.85	14.84		
$J_{14,15}$	9.90	11.11	11.11		
$J_{15,N}$	—	15.81	—		

^{a} s, Singlet; d, doublet; t, triplet; dd, doublet of doublets; bs, broad singlet.
^{b} In ppm. Numbering of carbons as in text.
^{c} Referenced to CD_2Cl_2, 5.32 ppm. Measured at 21°.
^{d} In Hz.
^{e} Low concentration of this isomer makes positive identification of all the resonances difficult.

atures, suggesting that rapid exchange of the N–H proton is occurring at room temperature.

$$\tag{1}$$

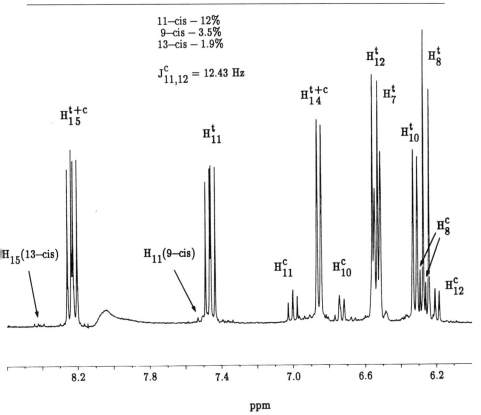

FIG. 2. Vinyl region of the 500-MHz ^{1}H NMR spectrum of **3** following photoisomerization. The resonances of the all-trans (**3**) and 11-cis (**5**) isomers are identified with the superscripts t and c, respectively.

Isomer Characterization by ^{1}H NMR

A tremendous advantage of ^{1}H NMR spectroscopy is that it allows for the identification of isomeric forms of retinylidene iminium salts in a single sample.[7] This is shown in Fig. 2, where the retinylidene iminium salt **3** has been photoisomerized and subsequently analyzed by ^{1}H NMR. Examination of this spectrum shows that it is much more complex than the spectrum of the all-trans isomer (Fig. 1). However, at 500 MHz, several well-resolved resonances arising from the 11-cis (**5**, major) and 9-cis (**6**, minor) isomers are present, as well as those of the all-trans species. Characterization of the 11-cis isomer is accomplished primarily through the location and multiplicity of the C-15–H and C-11–H reso-

[7] R. F. Childs and G. S. Shaw, *J. Am. Chem. Soc.* **110**, 3013 (1988).

nances and is described below. Identification of some of the resonances of the 9-cis isomer is accomplished after further irradiation of the same sample.

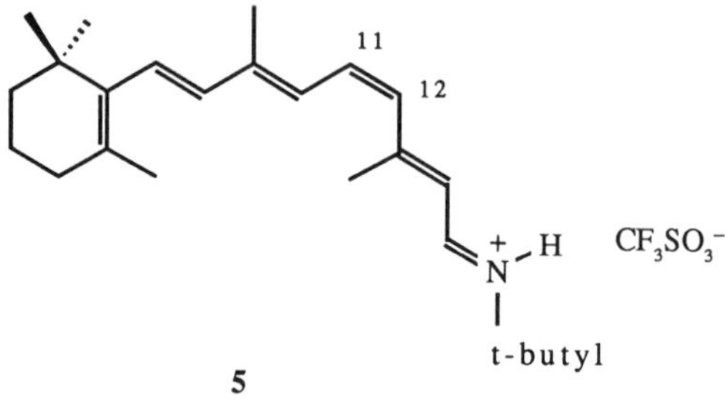

5

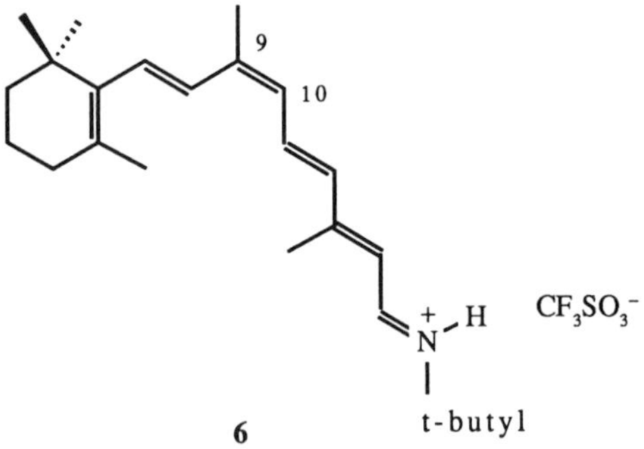

6

In Fig. 2, the doublet of doublets at 8.20 ppm arising from C-15–H of the all-trans isomer is now complicated by a second doublet of doublets at approximately the same chemical shift. This second resonance originates from C-15–H of the 11-cis isomer. As expected, isomerization of the remote C-11–C-12 bond has little effect on the C-15–H resonance or the coupling to N–H or C-14–H. In the C-11–H region, a new resonance belonging to the 11-cis isomer appears at 7.00 ppm. This is significantly upfield from its position in the all-trans isomer and is characteristic of a cis C-11–C-12 bond. Furthermore, the coupling constant across the C-11–C-12 bond ($J_{11,12}$ 12.4 Hz) is characteristic of a cis double bond. Similarly, the assignment of the remainder of the polyenylic portion of the 11-cis isomer can be carried out. Most noticeable are a large upfield shift of C-12–H and a large downfield shift of C-10–H relative to the all-trans isomer. Homonuclear decoupling experiments and measurement of the nuclear Overhauser effect have been used to confirm the assignments. The chemical shifts of **5** and **6** are given in Table I.

In the spectrum (Fig. 2), the *syn*-C=N and 13-cis isomers were not detected. However, they are easily identified by ^{1}H NMR using the C-15–H resonance.[7] In *n*-butyl retinylidene iminium salt derivatives, C-15–H of the all-trans isomer resonates at about 8.34 ppm ($J_{15,N}$ 15.7 Hz). In the 13-cis isomer this resonance appears about 0.1–0.2 ppm downfield and retains its characteristic doublet of doublets coupling pattern. In the all-*trans*-*syn*-C=N isomer, this resonance is located about 0.4 ppm downfield of the *anti*-C=N resonance. The *syn*-C=N resonance is also obvious from a collapse of the doublet of doublets coupling pattern to an apparent triplet ($J_{15,N} = J_{14,15} = 12$ Hz).

Solution ^{13}C NMR Spectroscopy

Solution ^{13}C NMR spectra are typically recorded at 62.9 MHz. Samples are prepared in deuterated dichloromethane to a concentration of about 30 mM and referenced to the central peak of the dichloromethane quintet at 53.8 ppm. The various resonances of the imines and iminium salts studied are assigned using a "J-modulated spin-echo" experiment with broadband ^{1}H decoupling.[8] This pulse sequence requires about the same amount of time as "normal" ^{13}C spectral acquisition and has the advantage of distinguishing between the various types of carbons based on the number of protons attached. Our experiments utilize a 90° pulse width of 30.5 μsec ($t_D = 7.5$ msec) and a relaxation delay (D1) of 2.0 sec. The sweep width is usually about 18,000 Hz with an acquisition time of 0.4 sec.

As with the ^{1}H NMR spectra, protonation of **2** to form the iminium salt **3** produces a general shift to lower field of the odd-numbered carbon resonances (Table II). However, because the frequency range for the ^{13}C nucleus is much greater than that for protons, the downfield shifts are much more dramatic in ^{13}C NMR spectroscopy. This is shown in Fig. 3, where a resonance such as C-13 in **3** is shifted about 20 ppm downfield of its position in **2**. A second noteworthy observation is that the even-numbered carbon resonances are shifted to higher field in the retinylidene iminium salt **3** compared to its neutral imine **2**.

The choice of the J-modulated spin-echo experiment for the assignment of the carbon resonances in these compounds proved to be critical in distinguishing between the C-13 and C-15 resonances. In retinylidene iminium salts, such as **4,** with an *N*-*n*-butyl group and a trichloroacetate counteranion, the C-15 resonance is at lower field than that of C-13 (Table

---

[8] S. L. Patt and J. N. Shoolery, *J. Magn. Reson.* **46**, 535 (1982).

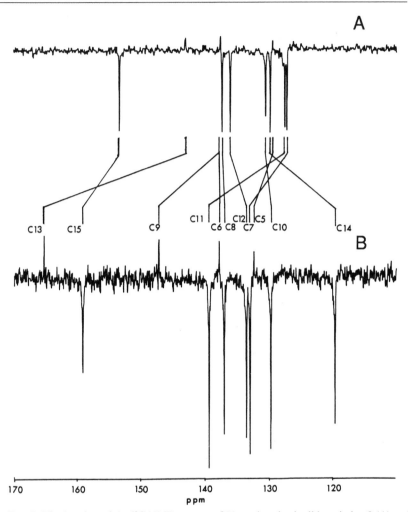

FIG. 3. Vinyl region of the ^{13}C NMR spectra of N-tert-butylretinylidene imine **2** (A) and its perchlorate salt **3** (B).

II). These resonances are reversed in **3** and are most likely a result of stronger hydrogen bonding in **4** between the N–H proton and its counter-anion. This is consistent with the ^{1}H NMR spectra, which suggest that a dynamic equilibrium, such as that in Eq. (1), is operating for a solution of **4** in dichloromethane. It has been shown that the addition of further trichloroacetic acid to this sample shifts the C-13 resonance downfield of C-15 presumably because it forces the equilibrium to the right.

TABLE II

^{13}C NMR CHEMICAL SHIFT^a DATA FOR RETINYLIDENE IMINES AND IMINIUM SALTS

	Compound				
Position	$\mathbf{2}^b$	$\mathbf{3}^b$ (solution)	$\mathbf{3}^d$ (solid)^e	$\mathbf{4}^b$ (solution)	$\mathbf{4}^d$ (solid)
C-1	34.22	34.66	36.69	34.21	35.47
C-2	37.74	40.15	45.15	39.46	40.43
C-3	19.29	19.55	19.92	19.13	20.84
C-4	33.10	33.74	34.78	33.27	35.47
C-5	129.64	132.44	136.76	131.74	131.69
C-6	137.78	137.97	140.11	137.45	140.03
C-7	127.37	132.90	134.29	131.74	131.69
C-8	137.52	137.18	136.76	137.19	136.84
C-9	137.78	146.95	147.60	145.24	143.99
C-10	130.76	129.92	129.66	129.60	130.37
C-11	127.70	139.02	140.11	137.19	140.03
C-12	136.25	133.67	133.41	133.63	132.37
C-13	143.21	164.57	167.16	163.27	162.75
C-14	130.01	119.74	120.34	120.35	122.54
C-15	153.50	159.02	160.98	164.64	166.90
C-16 C-17	28.75	28.59	29.90	28.76	31.52
C-18	21.43	21.95	22.32	21.77	25.41
C-19	12.58	13.40	12.84	13.11	13.43
C-20	12.88	14.83	15.53	13.90	17.20
C-1'	57.32	59.58	60.29	^{c}	54.75
C-2'	29.58	29.14	32.23	31.33	32.55
C-3'	—	—	—	19.87	20.84
C-4'	—	—	—	13.59	15.70

^{a} In ppm. Numbering of carbons as in text.
^{b} Referenced to CD_2Cl_2, 53.8 ppm. Measured at 21°.
^{c} Peak under solvent.
^{d} Referenced to adamantane, 29.50 and 38.57 ppm. Measured at 21°.
^{e} As the perchlorate salt.

Solid-State ^{13}C NMR Spectroscopy

Solid-state ^{13}C NMR spectra are obtained at 50.3 MHz using cross-polarization magic angle spinning on crystalline samples. Iminium salts are packed into alumina rotors and spun at approximately 4500 Hz. Contract and acquisition times of 3.0 msec and 0.3 sec, respectively, are used. A $\pi/2$ pulse width of 4 μsec is used to establish proton–carbon cross polarization. Spectra are referenced to adamantane.[9] Quaternary and methyl carbons are assigned using a delay without decoupling pulse se-

[9] W. L. Earl and D. L. VanderHart, *J. Magn. Reson.* **48**, 35 (1982).

TABLE III
CHEMICAL SHIFTS FOR C-5 POSITION OF
RETINAL DERIVATIVES IN SOLID STATE

Compound	6-s-cis	6-s-trans
all-*trans*-Retinal[a]	128.5	—
13-*cis*-Retinal[a]	126.7	136.8
all-*trans*-Retinoic acid[a]	127.8	135.9
3 (solid) = **7**	—	136.8

[a] From Ref. 1.

quence.[10] The assignment of the remaining carbon resonances is based on a comparison with solution ^{13}C data and literature values.

A comparison of the solid-state spectra of **3** and **4** with those obtained in solution (Table II) shows that most of the carbon resonances for each compound are very similar in chemical shift in the two phases. In **3**, however, there are notable differences in the resonance frequencies of C-2, C-5, C-6, C-13, and C-15. The changes in resonance position of C-13 and C-15 can be accounted for by a different cation–anion relationship in the solution and solid phases. However, this argument is not plausible for the C-2 and C-5 resonances, which are shifted nearly 5 ppm downfield in the solid spectrum of **3** compared to its solution spectrum. These chemical shift changes are consistent with the 6-s-trans conformation **(7)** being present in the solid state, whereas the 6-s-cis conformation **3** predominates in the solution phase. This observation is based on ^{13}C NMR spectral data, obtained by Harbison *et al.*,[1] of several retinoids where the three-dimensional structures had been determined X-ray crystallographically. A comparison of the C-5 chemical shift of **7** with the ^{13}C chemical shift for these retinoids is presented in Table III.

7

Acknowledgments

This work was supported by a grant and postgraduate scholarship (G.S.S.) from the Natural Sciences and Engineering Research Council of Canada.

[10] S. J. Opella and M. H. Frey, *J. Am. Chem. Soc.* **101**, 5854 (1979).

[12] Resonance Raman and Infrared Difference Spectroscopy of Retinal Proteins

By Friedrich Siebert

Introduction

Vibrational spectroscopy such as resonance Raman (RR) and infrared difference spectroscopy (IRD) has mostly been applied to retinal proteins having the retinal covalently bound to the protein via a protonated Schiff base. In all the cases studied, the side chain of a lysine constitutes the amino group. The vertebrate and invertebrate visual pigments and four different proteins located in the plasma membrane of the bacterium *Halobacterium halobium* belong to this class of proteins. The absorption of light by the chromophore causes its isomerization which, in turn, induces the protein structural changes responsible for the various functions of these systems. The proton pump bacteriorhodopsin, the chlorine pump halorhodopsin, the sensory pigments sensory rhodopsin and *P*-480 constitute the retinal proteins of *Halobacterium halobium*. Several reviews on these systems have been published.[1-4] As far as the visual pigments are concerned, the special volume (Vol. 13) of *Photobiochemistry and Photobiophysics* (1986) and the articles by Balogh-Nair and Nakanishi,[5] Ottolenghi,[2] and Sandorfy and Vocelle[6] provide a survey.

Vibrational spectroscopy can help to determine what are the structures of the chromophores in these pigments and in the intermediates of their photoreactions, what kind of interaction exists between the chromophore and the protein, and what are the structural changes evoked in the protein by the photoreaction. These topics are dealt with in two recent reviews on the application of RR spectroscopy[7,8] and in a review on the

[1] D. Oesterhelt and J. Tittor, *Trends Biochem. Sci.* **14,** 57 (1989).
[2] M. Ottolenghi, *in* "Advances in Photochemistry" (J. N. Pitts, G. S. Hammond, K. Gollnik, and D. Grosjean, eds.), p. 97. Wiley (Interscience), New York, 1980.
[3] W. Stoeckenius and R. A. Bogomolni, *Annu. Rev. Biochem.* **52,** 587 (1982).
[4] J. K. Lanyi, *Annu. Rev. Biophys. Biophys. Chem.* **15,** 11 (1986).
[5] V. Balogh-Nair and K. Nakanishi, *in* "New Comprehensive Biochemistry, Volume 3: Stereochemistry" (C. Tamm, ed.), p. 283. Elsevier Biomedical, Amsterdam, 1982.
[6] C. Sandorfy and D. Vocelle, *Can. J. Chem.* **64,** 2251 (1989).
[7] M. Stockburger, T. Alshuth, D. Oesterhelt, and W. Gärtner, *in* "Spectroscopy of Biological Systems" (J. H. Clark and R. E. Hester, eds.), p. 483. Wiley, New York, 1986.
[8] R. A. Mathies, S. O. Smith, and I. Palings, *in* "Biological Application of Raman Spectrometry, Volume 2: Resonance Raman Spectra of Polyenes and Aromatics" (T. G. Spiro, ed.), p. 59. Wiley, New York, 1987.

METHODS IN ENZYMOLOGY, VOL. 189

application of IRD spectroscopy.[9] Such problems may also apply to other retinal-containing proteins. The basic principles of the two methods are described here and some important applications discussed.

Resonance Raman Spectroscopy

Normal Raman scattering is a weak effect, and recording a complete spectrum usually requires many hours. However, if the molecule has an absorption band near the wavelength of the probing beam (usually a continuous wave laser), the scattered photon is emitted from the electronic excited state and the scattering cross section for vibrations of this molecule is increased by several orders of magnitude. Resonance Raman spectroscopy has, therefore, the unique advantage of selectivity in that, in the spectrum of such complex systems as retinal proteins, only vibrations of the chromophore will be reflected. In the retinal proteins, however, a photoreaction is evoked by the absorption of light. Resonance Raman scattering is, therefore, always connected with the generation of intermediates of the photoreaction, and special techniques have been developed to obtain spectra of these intermediates as well as of the initial state. The basic methods are described briefly.

Depending on the system being investigated, there are two basic techniques for overcoming these difficulties: namely, time-resolved technique and pump–probe technique. The first technique uses either a capillary flow system[10–12] or a rotating cell.[13,14] The main purpose of both methods is to continuously bring fresh sample into the laser beam. The laser power has to be low enough to accumulate negligible amounts of photoproducts during the short time the sample resides within the cross-section of the laser beam (usually a few microseconds). The pump–probe technique was first applied to rhodopsin by Oseroff and Callender.[15] In this application a photoequilibrium between three species of the photoreaction of rhodopsin can be established at 80 K (rhodopsin, bathorhodopsin, and isorhodopsin). By altering the wavelength of the "pump" laser beam, which usually

---

[9] M. S. Braiman and K. J. Rothschild, *Annu. Rev. Biophys. Biophys. Chem.* **17**, 541 (1988).

[10] R. H. Callender, A. Doukas, R. K. Crouch, and K. Nakanishi, *Biochemistry* **15**, 1621 (1976).

[11] R. A. Mathies, A. R. Oseroff, and L. Stryer, *Proc. Natl. Acad. Sci. U.S.A.* **73**, 1 (1976).

[12] R. A. Mathies, T. B. Freedman, and L. Stryer, *J. Mol. Biol.* **109**, 367 (1977).

[13] W. Kiefer and H. J. Bernstein, *Appl. Spectrosc.* **25**, 500 (1971).

[14] M. Stockburger, W. Klusmann, H. Gattermann, G. Massig, and R. Peters, *Biochemistry* **18**, 4886 (1979).

[15] A. R. Oseroff and R. H. Callender, *Biochemistry* **13**, 4243 (1974).

has the higher intensity, the composition of the photoequilibrium is changed. By tuning the wavelength of the "probe" laser beam to a position where one of the three species contributes most to the absorption spectrum, the RR scattering of this compound will selectively be enhanced. The probe beam should have a lower intensity in order not to alter the composition of the photoequilibrium. For systems which undergo a cyclic photoreaction, such as bacteriorhodopsin and halorhodopsin, the pump–probe technique can be applied at room temperature. The photostationary state will be dominated by the initial state and the intermediate with the slowest decay time.

The time-resolved and pump–probe techniques can, of course, be combined. The composition of intermediates can be altered by altering the power and wavelength of the pump laser. Again, by adjusting the probe beam, the RR scattering from the species of interest will be enhanced. In general, however, irrespective of the method employed, the RR spectrum will contain contributions from several species. Employing time-resolved techniques, the RR spectrum of the initial state can usually be obtained by illuminating the sample with a very weak probe beam only. This spectrum can then be used to deduce from the composite spectra the spectra of the intermediates of the photoreaction. By moving the focus of the probe beam from the focus of the pump beam, it is possible to collect spectra at times after the excitation of the sample by the pump beam. The time is given by the distance between pump and probe beam divided by the velocity of the flow. This is an additional method for altering the relative amounts of intermediates in the cross section of the probe beam. Intermediates arising several microseconds up to several milliseconds after excitation can be measured in this way. More technical details about the methods can be found in the review papers mentioned above.

One can obtain detailed molecular information by comparing the RR spectra of retinal proteins with those of model compounds. The basic investigation by the Mathies group has contributed greatly to our understanding of the vibrational spectra of isomers of retinal[16,17] and of protonated and unprotonated retinylidene Schiff bases.[18,19] An essential part of this study was the collaboration with the Lugtenburg group, providing the

[16] B. Curry, A. Broek, J. Lugtenburg, and R. A. Mathies, *J. Am. Chem. Soc.* **104,** 5274 (1982).
[17] B. Curry, I. Palings, A. Broek, J. A. Pardoen, P. P. J. Mulder, J. Lugtenburg, and R. A. Mathies, *J. Phys. Chem.* **88,** 688 (1984).
[18] S. O. Smith, A. B. Myers, R. A. Mathies, J. A. Pardoen, C. Winkel, E. M. M. Van den Berg, and J. Lugtenburg, *Biophys. J.* **47,** 653 (1985).
[19] B. Curry, I. Palings, J. A. Pardoen, J. Lugtenburg, and R. A. Mathies, *Adv. Infrared Raman Spectrosc.* **12,** 115 (1985).

[13]C- and [2]H-labeled retinals.[20,21] By developing an empirical molecular force field, almost all bands in the RR spectra could be assigned to specific vibrations of the molecules and characteristic bands for the various retinal isomers could be deduced. The effects of forming the Schiff base and its protonation have also been studied. Since retinals, as well as their protonated and unprotonated Schiff bases, are photolabile, the capillary flow technique proved to be essential.

In Fig. 1, the RR spectra of bacteriorhodopsin under various illumination conditions using the rotating cell technique are shown.[14] Spectrum (a) in Fig. 1 was collected with a weak probe beam of 514 nm only, reflecting the light-adapted species of bacteriorhodopsin only, BR568, which has an absorption maximum at 568 nm. By applying a broad pump beam of 514 nm and a coaxial probe beam of 450 nm, an almost pure spectrum of the long-lived intermediate M412 could be obtained [spectrum (b) in Fig. 1], having an absorption maximum at 412 nm. Only the shoulder at 1530 cm^{-1} indicates a small percentage of the initial state BR568. If the intensity of the probe beam is increased, spectrum (c) in Fig. 1 is obtained. This spectrum contains, in addition to BR568, the M412 intermediate, as can be seen from the band at 1567 cm^{-1}. If a pump beam of 450 is now applied, driving the M412 intermediate back to BR568, spectrum (d) (Fig. 1) is obtained.

These experiments demonstrate the principles of the pump–probe technique in combination with time-resolved techniques. The spectra (Fig. 1) are all dominated by a strong band between 1500 and 1600 cm^{-1}, caused by the ethylenic vibration of the retinal. Its position is correlated with the absorption maximum of the retinylidene Schiff base, owing to the dependency of the ethylenic force constants on π-electron delocalization.[22] The band at 1642 cm^{-1} (Fig. 1a) was shown to shift to 1624 cm^{-1} in ^2H_2O, demonstrating that it is caused by the C=N stretching vibration of the Schiff base and that the Schiff base is protonated. The vibrations of this group have been especially investigated, both theoretically and experimentally.[23–28] It appears unclear whether protonation increases or de-

[20] J. Lugtenburg, *Pure Appl. Chem.* **57,** 753 (1985).

[21] J. Pardoen, C. Winkel, P. Mulder, and J. Lugtenburg, *Recl. Trav. Chim. Pays-Bas* **103,** 135 (1984).

[22] M. E. Heyde, D. Gill, R. G. Kilponen, and L. Rimai, *J. Am. Chem. Soc.* **93,** 6776 (1971).

[23] H. Deng and R. H. Callender, *Biochemistry* **26,** 7418 (1987).

[24] H. S. R. Gilson, B. H. Honig, A. Croteau, G. Zarrilli, and K. Nakanishi, *Biophys. J.* **53,** 261 (1988).

[25] T. Baasov, N. Friedman, and M. Sheves, *Biochemistry* **26,** 3210 (1987).

[26] J. J. López-Garriga, G. T. Babcock, and J. F. Harrison, *J. Am. Chem. Soc.* **108,** 7131 (1986).

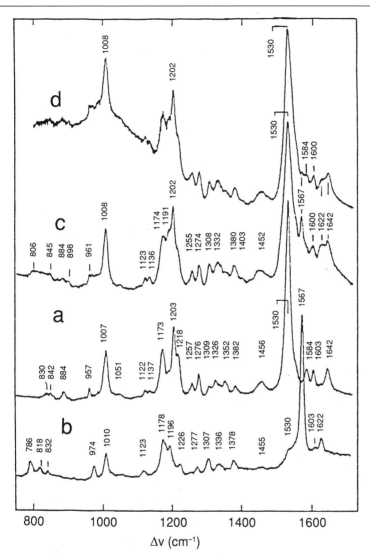

FIG. 1. Pump–probe resonance Raman spectra of bacteriorhodopsin at room temperature using a rotating cell system. (a) Spectrum of the initial state BR568 taken with a weak probe beam of 514 nm only; (b) spectrum of the long-lived intermediate M412, taken with a 514-nm pump beam a 457-nm probe beam; (c) spectrum obtained with strong probe beam of 514 nm, showing the presence of BR568 and M412; (d) as (c), but with an additional pump beam of 457 nm, driving the M412 intermediate back to BR568. [Reproduced from M. Stockburger, W. Klusmann, H. Gattermann, G. Massig, and R. Peters, *Biochemistry* **18,** 4886 (1979).]

creases the C=N force constant. Also, the contribution of the NH bending vibration to the C=N frequency is not completely understood. The understanding of this coupling is especially important, since it would provide a means to investigate the interaction of the protonated Schiff base group with its environment.

Figure 2 demonstrates the pump–probe technique at $-160°$, where the primary photoreaction of rhodopsin has been investigated.[29] Spectrum (A) in Fig. 2 was obtained with a probe beam of 585 nm. The sample was also illuminated with an all-lines argon laser pump beam. If the spectrum was recorded without the pump beam [spectrum (B), Fig. 2] only isorhodopsin was present. A pure rhodopsin spectrum was obtained by the capillary flow technique [spectrum (C), Fig. 2]. If appropriate amounts of spectra (B) and (C) are subtracted from spectrum (A), a pure spectrum of bathorhodopsin, the first intermediate which can be stabilized at low temperature, could be obtained. The downshift of the ethylenic mode (1545 versus 1536 cm^{-1}) is in accordance with the red-shifted absorption maximum of bathorhodopsin. The strong bands below 1000 cm^{-1} are of special interest. They could be assigned to hydrogen-out-of-plane (HOOP) vibrations of the retinal (i.e., the hydrogens move perpendicular to the polyene plane), and the high intensities were attributed to twists around single bonds of the polyene.[30,31] This means that the chromophore, already isomerized from 11-cis to all-trans, could not yet adopt a relaxed planar configuration owing to steric interaction with the protein, which is rigid at low temperature. A similar effect was observed for the 80 K photoproduct of bacteriorhodopsin, the K intermediate,[32] and interpreted in a similar way. A peculiarity in the bathorhodopsin spectrum is the anomalous decoupling of the 11- and 12-HOOP vibrations, causing the band at 920 cm^{-1}; it was attributed to a specific interaction with a negatively charged residue or dipole of the protein.

In recent illustrative applications of RR spectroscopy, demonstrating the selectivity of the method, the visual pigments in the retina of toad, anglefish, gecko, and bullfrog have been investigated.[33] In this case, the

[27] J. J. López-Garriga, G. T. Babcock, and J. F. Harrison, *J. Am. Chem. Soc.* **108,** 7241 (1986).

[28] J. J. López-Garriga, S. Hanton, G. T. Babcock, and J. F. Harrison, *J. Am. Chem. Soc.* **108,** 7251 (1986).

[29] G. Eyring and R. A. Mathies, *Proc. Natl. Acad. Sci. U.S.A.* **76,** 33 (1979).

[30] G. Eyring, B. Curry, A. Broek, J. Lugtenburg, and R. Mathies, *Biochemistry* **21,** 384 (1982).

[31] I. Palings, E. M. M. Van den Berg, J. Lugtenburg, and R. A. Mathies, *Biochemistry* **28,** 1498 (1989).

[32] M. Braiman and R. A. Mathies, *Proc. Natl. Acad. Sci. U.S.A.* **79,** 403 (1982).

[33] B. A. Barry and R. A. Mathies, *Biochemistry* **26,** 59 (1987).

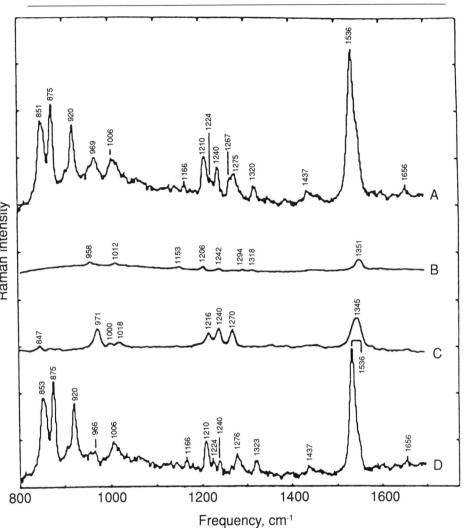

FIG. 2. Resonance Raman spectra of photostationary states of rhodopsin at −160°, demonstrating the pump–probe technique. (A) Spectrum with 585-nm probe beam in the presence of an all-lines argon laser pump beam; (B) as (A), but without the pump beam, reflecting a pure isorhodopsin spectrum; (C) rapid-flow RR spectrum of rhodopsin taken with 600-nm probe beam (from Ref. 11); (D) RR spectrum of pure bathorhodopsin obtained by subtracting the appropriate amounts of isorhodopsin (B) and rhodopsin (C) from spectrum (A). [Reproduced from G. Eyring and R. A. Mathies, *Proc. Natl. Acad. Sci. U.S.A.* **76**, 33 (1979).]

pump–probe technique at low temperature has been combined with the Raman microscope technique, to select the outer segment of the photoreceptor of interest. As in normal bathorhodopsin, in the low-temperature photoproducts of all pigments the unusual decoupling of the 11- and 12-HOOP vibrations is observed. This shows that the factor regulating the absorption maximum (ranging from 430 to 502 nm in this investigation) is not responsible for this special interaction with the protein. Investigations on the toad blue photoreceptor pigment (actually the isopigment, containing 9-*cis*-retinal) showed that the chromophore behaves like the 9-cis model compound in solution.[34] This was interpreted by the lack of two internal carboxyl groups, which in the red rod pigments provide a special interaction regulating the absorption maximum and altering the RR spectrum.

Resonance Raman spectroscopy has greatly contributed to our understanding of the photoreaction of bacteriorhodopsin, and the reader is referred to the reviews mentioned above. Two applications are of more general interest. The RR spectrum of vacuum-dried bacteriorhodopsin was measured, and it was concluded that water interacts with the Schiff base.[35] Support for this interaction was obtained by the observation that the $C=N$ stretching vibration of the Schiff base, being located near the bending vibration of water, is broader in H_2O than in ^2H_2O. Since protonation of the Schiff base requires a polar environment, the water molecules may serve this purpose. In a combination of protein chemistry methods and RR spectroscopy, it was possible to identify the retinal-binding lysine.[36] It is possible to recombine bacteriorhodopsin from its two chymotryptic fragments. One fragment was obtained from bacteriorhodopsin isolated from bacteria grown in a synthetic medium containing [ε-^{15}N]lysine, the other was from unmodified lysine. Only in the RR spectra of the recombinant in which lysine-216 was labeled was a downshift of the $C=N$ stretching vibration observed. This showed unequivocally that lysine-216 is the binding site. By recording the RR spectra of the M412 intermediate, a change of the binding site during the photoreaction could be excluded.

Infrared Difference Spectroscopy

At first sight, infrared spectroscopy would seem to be less suitable for the investigation of complex biological systems. Because the method is

[34] G. R. Loppnow, B. A. Barry, and R. A. Mathies, *Proc. Natl. Acad. Sci. U.S.A.* **86,** 1515 (1989).

[35] P. Hildebrandt and M. Stockburger, *Biochemistry* **23,** 5539 (1984).

[36] K. J. Rothschild, P. V. Argade, T. N. Earnest, K.-S. Huang, E. London, M.-J. Liao, H. Bayley, H. G. Khorana, and J. Herzfeld, *J. Biol. Chem.* **257,** 8592 (1982).

not selective, all groups of a protein contribute to the infrared absorbance, and it would be impossible to analyze the spectrum, which is constituted of many overlapping bands, in terms of single prosthetic groups or single amino acid side chains and their molecular changes. Only the vibrations of the amide group (amide I and amide II band) have been used to deduce information on the secondary structure. Infrared spectroscopy can, however, be rendered selective, that is, reflecting only functional groups, by forming the difference spectra between two different functional states. Only those groups which undergo molecular changes between these two states appear in the difference spectra, the remainder canceling each other by the subtraction. Infrared difference spectroscopy was first applied to retinal proteins (rhodopsin and bacteriorhodopsin) using time-resolved methods.[37-40] Somewhat later, static Fourier-transform infrared (FTIR) difference spectroscopy was applied to these systems.[41-46] With the development of faster scanning FTIR spectrophotometers, time-resolved FTIR spectroscopy became accessible.[47,48] The basic techniques of IRD are described here, with special emphasis on sample preparation and methods to assign bands to specific groups of the systems under investigation.

Infrared difference spectroscopy requires that the two spectra being subtracted differ only in the altered molecular states. In principle, it would be possible to record, for instance, spectra of a retinal-binding protein with and without retinal. The corresponding difference spectrum should reflect, in addition to the infrared bands of the retinal, the protein groups involved in binding. Since only a small percentage of the total system will contribute, the absorbance changes in the difference spectra will range from 10^{-4} to 10^{-2} at the most. Making the measurement of such small differences reliable requires, besides the different molecular states,

[37] F. Siebert and W. Mäntele, Biophys. Struct. Mech. 6, 147 (1980).
[38] F. Siebert, W. Mäntele, and W. Kreutz, Biophys. Struct. Mech. 6, 139 (1980).
[39] F. Siebert, W. Mäntele, and W. Kreutz, Can. J. Spectrosc. 26, 119 (1981).
[40] W. Mäntele, F. Siebert, and W. Kreutz, this series Vol. 88, p. 729.
[41] K. J. Rothschild, M. Zagaeski, and W. A. Cantore, Biochem. Biophys. Res. Commun. 103, 483 (1981).
[42] K. J. Rothschild and H. Marrero, Proc. Natl. Acad. Sci. U.S.A. 79, 4045 (1982).
[43] K. Bagley, G. Dollinger, L. Eisenstein, A. K. Singh, and L. Zimanyi, Proc. Natl. Acad. Sci. U.S.A. 79, 4972 (1982).
[44] F. Siebert and W. Mäntele, Eur. J. Biochem. 130, 565 (1983).
[45] F. Siebert, W. Mäntele, and K. Gerwert, Eur. J. Biochem. 136, 119 (1983).
[46] K. J. Rothschild, W. A. Cantore, and H. Marrero, Science 219, 1333 (1983).
[47] M. S. Braiman, P. L. Ahl, and K. J. Rothschild, Proc. Natl. Acad. Sci. U.S.A. 84, 5221 (1987).
[48] K. Gerwert, Ber. Bunsen-Ges. Phys. Chem. 92, 978 (1988).

samples almost identical in other respects. This is possible for soluble proteins, and investigations on enzyme–substrate interactions using IRD have been published[49]; however, this will be very difficult to realize for suspensions of membrane proteins. One has to realize that because water strongly absorbs infrared radiation in almost the total spectral range of interest, the concentrations have to be very high (of the order of 100 mg/ml) and the cuvette has to be very thin (4 μm). It is possible to make reproducible samples for soluble proteins under such conditions, but this is not feasible for suspensions owing to heterogeneity. Another method of sample preparation appears to be advantageous for membrane proteins: 100 to 200 μg of the membranes are dried onto one window of the infrared cuvette, rehydrated with the required amount of water, and sealed with a second window. Most IRD spectroscopy investigations on retinal proteins have been performed on such types of samples.

Infrared attenuated total reflection (ATR) spectroscopy has recently been applied to bacteriorhodopsin.[50,51] This method offers the advantage that, as the penetration depth of the infrared radiation is of the order of its wavelength only, samples which are deposited onto the ATR crystal can be bathed in aqueous solutions and the solutions can be exchanged in one experiment. Special precautions have to be taken to avoid detachment of the film from the crystal surface.

The difficulties in preparing highly reproducible samples of membrane proteins has limited investigations to photobiological systems. Here, the different states of the system can easily be evoked by irradiation with visible light. Intermediates of the photoreaction can be trapped by lowering the temperature or by adjusting the pH. The small absorbance changes in the difference spectra make it mandatory to employ FTIR spectroscopy, which exhibits a much higher sensitivity owing to the multiplex advantage and the larger energy through-put. Static IRD spectra are obtained by first recording a single-beam spectrum of the nonilluminated sample, then producing the desired intermediate, and subsequently recording the single-beam spectrum of the illuminated sample. By forming the logarithm of the ratio of the two single-beam spectra, the absorbance difference spectrum is obtained. Usually, to obtain a satisfactory signal-

[49] C. W. Wharton, R. S. Chittok, J. Austin, and R. E. Hester, in "Spectroscopy of Biological Molecules—New Advances" (E. D. Schmid, F. W. Schneider, and F. Siebert, eds.), p. 95. Wiley, New York, 1988.
[50] H. Marrero and K. J. Rothschild, FEBS Lett. 223, 289 (1987).
[51] H. Marrero and K. J. Rothschild, Biophys. J. 52, 629 (1987).

to-noise ratio, 100 to 1000 scans have to be coadded for one single-beam spectrum.

Figure 3 shows the static FTIR difference spectra of the photoreaction of rhodopsin.[52] The intermediates have been stabilized at low temperatures: the batho-, lumi-, metarhodopsin I, and metarhodopsin II intermediates at 80, 173, 240, and 273 K, respectively. The usual convention is that the bands of the initial state point downward and the bands of the photoproduct point upward. It is noteworthy that the strong bands below 1000 cm^{-1} of bathorhodopsin, which have been observed in the RR spectra and attributed to HOOP vibrations, appear also as strong bands in the infrared. In a recent investigation, this striking behavior was shown to be caused by a coupling of C–C modes to the HOOP vibrations if the retinal is twisted.[53] In the infrared, strong HOOP bands can also be taken as evidence of a twisted chromophore. Figure 3 shows that in the spectra of the later intermediates more and more protein molecular changes occur. This can be deduced from the increasing bands between 1600 and 1800 cm^{-1}.

One difficulty of IRD spectroscopy is discriminating between bands caused by the chromophore and those arising from the protein. Isotopic labeling of the chromophore is essential. A convenient method to clearly deduce the labeling effect is subtraction of the unlabeled from the labeled difference spectrum. Protein bands cancel in the subtraction, and bands of the chromophore which are affected by the specific labeling show up and can be assigned.[54,55]

The overlap of bands of the initial state and of the photoproduct is a common problem in difference spectroscopy. It can only be overcome if the absolute spectrum of one species is known. If the absorbance strength of bands of one species is much lower, an approximate absolute spectrum can be obtained.[54] Also, a comparison of the IRD spectra with corresponding RR spectra may be useful to deduce the true band positions in the difference spectra. These methods have been applied to obtain information on the geometry of the retinal in the intermediates of bacteriorhodopsin and rhodopsin and have been used to assign their C=N stretching frequencies.[54,55]

The main advantage of IRD is its capability to detect protein molecular

[52] U. M. Ganter, E. D. Schmid, D. Perez-Sala, R. R. Rando, and F. Siebert, *Biochemistry* **28**, 5954 (1989).
[53] K. Fahmy, M. F. Grossjean, F. Siebert, and P. Tavan, *J. Mol. Struct.* **214**, 257 (1989).
[54] K. Gerwert and F. Siebert, *EMBO J.* **5**, 805 (1986).
[55] U. M. Ganter, W. Gärtner, and F. Siebert, *Biochemistry* **27**, 7480 (1988).

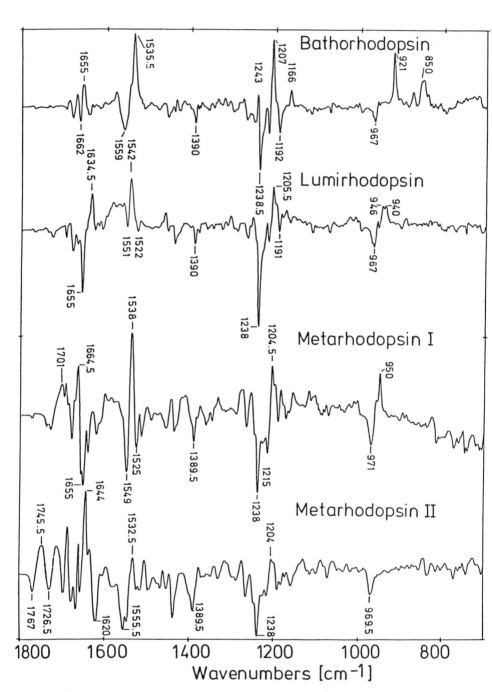

FIG. 3. Infrared difference spectra of the photoreaction of rhodopsin. Temperatures used to stabilize the intermediates were as follows: bathorhodopsin, 80 K; lumirhodopsin, 173 K; metarhodopsin I, 243 K; metarhodopsin II, 270 K and pH 5.5. [Reproduced from U. M. Ganter, E. D. Schmid, D. Perez-Sala, R. R. Rando, and F. Siebert, *Biochemistry* **28**, 5954 (1989).]

changes. In most cases, to assign bands unequivocally to specific groups ern methods of genetic engineering may provide additional tools by expressing nonbacterial systems in bacteria or by introducing point mutations. Such applications are now briefly discussed. of the protein, isotopic labeling is a prerequisite. Whereas for bacterial systems this is feasible, other systems pose greater difficulties. The mod-

Protonation changes of carboxyl groups during the photocycle of bacteriorhodopsin have been detected[41,56] and were later assigned by isotopic labeling to aspartic acids.[57,58] In the same way, molecular changes of tyrosines could be detected.[59-62] By the use of point mutations, some of the spectral changes could be attributed to a specific tyrosine in the amino acid chain. With the same technique, some of the aspartic acids could be assigned to specific positions in the peptide chain.[63] By incorporating deuterated lysine into bacteriorhodopsin, vibrational modes at which lysine participates could be detected.[64] It is reasonable to assume that these modes are caused by the retinal-binding lysine. The analysis may provide information on the reaction of lysine upon the isomerization of the retinal. In another application the role of water in rhodopsin was investigated. By exchanging $H_2^{18}O$ for $H_2^{16}O$, water molecules could be detected which undergo a slight change during the reaction of rhodopsin to bathorhodopsin. It was assumed that these water molecules are in the immediate neighborhood of the Schiff base and may help to stabilize its protonation state.

A prior RR investigation may provide a link between the retinal proteins discussed in this chapter and the retinal-binding proteins. In this investigation the pigments of lobster shell were examined. From the correlation of the C=C stretching frequency with the absorption maximum it

[56] F. Siebert, W. Mäntele, and W. Kreutz, *FEBS Lett.* **141**, 82 (1982).

[57] M. Engelhard, K. Gerwert, B. Hess, W. Kreutz, and F. Siebert, *Biochemistry* **24**, 400 (1985).

[58] L. Eisenstein, S.-L. Lin, G. Dollinger, K. Odashima, J. Termini, K. Konno, W.-D. Ding, and K. Nakanishi, *J. Am. Chem. Soc.* **109**, 6860 (1987).

[59] S.-L. Lin, P. Ormos, L. Eisenstein, R. Govindjee, K. Konno, and K. Nakanishi, *Biochemistry* **26**, 8327 (1987).

[60] K. J. Rothschild, P. Roepe, P. L. Ahl, T. N. Earnest, R. A. Bogomolni, S. K. Das Gupta, C. M. Mulliken, and J. Herzfeld, *Proc. Natl. Acad. Sci. U.S.A.* **83**, 347 (1986).

[61] P. Roepe, P. L. Ahl, S. K. Das Gupta, J. Herzfeld, and K. J. Rothschild, *Biochemistry* **26**, 6696 (1987).

[62] P. Roepe, P. Scherrer, P. L. Ahl, S. K. Das Gupta, R. A. Bogomolni, J. Herzfeld, and K. J. Rothschild, *Biochemistry* **26**, 6708 (1987).

[63] M. S. Braiman, T. Mogi, T. Marti, L. J. Stern, H. G. Khorana, and K. J. Rothschild, *Biochemistry* **27**, 8516 (1988).

[64] E. McMaster and A. Lewis, *Biochem. Biophys. Res. Commun.* **156**, 86 (1988).

was concluded that the color is dominated by excitonic coupling of carotenoids.[65-67] This example may provide some hints of how RR spectroscopy could also be applied to the retinal-binding proteins. Infrared difference spectroscopy may be more difficult to implement; however, owing to its capability to provide direct information on chromophore–protein interactions, it appears worthwhile.

[65] V. R. Salares, N. M. Young, P. R. Carey, and H. J. Bernstein, *J. Raman Spectrosc.* **6**, 282 (1977).
[66] V. R. Salares, N. M. Young, H. J. Bernstein, and P. R. Carey, *Biochemistry* **16**, 4751 (1977).
[67] V. R. Salares, R. Mendelsohn, P. R. Carey, and H. J. Bernstein, *J. Phys. Chem.* **80**, 1137 (1976).

[13] Analysis of Water-Soluble Compounds: Glucuronides

By Arun B. Barua

Introduction

Retinyl β-glucuronide and retinoyl β-glucuronide, the glucuronic acid conjugates of retinol and retinoic acid, respectively, occur naturally in bile,[1-3] several other tissues,[3-5] and human blood.[6,7] Unlike retinol (1–2.5 μM concentration), retinoic acid,[6-10] retinyl glucuronide, and retinoyl glucuronide occur in much smaller concentrations (1–3 nM) in human blood.[6,7] The concentration of these conjugates in other tissues is not known but appears to be low. Until recently, retinyl β-glucuronide (**I**) and retinoyl β-glucuronide (**II**) were not available in sufficient quantities to study their metabolism and possible role in the functions of vitamin A. This chapter describes (1) chemical synthesis of retinyl β-glucuronide and retinoyl β-glucuronide, and (2) extraction of the two compounds from

[1] R. D. Zachman, P. E. Dunagin, and J. A. Olson, *J. Lipid Res.* **7**, 3 (1966).
[2] P. E. Dunagin, E. H. Meadows, and J. A. Olson, *Science* **148**, 86 (1965).
[3] M. H. Zile, R. C. Inhorn, and H. F. DeLuca, *J. Biol. Chem.* **267**, 3537 (1982).
[4] K. Lippel and J. A. Olson, *J. Lipid Res.* **9**, 168 (1968).
[5] M. E. Cullum and M. H. Zile, *J. Biol. Chem.* **260**, 10590 (1985).
[6] A. B. Barua and J. A. Olson, *Am. J. Clin. Nutr.* **43**, 481 (1986).
[7] A. B. Barua, R. O. Batres, and J. A. Olson, *Am. J. Clin. Nutr.* **50**, 370 (1989).
[8] M. G. DeRuyter, W. E. Lambert, and A. P. DeLeenheer, *Anal. Biochem.* **98**, 402 (1979).
[9] A. P. DeLeenheer, W. E. Lambert, and I. Claeys, *J. Lipid Res.* **23**, 1362 (1982).
[10] J. L. Napoli, B. C. Pramanik, J. B. Williams, M. I. Dawson, and P. D. Hobbs, *J. Lipid Res.* **26**, 387 (1985).

serum, (3) analysis by high-performance liquid chromatography, and (4) characterization by mass spectrometry.

Synthesis

Synthesis of Retinyl β-Glucuronide. all-*trans*-Retinyl β-glucuronide (ROG) is synthesized by reaction of all-*trans*-retinol with methyl (tri-*O*-acetyl-1-bromoglucopyran)uronate (bromo sugar).[11] Retinol (4 g, oil) (prepared from all-*trans*-retinyl acetate by saponification[11]) is stirred with a magnetic stirrer at room temperature with the bromo sugar (3.2 g) (prepared from 6,3-glucuronolactone[11]) and silver carbonate (1.5 g) for 3 hr in the presence of butylated hydroxytoluene (100 mg). Diethyl ether (50 ml) is added, and the solution is filtered. The residue of silver carbonate is washed several times with ether, filtered, and the pooled filtrate is evaporated to dryness in a rotary evaporator. The oily residue is dissolved in methanol (25 ml), sodium methylate (1.2 g) is added, and the solution is refluxed for 10 min. The solution is diluted with water (25 ml) and neutralized with 10% acetic acid. After extraction with ethyl acetate (3 times, 50 ml each), the pooled extract is dried over anhydrous sodium sulfate, then evaporated to dryness in a rotary evaporator.

The residual oil is dissolved in a mixture of hexane/ether (1 : 1) (~5 ml) and subjected to column chromatography (CC) on 100 g of silica gel (for dry CC, Woelm Pharma, Eschwege, FRG) wet packed with hexane. Development with hexane containing 10–60% ether results in removal of

[11] A. B. Barua and J. A. Olson, *Biochem. J.* **244**, 231 (1987).

retinol and other products. ROG remains at the top of the column as an orange band. Development with dichloromethane containing 10% methanol results in movement of the ROG band. ROG is preceded by a dark-orange fraction. The eluates are collected in fractions of about 5 ml. All fractions showing absorption at 325 nm are pooled and then evaporated to dryness in a rotary evaporator. The oil is treated with methanol drop by drop until precipitation of ROG is complete. The solid is separated and dried. The yield is 2.1 g, or 90% based on the bromo sugar. ROG (100 mg) is dissolved in methanol and purified by HPLC on a Whatman Partisil 10 ODS-3 M9 column (9.4 × 50 cm) using methanol–water (4 : 1) at a flow rate of 4 ml/min. The elution time for all-*trans*-ROG is 17 min. Traces of polar impurities and cis isomers of ROG (<5%) separate well, and no α-anomer of ROG is seen.

Synthesis of Retinoyl β-Glucuronide. A three-step procedure for the preparation of retinoyl β-glucuronide (RAG) from retinoic acid was reported earlier.[12] The method has since been simplified to a two-step procedure.[13] Retinoic acid (RA) is converted to retinoyl fluoride (ROF) by treatment with diethylaminosulfur trifluoride.[12] The crystalline ROF (2.8 g) is dissolved in acetone (200 ml). Glucuronic acid (6 g) and sodium bicarbonate (6 g) are dissolved in water (100 ml) and added to the ROF solution. The mixture is stirred at room temperature for 16–24 hr. If the solution becomes turbid during stirring, water is added in small quantities until the solution is clear. Alternatively, if oily droplets of ROF appear, acetone is added until the solution is clear.

The solution is diluted with water (100 ml) and neutralized with 1 *N* acetic acid. The retinoids are extracted with ethyl acetate until the extract is colorless. The pooled extract is washed with water, dried over anhydrous sodium sulfate, and evaporated to dryness in a rotary evaporator. The residual oil is dissolved in hexane/diethyl ether (1 : 1, ~5 ml) and subjected to CC on silica gel (for dry CC) wet packed with hexane. Development with hexane containing 10–60% ether removes unconverted ROF and RA formed by hydrolysis of ROF. Development with dichloromethane containing 10% methanol results in elution of an unidentified brown-colored band followed immediately by RAG. The eluates are collected in fractions of about 5 ml. (A drop of each fraction is evaporated to dryness under argon, and the residue is dissolved in a few drops of water. If the residue is soluble and produces a yellow color, it is RAG.) All fractions containing RAG are pooled and evaporated to dryness in a rotary evaporator. The resultant yellow solid is dried under reduced pres-

---

[12] A. B. Barua and J. A. Olson, *J. Lipid Res.* **26**, 1277 (1985).
[13] A. B. Barua and J. A. Olson, U.S. Patent No. 4,855,463, Aug. 8, 1989.

sure. The yield is 1.5–2.5 g, or 30–60%. Pure all-*trans*-RAG is obtained by HPLC of a concentrated methanol solution on a Whatman ODS-3 M9 column using methanol–water (4 : 1) at a flow rate of 4 ml/min. The retention times of *cis*-RAG (3%), all-*trans*-RAG (90%), and an α-anomer of RAG (7%) are 13.2, 15.4, and 17 min (very broad), respectively.

Analysis of Retinoyl β-Glucuronide and Retinyl β-Glucuronide in Human Serum

RAG is easily hydrolyzed to retinoic acid in the presence of moderately strong acid or base.[6] ROG is unstable toward acid and is converted to an anhydroretinol-like compound. Earlier methods to quantitate retinoic acid[8–10] by using strong acid and base result in extensive hydrolysis of RAG.[6] Therefore, use of mineral acids and bases should be avoided, and care should be taken to prevent hydrolysis during extraction of the glucuronides.

Preparation of Standards. Pure all-*trans*-ROG and all-*trans*-RAG obtained by HPLC and crystalline all-*trans*-RA are dissolved in known volumes of methanol and the absorption spectra recorded. Using $E_{1 cm}^{1\%}$ values of 975 at 325 nm, 1065 at 360 nm, and 1480 at 350 nm for ROG, RAG, and retinoic acid, respectively, the concentrations of each retinoid are determined. Aliquots of the standard solution (0.5–20 ng) are injected into the HPLC system, and the peak areas are determined by use of an integrator connected to the detector. Standard curves are made for each retinoid.

Extraction of Retinoids from Serum. The extraction procedure is essentially the same as described earlier.[6,7] All operations are carried out with precooled solvents at 4° under yellow light. Serum (1 ml), taken in a disposable glass culture tube (13 × 100 mm), is treated with ethanol (90–100%) (2 ml) containing 0.1% (w/v) butylated hydroxytoluene and with ethyl acetate (2 ml). After vortexing for 30 sec and centrifuging at 1500 rpm for 30 sec, the supernatant solution is pipetted out and saved. The pellet is broken and extracted with ethyl acetate (2 times, 1 ml each). The first supernatant is diluted with water (2 ml), vortexed, and centrifuged as above. The upper organic phase is saved, and the aqueous phase is treated with 1 drop (50 μl) of 10% (v/v) acetic acid in water. The aqueous phase is extracted with the ethyl acetate (pellet extract, 2 ml) and once with hexane (1 ml) by vortexing and centrifuging as just described. All organic solvent phases are pooled, water (1 ml) is added, and the mixture vortexed and centrifuged as before. The organic phase is carefully pipetted out and evaporated to dryness under a stream of argon. The residue is dissolved in methanol–dichloromethane (4 : 1, v/v) (100 μl). Aliquots of

50–100 μl corresponding to 0.5–1 ml of the serum are analyzed immediately by high-performance liquid chromatography (HPLC).

High-Performance Liquid Chromatographic Analysis of Serum Extract

For the estimation of endogenous retinoid glucuronides and retinoic acid, but not retinol, in serum, an isocratic reversed-phase HPLC system[6,7] is preferred. The isocratic method gives a stable baseline and facilitates accurate quantitation of these minor retinoids in serum. Furthermore, a sensitive detector connected to a sensitive integrator are required in order to see the peaks arising from 1–8 ng of the glucuronides of vitamin A and retinoic acid present in 0.5–1 ml of serum extract. When radiolabeled retinoids are analyzed, and when the quantitation of the amount of radioactivity associated with each peak, rather than the area of the absorption peak, is the primary goal, gradient HPLC[13] is the method of choice. The advantage of gradient HPLC lies in the analysis of a whole range of very polar (4-oxoretinoic acid) to least polar (retinyl esters) retinoids in one run in 30 min.

Isocratic Method. The reversed-phase isocratic HPLC apparatus used in the author's laboratory consists of a Waters Associates (Milford, MA) Model 510 pump, an ISCO (Lincoln, NE) Model V^4 absorbance detector set at 325 nm, and a Rheodyne (Cotati, CA) injector. A Waters Resolve 5-μm spherical C_{18} column (3.9 × 15 cm) preceded by a Upchurch (Oak Harbor, WA) C-130 guard column is used with a solvent system of methanol–water (68 : 32, v/v) containing 10 mM ammonium acetate (solvent A) at a flow rate of 1.5 ml/min. An Upchurch Model U-446 back-pressure regulator is useful at the detector exit to prevent solvent degassing in the detector cell. A Shimadzu (Kyoto, Japan) Model CR3-A or CR4-A integrator is used to record the chromatograms at an initial setting of attenuation 3 (0.008 AUFS). The chromatograms are saved, and, if necessary, at the end of analysis, they are reanalyzed by changing the parameters of attenuation, slope, etc. Under these conditions, the retention times of RAG, ROG, and RA are 7.7, 9.2, and 16.2 min, respectively (Fig. 1A).[14] The column is to be flushed with methanol (100%) after every three serum assays in order to elute retinol, which otherwise will interfere with the assay. The separation of RAG, ROG, and RA in human serum by isocratic HPLC is shown in Fig. 1B.

Gradient Method. For reversed-phase gradient HPLC, in addition to the equipment mentioned above, a second pump and a solvent programmer are required. The solvent system consists of a linear gradient of

---

[14] A. B. Barua, R. O. Batres, and J. A. Olson, *Biochem. J.* **252**, 415 (1988).

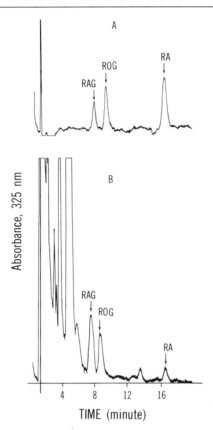

TIME (minute)

FIG. 1. Chromatogram obtained with reversed-phase isocratic HPLC of (A) a standard mixture of all-*trans*-retinoyl β-glucuronide (RAG), all-*trans*-retinyl β-glucuronide (ROG), and all-*trans*-retinoic acid and (B) extract of human serum (1 ml). Separation was achieved on a Waters Resolve 5-μm spherical C₁₈ column (3.9 mm i.d. × 15 cm) eluted with methanol–water (68:32) containing 10 mM ammonium acetate at a flow rate of 1.5 ml/min; detection wavelength: 325 nm. [Reprinted with permission from *American Journal of Clinical Nutrition*, American Society for Clinical Nutrition. From Barua *et al.*[14]]

solvent A to solvent B (methanol–dichloromethane, 4:1, v/v) over 20 min at a flow rate of 1 ml/min. At the end of the program, solvent B is run for 10 min more, then the gradient is changed to initial conditions over a 5-min period. After allowing 10 min for the column to equilibrate fully to the initial conditions, the next injection is made. A chromatogram obtained with standard retinoids by the reversed-phase gradient HPLC is shown in Fig. 2. Gradient HPLC is useful to study the metabolism of labeled retinoids.[14]

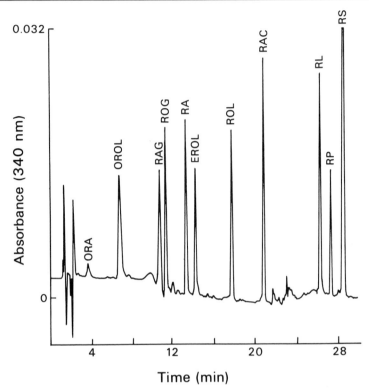

Time (min)

FIG. 2. Chromatogram obtained with reversed-phase gradient HPLC of a standard mixture of retinoids. Separation was achieved on a Waters Resolve 5-μm spherical C_{18} column (3.9 mm i.d. × 15 cm). A linear gradient of methanol–water (68:32) containing 10 mM ammonium acetate for 20 min to methanol–dichloromethane (4:1) at a flow rate of 1 ml/min was used. Abbreviations: ORA, 4-oxoretinoic acid; OROL, 4-oxoretinol; RAG, retinoyl β-glucuronide; ROG, retinyl β-glucuronide; RA, retinoic acid; EROL, 5,6-epoxyretinol; ROL, retinol; RAC, retinyl acetate; RL, retinyl linoleate; RP, retinyl palmitate; RS, retinyl stearate. Detection wavelength: 340 nm.

Mass Spectrometry

It is not uncommon that two compounds with different structures coelute during HPLC. Therefore, besides comparing the retention time of the compound under investigation with a standard compound, often it is necessary to confirm the identity of the compound by determining the mass spectrum. In order to increase the volatility and to obtain a suitable parent ion, it is also necessary to derivatize most of the retinoids.

Methylation.[6] The extract from 6–7 ml of serum is dissolved in methanol (500 μl) and treated with a solution of diazomethane (~5 mg in 3 ml of diethyl ether) at ice temperature (4°). The solution is kept at room temperature for 30–40 min, and then the solvent, along with the excess diazomethane (which is a very poisonous gas), is carefully removed under argon in a hood. The residue is either converted to the trimethylsilyl derivative (see below) or dissolved in methanol (100 μl) for HPLC analysis to isolate methylated retinoyl glucuronide. A Waters 5-μm Resolve column and a solvent mixture of methanol–water (7 : 3) containing 10 mM ammonium acetate are used at a flow rate of 1.5 ml/min for 10 min, followed by a linear gradient to methanol–water (9 : 1) over a 15-min

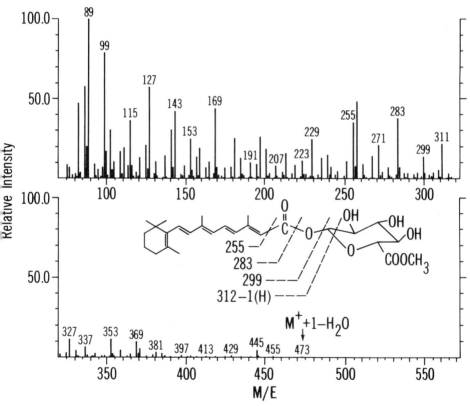

FIG. 3. Mass spectrum of endogenous human serum retinoyl β-glucuronide after methylation with diazomethane. The spectrum was obtained with a Finnigan Model 4000 GC–MS instrument by desorption chemical ionization using methane or isobutane as the reagent gas. [Reprinted with permission from *American Journal of Clinical Nutrition*, American Society for Clinical Nutrition. From Barua and Olson[6].]

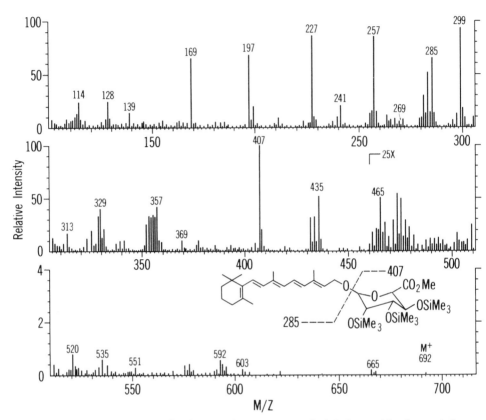

FIG. 4. Mass spectrum of endogenous human serum retinyl β-glucuronide after methylation with diazomethane and silylation with 1-(trimethylsilyl)imidazole. The spectrum was obtained with a Finnigan Model 4000 GC–MS instrument by desorption chemical ionization using isobutane as the reagent gas. [Reprinted with permission from *American Journal of Clinical Nutrition*, American Society for Clinical Nutrition. From Baruq *et al.*[14]]

period. Methylated RAG elutes at 21.7 min, and methyl retinoate at 30.3 min. The peak carrying methylated RAG is collected, an equal volume of water is added, and the retinoid is extracted with diethyl ether (3 times, 1 ml each) by vortexing and centrifuging. The solvent from the extract is removed under argon. The residue is dissolved in 50–100 μl of methanol and analyzed by mass spectrometry.

Trimethysilylation.[7] Formation of a silyl derivative is helpful for several reasons: first, the mass of the compound is considerably increased, thereby allowing easier manipulation of very small amounts; second, the mass spectra of the silylated compounds often have diagnostic fragmentation patterns that are clearer than the fragmentation pattern of the underivatized compound; and, third, the silyl derivative can be separated easily from other impurities by HPLC. Silylation is achieved by treating the methylated retinoid with 1-(trimethylsilyl)imidazole (TMS-Im). The residue of the methylated retinoid mixture is dissolved in pyridine (100 µl) to which TMS-Im (100 µl) is added at ice temperature (4°). The mixture is kept at room temperature for 15 min. Water (1 ml) is added, and the retinoids are extracted with hexane (3 times, 1 ml each) by vortexing and centrifuging. The pooled hexane extract is evaporated under argon, and the residue is dissolved in hexane (50 µl) to which methanol (100 µl) is added. On a Waters 5-µm Resolve column using methanol at a flow rate of 0.5 ml/min, the trimethylsilyl derivative of methylated ROG elutes at 11 min. This broad peak is collected in small fractions, and each fraction is evaporated under argon to dryness, dissolved in hexane (50–100 µl), and analyzed by mass spectrometry.

Mass Spectrum. A satisfactory mass spectrum of methylated retinoic acid is obtained in either the electron-impact (EI) mode[9] or chemical ionization (CI) mode[10] using an appropriate reagent gas. Although underivatized retinoid glucuronides do not give satisfactory mass spectra in either the EI or CI mode, reasonably satisfactory mass spectra of partially derivatized retinoyl glucuronide and fully derivatized retinyl glucuronide can be obtained in the desorption chemical ionization (DCI) mode. The mass spectra of methylated retinoyl β-glucuronide and that of the trimethylsilyl derivative of the methylated retinyl β-glucuronide are shown in Figs. 3 and 4. The parent peak (molecular ion) is absent or very weak in both the glucuronides. However, peaks arising from fragmentation of the retinoid moiety are quite intense and distinctive, thus supporting the identity of the compounds.

Acknowledgments

I am grateful to Professor James Allen Olson for introducing me to this project on water-soluble forms of vitamin A and for his help and constructive criticism in preparing the manuscript. This work was supported by the Competitive Research Grants Program, Science and Education Administration, U.S. Department of Agriculture (84-CRCR-1-1418 and 87-CRCR-1-2320), by the National Institutes of Health (DK 32793 and DK 39733), by the Allen Whitfield Memorial Cancer Fund, and by the Iowa Agriculture and Home Economics Experiment Station (Project No. 2534).

[14] Determination of Retinoids in Plasma by High-Performance Liquid Chromatography and Automated Column Switching

By RONALD WYSS

Introduction

Automation of bioanalytical methods has gained in importance, and not only for the routine laboratory. In this context, sample pretreatment before chromatographic separation is one of the most critical steps. An elegant solution of this problem in the field of drug determination in biological fluids is the direct injection of plasma samples using on-line solid-phase extraction and automated column switching. Such a system has been described by Roth et al.[1] using a C_{18} bonded phase as the precolumn material and backflush elution. This column-switching technique is especially interesting for the determination of retinoids, not only because of the high degree of automation, but also because of the total protection from light during analysis of the photosensitive retinoids. Furthermore, trace enrichment in the precolumn allows the development of highly sensitive methods.

The advantages of this technique were demonstrated by methods for the determination of isotretinoin (13-*cis*-retinoic acid), tretinoin (all-*trans*-retinoic acid), and their 4-oxo metabolites,[2] as well as etretinate, acitretin, and 13-*cis*-acitretin,[3] in plasma by high-performance liquid chromatography (HPLC) using automated column switching. A similar method was reported by Creech Kraft et al.[4] for retinoic acids, their 4-oxo metabolites, and retinol in plasma, amniotic fluid, and embryos. Early problems with this technique, such as stability of the analytes in the biological matrix in the autosampler before injection or recovery problems owing to the high protein binding of the retinoids, are now more or less solved.[5] Recovery problems, which may be due to the binding of retinoids to

[1] W. Roth, K. Beschke, R. Jauch, A. Zimmer, and F. W. Koss, *J. Chromatogr.* **222**, 13 (1981).

[2] R. Wyss and F. Bucheli, *J. Chromatogr.* **424**, 303 (1988).

[3] R. Wyss and F. Bucheli, *J. Chromatogr.* **431**, 297 (1988).

[4] J. Creech Kraft, C. Echoff, W. Kuhnz, B. Löfberg, and H. Nau, *J. Liq. Chromatogr.* **11**, 2051 (1988).

[5] R. Wyss and F. Bucheli, *J. Chromatogr.* **456**, 33 (1988).

different plasma proteins,[6] were overcome by careful choice of the injection solutions.

This chapter describes two HPLC methods developed in the author's laboratory. Best recoveries are obtained by diluting plasma with 9 mM sodium hydroxide–acetonitrile (8:2) or by precipitating proteins with ethanol, for the isotretinoin and etretinate series, respectively. After injection of 0.5 ml of the sample onto a precolumn, proteins and polar components are washed out with 1% ammonium acetate and 10–20% acetonitrile. After valve switching, the retained retinoids are transferred to the analytical column in the backflush mode, separated by gradient elution, and detected at 360 nm by UV detection.

Materials and Reagents

Tetrahydrofuran, 2-propanol, acetic acid, and ammonium acetate (all puriss. p.a.) were obtained from Fluka (Buchs, Switzerland), ethanol (HPLC grade) and sodium hydroxide (Titrisol) from Merck (Darmstadt, FRG), and acetonitrile (HPLC grade S) from Rathburn (Walkerburn, UK). Reference compounds and internal standards were provided by Hoffmann-La Roche (Basel, Switzerland) and were kept under argon at −20°.

Solutions and Standards

The preparation of plasma standards and the dilution or deproteination of the samples are performed under diffuse light conditions. Stock solutions of isotretinoin, tretinoin, and their 4-oxo metabolites are prepared in amberized volumetric flasks by dissolving 10 mg in 10 ml of ethanol. Stock solutions of etretinate, acitretin, and 13-*cis*-acitretin are prepared in the same manner by dissolving 10 mg in 2 ml of tetrahydrofuran and making up to 20 ml with ethanol. Appropriate amounts of the stock solutions are combined and diluted with ethanol. These solutions are used as plasma standards by diluting 0.2 ml with blank plasma to 20 ml, yielding concentrations of 2–1000 or 2–2000 ng/ml of plasma. The plasma standards are divided into portions of 1.2 ml and stored at −20°. For the determination of isotretinoin and metabolites, an internal standard solution is prepared containing 100 ng/ml of acitretin in 9 mM sodium hydrox-

[6] A. Vahlquist, G. Michaelsson, A. Kober, I. Sjöholm, G. Palmskog, and U. Pettersson, *in* "Retinoids: Advances in Basic Research and Therapy" (C. E. Orfanos, O. Braun-Falco, E. M. Farber, C. Grupper, M. K. Polano, and R. Schuppli, eds.), p. 109. Springer-Verlag, Berlin and New York, 1981.

ide–acetonitrile (8 : 2, v/v). For etretinate and metabolites the concentrations of the two internal standards used, isotretinoin and Ro 12-7554 [ethyl all-*trans*-9-(2,6-dichloro-4-methoxy-*m*-tolyl)-3,7-dimethyl-2,4,6,8-nonatetraenoate], are 100 ng/ml each in ethanol. These internal standard solutions are freshly prepared prior to use from stock solutions, which may be stored for several months at 4°.

Chromatographic System

A schematic representation of the column-switching system is given in Fig. 1. An HPLC pump 420 (P1; Kontron, Zurich, Switzerland) delivers mobile phase M1, which is used as the purge solvent at a flow rate of 1.5 ml/min. Aliquots (0.5 ml) are injected by a WISP 712 automatic sample injector with cooling module (I1; Waters, Milford, MA) onto the precolumn (PC). In order to inject sample volumes larger than 200 μl, the autoinjector is used with a 1-ml syringe, the auxiliary sample loop, and a syringe motor rate of 1.85 μl/sec. The UV detector D1 (Spectroflow 773; Kratos, Ramsey, NJ), operating at 240 nm, together with a W+W recorder 320 (R; Kontron; sensitivity 10 mV, chart speed 0.5 cm/min), is used during method validation to monitor the removal of plasma components from the precolumn during the purge step; they are not needed for

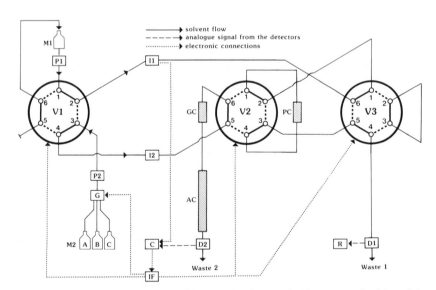

FIG. 1. Schematic representation of the HPLC column-switching system. Position of the valves: V1 = T5, V2 = T4, and V3 = T8 (see text for further details). (Reprinted from Ref. 5, by permission of Elsevier Science Publishers.)

routine analysis. Pump P2 with a low-pressure gradient system (G; Spectroflow 400 solvent delivery system and 430 gradient former, Kratos) delivers the gradient mobile phase M2 (flow rate 1 ml/min).

A manual injector (I2; Model 7125 with a 500-μl loop; Rheodyne, Cotati, CA), situated between pump 2 and valve 2, is used for direct injection onto the analytical column (e.g., for recovery experiments). Detection of the eluted compounds is carried out at 360 nm with a UV detector (D2; Spectroflow 783, Kratos; rise time 1 sec, range 0.01 AUFS), and integration is performed by means of a computing integrator (C; Model SP 4200; Spectra-Physics, San Jose, CA; sensitivity 8 mV, chart speed 0.5 cm/min). The gradient former (G) and the three air-actuated switching valves (V1–V3; Model 7000A, Rheodyne), the latter connected to three solenoid valves (Model 7163, Rheodyne), are controlled by the external time events of the computing integrator (C). To achieve compatibility, an interface (IF), produced in our electronic workshop, was placed between the integrator output and the solenoid valve input. The positions of the valves in Fig. 1 are as follows: V1 = T5 (alternative flow T6), V2 = T4 (T3), and V3 = T8 (T7).

Columns and Mobile Phases

A guard column between V1 and I1 (not shown in Fig. 1) and the precolumn PC (both 14 × 4.6 mm i.d.; Bischoff-Analysentechnik, Leonberg, FRG) are dry-packed with Bondapak C_{18} Corasil, 37–50 μm (Waters), and used with sieves (3 μm) without glass fiber filters to avoid column blocking. The analytical column (AC) for isotretinoin and its metabolites consists of two columns (125 × 4 mm i.d.; Hibar type, Merck), linked by a sleeve nut (Merck), and for etretinate and metabolites one column of the same type. A guard column (GC; 30 × 4 mm i.d.; Merck) is also linked to the analytical column using a sleeve nut. The AC and GC are packed with Spherisorb ODS 1, 5 μm (Phase Separations, Queensferry, UK), using a slurry technique.

The following mobile phases are used for isotretinoin and its metabolites: M1, 1% ammonium acetate–acetonitrile (9 : 1, v/v); M2, (A) 4 ml of 10% ammonium acetate, 400 ml of double-distilled water, 600 ml of acetonitrile, and 30 ml of acetic acid, (B) 4 ml of 10% ammonium acetate, 146 ml of double-distilled water, 850 ml of acetonitrile, and 10 ml of acetic acid. Mobile phases used for etretinate and metabolites are as follows: M1, 1% ammonium acetate and 1% acetic acid–acetonitrile (8 : 2, v/v); M2, (A) 4 ml of 10% ammonium acetate, 300 ml of double-distilled water, 700 ml of acetonitrile, and 3 ml of acetic acid, (B) 4 ml of 10% ammonium acetate, 146 ml of double-distilled water, 850 ml of acetonitrile, and 1 ml of acetic acid. All mobile phases are degassed with helium prior to use.

Analytical Procedure

Isotretinoin

Plasma (0.5 ml) is mixed with 0.75 ml of 9 mM sodium hydroxide–acetonitrile (8 : 2, v/v), containing the internal standard. After centrifugation (3 min at 1500 g), 0.5 ml is injected. The samples are kept at 10° in the autosampler before injection. The total sequence of automated sample analysis requires 37 min and includes the following five steps.

Step A (0–7 min, V1 = T5, V2 = T4, V3 = T8): Injection of the sample onto the PC. Proteins and polar components are washed out to waste 1. The GC and AC are equilibrated with M2 (100% A).

Step B (7–10 min, V1 = T5, V2 = T4, V3 = T7). The PC is purged in the backflush mode by M1.

Step C (10–29 min, V1 = T5, V2 = T3, V3 = T7). M1 passes directly to waste 1. The retained components are transferred from the PC to the GC/AC in the backflush mode by the gradient M2 from 100% A to 70% A (10–16 min), 70% A to 0% A (16–21 min), and 100% B (21–32 min).

Step D (29–32 min, V1 = T6, V2 = T3, V3 = T7; 32–32.9 min, V1 = T6, V2 = T4, V3 = T8). While M1 is running in a recycling mode, the capillaries between I1 and D1 are purged with M2 (100% B) to prevent any memory effects during the next injection. There is no flow through the GC and AC during this period.

Step E (32.9–37 min, V1 = T5, V2 = T4, V3 = T8). M2 is changed from 0% A to 100% A in 0.1 min, and the GC/AC and PC are reequilibrated with M2 and M1, respectively.

Etretinate

To 0.5 ml plasma, 1 ml of the ethanolic internal standard solution is added for protein precipitation. After vortex mixing and standing for 15 min in a refrigerator at 4°, the vial is centrifuged (3 min at 1800 g), the supernatant transferred to the autosampler vial, and 0.5 ml injected. The samples are kept at 20° in the autosampler before injection. The total sequence of automated sample analysis requires 34 min and is identical to that described above for isotretinoin, with the following exceptions:

Step C (10–26 min). Gradient M2 from 100% A to 0% A (10–18 min) and 100% B (18–29 min).

Step D (26–29 min, V1 = T6, V2 = T3, V3 = T7; 29–29.9 min, V1 = T6, V2 = T4, V3 = T8).
Step E (29.9–34 min).

Calibration and Calculations

Together with the unknown and quality control samples, seven plasma standards, distributed over the whole set of samples, are processed as described above. As isotretinoin, tretinoin, and their metabolites can be found as endogenous peaks in blank plasma samples (~0.5–4 ng/ml), the calibration is performed by standard addition of known amounts. Calibration curves ($y = a + bx$) for all four substances are obtained by weighted linear least-squares regression (weighting factor $1/y^2$) of the peak height ratios, using the internal standard acitretin, versus the concentration. The concentration intercepts ($x = -a/b$) are calculated and added to the weighed amounts. The calibration curves with corrected concentrations are used to interpolate unknown concentrations in the biological samples from the peak height ratios.

For etretinate and its metabolites, no endogenous substance or interference is found, and the intercepts are not significantly different from zero. Therefore, the same calibration procedure, but without intercept correction, is used with Ro 12-7554 as internal standard for etretinate, and isotretinoin for acitretin and 13-cis-acitretin.

Separation of Isotretinoin/Tretinoin or Etretinate and Their Metabolites

In principle, only one switching valve (V2 in Fig. 1) is needed for direct injection of plasma samples onto a precolumn and subsequent separation on an analytical column. A second valve (V1) is used for purging the steel capillaries between the autosampler and the detector 1. The installation of a third valve (V3) allows forward- and backflush purging of the precolumn, which delays clogging of the precolumn and prevents deterioration of the analytical column. These effects can occur by adsorption of proteins and solid particles on the sieves on the top of the precolumn and, after valve switching, by transfer to the analytical column. The three-valve technique allows an increase in injections onto the precolumn and proves to be very effective after the injection of dirty samples, such as tissue homogenates.

The addition of acetonitrile to the injection solution and to mobile phase 1 improves recoveries of the retinoids. However, an adequate recovery for etretinate could only be obtained by protein precipitation with ethanol.[5] Sodium hydroxide is used to stabilize isotretinoin, tretinoin, and their metabolites in the autosampler vials. Although the use of an internal standard in a column-switching technique is not absolutely necessary as far as precision is concerned, it may be helpful in reducing problems associated with autosampler malfunction, incomplete recovery, instability, etc. Two internal standards are used in one method because of the different chemical properties of etretinate, which is an ester, and acitretin, which is an acid.

Chromatograms of plasma standards and the corresponding blank plasma samples are shown in Fig. 2 for isotretinoin and metabolites and in

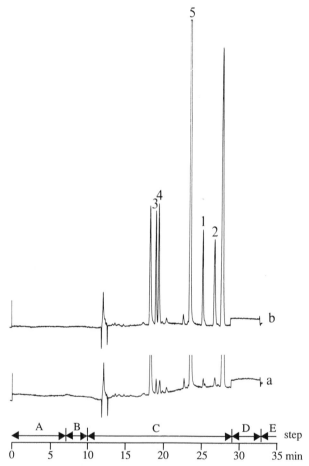

FIG. 2. Chromatograms of human plasma samples. (a) Blank plasma sample containing the internal standard 5 and endogenous levels of 1.9–2.8 ng/ml of 1–4; (b) blank plasma sample spiked with 20 ng/ml of 1–4 and 150 ng/ml of 5. Peak identification: 1, isotretinoin; 2, tretinoin; 3, 4-oxoisotretinoin; 4, 4-oxotretinoin; 5, acitretin (internal standard).

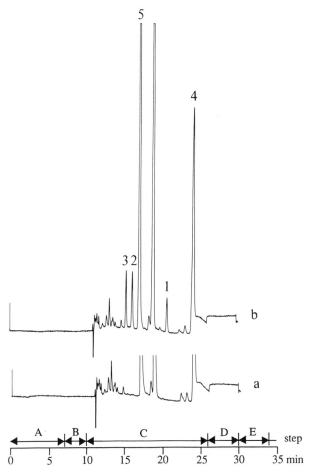

FIG. 3. Chromatograms of human plasma samples. (a) Blank plasma sample containing the internal standards 4 and 5; (b) blank plasma sample spiked with 20 ng/ml of 1–3 and 200 ng/ml of 4 and 5. Peak identification: 1, etretinate; 2, acitretin; 3, 13-*cis*-acitretin; 4, Ro 12-7554 (internal standard for 1); 5, isotretinoin (internal standard for 2 and 3).

Fig. 3 for etretinate and metabolites. Because of the big difference in polarity, a gradient was used for the simultaneous determination of the drugs and their metabolites. Spherisorb ODS 1 (5 μm) has proved to be the best packing material regarding isomer separation. For the difficult separation of 13-*cis*- from all-*trans*-4-oxoretinoic acid, two coupled analytical columns are used. As a precaution, the precolumn is replaced every day (after 40–50 injections), even when no deterioration is observed.

Characteristics of Methods

Limit of Quantification. The quantification limit is 2 ng/ml for all substances. Intraassay precision (*n* = 5–7) at this concentration is 1.8–9.5% for isotretinoin, tretinoin, and their 4-oxo metabolites and 6.6–12.4% for etretinate and its metabolites. Real detection limits, defined by a signal-to-noise ratio of approximately 3 : 1, are in the range 0.5–1 ng/ml.

Linearity. The methods are linear in the range 2–2000 and 2–1000 ng/ml for isotretinoin and etretinate, respectively.

Reproducibility. The precision and accuracy of the methods have been investigated in the range 20–1000 ng/ml in interassay studies (*n* = 6–9). The coefficients of variation are 2.3–3.2 and 1.4–2.7% for the isotretinoin and etretinate series, respectively. The inaccuracy is not more than 4.7%.

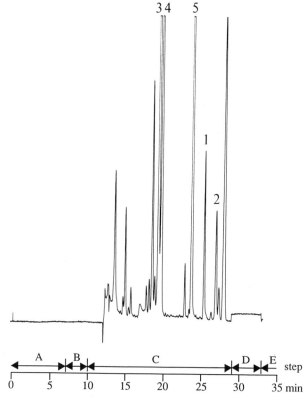

FIG. 4. Chromatogram of a patient plasma sample taken 24 hr after a daily dose of 1 mg/kg isotretinoin over 2 weeks. Peaks and concentrations: 1, isotretinoin 39.0 ng/ml; 2, tretinoin 28.2 ng/ml; 3, 4-oxoisotretinoin 253 ng/ml; 4, 4-oxotretinoin 154 ng/ml; 5, acitretin (internal standard).

Recovery. Recoveries are difficult to determine with the column-switching technique, because no off-line extraction is carried out. Under these circumstances, recovery is normally determined by injecting the analyte directly onto the analytical column using an aqueous solution. However, retinoids are not soluble in water, and, therefore, 50% methanol or ethanol has to be used. Under these conditions, recoveries of 88–98 and 63–80% are obtained for isotretinoin and metabolites and etretinate and metabolites, respectively. Systematic investigations to improve the recovery of very lipophilic retinoids such as etretinate have been described.[5]

Application of Methods

The methods described have been successfully applied to the analysis of more than 1000 human and animal plasma samples. Figure 4 shows a chromatogram of a plasma sample from a patient that demonstrates the need for a good separation, especially for the 4-oxo metabolites. The methods have also been used for the determination of other similar retinoids and could be adapted to other biological fluids and even tissue homogenates.

Acknowledgments

I thank Mr. F. Bucheli for assistance in the development of these methods, Mr. H. Suter for the drawings, and Drs. D. Dell and J. Burckhardt for correction of the manuscript.

[15] Simultaneous, High-Performance Liquid Chromatographic Analysis of Retinol, Tocopherols, Lycopene, and α- and β-Carotene in Serum and Plasma

By Lawrence A. Kaplan, Judith A. Miller, Evan A. Stein, and Meir J. Stampfer

Introduction

The analysis of plasma (or serum) for retinol, tocopherols, lycopene, and α- and β-carotene is a complex analytical problem. The laboratory performing these analyses must have a good working knowledge of liquid chromatography and pay strict attention to basic analytical techniques. The quality assurance group for micronutrients, sponsored by the National Cancer Institute (NCI) and the National Institutes for Standards and Technology (NIST; formerly the National Bureau of Standards, NBS), has shown that the laboratories which perform these analyses best seem to be those which pay sufficient attention to these concerns. Good analytical performance does not appear to depend absolutely on any specific type of technique. The material provided in this chapter describes one method in detail that has been in use for many years.[1,2]

Preparation of Working Standards

The single most critical element in this analysis is proper standardization, which begins with an accurate assignment of the concentration of each stock standard to be used. This assignment should not be based on gravimetric preparation of the standards, but must be carried out by spectrophotometric analysis.

Stock Standards

All solvents should be high-performance liquid chromatography (HPLC) or spectrograde quality. Preparation of all standards is performed in a darkened room with a red safety light.

Each standard is prepared to a target concentration based on the weight listed on the commercially available vials of the compounds. Lycopene (to 100 μg/ml), α-carotene (to 10 μg/ml), β-carotene (synthetic, all-*trans*; to 1 mg/ml), and retinyl palmitate (to 40 mg/ml) were purchased from Sigma Chemical Company (St. Louis, MO) and were prepared in spectrograde chloroform. Retinol (to 1 mg/ml), retinyl acetate (to 1 mg/ml), α- and γ-tocopherol (each to 0.5 mg/ml), and α-tocopheryl acetate were purchased from Eastman Kodak Company (Rochester, NY) and prepared in 100% ethanol.

The purity of the standards should be validated before they are used. At a minimum, each stock solution should be injected into the chromatographic system and the chromatogram monitored at the appropriate wavelength listed in Table I. Significant peaks (>2% of total) other than the one for the standard should indicate that the sample is not suitable for a standard. Since this validation does not check for impurities that are totally retained on the stationary phase of the column, one should also compare the spectrum of each lot of standard with the published spectrum of the pure compound.

[1] N. Katrangi, L. A. Kaplan, and E. A. Stein, *J. Lipid Res.* **25**, 400 (1984).
[2] L. A. Kaplan, J. A. Miller, and E. A. Stein, *J. Clin. Lab. Anal.* **1**, 147 (1987).

The actual concentration of each stock standard is determined by carefully diluting each stock solution with an appropriate diluent and measuring the absorbance in a dual-beam spectrophotometer (Perkin-Elmer, Norwalk, CT). The diluents, wavelengths, and molar absorptivities (or $E_{1\,cm}^{1\%}$) of the compounds are listed in Table I. These spectrophotometrically assigned values are used in the subsequent calculations of the standard curve described below.

Secondary Stock Solutions

Secondary stock solutions are prepared to target concentrations from stock solutions by diluting β-carotene to 20 μg/ml, retinol and retinyl acetate to 100 μg/ml, and retinyl palmitate to 400 μg/ml with 100% ethanol. The actual concentrations of these solutions are calculated from the spectrophotometrically assigned values of the stock solutions.

TABLE I

SPECTROPHOTOMETRIC CHARACTERISTICS OF RETINOIDS, TOCOPHEROLS, AND
CAROTENOIDS USED TO STANDARDIZE THE HPLC ASSAY[a]

Analyte	Solvent	Wavelength (nm)	Molar absorptivity (liter mol⁻¹ cm⁻¹)	$E_{1\,cm}^{1\%}$	Ref.[b]
Retinol	Ethanol	325	52,995	—	1
Retinyl acetate	2-Propanol	326	50,260	—	2
Retinyl palmitate	2-Propanol	326	50,390	—	2
α-Tocopherol	Ethanol	292	—	75.8	3
γ-Tocopherol	Ethanol	298	—	91.4	3
Cryptoxanthin	Petroleum ether	451	—	2160	2
Zeaxanthin	Petroleum ether	452	—	2350	2
Canthoxanthin	Petroleum ether	466	186,000	2200	2
Lycopene	Petroleum ether	472	—	—	2
α-Carotene	Petroleum ether	444	137,000	2800	4
β-Carotene	Petroleum ether	450	—	—	1

[a] From L. A. Kaplan, J. A. Miller, and E. A. Stein, *J. Clin. Lab. Anal.* **1**, 147 (1987).
[b] Key to references: (1) W. Vetter, G. Englert, N. Rigassi, and U. Schwieter, *in* "Carotenoids" (O. Isler, ed.), p. 190. Halsted, New York, 1971. (2) O. A. Roels and S. MacLadern, *in* "The Vitamins: Chemistry, Physiology, Pathology, Methods" (P. Gregory and W. N. Pearson, eds.), Vol. 6, 2nd Ed., p. 139. Academic Press, New York, 1967. (3) P. Schudel, H. Mayer, and O. Isler, *in* "The Vitamins: Chemistry, Physiology, Pathology, Methods" (W. H. Sebrell and R. S. Harris, eds.), Vol. 5, 2nd Ed., p. 168. Academic Press, New York, 1967. (4) U. Schwieter and O. Isler, *in* "The Vitamins: Chemistry, Physiology, Pathology, Methods" (W. H. Sebrell, Jr., and R. S. Harris, eds.), Vol. 1, 2nd Ed., p. 113. Academic Press, New York, 1967.

Working Combined Internal Standard I

For the preparation of a serum-base standard curve, the combined internal standard solution is prepared in 100% ethanol by diluting the secondary stock solutions of retinyl acetate and retinyl palmitate to 2 and 4 μg/ml, respectively, and the stock α-tocopheryl acetate to 100 μg/ml.

Working Combined Internal Standard II

For the routine analysis of plasma or serum samples, a combined solution of internal standards is prepared by diluting the secondary stock solutions of retinyl acetate and retinyl palmitate to 0.5 and 1 μg/ml, respectively, and the stock α-tocopheryl acetate to 25 μg/ml; all these compounds were diluted with 100% ethanol.

Working Combined Standard

Employing convenient, accurate volumes, the stock or secondary stock solutions of the analytes are combined and diluted to yield a working combined standard. The *approximate* target concentrations were as follows: lycopene, 2 μg/ml; α-carotene, 0.2 μg/ml; β-carotene, 0.8 μg/ml; retinol 2 μg/ml; and α- and γ-tocopherol, each to 40 μg/ml. The *actual* concentration of each analyte is calculated from the spectrophotometrically assigned value of the stock solutions as described above.

Preparation of Standard Curves

Since the tocopherols, lycopene, and carotene are transported in plasma in lipoprotein particles,[3,4] one major concern of the assay is to be able to extract these compounds from the plasma matrix. Since the lipoprotein content of samples will differ, it is possible that the efficiency of extraction of these compounds may also vary. To correct for this, our standard curve is based on internal standards that have extraction and chromatographic properties similar to those of the analyte and on standards that are added to a matrix of pooled serum or plasma.

Clear, nonicteric, nonhemolysed human serum or plasma is pooled and centrifuged to remove fibrin clots. One milliliter of the pool is added to each of five pairs of 16 × 100 mm culture tubes with Teflon-sealed screw caps. Five spiked serum standards are prepared by adding increasing volumes of the working combined standard, 250 μl of working internal

[3] W. A. Behrens, J. N. Thompson, and R. Malene, *Am. J. Clin. Nutr.* **35,** 691 (1982).
[4] B. A. Underwood, *in* "The Retinoids" (M. B. Sporn, A. B. Roberts, and D. S. Goodman, eds.), Vol. 2, p. 11. Academic Press, New York, 1984.

standard I, and sufficient volumes of 100% ethanol to make the final volume of each tube 2 ml (or 0.6 ml for extractions of smaller sample volume; see values in parentheses below). This is summarized in the following tabulation.

Additive (in order of addition)	Volume added (ml)				
Serum pool	1	1	1	1	1
	(0.3)	(0.3)	(0.3)	(0.3)	(0.3)
Working combined standard	0	0.1	0.2	0.4	0.6
		(0.025)	(0.050)	(0.10)	(0.20)
Working internal standard I	0.25	0.25	0.25	0.25	0.25
	(0.075)	(0.075)	(0.075)	(0.075)	(0.075)
Ethanol (100%)	0.75	0.65	0.55	0.35	0.15
	(0.225)	(0.20)	(0.175)	(0.125)	(0.025)

It is important to continually vortex mix the sample while slowly adding the ethanolic solutions. The actual concentrations of added analyte in each of the serum standards is, of course, calculated from the spectrophotometric-based estimates of the analyte concentrations assigned to the working combined standard. The standard curve can be extended by adding larger volumes of working combined standard or by adjusting the concentrations of the analytes in the working combined standard. A modified standard curve can be prepared in order to analyze smaller sample volumes. The italicized numbers in parentheses indicate the volumes employed for 0.3 ml of sample. Currently we employ 250 μl of sample, and others use as little as 100 μl of sample. The serum standards are then extracted and analyzed as described below.

Sample Extraction

All extractions are also performed at room temperature and under a red safety light. One milliliter of serum or plasma sample is placed in an aluminum foil-covered 16 × 100 mm culture tube with a Teflon-sealed screw cap. For patient samples, 1 ml of working combined internal standard II is slowly added to each tube while rapidly vortex mixing. For extractions using 0.3 ml of patient sample, 0.3 ml of internal standard solution is used. Three milliliters of HPLC-grade hexane or petroleum ether (hexane is preferred) is added to each tube, which is then capped and vortexed mixed for 2 min at room temperature. We usually mix four tubes at a time, two tubes per hand. This vortexing step is probably the most critical manipulation of the assay and is the greatest source of intra- and intertechnologist variation. The mixing must be uniform and sufficiently vigorous to disrupt the precipitated lipoprotein particles. This is especially true for the second extraction when the pelleted protein must be dispersed. The tubes are then centrifuged at 2500 g for 5 min at room temperature. As much of the upper phase as possible is then carefully transferred to an aluminum foil-covered 13 × 100 mm test tube. Two milliliters of extraction solvent is added to the culture tubes and the extraction repeated. The combined supernatants are evaporated at 37° under a stream of N_2 and the residue reconstituted in 200 μl of HPLC mobile phase (see below). The reconstituted samples are transferred to a conical minivial and placed in an amber-colored autosampler vial (Waters, Associates, Division of Millipore, Milford, MA). If many analyses are being performed in a single run, it is preferable to reconstitute extracted samples in staggered groups so as to minimize the time the samples are reconstituted before analysis.

HPLC Analysis

The HPLC system initially consisted of the following equipment purchased from Waters: an automated sample injector (WISP 710B) and M6000 pump (or equivalent) and the Models 480 and 450 variable-wavelength detectors. More recently we have also used the Waters 490 Multiwavelength Detector satisfactorily; the exact model of a spectrophotometer is not critical for this analysis. The spectrophotometer(s) is set to monitor the carotenoids at 460 nm (AUFS = 0.005) and the retinoids and tocopherols at 292 nm (AUFS = 0.05). When using a photodiode-array spectrophotometer, it is possible to monitor each analyte at its wavelength maximum. However, at the serum levels usually seen for these analytes, the increased sensitivity thus obtained is not an important factor in the analysis.

The chromatographic analysis is performed on a 5-μm particle C_{18} (octadecyl) Biophase ODS column (Bioanalytical Systems, Inc., West Lafayette, IN). A guard column of C_{18} material (Waters) is placed immediately before the analytical column and changed every 40–50 injections. The mobile phase, consisting of acetonitrile, chloroform, 2-propanol, and water (78 : 16 : 3.5 : 2.5, v/v) is pumped at a flow rate of 2 ml/min. The effluent from the column is recirculated into a mobile-phase reservoir, and fresh, degassed and filtered solvent is prepared weekly. Data from the NIST/NCI Micronutrient Quality Control Program suggests that good performance may not depend on the type of analytical column and mobile phase employed. Chromatograms are recorded on two HP 3390A (or equivalent) integrater-recorders (Hewlett Packard, Palo Alto, CA). The recorder chart speeds are set at 0.5 cm/min.

Extracts of serum or plasma are routinely analyzed by injecting 40 μl of extract into the chromatographic system. The chromatography is run until the retinyl palmitate peak is eluted ($\sim$28 min). There are essentially no other peaks detectable at 460 or 292 nm after this time. Figures 1 and 2 demonstrate the chromatograms usually observed for standards and serum samples.

Calculation of Results

Analyte concentrations in unknown samples are determined from a standard curve of the peak height (or the peak area) ratios of the analyte/internal standard plotted versus the concentration of the analyte added to the serum pool used to prepare the standard curve.[2] Retinyl acetate is the

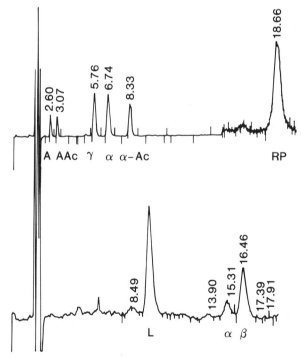

FIG. 1. Chromatography of standards and internal standards. (Top, 292 nm detector): retinol (A), retinyl acetate (AAc), γ-tocopherol (γ), α-tocopherol (α), α-tocopheryl acetate (α-Ac), and retinyl palmitate (RP) (recorder sensitivity changed at 15 min to detect RP); (bottom, 450 nm detector): lycopene (L), α-carotene (α), and β-carotene (β).

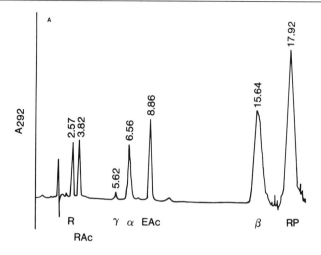

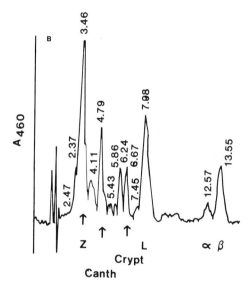

FIG. 2. Chromatography of a sample from a patient with elevated β-carotene levels. (A) Detection at 292 nm: retinol (R), retinyl acetate (RAc), γ-tocopherol (γ), α-tocopherol (α), α-tocopheryl acetate (EAc), β-carotene (β), and retinyl palmitate (RP). (B) Zeaxanthin region of chromatogram (Z), canthoxanthin region of chromatogram (Canth), cryptoxanthin (Crypt, 6.67 min), α-carotene (α), and β-carotene (β). Lycopene is the peak eluting at 7.98 min in this chromatogram.

internal standard for retinol, α-tocopheryl acetate, the internal standard for the tocopherols and lycopene, and retinyl palmitate, the internal standard for α- and β-carotene. We have found that using α-tocopheryl acetate as the internal standard for α- and β-carotene does not adversely effect the accuracy and precision of this assay, and our assay is currently being performed with α-tocopheryl acetate as the only internal standard.

The peak height ratios for each analyte to its internal standard are plotted against the final analyte concentration for the spike to pool standards. Regression analysis of the data is performed for each analyte, and the Y (vertical) axis intercept value is subtracted from each peak height value for each analyte to correct for the contribution of endogenous analyte in the pool. Regression analysis of the corrected peak height ratio versus the concentrations of the spiked analytes yields a regression formula for calculating the analyte concentration in patient samples:

$$\text{Concentration of analyte} = PHR/m$$

where PHR is the corrected peak height ratio of the analyte to its internal standard and m is the slope of the regression curve. If the regression coefficient for this analysis is below 0.99, the entire analysis is repeated. Although the regression calculation yields a small intercept value, for computational purposes, this is assumed to be zero. Unusually large intercept values are indicative of a poor standard curve. We have employed peak areas as well as peak heights for these calculations without finding a discernible effect in the results of our analysis.

Notes on the Analysis

1. One technologist can comfortably extract and set up for analysis 20 to 30 samples plus controls in a single working day. We have connected the HPLC system to a timer so that it shuts down after the analysis time has been completed. Given time for calculation of results, a technologist can reasonably complete 90–150 analyses in a week.

2. A standard curve is prepared after allowing a new HPLC column to equilibrate with mobile phase for several days. No further standardization of the column is performed, nor do we feel that one is needed if the column is performing properly. Column performance is monitored by close observation of the peak shapes and retention times of the analytes, column pressure, and the results of analysis of quality control pools that are analyzed with each analytical run. Significant changes from baseline values indicate the need to replace the column with a new one.

TABLE II
QUALITY ASSURANCE DATA FOR RETINOL, TOCOPHEROLS, AND CAROTENOID ANALYSIS[a]

Analyte	N	Mean (μg/liter)	S.D. (μg/liter)	%CV
Between-run comparison				
Retinol	81	524	42.6	8.2
γ-Tocopherol	81	1.44[b]	0.16[b]	11.1
α-Tocopherol	81	5.60[b]	0.74[b]	13.2
Lycopene	98	108	17	16
α-Carotene	84	21	2	9.5
β-Carotene	78	72	10	13.9
Within-run comparison				
Retinol	12	635	24	3.8
γ-Tocopherol	12	4.07[b]	0.20[b]	4.9
Lycopene	12	80	5.6	6.9
α-Carotene	11	16	2.4	15.0
β-Carotene	12	67	4.7	7.0

[a] From L. A. Kaplan, J. A. Miller, and E. A. Stein, *J. Clin. Lab. Anal.* **1**, 147 (1987). CV, Coefficient of variation.
[b] mg/liter concentrations.

3. Column life can vary with each lot of stationary phase. Nevertheless, in our experience we typically achieve from 200 to 450 sample injections per column.

4. The within-run and between-day imprecision of this method was estimated by repeated analysis of a commercially available quality control pool (aliquoted and frozen at $-70°$.[2] Table II presents data accumulated for 4 months. This level of imprecision is a reasonable estimate of what might be expected when this assay is performed almost daily throughout the year and employing two or three different technologists over this time period.

5. By participating in the NIST/NCI-sponsored Quality Assurance Program, we have been able to compare the performance of our assay with those of other methods. The blinded quality assurance specimens sent to participating laboratories ranged in concentration from 0.3 to 1.2 μg/ml for retinol, 4.9 to 12.6 μg/ml for α-tocopherol, and 0.08 to 1.6 μg/ml for β-carotene. For the last three trials (13 samples) the average percent bias of this method from the assigned values were as follows: retinol, $+2.8\%$; α-tocopherol, -6.7%; and β-carotene, -0.12%. (The NIST laboratory showed a negative bias from the assigned values for α-tocopherol.) The NIST considers a percentage bias of the assigned value of 0 to 5% to

TABLE III
PLASMA NUTRIENT MEANS[a] BY AGE AND SEX[b]

Number and nutrient	Age group				
	18–24	25–34	35–44	45+	All
Number					
Male	24	51	30	32	137
Female	40	75	37	41	193
Retinol (μg/liter)					
Male	520 (230–760)	600 (310–830)	660 (430–990)	710 (450–1030)	630 (380–930)
Female	480 (310–780)	530 (320–840)	490 (300–1000)	610 (360–980)	530 (320–800)
	$p < 0.02$	$p < 0.007$	$p < 0.003$	$p < 0.03$	$p < 0.001$
α-Tocopherol (mg/liter)					
Male	7.4 (3.9–11.4)	9.5 (6.2–15.8)	9.1 (4.2–19.3)	10.4 (5.2–19.5)	9.3 (5–16.2)
Female	9.0 (4.3–19.0)	9.1 (5.8–15.1)	9.0 (5.1–13.7)	10.9 (5.2–17.4)	9.4 (5–17.3)
γ-Tocopherol (mg/liter)					
Male	1.3 (0.35–3.0)	1.4 (0.46–2.6)	2.1 (1.2–4.3)	2.0 (0.56–4.8)	1.7 (0.59–3.5)
Female	1.3 (0.3–2.7)	1.4 (0.6–3.1)	1.7 (0.3–3.1)	1.7 (0.6–3.6)	1.6 (0.5–3.1)
α-Carotene (μg/liter)					
Male	29 (11–79)	56 (10–197)	45 (5–273)	37 (3–156)	45 (6.8–106)
Female	81 (15–308)	72 (11–212)	68 (11–177)	77 (10–164)	74 (12–213)
β-Carotene (μg/liter)					
Male	107 (34–235)	215 (31–768)	172 (43–546)	210 (31–946)	185 (36–590)
Female	273 (61–639)	320 (61–866)	260 (52–618)	411 (56–987)	318 (61–745)
	$p < 0.002$	$p > 0.1$	$p > 0.1$	$p < 0.001$	$p < 0.001$
Lycopene (μg/liter)					
Male	270 (150–650)	388 (150–730)	296 (127–677)	280 (81–550)	339 (126–652)
Female	413 (140–740)	339 (130–630)	378 (135–774)	318 (63–705)	357 (125–641)
	$p < 0.001$	$p < 0.05$	$p < 0.05$	$p < 0.005$	$p < 0.001$

[a] 5th–95th percentiles in parentheses; p values from t statistic comparing log-transformed means, men versus women.
[b] L. A. Kaplan, E. A. Stein, W. C. Willett, M. J. Stampfer, and W. S. Stryker, *Clin. Physiol. Biochem.* **5**, 297 (1987).

TABLE IV
PLASMA NUTRIENT LEVELS[a]

Sex and nutrient	Percent points for plasma nutrient levels				
	10	25	50	75	90
Sexes combined					
Retinol (μg/liter)	360	440	550	670	780
α-Tocopherol (mg/liter)	6	7.2	8.7	10.8	13.3
γ-Tocopherol (mg/liter)	0.7	1.1	1.5	2.0	2.7
α-Carotene (μg/liter)	13	24	41	78	132
β-Carotene (μg/liter)	63	110	190	332	546
Lycopene (μg/liter)	156	224	333	446	587
Males					
Retinol (μg/liter)	430	510	620	720	810
α-Tocopherol (mg/liter)	5.9	7.0	8.7	10.7	13.1
γ-Tocopherol (mg/liter)	0.8	1.1	1.5	2.0	2.7
α-Carotene (μg/liter)	11	16	33	48	83
β-Carotene (μg/liter)	52	78	141	196	361
Lycopene (μg/liter)	148	216	328	421	547
Females					
Retinol (μg/liter)	340	410	500	620	740
α-Tocopherol (mg/liter)	6.0	7.4	8.7	10.8	13.5
γ-Tocopherol (mg/liter)	0.7	1.1	1.4	2.0	2.6
α-Carotene (μg/liter)	18	32	57	97	145
β-Carotene (μg/liter)	85	147	254	407	602
Lycopene (μg/liter)	162	229	334	464	601

[a] From L. A. Kaplan, E. A. Stein, W. C. Willett, M. J. Stampfer, and W. C. Stryker, *Clin. Physiol. Biochem.* **5**, 297 (1987).

represent exceptional performance, >5 to 10% to be acceptable performance, and >10 to 20% to be marginally acceptable performance.

6. Because proper standardization of the method is the most critical factor for yielding good results, we have employed special pools from the NIST as a quality assurance check for a new standard curve. A standard curve is accepted only when the results from the analysis of these three pools fall within the assigned limits. If these results are not obtained, the standard curve is repeated. The HPLC procedure described in this chapter is one of five independent assays used to assign target values for these quality assurance pools which can now be purchased from the NIST.[5]

[5] National Institute of Standards and Technology. Standard Reference Material 968, Fat-Soluble Vitamins in Human Serum, 1989.

Results

The HPLC method described above was used to analyze plasma from 330 nonfasting individuals serving as generally healthy controls for a separate study.[6] Individuals with diabetes or cardiovascular disease were not excluded from this study, and none of the women were pregnant. The plasma from venous blood collected with EDTA was stored at $-70°$ until shipment (on dry ice) and analysis could be performed. Table III lists, by age, the mean and 95th percentile ranges for males and females for the analytes originally measured by this HPLC assay. Table IV lists the level of these analytes at various percentiles for males and females of all ages.

[6] L. A. Kaplan, E. A. Stein, W. C. Willett, M. J. Stampfer, and W. C. Stryker, *Clin. Physiol. Biochem.* **5**, 297 (1987).

[16] High-Performance Liquid Chromatography of Aromatic Retinoids and Isotretinoin in Biological Fluids

By H. Bun, N. R. Al-Mallah, C. Aubert, and J. P. Cano

Introduction

The thermal instability of retinoids excludes gas–liquid chromatographic procedures for their determination. Several analytical methods involving normal- and reversed-phase high-performance liquid chromatography (HPLC) for the quantification of isotretinoin and/or etretinate and some of their metabolites in biological fluids have been described.[1-4] These methods are time-consuming because they require a cumbersome extraction procedure followed by evaporation of the organic layer. They are also inadequate as they do not separate all the metabolites and do not determine with enough sensitivity the major metabolites in the low nanogram range. We describe here a rapid and specific method for analysis of aromatic retinoids or isotretinoin and their major metabolites in plasma by

[1] J. G. Besner, S. Meloche, R. Leclaire, P. Band, and S. Mailhot, *J. Chromatogr.* **231**, 467 (1982).
[2] U. Paravicini and A. Busslinger, *J. Chromatogr.* **276**, 359 (1983).
[3] C. J. L. Buggé, L. C. Rodriguez, and F. M. Vane, *J. Pharm. Biomed. Anal.* **3**, 269 (1985).
[4] P. Jakobsen, F. G. Larsen, and C. G. Larsen, *J. Chromatogr.* **415**, 413 (1987).

METHODS IN ENZYMOLOGY, VOL. 189

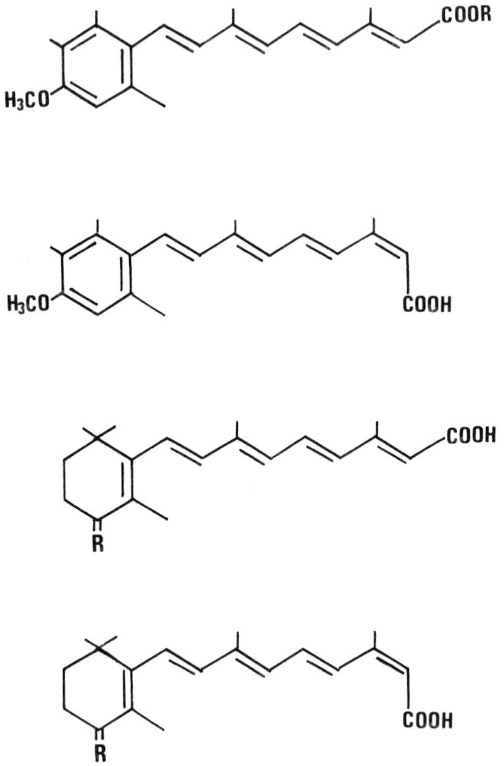

R	Compounds
CH_2CH_3	Etretinate Ro 10-9359
H	Acitretin Ro 10-1670
H	13-*cis*-Acitretin Ro 13-7652
H, H	Tretinoin Ro 1-5488
H, H	Isotretinoin Ro 4-3780
O	4-OXO-13-*cis*-RETINOIC ACID Ro 22-6595

FIG. 1. Chemical structures of etretinate, isotretinoin, and their main metabolites.

HPLC (Fig. 1).[5,6] This method was applied in a pharmacokinetic study and in monitoring patients receiving these compounds.

Experimental

Extraction Procedure. Volumes of 0.5–2 ml of plasma and 0.1 ml of phosphate buffer (pH 7) are added to a suitable volume of the internal standard solution (tretinoin for aromatic retinoids, 13-*cis*-acitretin for isotretinoin) that has been evaporated to near dryness in tapered 10-ml yellow-amber tubes under a stream of nitrogen. The mixture is extracted for

[5] N. R. Al-Mallah, H. Bun, P. Coassolo, C. Aubert, and J. P. Cano, *J. Chromatogr.* **421,** 177 (1987).

[6] N. R. Al-Mallah, H. Bun, and A. Durand, *Analytical Lett.* **21,** 1603 (1988).

5 min with 2 ml of a diethyl ether–ethyl acetate (50 : 50, v/v). The tubes are gently shaken on a vortex mixer to avoid emulsion. After centrifugation for 10 min at 2000 g, the organic phase is transferred to the same type of yellow-amber tubes, then evaporated to dryness under a stream of nitrogen. The residue is dissolved in 30–100 μl of methanol and transferred to an injection vial for HPLC determination. A 25-μl sample of each extract is injected onto the column.

HPLC. HPLC analyses are performed with a variable-wavelength UV detector (350 nm), Nucleosil C_{18} columns (25 cm × 4.6 mm i.d.) with 5-μm particles at ambient temperature. The mobile phase is methanol–1% aqueous acetic acid (85 : 15, v/v) at a flow rate of 1.5 ml/min. Retention times of these compounds are shown in Fig. 2.

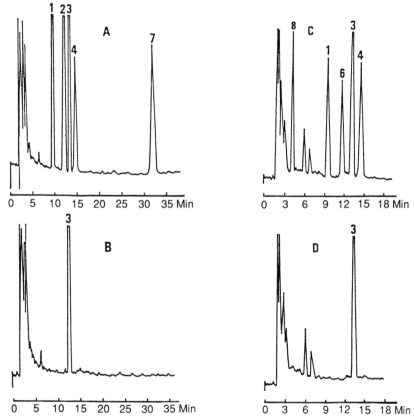

FIG. 2. Chromatograms obtained after extraction of (A) control plasma spiked with 13-*cis*-acitretin (Peak 1) acitretin (2), tretinoin (4), and etretinate (7); (B) control plasma; (C) control plasma spiked with 4-oxo-13-*cis*-retinoic acid (8), 13-*cis*-acitretin (1), isotretinoin (6), and tretinoin (4); and (D) control plasma.

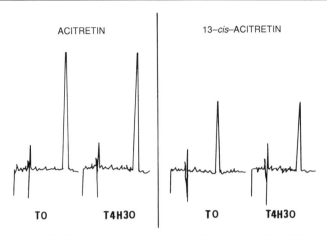

ACITRETIN   13–*cis*–ACITRETIN

T0   T4H30   T0   T4H30

FIG. 3. Chromatograms obtained working under yellow light.

Results

Precision and Accuracy. Intra- and interday precision (relative standard deviation) and accuracy (difference between the determined and expected concentrations) were calculated for 5 to 420 ng/ml plasma concentrations of these compounds. Results were acceptable within these concentration ranges; reproducibility and accuracy were under 11%.

Limit of Detection. The limit of detection (signal-to-noise ratio of about 3 or 4) was approximately 2 ng/ml. Near this detection limit the intra- and interassay reproducibility and accuracy were better than 25%.

Linearity. Linearity of regression graphs was satisfactory for plasma concentrations of about 2 to 1200 ng/ml of these compounds. The correlation coefficient ranged from 0.9878 to 0.9999, and the intercepts of the calibration graphs did not differ significantly from zero.

Stability. The stability of the working solutions of these compounds was tested under natural light, normal artificial light, yellow light, and complete darkness, for periods of up to 5 hr, at 20°. These compounds were stable under yellow light and in darkness, but they were very unstable under other light conditions (Figs. 3 and 4). The spiked plasma samples were stable during storage (from 0 to 24 hr at ambient temperature and from 0 to 90 days at −20°).

Extraction Efficiency. Recovery of these compounds from human plasma was in the range of 80–99%, regardless of the type of compound and its concentration.

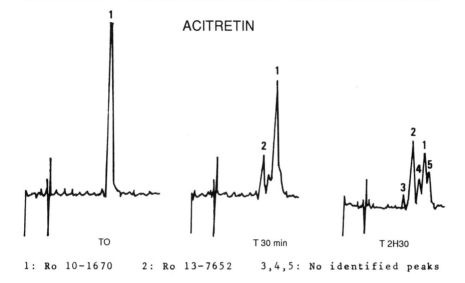

ACITRETIN

TO   T 30 min   T 2H30

1: Ro 10-1670   2: Ro 13-7652   3,4,5: No identified peaks

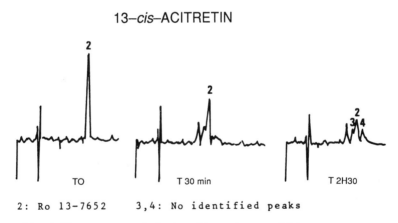

13–*cis*–ACITRETIN

TO   T 30 min   T 2H30

2: Ro 13-7652   3,4: No identified peaks

FIG. 4. Chromatograms obtained working under natural light.

Discussion and Conclusion

Administration of retinoids to patients is frequently associated with other drugs such as antidepressants, benzodiazepines, or psoralen. Some of the drugs in these classes were tested for possible interference. No interference was observed for any of these drugs. For endogenous plasma

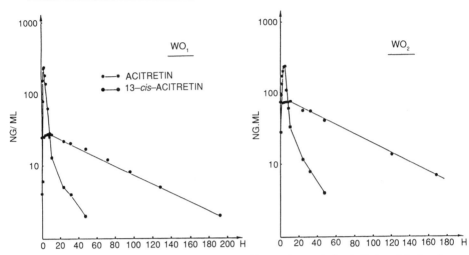

FIG. 5. Plasma concentrations of acitretin and its 13-cis metabolite in Patient 2 during the first washout (WO$_1$) after 42 days with daily oral doses of 30 mg acitretin and during the second washout (WO$_2$) after 290 days with daily oral doses of 30 mg acitretin.

components, the specificity was satisfactory. Acitretin and isotretinoin have the same retention time, about 10 min. Analytical interference does not pose a problem because isotretinoin was not administered concomitantly with etretinate or acitretin. The HPLC method proposed was used successfully for determination of these compounds in skin,[7] and it was applied to pharmacokinetic and drug monitoring studies in patients receiving single- or multiple-dose administrations of these compounds (Fig. 5).

[7] P. Laugier, P. Berbis, C. Brindley, H. Bun et al., Skin Pharmacol. **2**, 181 (1989).

[17] Determination of Retinol, α-Tocopherol, and β-Carotene in Serum by Liquid Chromatography

By WILLIAM A. MacCREHAN

Introduction

Serum or plasma are the most convenient samples for assessing subject nutritional status for retinol (vitamin A), α-tocopherol (vitamin E), and β-carotene (provitamin A). Normal serum values range from 200 to

800 ng/ml for retinol, 5 to 25 μg/ml for α-tocopherol, and 10 to 2000 ng/ml for β-carotene. Because of the complexity of the serum matrix and the sensitivity required for these analytes, liquid chromatography (LC) has become the method of choice for their determination,[1] completely displacing the formerly used, and less reliable, colorimetric methods.

Recovering Lipophilic Vitamins. β-Carotene and α-tocopherol are compartmentalized in the lipoprotein fraction of serum, whereas a specific protein, retinol-binding protein, has been found to bind retinol. Although vitamin determination by the direct injection of whole serum might seem to be an attractive approach, the precipitation of serum protein upon injection into hydroorganic mobile phases produces LC column pressure problems. Additionally, these vitamins are poorly recovered because of their protein encapsulation. Therefore, for the determination of these lipophilic analytes, a solvent extraction is always performed prior to analysis. The proteins are precipitated optimally with ethanol and extracted with hexane. We have found that, because of some loss of hexane to the ethanol/precipitate layer, two sequential hexane extractions provide the highest and most reproducible recoveries. Whereas a single serum extraction provides a recovery of the internal standard of 81 ± 15% (1 S.D.), two extractions provide a recovery of 97 ± 3.0%.[2]

Assay Procedure

Vitamin Extraction. For the extraction,[2] a serum volume of 250 μl is placed in a 13-ml centrifuge tube and briefly vortex mixed with 25 μl of an ethanol solution of the internal standard, tocol (see below) with a concentration of approximately 80 μg/ml. The proteins are then precipitated by addition of 250 μl of ethanol containing 20 μg/ml of butylated hydroxytoluene (BHT), an antioxidant. The sample is then extracted with 1.5 ml of hexane by vortex mixing for 1 min. After 1.0 ml of the upper layer is removed, a second 500 μl of hexane is added, and the sample is vortex mixed for 1 min. A second 300 μl of upper layer is combined with the first extract in a sample vial with a two-holed polyethylene stopper. Inert gas is then gently blown over the sample to remove the hexane. The sample is reconstituted with 250 μl of the ethanol/BHT using ultrasonic agitation. Occasionally, this final extract is cloudy, and centrifugation is required. For LC analysis, the extract is placed in a glass sample vial, and 25 μl is injected into the chromatograph.

---

[1] G. L. Catignani, J. G. Bieri, W. J. Driskell, *et al., Clin. Chem.* 29, 708 (1983).
[2] W. A. MacCrehan and E. Schönberger, *Clin. Chem.* 33, 1585 (1987).

Storage. All three vitamins are subject to easy oxidation by atmospheric oxygen. Consequently, the choice of storage conditions for the sera and extracts are crucial to ensure the accuracy of the determination. No losses of the three analytes are observed for short-term (hours to days) storage on ice, under subdued light. Control sera stored in glass ampoules at $-80°$ have shown excellent stability for 3 years. Ethanol/ BHT solutions of standards and serum extracts show small losses of all-*trans*-retinol and all-*trans*-β-carotene via oxidation and isomerization when stored at room temperature for periods greater than 6 hr, or when allowed to sit for several minutes unprotected from fluorescent room light.

Standardization. Although α-tocopherol can be obtained in high purity from commercial sources, all-*trans*-retinol and all-*trans*-β-carotene compounds typically show relatively low and variable purity. In addition to a small percentage of geometric isomers, these materials also contain 10–50% oxidized impurities.

Calibration of the chromatographic instrumentation for determination of these vitamins consists of spectrophotometric measurement of the absorbance of ethanolic calibration solutions using the following absorptivity values (dl/g · cm): all-*trans*-retinol, $\varepsilon_{324\,nm} = 1850$; α-tocopherol, $\varepsilon_{292\,nm} = 75.8$; and all-*trans*-$\beta$-carotene, $\varepsilon_{452\,nm} = 2620$. The concentrations are then corrected for the impurities that contribute to the absorbance by chromatographing the standard solutions using absorbance detection at the same wavelength used for the absorptivity measurement. The concentration of vitamin in the standard is given by Eq. (1):

$$\begin{array}{c} \text{Concentration} \\ \text{of standard} \\ \text{solution} \end{array} = \frac{\text{absorbance}}{\begin{array}{c}\text{path length} \times \text{absorptivity} \\ \text{(from spectrophotometer)}\end{array}} \times \frac{\text{area vitamin peak}}{\begin{array}{c}\text{total area under all peaks} \\ \text{(from chromatograph)}\end{array}}$$

(1)

In order to correct for recoveries during the serum extraction, an internal standard should be added to the serum prior to commencing the extraction process. Tocol, a compound similar to α-tocopherol, but lacking the three methyl groups on the phenolic ring, is commercially available in high purity and elutes in the middle of the multivitamin separation.[2] Although some workers have advocated the use of individual internal standards for each vitamin (i.e., retinyl acetate for A, α-tocopheryl acetate for E, and echinenone for carotene), as well as the use of standard solutions added to serum for calibration, there is little evidence that these standards become incorporated into the lipoproteins in a manner similar to the natural vitamins. The recovery of the "spike" primarily reflects the volume recovery of the extractant phase. Therefore, the measurement of a single internal standard (tocol) suffices for correction of the recovery of the extracted phase in each sample.

After establishing the linearity of a multipoint calibration curve, a single ethanol calibrant solution of the three vitamins may be used. Because the vitamin concentrations are limited to a comparatively narrow range, this single-point calibration approach may be used without significant loss in accuracy. Imprecision in the determination is primarily a result of the irreproducibility of the extraction. A convenient working ethanolic calibrant contains approximately 500 ng/ml for retinol, 8 μg/ml for both tocol and α-tocopherol, 200 ng/ml for β-carotene, and 20 μg/ml of BHT. Such standards are stable in sealed glass ampoules stored at −4° for several months; however, care must be taken to warm the solutions to room temperature and agitate to ensure redissolution of the β-carotene before use.

When possible, new analytical procedures should be validated through the use of a matrix-matched control material. A standard reference material (SRM 968, fat-soluble vitamins in human serum) has been prepared by NIST (the National Institute of Standards and Technology), with values certified for retinol, α-tocopherol, and β-carotene. This reference material should be used in the development and quality assurance of analytical methods for these vitamins in serum.

Reversed-Phase Separation of Retinol and β-Carotene Isomers

Approaches. Because of the differences in physiological activity of the geometric isomers of retinol, tocopherol, and carotene, it is imperative that they be separated in samples of clinical importance. It is generally accepted that the most powerful approach to isomer separation of these species has been normal-phase LC, because the liquid–solid adsorption process possesses selectivity for geometric isomers.[3] However, the reproducibility of normal-phase separations is generally poor because of the sensitivity of analyte retention to small amounts of strongly adsorbed water in the samples and mobile phase. Thus, reversed-phase separations are generally favored for analysis of biological samples. Although octadecylsilyl-modified silica (C_{18}) stationary phases are most frequently used for separation of these analytes, not all C_{18} columns show the same selectivity for resolution of the geometric isomers. Both all-*trans*-retinol and all-*trans*-β-carotene are rigid linear molecules with modeled length by width dimensions of 18.1 × 7.5 and 32.9 × 8.9 Å, respectively.[4] Most

[3] B. Stancher and F. Zonta, *J. Chromatogr.* **287,** 353 (1984).
[4] N. Craft, "Chromatographic Methods for the Separation of β-Carotene Isomers." NCI–NIST Micronutrient Analysis Workshop, Gaithersburg, Maryland, November 3, 1988.

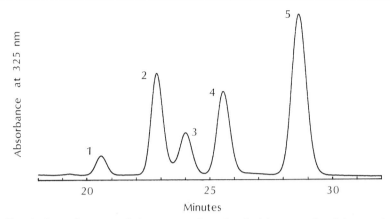

FIG. 1. Isocratic reversed-phase separation of retinol isomers. Conditions: column, Vydac 201TP54 5 μm; solvent, methanol/n-butanol/water (65 : 10 : 25) containing 10 mM ammonium acetate, pH 3.2; flow rate, 1.0 ml/min. Retinol isomer peak identities: 1, di-*cis*-; 2, 11-*cis*-; 3, 9-*cis*-; 4, 13-*cis*-; 5, all-*trans*-retinol. (Reproduced from MacCrehan and Schönberger.[5])

commercial C_{18} silica columns have a pore diameter of only 60 to 100 Å, which is further occluded by the 5 to 15 Å thickness of the bonded alkyl silane phase. Thus, the analyte molecules, particularly β-carotene,[4] may be partially excluded from interacting with the C_{18} groups inside the pores of most commercial columns. Wide-pore (300 Å) C_{18} phases do not exclude these rigid analytes from the pores and thus show much better selectivity for the discrimination of the retinol/carotene geometric isomers.[5,6]

Retinol Isomer Separation. Figure 1 shows the isocratic reversed-phase separation of the 11-, 9-, and 13-cis isomers of retinol from all-*trans*-retinol in less than 30 min.[5] Addition of n-butanol to the mobile phase was crucial in achieving this resolution, presumably because this long-chain alkyl alcohol wets the C_{18} phase, changing the interfacial structure. Even in separations providing a retinol retention time of 5 min, the all-*trans*-retinol may still be baseline separated from the three commonly occurring cis isomers, which coelute and appear as a single peak.[2]

β-Carotene Isomer Separation. Although all-*trans*-β-carotene is relatively easily separated from its all-trans α isomer[7] using narrow pore (60 to 100 Å) columns derivatized with monomeric C_{18} phases, these columns

[5] W. A. MacCrehan and E. Schönberger, *J. Chromatogr. Biomed. Appl.* **417,** 65 (1987).
[6] F. W. Quackenbush, *J. Liq. Chromatogr.* **10,** 643 (1987).
[7] J. G. Bieri, E. D. Brown, and J. C. Smith, *J. Liq. Chromatogr.* **8,** 473 (1985).

do not provide complete resolution of the cis/trans isomers. On the other hand, a wide (300 Å) pore column derivatized with a polymeric C_{18} phase can resolve all-trans from its commonly occurring cis isomers, 9- and 13-cis,[4,6] which elute following the all-trans isomer. Evidently, the wide pores of this column provide more geometric opportunity for the rigid β-carotene isomers to interact with the stationary phase. Additionally, the high carbon load of the polymeric C_{18} phase provides the increased retention and ligand phase thickness[4] required to resolve geometric isomers. The cis isomers, with their bent shape, may interact more strongly with the large curved pores of the Vydac column than does the straight all-trans. This might account for the longer retention of the cis isomers in the separation.

On-Column β-Carotene Losses. One difficulty encountered in β-carotene separations are losses on the analytical column. Losses are dependent on the composition of the frits used in the ends of the column and the mobile phase composition at the time of injection. Using the most common LC frit material, type 316 stainless steel, losses in peak area of 50% for β-carotene are observed with a mobile phase containing 15% water. Substituting porous Teflon,[2] titanium, or Hastelloy C,[8] which are more inert, for the stainless steel frit of the column greatly decreases the problem. The β-carotene is lost either by quasi-irreversible precipitation on the stainless steel frit[2] or by rapid decomposition through oxidation in the presence of the frit material.[8] Analytical columns with inert frits can be purchased from commercial manufacturers through special orders. Despite some nonlinearity in the β-carotene calibration curve in the low concentration range, resulting from the on-column loss with stainless steel frits, fairly reproducible peak areas are observed for a given column and mobile phase. Recalibration is required when either are changed.

Gradient Multivitamin Separation. Isocratic reversed-phase separations have been published for the simultaneous determination of retinol, α-tocopherol, and β-carotene.[9,10] However, because of the relatively short retention characteristics of retinol compared to β-carotene, it is very difficult to find elution conditions that provide both a reasonable elution time for β-carotene and baseline resolution of all-*trans*-retinol and α-tocopherol from coeluting serum components and their geometric isomers.[2] Because of this difficulty, some laboratories use two different isocratic separations, one for determining retinol and α-tocopherol and a second with a stronger mobile phase for the determination of β-caro-

[8] D. W. Nierenberg and D. C. Lester, *J. Nutr. Growth Cancer* **3**, 215 (1986).
[9] D. B. Milne and J. Botnen, *Clin. Chem.* **32**, 874 (1986).
[10] K. W. Miller and C. S. Yang, *Anal. Biochem.* **145**, 21 (1985).

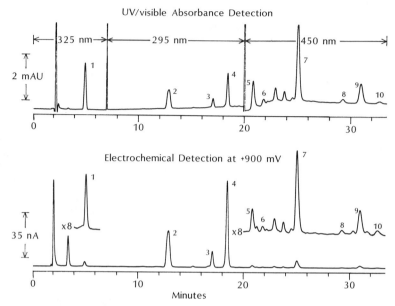

FIG. 2. Gradient-elution reversed-phase separation of vitamins in serum. Peak identities: 1, all-*trans*-retinol; 2, tocol; 3, γ-tocopherol; 4, α-tocopherol; 5, lutein; 6, zeaxanthin; 7, cryptoxanthin; 8, all-*trans*-α-carotene; 9, all-*trans*-β-carotene; 10, 9-*cis*-β-carotene. (Reproduced from MacCrehan and Schönberger.[2])

tene.[11] We have chosen to use a single multivitamin separation with gradient solvent elution on a wide-pore, polymerically bonded C_{18} column.[2] The separation shown in Fig. 2 allows simultaneous determination of all three analytes in a single chromatographic run. A Vydac 201TP54 (5 μm, 0.46 × 25 cm) column is used. Solvents A (15 : 75 : 10) and B (2 : 88 : 10), water/methanol/*n*-butanol, contain 50 mmol/liter ammonium acetate buffer, pH 5.5. The gradient employed is solvent A for 3 min, linear to B over 15 min, hold at B for 17 min, and then return. Under these conditions the following compounds are baseline separated: all-*trans*-retinol from all of its cis isomers, α- from γ-tocopherol (however, the optical isomers of α-tocopherol are not separated), and all-*trans*-β-carotene from its cis and all-*trans* α isomers. Retinyl palmitate may also be determined by monitoring the absorbance at 325 nm at a time of 31 to 33 min. The analyte compounds are also well separated from other serum sample constituents such as phthalate plasticizers, xanthophylls, and other carotenoids.

[11] N. E. Craft, E. D. Brown, and J. C. Smith, *Clin. Chem.* **34,** 44 (1988).

Liquid Chromatography Detection of Vitamins

Ultraviolet/Visible Absorbance Detection. Because of the comparatively high absorptivities of retinol and β-carotene at favorable wavelengths, UV/visible absorbance detection is an attractive method for the LC measurement of these analytes. Although the absorptivity for α-tocopherol is much smaller, normal serum concentrations are such that it, too, may be well detected. In recent years, relatively inexpensive, wavelength-programmable LC absorbance detectors have become available. These new detectors provide two crucial advantages for vitamin measurements: (1) very small absorbances can be measured, with absorbance noise levels as low as 1×10^{-5} absorbance units, providing high sensitivity; and (2) the wavelength can be optimized for each analyte during the LC separation, greatly increasing both sensitivity and selectivity. This is a significant improvement over detectors found in the first generation of LC equipment, with fixed wavelengths via filters and/or phosphors to achieve compromise wavelengths such as those based on a mercury light source and phosphor at 280 nm. The baseline absorbance noise in this old design is typically a factor of 50 greater than that in the better modern detectors with deuterium sources. The noise contributes to difficulty in the measurement of the peak height or area (especially by microprocessor-based integrators), contributes uncertainty to the determination, and requires the use of a larger sample injection onto the LC column.

Some methods for the simultaneous determination of retinol and tocopherol suggest detection at 280 nm.[1] In addition to a poor signal for retinol (λ_{max} of 325 nm) at 280 nm, interference from phthalate plasticizers, used in containers for serum collection and handling, has been observed. If a compromise wavelength must be chosen for vitamin A and E determination with a nonprogrammable variable-wavelength detector, then 300 nm is recommended, providing adequate signal and better immunity to interferences from low-wavelength UV absorbers.

Diode-array detectors can provide complete absorbance measurements over an entire spectral range during the LC separation, a powerful approach to evaluate the presence of coeluting serum components. Although useful in the development of a new analytical method, these detectors have an absorbance noise level 5 to 100 times greater than that of the best fixed-wavelength detectors, and thus they may not be optimal for routine determinations at fixed wavelength. Wavelength-programmed fluorescence detection may also be used for the simultaneous detection[12] of retinol using 325 nm for λ_{ex} and 390 nm for λ_{em} and α-tocopherol using

[12] A. T. R. Williams, *J. Chromatogr.* **341**, 198 (1985).

TABLE I
DETECTION LIMITS FOR ANALYTES

	Concentration (μg/liter)	
Analyte	Absorbance	Electrochemical
all-*trans*-Retinol	6.0 (150)	4.1 (103)
α-Tocopherol	96 (2400)	0.65 (16)
all-*trans*-β-Carotene	29 (730)	2.1 (53)

a Numbers in parentheses indicate pg of analyte in 25-μl injection volume.

295 nm for λ_{ex} and 480 nm for λ_{em}; however, β-carotene is nonfluorescent and may not be determined with this detector.

The top chromatogram in Fig. 2 shows wavelength-programmed absorbance detection of retinol, α-tocopherol, and β-carotene. Note that the analytes and their major isomers are baseline resolved from other serum components using the selectivity of the wavelength-programmed absorbance detection. The detection limits given in Table I are excellent under these conditions, allowing facile determination of even low serum levels of the analytes.

Amperometric Electrochemical Detection. Amperometric detection can also be used for the serum determination of these three vitamins,[2] as shown in Fig. 2. All three analytes and their isomers can be detected by their oxidation at a glassy carbon electrode using the following optimal plateau potentials: +1050 mV for retinol, +900 mV for tocol, +750 mV for α-tocopherol, and +700 mV for β-carotene, when measured in 50 mmol/liter, pH 5.5, buffer versus an Ag/AgCl, 3 mol/liter KCl reference electrode. For serum determinations, a compromise potential of +900 mV should be employed. The selectivity and sensitivity for retinol and β-carotene are not very much different for this detector than the wavelength-programmed absorbance detector, but amperometric detection provides somewhat better selectivity and much better detection limits for α-tocopherol (see Table I). It should be noted that although the electrochemical detector requires the addition of a dilute supporting electrolyte (i.e., 50 mmol/liter buffer), and more frequent calibration (at least twice daily) owing to drift in the analyte response factors, it produces analytical results that are not significantly different than the absorbance detector.[2]

Equipment Note. Certain commercial equipment, instruments, or materials are identified in this report to specify adequately the experimental procedure. Such identification does not imply recommendation or endorsement by the National Institute of Standards and Technology, nor does it imply that the materials or equipment identified are necessarily the best available for the purpose.

[18] Separation of Retinyl Esters and Their Geometric Isomers by Isocratic Adsorption High-Performance Liquid Chromatography

By Hans K. Biesalski

Introduction

Vitamin A is stored as its fatty acid ester primarily in the liver. The large reserves in the liver guarantee that supplies to all tissues dependent on vitamin A are constant, even when dietary intake is low. Retinyl esters are also located in various concentrations and esterified with various fatty acids in the pigment epithelium of the eye[1,2] and the mucous membranes of the intestines.[3,4] The significance of these peripheral retinyl ester stores is still unclear. In the pigment epithelium, they appear to protect the tissue for a restricted period against a lack of supplies. Accordingly, a close relationship exists between the amount of vitamin A required by a specific tissue and the concentration of the retinyl esters located there. Thus, determining retinyl esters in various tissues can provide information about the dependency of such tissues on an adequate supply of vitamin A. Such an analysis can also be performed on bioptic material, for example, taken from neoplasms, and may help to assess high-dosage courses of therapy with natural vitamin A[5] that are often prescribed. It has therefore been necessary to develop a method for quantitative and qualitative analysis of different retinyl esters in low concentrations from a wide variety of tissue samples.

So far, standard techniques[6–9] have only partially fulfilled the require-

[1] N. I. Krinsky, *J. Biol. Chem.* **23**, 881 (1958).
[2] J. S. Andrews and S. Futterman, *J. Biol. Chem.* **239**, 4073 (1964).
[3] J. Ganguly, *Vitam. Horm.* (*N.Y.*) **18**, 387 (1960).
[4] H. S. Huang and D. S. Goodman, *J. Biol. Chem.* **240**, 2192 (1965).
[5] G. E. Goodman, D. S. Alberts, and D. L. Ernest, *J. Clin. Oncol.* **1**, 394 (1983).
[6] C. D. B. Bridges, *Vision Res.* **15**, 1311 (1976).
[7] R. Alvarez, C. D. B. Bridges, and S. Fong, *Invest. Ophthalmol.* **20**, 304 (1981).
[8] A. C. Ross, this series, Vol. 123, p. 68.
[9] H. Steuerle, *J. Chromatogr.* **206**, 319 (1986).

ments imposed on trace-analytical high-performance liquid chromatography (HPLC) techniques,[10] namely, extraction without major concentrating and derivatization steps, transfer to isocratic separation without the need for further concentrating stages, and separation with short retention times for obtaining sharp peaks. In particular, transferring the extract without the need for concentrating techniques (i.e., from a nonpolar extraction phase to a corresponding mobile phase) has proved to be a difficult problem. A recently developed method, however, has made it possible for the first time to qualitatively and quantitatively determine different retinyl esters in the smallest of tissue samples, such as tracheal mucous membranes, inner ear, testicles, and other tissues, by isocratic separation on a pure silica column.[11,12]

Equipment

The HPLC system employed consists of a Perkin-Elmer (Norwalk, CT) pump (Series 1), a UV detector (LC-85, set at 325 nm), and a fluorescence detector (650-10S, set at 342 nm excitation and 475 nm emission, slit 10/10 nm), in addition to a Rheodyne (Cotati, CA) injector (Model 7125) and a Perkin Elmer recorder (561) connected to a data evaluation system (Contron Data P 450). The separation is performed on a 3-μm Spherisorb silica column (25 × 4.6 mm) connected to a short precolumn (25 × 4.6 mm) packed with the same material (Bischoff, Leonberg, FRG). The same manufacturer supplies other columns with 3-μm material (Shandon 005 Hypersil, Nucleosil 100 silica) that are just as suitable for use with short precolumns (packed with appropriate material).

Reagents. In line with the requirements placed on a trace-analytical method, extraction and separation are performed on comparable solvents. *n*-Hexane (analytical grade) is used for extraction, and a 98.5 : 1.5 (v/v) mixture of *n*-hexane and diisopropyl ether is used as the mobile phase for isocratic separation. Eluents of spectroscopic quality are recommended for very sensitive measurements (e.g., *n*-hexane Uvasol, Merck, Darmstadt, FRG). They increase the sensitivity by reducing noise. At the same time, these solvents contain less water, a fact which is of major consequence for separation on adsorption columns. The water content is reduced by first passing the mobile phase through a glass column (500 × 50 mm) filled with Al_2O_3, which adsorbs water and perox-

[10] L. R. Snyder and J. J. Kirkland, "Introduction to Modern Liquid Chromatography," 2nd Ed. Wiley, New York, 1979.
[11] H. K. Biesalski, *Int. J. Vitam. Nutr. Res.* **54**, 113 (1984).
[12] H. K. Biesalski and H. Weiser, *J. Clin. Chem. Clin. Biochem.* **27**, 65 (1989).

ides. The eluent is thus predried, and contamination of the stationary phase with water can be prevented.

Recycling System. As the method was being developed, it turned out that reproducible resolution with short retention times can only be achieved permanently if the conditions for the mobile and stationary phases can be held constant. When a nonpolar, binary mobile phase is used in conjunction with an adsorption column, it must always be borne in mind that the separating properties of the column can change through adsorption of water from the mobile phase; this adsorption is difficult to monitor.[13] The use of a permanently dry eluent allows constant drying and thus constant adsorption homogeneity of the stationary phase.[14] Consequently, it is possible to achieve acceptable resolution of the retinyl esters with short retention times. In order to guarantee that the mobile and stationary phases remain dry over a protracted period, we developed a closed recycling system from simple, commercially available glass materials (Fig. 1). We have achieved constant separating efficiency for the sensitive silica columns for 2 years (1000 separations). Reproducible separations are achieved after the columns have been allowed to equilibrate for 6 hr using a predried solvent. If the retention times have changed, as a result of contamination by water from those extraction phases which are not so readily dried, a reequilibration time of 1 hr suffices.

Standards. The qualitative and quantitative determinations are performed relative to external standards (all-*trans*-, 9-*cis*-, 11-*cis*-, 13-*cis*-retinyl palmitate; all-*trans*-, 11-*cis*-stearate; all-*trans*-oleate; all-*trans*-palmitoleate; all-*trans*-linoleate), all of which were kindly supplied by Hoffmann-La Roche (Basel, Switzerland). Stock solutions of the external standards are prepared by dissolving 25 mg of the retinyl esters in 100 ml of 2-propanol in brown flasks. Aliquots of these solutions are diluted with 2-propanol until an absorption of approximately 0.5 is obtained relative to 2-propanol at 328 nm. The concentrations are calculated from the specific absorption quotient for retinyl palmitate, which is 940 in ethanol measured at 328 nm. These stock solutions are then diluted with *n*-hexane to prepare series of reference solutions (10–100 μg/ml). The stock solutions are stable up to 6 weeks when stored at $-28°$ in the dark. The actual calibration solutions should be freshly prepared each time. The purity of the stock solutions can be determined by recording the derivative spectrum[15] or by saponification and measurement of the molar absorption coefficient for retinol.[16]

[13] R. P. W. Scott and S. Traiman, *J. Chromatogr.* **196**, 193 (1980).
[14] J. P. Thomas, A. Brun, and J. P. Bouning, *J. Chromatogr.* **172**, 107 (1979).
[15] U. Wellner and H. K. Biesalski, *Lecture Notes Med. Inf.* **25**, 579 (1985).
[16] A. C. Ross, *Anal. Biochem.* **115**, 324 (1981).

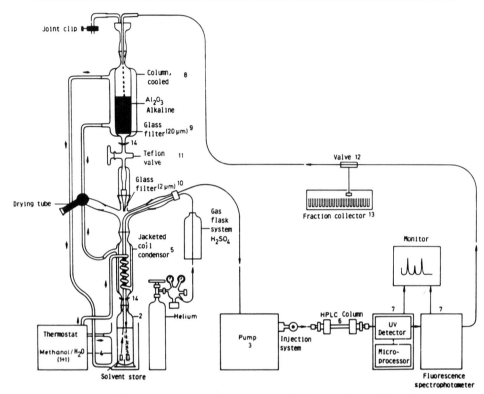

FIG. 1. Recycling and moisture control system (built from readily available glass material). After being mixed, the components of the mobile phase are passed over a drying column (8) containing alumina in a brown glass flask (2). Through this "adsorptive filtration," the water from the mobile phase is removed. The dry solvent is aspirated from the flask into the HPLC pump (3). Before this, the mobile phase is degassed in an ultrasonic bath; during measurement it is degassed with helium, which is passed through a gas flask system containing sulfuric acid. During the chromatographic separation process the storage flask and the recycling system are thermostatted (usually at 14°) (4). The thermostatted jacketed coil condensor (5), which is mounted at the storage flask, prevents one or both eluents of the mobile phase from condensing. After flowing through the column (6) and the detector (7) the mobile phase reaches a dropping funnel, which is connected to the thermostatted system and contains dried neutral alumina (8). The dropping funnel is sealed with a 20-μm fritted glass disk (9), to which a separate shell containing a 2-μm fritted glass disk is connected (10), which can be easily separated and purified when contaminated with alumina particles. At the exit of the dropping funnel, a three-way stopcock (11) controls the flow of the eluent, so that the alumina of the solid-phase is always covered with the eluent. An opening to the atmosphere is provided via a drying tube, which is filled with blue silica gel. Between the detector and the dropping funnel is a fraction collector with a three-way valve which allows collection of the separated compounds either for further analysis or to prevent overloading of the aluminum column.

Quantitative and Qualitative Analysis

Quantitative trace analysis is best performed by determining the peak height. This permits greater sensitivity because the retention times are reduced (increased signal-to-noise ratio). At the same time hard to eradicate interferences (e.g., fluctuations in pump pulses) that affect only peak area do not exert any influence.

Qualitative analysis entails comparing the retention times and the UV and fluorescence spectra in the stopped-flow mode with those of the external standards. Cochromatographing impurities can be excluded by determining the so-called absorption quotient,[15] which can be calculated automatically by suitably equipped UV detectors. If two detectors are connected in series, the detector quotient method[5] can be used to determine cochromatographing isomers of the retinyl esters. This type of cochromatography can arise with the compounds listed in Table I in the system described here. A prerequisite for using the quotient of the signals is that constant conditions must be maintained for determining the quotients of the external standards for comparison with the sample. In this case, the quotients listed in Table II are obtained. If cochromatography occurs as shown in Table III, the quotients change and thus permit identification of cochromatographing retinyl esters. (See Figs. 2 and 3.)

Preparation of Samples. In order to be able to work under traceanalytical conditions, the extraction procedure must be restricted to a minimum number of stages, and light and oxygen must be excluded as far as possible. The retinyl esters are extracted from tissue samples by freezing the samples (which should be as small as possible) immediately in liquid nitrogen and then freeze-drying them at $-55°$ and 3×10^{-3} torr for 24–48 hr according to their weight. Freeze-dried samples weighing more than 100 mg are pulverized in a mortar. From 5–10 mg of the powder is weighed out and extracted. The extraction procedure consists of covering the tissue powder with 4 ml of *n*-hexane, homogenizing for 1 min with an Ultraturrax (Ika Werk, Stauffen, FRG) and subsequently centrifuging at 5000 rpm for 2 min. A sample of the clear hexane supernatant liquid is

TABLE I
COCHROMATOGRAPHY OF DIFFERENT
RETINYL ESTERS

Compound	Cochromatography with
13-*cis*-Retinyl oleate	11-*cis*-Retinyl palmitate
11-*cis*-Retinyl oleate	9-*cis*-Retinyl stearate
9-*cis*-Retinyl oleate	all-*trans*-Retinyl stearate

TABLE II
MEAN VALUES OF TYPICAL DETECTOR QUOTIENTS OF
RETINYL ESTER ISOMERS[a]

	Isomer			
Parameter	13-cis	11-cis	9-cis	all-trans
Mean	1.78	0.374	1.75	2.81
S.D.	0.07	0.01	0.08	0.07
CV^b	3.9%	2.8%	4.5%	2.4%
n^c	15	15	15	21

[a] Fluorescence: UV peak height.
[b] CV, Coefficient of variance.
[c] n, Number of repeated samples.

injected onto the HPLC column. Smaller samples weighing less than 1 mg (dry weight) are transferred after the freeze-drying stage to a glass Micro-potter (Kontes, Michigan) in a volume of 100–200 μl and ground dry with a glass pestle. The tissue is coated with 50 μl of n-hexane and ground further with the glass pestle or comminuted with an ultrasonic rod (Braun, Melsungen, FRG). Samples are then centrifuged in an Eppendorf micro-fuge (Eppendorf, Hamburg, FRG) for 2 min at 7000 rpm. The organic supernatant layer is either removed by means of a 5- or 10-μl Eppendorf micropipette or aspirated by means of a Hamilton injection syringe and then injected into the HPLC system.

Precautions. During quantitative and qualitative trace analysis, all conditions must be kept constant in order to achieve reproducible and reliable results. This applies particularly to all parameters that affect peak height by band broadening or shifting of retention times. For this reason, the concentration ranges and injection volumes of the samples should approximate those of the calibration curves. Samples from tissues containing a high concentration of retinyl esters should therefore be diluted appropriately.

TABLE III
ALTERATION OF DETECTOR QUOTIENTS OWING TO COCHROMATOGRAPHY OF
DIFFERENT RETINYL ESTERS

Standard	Typical quotient range	Cochromatographing peak	Quotient alteration
11-cis (16:0)	0.35–0.4	13-cis (18:1)	>0.4
9-cis (18:0)	1.65–1.85	11-cis (18:1)	<1.6
all-trans (18:0)	2.75–2.9	9-cis (18:1)	<2.7

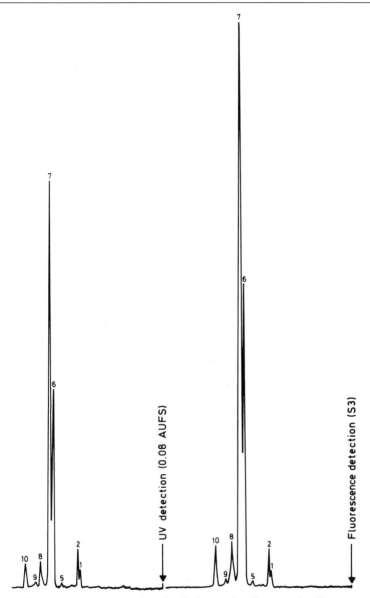

FIG. 2. Chromatographic separation of retinyl esters in a liver sample. Chromatographic conditions were as described in the text. Peak identification: (1) 13-*cis*-retinyl stearate; (2) 13-*cis*-retinyl palmitate; (3) 11-*cis*-retinyl stearate; (4) 11-*cis*-retinyl palmitate; (5) 9-*cis*-retinyl palmitate; (6) all-*trans*-retinyl stearate; (7) all-*trans*-retinyl palmitate; (8) all-*trans*-retinyl oleate; (9) all-*trans*-retinyl palmitoleate; (10) all-*trans*-retinyl linolate.

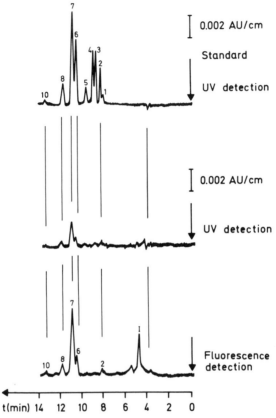

FIG. 3. Separation of retinyl esters in a small tissue sample (organ of Corti of the guinea pig inner ear). Peak identifications as in Fig. 2.

Determination of Retinyl Esters in Different Tissues

The distribution of different fatty acid esters of retinol was determined in different tissues in normal nourished guinea pigs (Table IV). This qualitative and quantitative distribution shows some differences between selected tissues. The highest concentration in all cases is that of retinyl palmitate, followed by retinyl stearate and retinyl oleate. Exceptions to this are all tissues with mucous membranes where retinyl palmitoleate was not detectable and a relative high concentration of retinyl linoleate was measured. Especially in the mucous membranes of the respiratory tract (lung, trachea, nasal mucosa), a higher relative concentration of retinyl oleate becomes obvious. In sensory tissues, however, retinyl palmitate and retinyl stearate show the highest concentration whereas the other esters are only detectable in trace amounts. We assume that the

TABLE IV
DISTRIBUTION OF RETINYL ESTERS IN BIOLOGICAL SAMPLES[a]

Sample	Number	Concentration of retinyl ester[b] (μg/g dry weight)				
		16:0	18:0	18:1	16:1	18:2
Liver	14	105.3 ± 31.5	45.6 ± 9.2	12.7 ± 2.8	4.8 ± 0.9	16.7 ± 5.5
Kidney	6	1.31 ± 0.65	0.72 ± 0.30	0.11 ± 0.08	0.02 ± 0.01	0.15 ± 0.06
Testicle	7	5.72 ± 4.20	2.88 ± 1.95	0.71 ± 0.43	0.06 ± 0.02	1.22 ± 0.75
Epididymis	7	2.35 ± 0.94	0.81 ± 0.44	0.27 ± 0.11	n.d.[c]	0.72 ± 0.51
Spermatic cord	9	3.90 ± 1.55	1.15 ± 0.37	0.31 ± 0.20	n.d.	1.95 ± 0.69
Trachea	6	0.18 ± 0.11	0.13 ± 0.07	0.11 ± 0.05	n.d.	0.09 ± 0.04
Lung	12	0.42 ± 0.31	0.23 ± 0.11	0.19 ± 0.12	n.d.	0.21 ± 0.09
Nasal mucosa	9	0.58 ± 0.40	0.11 ± 0.10	0.14 ± 0.08	n.d.	0.17 ± 0.11
Tongue	10	1.81 ± 0.68	0.97 ± 0.46	0.28 ± 0.21	0.08 ± 0.06	0.14 ± 0.11
Pigment epithelium	8	21.92 ± 6.35	7.89 ± 2.21	1.10 ± 0.32	0.08 ± 0.05	0.91 ± 0.37
Inner ear, organ of Corti	12	4.8 ± 0.3	1.6 ± 0.2	0.8 ± 0.4	n.d.	0.04 ± 0.1

[a] Guinea pig.

[b] 16:0, Retinyl palmitate; 18:0, retinyl stearate; 18:1, retinyl oleate; 16:1, retinyl palmitoleate; 18:2, retinyl linoleate.

[c] n.d., Not determined.

distribution of different fatty acid esters of retinol depends on the availability of typical fatty acids within the tissues.

From these results we conclude that vitamin A dependent tissues are able to store the vitamin in or near the target cells. In cases of a period of limited vitamin A supply, with lowered retinol plasma values, or a suddenly increased cellular demand, the hydrolysis of the retinyl esters will result in an intracellular release of retinol. Consequently, a decrease of the intracellular retinol concentration due to a decreased plasma retinol level or an increased cellular demand, could be avoided over a period of time depending on the amount of retinyl esters at the target cell side. If the retinyl esters are stored in typical storing cells near the target cells as was demonstrated for the lung,[17] the transport of the hydrolysis product retinol to the target cell could be mediated by cytosolic retinol binding protein. The repletion of the peripheral storage sites with retinyl esters, which is possible by intravenous administration of retinyl esters,[18] provides a direct way to supply the vitamin to peripheral, vitamin A dependent tissues, especially in cases of protein deficiency with impaired RBP-synthesis.

[17] T. Okabe, H. Yorifuji, E. Yamada, and F. Takaku, *Exp. Cell. Res.* **154,** 125 (1984).

[18] T. Gerlach, H. K. Biesalski, B. Haeussermann, and K. H. Baessler, *Am. J. Clin. Nutr.* **50,** 1029 (1989).

Section II

Receptors, Transport and Binding Proteins

[19] Purification and Properties of Plasma Retinol-Binding Protein

By WILLIAM S. BLANER and DEWITT S. GOODMAN

Introduction

Retinol is mobilized from retinoid stores in the liver bound to its specific plasma transport protein, retinol-binding protein (RBP). RBP, first isolated by Kanai et al. in 1968,[1] is comprised of a single polypeptide chain with a molecular weight of close to 21,000 and has a single binding site for one molecule of all-trans-retinol. In the circulation, RBP forms a 1 : 1 molar complex with another plasma protein, transthyretin (TTR). RBP has been extensively studied, and much information regarding its physicochemical properties, chemical structure, gene structure, synthesis, secretion, and physiological roles is now available. These topics have been reviewed extensively elsewhere.[2,3]

The first portion of this chapter describes some of the physicochemical properties of RBP with an emphasis on those properties which are important for the purification of RBP. The middle and main portion of this chapter describes, in detail, procedures which employ conventional column chromatography for the purification of rat RBP and human RBP from plasma. The procedures for the purification of RBP from these two species are similar. Both procedures initially isolate RBP as a complex with TTR and separate the RBP from the TTR in a final gel-filtration step. For the purification of rat and human RBP, some representative column elution profiles have been provided. At the end of the chapter a brief discussion is provided concerning RBP gene structure and expression.

Physicochemical Properties of Retinol-Binding Protein

In order to purify RBP, it is necessary to be familiar with some of its major physicochemical properties. Both human and rat RBP are isolated from plasma as polypeptides, each containing 182 amino acids.[4,5] The

[1] M. Kanai, A. Raz, and D. S. Goodman, J. Clin. Invest. 47, 2504 (1968).
[2] D. S. Goodman, in "The Retinoids" (M. B. Sporn, A. B. Roberts, and D. S. Goodman, eds.), Vol. 2, p. 41. Academic Press, Orlando, Florida, 1984.
[3] W. S. Blaner, Endocr. Rev. 10, 308 (1989).
[4] L. Rask, H. Anundi, and P. A. Peterson, FEBS Lett. 104, 55 (1979).
[5] J. Sundelin, B. C. Laurent, H. Anundi, L. Trägårdh, D. Larhammar, L. Björck, U. Eriksson, B. Åkerström, A. Jones, M. Newcomer, P. A. Peterson, and L. Rask, J. Biol. Chem. 260, 6472 (1985).

METHODS IN ENZYMOLOGY, VOL. 189

three-dimensional structure of human holo-RBP, as determined by X-ray crystallography, shows that RBP consists of a single globular domain approximately 40 Å in diameter and is composed of an amino-terminal coil, a β-sheet core, an α helix, and a carboxy-terminal coil.[6] The β-sheet core of RBP consists of a β-barrel structure composed of eight antiparallel β strands which are arranged in two stacked orthogonal β sheets. Retinol is encapsulated within the two orthogonal β sheets with the β-ionone ring of the retinol positioned deep within the β barrel and the isoprene tail of the retinol projecting out along the barrel axis to near the surface of the protein. The primary sequences of human, rat, and rabbit RBP have been compared and found to be very similar.[5] Rat RBP has an 86% sequence similarity with human RBP, and rabbit RBP has a 92% similarity with human RBP. The amino acid differences between human, rat, and rabbit RBP are not equally distributed throughout the entire polypeptide chain. Three stretches along the polypeptide chain are totally conserved (residues 1–20, 69–98, and 126–141), and two short stretches contain many substitutions (residues 145–151 and 175–179).

Human RBP has a sedimentation coefficient ($s_{20,w}$) of 2.13 and an estimated Stokes radius of 20.5 Å.[7] No bound lipid (other than retinol) has been detected in purified RBP,[1] and purified RBP has been reported to contain very little or no carbohydrate.[7,8] RBP from mammalian species migrates with α_1 mobility on electrophoresis and displays charge microheterogeneity on native polyacrylamide gel electrophoresis[7,8] or native isoelectric focusing gel electrophoresis.[9] In these electrophoresis systems, three or four different immunologically identical RBP bands have been observed.[7–9] The ultraviolet absorption spectrum of holo-RBP in solution has two absorbance maxima, at 280 nm (arising from certain amino acids of the protein) and at 330 nm (arising from the protein-bound retinol). For holo-RBP, the two absorbance peaks are of approximately the same magnitude ($A_{330}:A_{280}$ ratio of ~1.0). The $E_{1\,cm}^{1\%}$ of RBP has been estimated to be 19.4.[1] In solution, the retinol–RBP complex is highly fluorescent, exhibiting an excitation maximum at 334 nm and an emission maximum at 463 nm.[10,11] The intensity of fluorescence of retinol bound to RBP is highly enhanced (by an order of magnitude) and the fluorescence of bound retinol displays a blue shift (~15–20 nm) when compared to

[6] M. E. Newcomer, A. Liljas, J. Sundelin, U. Eriksson, L. Rask, and P. A. Peterson, *EMBO J.* **3**, 1451 (1984).

[7] A. Raz, R. Shiratori, and D. S. Goodman, *J. Biol. Chem.* **245**, 1903 (1970).

[8] P. A. Peterson, *J. Biol. Chem.* **246**, 34 (1971).

[9] P. A. Peterson and I. Berggård, *J. Biol. Chem.* **246**, 25 (1971).

[10] P. A. Peterson and L. Rask, *J. Biol. Chem.* **246**, 7544 (1971).

[11] D. S. Goodman and R. B. Leslie, *Biochim. Biophys. Acta* **260**, 670 (1972).

retinol dissolved in a variety of organic solvents.[11,12] RBP forms a stable protein–protein complex with TTR, with an association constant of the order of 10^6 to 10^7.[2] The interaction of RBP with TTR is very sensitive both to ionic strength and to pH.[13,14] At an ionic strength of 2 mM or lower, complex formation between RBP and TTR does not occur, and RBP–TTR affinity decreases rapidly above pH 9 and below pH 5.

Purification of Rat Retinol-Binding Protein

Several procedures for the purification of rat RBP have been reported in the literature.[15-19] These procedures employ various combinations of ion-exchange chromatography, gel-filtration chromatography, TTR affinity chromatography, and preparative polyacrylamide gel electrophoresis to obtain purified RBP. The purification scheme described below has been used successfully and repeatedly in our laboratory for the isolation of rat RBP.[19] This purification scheme for rat RBP involves fractionation of plasma (or serum) by chromatography on a DEAE-cellulose anion-exchange column followed by two gel-filtration steps on Sephadex G-100. With one additional gel-filtration step, it is possible to obtain pure rat TTR in addition to pure rat RBP. This scheme is relatively simple to carry out and provides pure RBP in high yield. The recovery of rat RBP routinely ranges between 30 and 35% (a yield of 10–15 mg pure RBP per liter plasma). The isolated RBP consists of a mixture of holo- and apo-RBP, with approximately 80–85% of the purified RBP being in the holo form, as judged by the $A_{330}:A_{280}$ ratio (i.e., the $A_{330}:A_{280}$ ratio is between 0.8 and 0.85).

All procedures described below should be carried out at 4°. Attempts also should be made to protect the light-sensitive retinol, bound to RBP, from prolonged exposure to light. Although it is not possible to eliminate all exposure to light, it is recommended that all column fractions, concentrated fractions, and pools be wrapped with aluminum foil to decrease their exposure to light during processing.

[12] S. Futterman, D. Swanson, and R. E. Kalina, *Invest. Ophthalmol.* **14**, 125 (1975).
[13] P. A. Peterson, *J. Biol. Chem.* **246**, 44 (1971).
[14] P. P. van Jaarsveld, H. Edelhoch, D. S. Goodman, and J. Robbins, *J. Biol. Chem.* **248**, 4698 (1973).
[15] Y. Muto and D. S. Goodman, *J. Biol. Chem.* **247**, 2533 (1972).
[16] P. A. Peterson, L. Rask, L. Östberg, L. Anderson, F. Kamwendo, and H. Pertoft, *J. Biol. Chem.* **248**, 4009 (1973).
[17] A. R. Poole, J. T. Dingle, A. K. Mallia, and D. S. Goodman, *J. Cell Sci.* **19**, 379 (1975).
[18] B. W. McGuire and F. Chytil, *Biochim. Biophys. Acta* **621**, 324 (1980).
[19] M. Kato, K. Kato, and D. S. Goodman, *J. Cell Biol.* **98**, 1696 (1984).

Step 1. Preparation of Rat Plasma for Purification of RBP

For the chromatography procedures described below, approximately 1 liter of rat plasma (or serum) must first be dialyzed against 20 mM imidazole–acetate buffer, pH 6.0. The initial dialysis should be against 20 liters of this buffer, and the dialysis buffer (20 liters) should be changed at least 3 times over a period of 3 days. After dialysis, the plasma is centrifuged for 30 min at 20,000 g and 4° to remove a large quantity of precipitated protein. The clear supernatant is removed, and its pH and conductivity are measured to ensure that they are the same as the values seen for the 20 mM imidazole–acetate buffer (pH 6.0 and conductivity of 1.7 mS). If the supernatant and the dialysis buffer are not found to be identical, further dialysis (and another centrifugation step) is required. When the pH values and conductivities are identical, the supernatant can be either applied to the DEAE-cellulose column (Step 2 below) or stored at 4°, protected from light, for up to 24 hr before application to the DEAE-cellulose column.

Step 2. Ion-Exchange Chromatography on DEAE-Cellulose

The dialyzed plasma, processed as described in Step 1, is applied to an 8 × 110 cm column of DEAE-cellulose (Whatman DE-52, Whatman International Ltd., Maidstone, UK). It is essential that this DEAE-cellulose column first be equilibrated with 20 mM imidazole–acetate buffer, pH 6.0 (as judged by the identities of the pH values and conductivities of the affluent and effluent buffer), prior to applying the dialyzed plasma to the column. On application of the sample to the column, collection of fractions, each containing approximately 25 ml, is begun. After the dialyzed plasma has fully entered the column, the anion-exchange resin is washed with approximately 5 liters of the 20 mM imidazole–acetate buffer, pH 6.0. The column is then eluted with the same buffer containing a linear gradient of NaCl concentration, from 0 to 90 mM. This elution gradient is constructed with 8 liters of 20 mM imidazole–acetate buffer, pH 6.0, and 8 liters of 20 mM imidazole–acetate, pH 6.0, containing 90 mM NaCl. The RBP begins to elute from the DEAE-cellulose when the conductivity of the elution buffer reaches 4.0 mS. Throughout the elution procedure, the flow rate of the column is kept constant at approximately 250 ml/hr. The presence of RBP in the fractions is assessed by measuring the fluorescence of the retinol–RBP complex (excitation 330 nm, emission 465 nm). Figure 1 provides a representative elution profile for the DEAE-cellulose chromatography procedure. The first fluorescence peak (labeled with a bar in Fig. 1) contains the RBP (which is eluted here as a complex with TTR). The latter fluorescence peaks do not contain RBP.

The RBP-containing fractions are pooled and concentrated in an Ami-

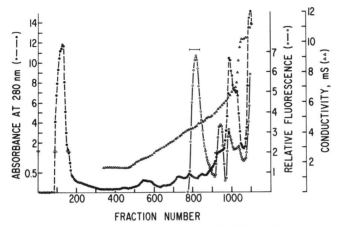

FRACTION NUMBER

FIG. 1. Elution profile of rat RBP from a 8 × 110 cm DEAE-cellulose column. The RBP-containing fractions are indicated by the bar. Details of the chromatography conditions are given in the text.

con concentrator (Amicon Corporation, Danvers, MA), equipped with an Amicon YM5 membrane, to approximately 20 ml. The concentrated RBP-containing sample is dialyzed against 3 changes of 4 liters of 50 mM Tris-HCl, pH 8.4, in order to prepare it for the next step.

Step 3. Sephadex G-100 Gel Filtration

The RBP-containing sample, prepared as described in Step 2, is carefully applied to a 5 × 100 cm column of Sephadex G-100 Superfine (Pharmacia LKB Biotechnology, Piscataway, NJ) previously equilibrated with 50 mM Tris-HCl, pH 8.4. Chromatography is carried out in 50 mM Tris-HCl, pH 8.4, at a flow rate of 20 ml/hr, and fractions containing approximately 10 ml are collected. The RBP elutes from this column as a complex with TTR at a molecular weight corresponding to approximately 75,000. The presence of RBP in the eluted fractions is determined by measurement of the retinol fluorescence of the holo-RBP (excitation 330 nm, emission 465 nm). The fractions found to contain RBP (and TTR) are pooled and concentrated in an Amicon concentrator with a Amicon YM5 membrane to a final volume of 4 ml before further purification.

Step 4. Sephadex G-100 Gel Filtration in 3 M Urea

In order to separate the RBP from TTR, it is necessary to disrupt the RBP–TTR complex; this is conveniently done with 3 M urea. For this purpose, the RBP-containing concentrate, prepared in Step 3, is diluted with an equal volume of a solution containing 6 M urea in water. This diluted RBP solution is allowed to remain for 18 hr at 4°. The 3 M urea solution of RBP (and TTR) is applied to a 5 × 100 cm column of Sephadex G-100 Superfine, equilibrated with 3 M urea in 25 mM Tris-HCl, pH 8.4. Chromatography is carried out with a solution of 25 mM Tris-HCl, pH 8.4, containing 3 M urea, at a flow rate of approximately 10 ml/hr, and fractions containing 7 ml are collected. In the 3 M urea buffer, the RBP elutes as free RBP, not complexed to TTR, with a molecular weight corresponding to 21,000. The RBP-containing fractions are identified by measurement of the retinol fluorescence of the holo-RBP.

Figure 2 shows a representative elution profile for a 3 M urea–Sephadex G-100 column. Normally, three pools (labeled Pools 1, 2, and 3 in Fig. 2) are made from the fractions collected from the 3 M urea–Sephadex G-100 column. Pool 1, consisting of fractions from the upper half of the ascending limb of the retinol fluorescence peak through the end of the descending limb (fractions 110–122 in Fig. 2), contains homogeneously purified RBP. Pool 2, consisting of the fractions comprising the lower half of the ascending limb of the retinol fluorescence peak (fractions 98–109), usually also contains homogeneously purified RBP; however, in some preparations some TTR will be present in this pool. Pool 3 consists of approximately 200 ml of buffer eluting immediately before the start of

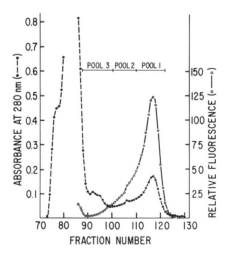

Fig. 2. Elution profile of rat RBP from a 5 × 100 cm Sephadex G-100 Superfine column eluted with buffer containing 3 M urea. Three pools (1, 2, and 3) are made from the fractions as indicated by the bars over the fractions which compose each pool. The contents of each pool are described in the text. Details of the conditions of chromatography are given in the text.

retinol fluorescence peak (fractions 86–97) and contains the majority of the TTR eluting from this column. Each pool is concentrated in an Amicon concentrator with a Amicon YM5 membrane to approximately 5 ml and dialyzed against water (if the pools are to be lyophilized and stored) or an appropriate buffer based on experimental needs.

The purity of each pool is checked by sodium dodecyl sulfate (SDS)–polyacrylamide electrophoresis (SDS–PAGE). RBP migrates as a single protein band at 21 kDa. On SDS–PAGE, TTR migrates as a major protein band at 14 kDa (monomeric TTR) and a second, much less prominent protein band at 28 kDa (dimeric TTR).

In some preparations, Pool 2 may contain some contaminating TTR, and the RBP in this pool will require further purification. If further purification of Pool 2 is necessary, the pool should be rechromatographed on an appropriately scaled-down column of 3 M urea–Sephadex G-100. When necessary, rechromatography of Pool 2 will provide additional homogeneously purified RBP.

Pool 3 contains TTR and, usually, low levels of several contaminating (and unidentified) proteins. Purified TTR can be obtained through the use of an additional gel-filtration step on a 1 × 100 cm column of Sephadex G-100 Superfine, equilibrated and run in 2 mM Tris-HCl buffer, pH 8.4. Using this further purification step for TTR, approximately 20 to 25 mg TTR can be recovered from 1 liter of rat plasma.

Purification of Human Retinol-Binding Protein

A number of different procedures for the purification of human RBP have been reported in the literature.[1,8,20–24] Most of the reported purification schemes are similar in that they employ combinations of ion-exchange chromatography and gel filtration to achieve RBP purification. In these schemes, RBP has been isolated from serum or plasma by first isolating the RBP–TTR complex and then dissociating the complex into its two components. Procedures have also been described which employ affinity chromatography to isolate human RBP, utilizing either human TTR coupled to Sepharose[21] or retinoic acid linked to Sepharose.[23] One recently published procedure reports the purification of RBP using a sequence of procedures consisting of ammonium sulfate fractionation, hy-

[20] A. Raz and D. S. Goodman, *J. Biol. Chem.* **244**, 3230 (1969).
[21] A. Vahlquist, S. F. Nilsson, and P. A. Peterson, *Eur. J. Biochem.* **20**, 160 (1971).
[22] H. Haupt and K. Heide, *Blut* **24**, 94 (1972).
[23] G. Fex and B. Hansson, *Biochim. Biophys. Acta* **537**, 358 (1978).
[24] Y. Muto, Y. Shidoji, and Y. Kanda, this series **81**, 840 (1982).

drophobic interaction chromatography on Phenyl-Sepharose, and gel-filtration chromatography on Sephadex G-50.[25] Procedures have also been reported for the isolation of RBP from the urine of patients with tubular proteinuria.[9,21,26]

The purification scheme described below for the isolation of human plasma RBP employs two ion-exchange chromatography steps and two gel-filtration steps, one of which separates RBP from TTR. An additional chromatography step is required to obtain purified TTR. The scheme is relatively easy to carry out and yields approximately 15 mg of purified RBP and 25 mg of purified TTR from 1 liter of human plasma. About 85% of the purified RBP obtained is in the holo form, as judged by the $A_{330} : A_{280}$ ratio. As described above for the purification of rat RBP, all procedures should be carried out at 4°, and precautions should be taken to minimize the exposure of the RBP-containing fractions to light.

Step 1. Preparation of Human Plasma for Purification of RBP

Approximately 1 liter of human plasma is dialyzed against 20 liters of 50 mM sodium phosphate buffer, pH 7.6, containing 75 mM NaCl. The dialysis buffer is changed at least 3 times during a period of 3 days. After dialysis the plasma is centrifuged for 30 min at 20,000 g to remove precipitated protein. The conductivity and pH of the supernatant should be identical to that of the dialysis buffer (pH 7.6, conductivity of 12.5 mS). If the pH values and conductivities are identical, the sample is ready for application to the DEAE-Sepharose column; otherwise, the plasma should be dialyzed further.

Step 2. DEAE-Sepharose Chromatography

A 4 × 50 cm column of DEAE-Sepharose (Pharmacia LKB Biotechnology) is equilibrated with 50 mM sodium phosphate buffer, pH 7.6, containing 75 mM NaCl, prior to the start of chromatography. After application of the dialyzed plasma to this equilibrated column, the column is washed with 2 liters of the same buffer solution. The RBP, which under these conditions is retained by the resin, is eluted from the column with a linear gradient of increasing NaCl concentration. The gradient is constructed with 2.5 liters of 50 mM sodium phosphate buffer, pH 7.6, containing 75 mM NaCl and 2.5 liters of 50 mM sodium phosphate buffer, pH 7.6, containing 600 mM NaCl. The column is run at a flow rate of about

[25] R. Berni, S. Ottonello, and H. L. Monaco, *Anal. Biochem.* **150**, 273 (1985).
[26] G. Fex and B. Hansson, *Eur. J. Biochem.* **94**, 307 (1979).

120 ml/hr, and fractions containing approximately 20 ml are collected. The RBP elutes from the DEAE-Sepharose as the conductivity approaches 20 mS. Ceruloplasmin, a blue-colored protein, elutes from the column shortly before RBP. The RBP-containing fractions are identified by measurement of the fluorescence of retinol bound to RBP (see above). These fractions are pooled and concentrated in an Amicon concentrator with an Amicon YM5 membrane to a final volume of approximately 10 ml.

Step 3. Sephadex G-100 Gel Filtration

The concentrated RBP-containing sample, prepared in Step 2, is applied to a 5 × 100 cm column of Sephadex G-100 Superfine, equilibrated with 50 mM Tris-HCl buffer, pH 8.4. Chromatography is carried out with 50 mM Tris-HCl buffer, pH 8.4, at a flow rate of about 10 ml/hr. Fractions of 8 ml are collected from the column, and the presence of RBP in the fractions is assessed by measurement of the fluorescence of the holo-RBP (excitation at 330 nm, emission at 465 nm). Figure 3 provides a representative elution profile for this column. The bar in Fig. 3 indicates the fractions which contain RBP (as the RBP–TTR complex). These RBP-containing fractions are pooled, and sufficient solid NaCl is added to bring the final NaCl concentration of the pool to 75 mM.

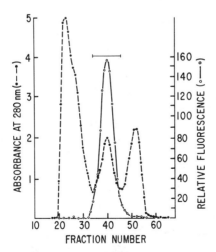

Fig. 3. Elution profile of human RBP from a 5 × 100 cm Sephadex G-100 Superfine column run in the presence of 50 mM Tris-HCl buffer, pH 8.4. The bar indicates the fractions containing the RBP–TTR complex which are pooled for subsequent purification. Details of the conditions of chromatography are given in the text.

Step 4. DEAE-Cellulose Chromatography

A 2.5 × 50 cm column of DEAE-cellulose (DE-52, Whatman) is equilibrated with 50 mM Tris-HCl buffer, pH 8.4, containing 75 mM NaCl. To this column, the RBP-containing pool, prepared in Step 3, is added and allowed slowly to enter the column. After the sample fully enters the column, the column is washed with 500 ml of 50 mM Tris-HCl buffer, pH 8.4, containing 75 mM NaCl. The RBP, which binds to the DEAE-cellulose, is eluted from the column with a linear gradient of increasing NaCl concentration. The gradient is constructed with 1 liter of 50 mM Tris-HCl buffer, pH 8.4, containing 75 mM NaCl and 1 liter of 50 mM Tris-HCl buffer, pH 8.4, containing 450 mM NaCl. The column is allowed to flow at a rate of approximately 50 ml/hr. RBP elutes from the anion-exchange resin as the conductivity of the buffer approaches 17 mS. The column fractions, about 10 ml each, are screened for retinol fluorescence (excitation at 330 nm, emission at 465 nm) to identify the RBP-containing fractions. Fractions found to contain RBP are pooled and concentrated to 5 ml in an Amicon concentrator with an Amicon YM5 membrane.

At this point in the purification, the RBP is still present as a complex with TTR. In order to dissociate the protein–protein complex, the RBP-containing sample is dialyzed against 4 liters of 0.25 mM potassium carbonate–potassium borate buffer adjusted to pH 10 with a few drops of concentrated NH$_4$OH. The dialysis is carried out exhaustively over 3 days with at least 4 changes, of 4 liters each, of the dialysis buffer.

Step 5. Sephadex G-100 Gel Filtration

The purification of human RBP is completed by separating RBP from TTR by gel filtration of the dissociated complex. The dialyzed RBP sample prepared in Step 4 is applied to a 2.5 × 100 cm column of Sephadex G-100 Superfine previously equilibrated with 0.25 mM potassium carbonate–potassium borate buffer, pH 10. Gel filtration is carried out with the same 0.25 mM potassium carbonate–potassium borate buffer, pH 10. The column is eluted at a rate of 10 ml/hr and fractions containing 5 ml are collected. The presence of RBP in the fractions is monitored, as before, by measuring the fluorescence of the retinol bound to RBP.

The RBP-containing fractions are analyzed individually for purity by SDS–PAGE. Usually, the fractions with detectable retinol fluorescence contain homogeneously purified RBP. Occasionally, however, a small amount of TTR is still found in some of the fractions which are present at the start of the ascending limb of the peak of retinol fluorescence (RBP); thus, it is advisable to check individual fractions for purity before the fractions are pooled.

The large peak of protein (as determined by absorbance at 280 nm) immediately preceding the peak of retinol fluorescence (RBP) contains TTR. Should it be necessary, the TTR in this pool can be purified to homogeneity by chromatography on Cibacron Blue F3-GA agarose (Bio-Rad Laboratories, Richmond, CA) according to published procedures.[27] Briefly, the TTR-containing pool is dialyzed against 30 mM sodium phosphate buffer, pH 7.0, and applied to a 2 × 10 column of the Cibacron Blue F3-GA agarose previously equilibrated with the same buffer. On this resin, TTR is not retained, whereas the contaminating proteins bind tightly to the resin.

Purification of Retinol-Binding Protein from Other Species

Retinol-binding protein has been isolated from many mammalian species in addition to humans and rats. Purification procedures for RBP from the cow,[28,29] ox,[30] rabbit,[31] pig,[32] dog,[33] and cynomolgus monkey[34] have been published. In addition, RBP has been purified from chicken plasma[35,36] and from plasma of young yellowtail tuna.[37] In general, the isolation schemes used for purification of RBP from these various species employed various combinations of ion-exchange, gel-filtration, and TTR affinity chromatography. The reported RBP purification procedures are, in general, described in detail, and the reader is referred to the original reports if more information regarding a procedure is needed.

Retinol-Binding Protein Gene Structure and Expression

Much information is now available regarding the structure of the RBP gene and its expression. Both the human[38,39] and rat[5,40] genes for RBP

[27] E. Gianazza and P. Arnaud, *Biochem. J.* **201**, 129 (1982).
[28] J. Heller, *J. Biol. Chem.* **250**, 6549 (1975).
[29] G. Fex and R. Lindgren, *Biochim. Biophys. Acta* **493**, 410 (1977).
[30] J. Glover, *Vitam. Horm. (N.Y.)* **31**, 1 (1973).
[31] Y. Muto, M. Nakanishi, and Y. Shidoji, *J. Biochem.* **79**, 775 (1976).
[32] L. Rask, *Eur. J. Biochem.* **44**, 1 (1974).
[33] M. D. Poulik, D. Farrah, G. H. Malek, C. J. Shinnick, and O. Smithies, *Biochim. Biophys. Acta* **412**, 326 (1975).
[34] A. Vahlquist and P. A. Peterson, *Biochemistry* **11**, 4526 (1972).
[35] S. Moukady and M. Tal, *Biochim. Biophys. Acta* **336**, 361 (1974).
[36] T. Abe, Y. Muto, and N. Hosaya, *J. Lipid Res.* **16**, 200 (1975).
[37] Y. Shidoji and Y. Muto, *J. Lipid Res.* **18**, 679 (1977).
[38] C. D. D'Onofrie, V. Colantuoni, and R. Cortese, *EMBO J.* **4**, 1981 (1985).
[39] V. Colantuoni, V. Romano, G. Bensi, C. Santoro, F. Costanzo, G. Raugei, and R. Cortese, *Nucleic Acids Res.* **11**, 7769 (1983).
[40] B. C. Laurent, M. H. L. Nilsson, C. O. Båvik, T. A. Jones, J. Sundelin, and P. A. Peterson, *J. Biol. Chem.* **260**, 11476 (1985).

have been cloned and characterized. The human gene was found to contain approximately 10 kilobases (kb) of genomic DNA and to consist of six exons and five introns.[38,39] The exons are of relatively short length, with exons 1 to 5 consisting of 69, 123, 137, 107, and 213 nucleotides, respectively. Intron size, however, was found to be quite variable. The lengths of the first three introns are, respectively, 0.131, 0.110, and 0.165 kb, whereas intron 4 consists of approximately 8 kb and intron 5 of at least 1.2 kb. RBP is synthesized initially in liver as a larger pre-RBP molecule, which is processed cotranslationally, with removal of a signal peptide of 16 amino acids, to the mature and subsequently secreted protein.[39,41] Interestingly, the amino acid sequence for mature human RBP determined by nucleotide sequencing was found to consist of 184 amino acids[39] rather than 182 as determined by protein sequencing techniques with the isolated protein. The basis for this discrepancy is not known. The 5'-flanking region of the human RBP gene contains at least three control elements, a nontissue-specific enhancer which was found to activate promoters in a variety of cell lines, a negative cis-activating element which is proposed to bind a repressor molecule, and a promoter element.[42] How these controlling elements interact to regulate RBP gene expression is not presently clear and requires further investigation.

The gene for rat RBP spans 6.9 kb and, like the human gene, consists of six exons and five introns.[40] The rat gene also resembles the human gene in that the exons are of relatively short lengths, whereas intron size is variable and ranges from 78 base pairs to 4.4 kb (intron 4); introns account for more than 86% of the rat RBP gene. Interestingly, all of the translated exons correspond closely to discrete structural elements of RBP.[40] Splice junctions for these exons occur at residues on the surface of the RBP molecule. All introns, with the exception of intron 4, which occurs at a β bend marking a directional change, fall at structural boundaries defined by β strands. The amino acid sequence of rat RBP determined from the nucleotide sequence indicates that rate RBP contains 183 amino acids,[40] although the amino acid sequence determined from the purified protein contains only 182 amino acids.[5] As with human RBP, it is not clear if this difference is physiologically meaningful.

With the isolation of cDNA clones for RBP,[5,38–40,43] it has become possible to explore the tissue-specific expression of the RBP gene and its regulation. It is now clear that RBP is expressed not only in liver, but also

[41] D. R. Soprano, C. B. Pickett, J. E. Smith, and D. S. Goodman, *J. Biol. Chem.* **256**, 8256 (1981).

[42] V. Colantuoni, A. Pirozzi, C. Blance, and R. Cortese, *EMBO J.* **6**, 631 (1987).

[43] D. R. Soprano, M. L. Wyatt J. L. Dixon, K. J. Soprano, and D. S. Goodman, *J. Biol. Chem.* **263**, 2934 (1988).

in a variety of extrahepatic tissue. Soprano *et al.* showed by Northern blot analysis that RBP mRNA was present in a large number of extrahepatic tissues in the rat.[44] The kidneys were found to contain RBP mRNA at a level of 5–10% of that of the liver. The lungs, spleen, brain, heart, and skeletal muscle were also found to contain low levels of RBP mRNA, at levels which ranged from 1 to 3% of the levels found in the liver. Other studies have demonstrated by *in situ* hybridization and RNase protection assay that RBP mRNA is present in perinephric and epididymal adipose tissue.[45] The RBP mRNA levels in epididymal adipose tissue were found to be approximately 20% of those of the liver. Recently, RBP mRNA has been reported to be present in the retinal pigment epithelium of the rat eye.[46]

Using the technique of *in situ* hybridization, Makover *et al.* have explored the expression of the gene for RBP in the rat kidney.[45] RBP mRNA was found to be specifically localized in the S_3 segment of the proximal tubules of the kidney, which are present in the anatomic region known as the outer stripe of the medulla. In contrast, immunoreactive RBP protein is localized in the proximal convoluted tubular cells of the renal cortex,[19] where serum RBP is presumably filtered and reabsorbed. Thus, RBP mRNA in the kidney is localized in an anatomic region different from that of immunoreactive RBP. The physiological significance of this is not understood.

RBP mRNA has also been localized in fetal rat tissues. By Northern blot analysis, Soprano *et al.* showed that both RBP mRNA and TTR mRNA are present in the visceral yolk sac endoderm and fetal liver at 14 days of gestation.[47] RBP and TTR mRNA levels in the visceral yolk sac were found to remain constant from 14 to 20 days of gestation, averaging 58 and 51%, respectively, of adult liver levels of these transcripts. Immunoprecipitable RBP and TTR were found to be synthesized and secreted by explant cultures of the visceral yolk sac tissue. More recent studies by Makover *et al.*,[48] using *in situ* hybridization, have shown that both RBP mRNA and TTR mRNA are present in the visceral endoderm as early as 7 days of gestation, and the transcripts are strongly and in-

[44] D. R. Soprano, K. J. Soprano, and D. S. Goodman, *J. Lipid Res.* **27,** 166 (1986).
[45] A. Makover, D. R. Soprano, M. L. Wyatt, and D. S. Goodman, *J. Lipid Res.* **30,** 171 (1989).
[46] R. L. Martone, E. A. Schon, D. S. Goodman, D. R. Soprano, and J. Herbert, *Biochem. Biophys. Res. Commun.* **157,** 1078 (1988).
[47] D. R. Soprano, K. J. Soprano, and D. S. Goodman, *Proc. Natl. Acad. Sci. U.S.A.* **83,** 7330 (1986).
[48] A. Makover, D. R. Soprano, M. L. Wyatt, and D. S. Goodman, *Differentiation* **40,** 17 (1989).

creasingly expressed in the visceral endoderm and visceral yolk sac endoderm from days 7 to 13 of gestation. It was suggested[47] that visceral yolk sac-derived RBP may function in the transport and delivery of retinol from maternal blood to the developing fetus. In line with this hypothesis, we have recently found RBP mRNA to be present at significant levels in the human placenta (W. S. Blaner *et al.*, 1990, unpublished).

In vitro studies have shown that RBP gene expression can be induced in F9 embryonal carcinoma cells which have been induced to differentiate to embryoid bodies.[49] Using *in situ* hybridization, it was possible to show that RBP mRNA was specifically localized in the outer layer of cells (which are visceral endodermlike cells) of the embryoid bodies. Undifferentiated F9 cells or F9 cells induced to differentiate to parietal endoderm were not found to contain RBP mRNA.

The regulation and the functional significance of the expression of the RBP gene in extrahepatic tissues has become an area of active interest and importance. It is now clear that the RBP gene (in the rat) is expressed (as evidenced by the presence of RBP mRNA) in many tissues, including kidney, lungs, spleen, brain, stomach, heart, skeletal muscle, large intestine, small intestine, testes, pancreas, eye, fat (both perinephric and epididymal), and the visceral yolk sac. It has been hypothesized[44,45] that RBP synthesized in extrahepatic tissues may play an important role in the transport and recycling of retinol in the body. This hypothesis provides a challenge for future research on RBP synthesis, functions, and physiology.

Acknowledgments

The authors wish to acknowledge the support of National Institutes of Health Grants DK05968 and HL21006 (SCOR).

[49] D. R. Soprano, K. J. Soprano, J. L. Wyatt, and D. S. Goodman, *J. Biol. Chem.* **263**, 17897 (1988).

[20] Interstitial Retinol-Binding Protein: Purification, Characterization, Molecular Cloning, and Sequence

By SHAO-LING FONG and C. D. B. BRIDGES

Introduction

Interstitial or interphotoreceptor retinol-binding protein (IRBP) is a large, elongated glycoprotein that is synthesized and secreted by the photoreceptor cells.[1-10] The average molecular weight in vertebrates is 134,200, except in teleosts, where it is about one-half of this value.[11] IRBP is found in the interphotoreceptor matrix of the eye,[5,12] where its major function is believed to be the transport of 11-*cis*- and all-*trans*-retinoids between the neural retina and retinal pigment epithelium.[13] We have purified and characterized human IRBP, localized its gene to chromosome 10, and obtained its complete cDNA sequence.[4,14,15] The translated human cDNA sequence has been aligned with the amino acid sequences of tryptic peptides from bovine IRBP.[15]

The first human IRBP clone (H.4 IRBP) was obtained by screening a

[1] A. J. Adler, C. D. Evans, and W. F. Stafford III, *J. Biol. Chem.* **260**, 4850 (1985).

[2] A. J. Adler, W. F. Stafford III, and H. S. Slayter, *J. Biol. Chem.* **262**, 13198 (1987).

[3] S.-L. Fong, G. I. Liou, R. A. Landers, R. A. Alvarez, and C. D. B. Bridges, *J. Biol. Chem.* **259**, 6534 (1984).

[4] S.-L. Fong, G. I. Liou, R. A. Landers, R. A. Alvarez, F. Gonzalez-Fernandez, P. A. Glazebrook, D. M. K. Lam, and C. D. B. Bridges, *J. Neurochem.* **42**, 1667 (1984).

[5] F. Gonzalez-Fernandez, R. A. Landers, P. A. Glazebrook, S.-L. Fong, G. I. Liou, D. M. K. Lam, and C. D. B. Bridges, *J. Cell Biol.* **99**, 2092 (1984).

[6] T. M. Redmond, B. Wiggert, F. A. Robey, N. Y. Nguyen, M. S. Lewis, L. Lee, and G. J. Chader, *Biochemistry* **24**, 787 (1985).

[7] J. C. Saari, D. C. Teller, J. W. Crabb, and L. Bredberg, *J. Biol. Chem.* **260**, 195 (1985).

[8] G. I. Liou, C. D. B. Bridges, S.-L. Fong, R. A. Alvarez, and F. Gonzalez-Fernandez, *Vision Res.* **22**, 1457 (1982).

[9] L. Carter-Dawson, R. A. Alvarez, S.-L. Fong, G. I. Liou, H. G. Sperling, and C. D. B. Bridges, *Dev. Biol.* **116**, 431 (1986).

[10] T. van Veen, A. Katial, T. Shinohara, D. J. Barrett, B. Wiggert, G. J. Chader, and J. M. Nickerson, *FEBS Lett.* **208**, 133 (1986).

[11] C. D. B. Bridges, G. I. Liou, R. A. Alvarez, R. A. Landers, A. M. Landry, Jr., and S.-L. Fong, *J. Exp. Zool.* **239**, 335 (1986).

[12] A. Bunt-Milam and J. Saari, *J. Cell. Biol.* **97**, 703 (1983).

[13] Z.-S. Lin, S.-L. Fong, and C. D. B. Bridges, *Vision Res.* **29**, 1699 (1989).

[14] G. I. Liou, S.-L. Fong, J. Gosden, P. van Tuinen, D. H. Ledbetter, S. Christie, D. Rout, S. Bhattacharya, R. G. Cook, Y. Li, C. Wang, and C. D. B. Bridges, *Somatic Cell Mol. Genet.* **13**, 315 (1987).

[15] S.-L. Fong and C. D. B. Bridges, *J. Biol. Chem.* **263**, 15330 (1988).

METHODS IN ENZYMOLOGY, VOL. 189

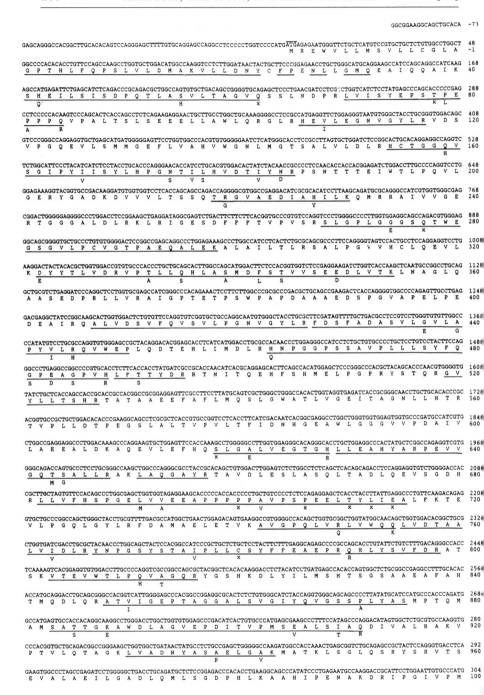

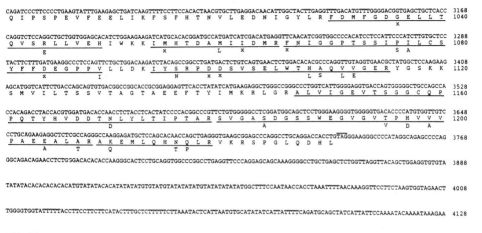

FIG. 1. (*continued*)

human retina λgt10 cDNA library with a 258-base-pair bovine probe (B23) obtained from a bovine retina λgt11 expression library. H.4 IRBP was then employed as a probe to enable isolation and purification of three overlapping human cDNA clones (H8, H12, and H18) from the same human retina λgt10 cDNA library.[15] The complete cDNA sequence of 4230 bases (Fig. 1) was derived from the sequences of H.12, H.4, and H.18.

Purification of Bovine IRBP

Because human eyes are not available in large quantities,[4] most biochemical studies have been carried out on the bovine protein. Bovine IRBP is prepared from 100–200 eyes according to Fong *et al.*[3,16] The interphotoreceptor matrix preparation is dialyzed and loaded on a DEAE-

[16] S.-L. Fong, G. I. Liou, and C. D. B. Bridges, this series, Vol. 123, p. 102.

FIG. 1. Nucleotide and deduced amino acid sequences of human IRBP. The amino-terminal sequence obtained for the mature protein is underlined from amino acid residues 1–32 (gaps denote uncertain residue identification). The tryptic peptide sequences of bovine IRBP corresponding to the deduced human IRBP sequence are also underlined; mismatches are indicated below the underline. Uncertain residues identified by a gas-phase protein sequencer or by mismatches between the peptide sequences and the deduced bovine IRBP sequence [D. E. Borst, T. M. Redmond, J. E. Elser, M. A. Gorda, B. Wiggert, G. J. Chader, and J. M. Nickerson, *J. Biol. Chem.* **264,** 1115 (1989)] are indicated by an x below the underline. The translation initiation site and the stop codon (TAG) are overlined.

cellulose column. The eluted fractions containing IRBP are pooled and passed through a concanavalin A column, which binds IRBP. Fractions eluted with methyl α-D-mannopyranoside are concentrated in the presence of 1% octyl β-D-glucopyranoside. Sugar and detergent are removed by a final gel-filtration step on Sepharose CL-4B.

Preparation and Characterization of Tryptic Peptides of Bovine IRBP

Purified bovine IRBP (23.2 mg) is reduced under a nitrogen barrier with 0.2 g 2-mercaptoethanol in 15 ml of 0.6 M Tris-HCl (pH 8.6), 6 M guanidine-HCl, and 0.2% EDTA.[17] The sulfhydryl groups of cysteine residues are then modified by stirring for 15 min with 476 mg iodoacetic acid in 2 ml of 1 N NaOH. After dialysis against water, the carboxymethylated IRBP is resuspended in 5 ml of 0.2 N N-ethylmorpholinoacetic acid buffer (pH 8.2). Trypsin digestion is carried out by incubating at 37° with two consecutive additions of 1.5 and 2% (w/w) trypsin solutions at 2-hr intervals. At the end of the incubation, the tryptic peptides are separated on a Sephadex G-50 column (1 × 100 cm) and eluted with 0.1 M NH$_4$HCO$_3$. Six 280-nm absorption peaks (T1 through T6) are pooled as illustrated by the typical chromatograph in Fig. 2A.

Each pooled peak is lyophilized and dissolved in 0.1% trifluoroacetic acid. Aliquots of each peak (except T1) are freed of particulate matter by centrifugal filtration (Rainin, Woburn, MA, 0.2-μm cellulosic membranes) and injected into an FPLC (fast purification liquid chromatography) system based on a Pharmacia reversed-phase Pep RPC HR 5/5 column. Gradient chromatography is carried out with two Altex-Beckman Model 100 pumps, an Altex 400 solvent programmer, and an Altex-Hitachi Model 155-10 variable-wavelength detector. The effluent should be continuously monitored at 280 nm.

A typical chromatogram obtained by FPLC of T3 is shown in Fig. 2B. Five to ten percent of each fraction derived from each FPLC peak is loaded on a gas-phase protein sequencer (Applied Biosystems, Foster City, CA) for amino acid sequence determination. In our experience, most of the peaks from the FPLC reversed-phase column contain more than one peptide. However, many of them are present in different relative amounts and can therefore be sequenced directly because the different amounts of phenylthiohydantoin (PTH)-amino acids produced in each cycle of Edman degradation can be used to distinguish different peptides. Thus, peaks R1 and R3 in Fig. 2B are found to contain two peptides each. Peaks R4, R5, and R6 contained single peptides. If the amounts of two or more peptides are about the same, however, further separation on an-

---

[17] A. M. Crestfield, S. Moore, and W. H. Stein, *J. Biol. Chem.* **238**, 622 (1963).

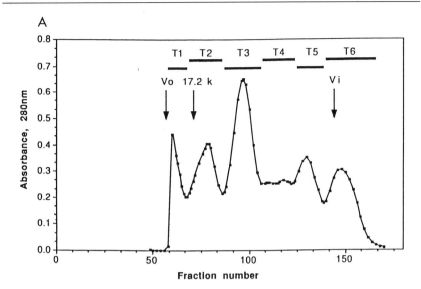

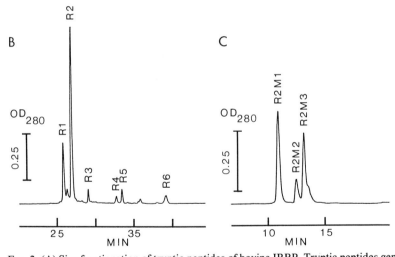

FIG. 2. (A) Size fractionation of tryptic peptides of bovine IRBP. Tryptic peptides generated from reduced carboxymethylated bovine IRBP were loaded on a Sephadex G-50
column (1 × 100 cm) and eluted with 0.1 N NH$_4$HCO$_3$. Fractions of 1 ml were collected. V_o,
void volume; V_i, included volume. (B) Reversed-phase chromatography of T3 on a Pep RPC
HR 5/5 column. Eluent A, 0.1% trifluoroacetic acid in water; eluent B, 0.1% trifluoroacetic
acid in acetonitrile. Flow rate: 0.7 ml/min. Linear gradient 0–50% (B in A) in 30 min. (C)
Ion-exchange chromatography of R2 on a Mono Q column. Eluent A, 20 mM Tris-HCl, pH
7.8; eluent B, 1 M NaCl in eluent A. Flow rate: 1 ml/min. Concave gradient, 0–50% (B in A)
in 15 min, exponent m = 3.

TABLE I
AMINO ACID SEQUENCES OF 10 BOVINE IRBP TRYPTIC PEPTIDES
PURIFIED FROM FRACTION T3

Peptide	Sequence[a]
R1.1.	IYNRPXXSVSELWTLSQLEGER
R1.2[b,c]	LXXTSALVLXL
R2M1	SVGAADGSSWEGVGVVPDVAVPAEAALTR
R2M2	ALVIGEVTSGGCQPPQTYHVDDTDLYLTIPTAR
R2M3	SLGPLGEGSXTWEGSGVLPCVGTPAEQALEK
R3.1[b]	SQEILSISDPQTLAHVLTAGVX
R3.2[b,c]	XASTGEXXXLAGVEPDXXVPM
R4[b]	XNIGGPTSSISALCSYFFXEGPPI
R5	EYYTLVDRVPALLSHLAAMDLSSVVSEDDLVTK
R6[b]	FDSFADASVLEVLGPYILHQVWEP

[a] X denotes residues not identified during peptide sequencing or mistmatched residues between the peptide sequence and the deduced amino acid sequence for bovine IRBP.
[b] These sequences were not completed to the terminal R or K.
[c] The sequence is not underlined in Fig. 1.

other column is necessary. We have found that Mono Q columns are a good choice for this purpose. The Mono Q column is packed with a hydrophilic polymer containing quaternary amine groups that remain equally charged over a wide pH range of 2–12. When the major peak (R2) from the reversed-phase column is chromatographed on this anion-exchange column, three peptides are easily resolved, namely, R2M1, R2M2, and R2M3. The sequences of 10 tryptic peptides obtained from T3 are shown in Table I.

The same strategy may be applied to the other T fractions. The amino-terminal sequence of the protein can be obtained directly from purified bovine IRBP[18,19] (see underlined amino acid residues 1–32 in Fig. 1). During gel-filtration chromatography, T1 elutes in the void volume. This corresponds to a molecular weight of 30,000 or larger, which cannot be analyzed on the Pep RPC column, which has an exclusion limit of 6,000.

Molecular Biology

In our original work,[8] a bovine retina cDNA library in the expression vector λgt11 provided by Drs. Khorana and Oprian (Massachusetts Insti-

[18] S.-L. Fong, R. G. Cook, R. A. Alvarez, G. I. Liou, and C. D. B. Bridges, *FEBS Lett.* **205**, 309 (1986).
[19] T. M. Redmond, B. Wiggert, F. A. Robey, and G. J. Chader, *Biochem. J.* **240**, 19 (1986).

tute of Technology, Cambridge, MA) was screened with polyclonal rabbit anti-bovine IRBP immunoglobulin G (IgG) and ^{125}I-labeled protein or horseradish peroxidase-labeled goat anti-rabbit IgG (Bio-Rad, Richmond, CA). The procedure used was described by Young and Davis[20] and Huynh et al.[21] After the λ phage was absorbed to the Y1090 host cell, 10[5] plaque-forming units (pfu) was plated on a 150-mm plate. IPTG-containing nitro-cellulose filters were overlaid on the plate and saturated with 20% fetal calf serum and 5% nonfat milk. Following incubation with anti-bovine IRBP IgG, the filter was treated with either ^{125}I-labeled protein A or horseradish peroxidase-labeled goat anti-rabbit IgG.

When the human retina cDNA λgt10 library (provided by Dr. Jeremy Nathans, Johns Hopkins School of Medicine, Baltimore, MD) was screened, nick-translated, ^{32}P-labeled cDNA fragments (bovine probe B23 and H.4 IRBP) were used as probes. The procedure of Maniatis et al.[22] was followed. Phage DNA from positive, isolated clones was purified by the plate lysate method using LambdaSorb (Promega Biotec, Madison, WI). The insert was removed from the phage by EcoRI, purified by agarose gel electrophoresis, and subcloned into M13mp18 and M13mp19 for sequencing by the chain-termination method[23] using the procedure of Dale et al.[24]

[20] R. A. Young and R. W. Davis, Science 222, 778 (1983).
[21] T. V. Huynh, R. A. Young, and R. W. Davis, in "DNA Cloning, A Practical Approach" (D. M. Glover, ed.), Vol. 1. IRL Press, Oxford and Washington, D.C., 1985.
[22] T. Maniatis, E. F. Fritsch, and J. Sambrook, "Molecular Cloning: A Laboratory Manual." Cold Spring Harbor Laboratory, Cold Spring Harbor, New York, 1982.
[23] F. Sanger, S. Nicklen, and A. R. Coulson, Proc. Natl. Acad. Sci. U.S.A. 74, 5463 (1979).
[24] R. M. K. Dale, B. A. McClure, and J. P. Houchins, Plasmid 13, 31 (1985).

[21] Purification and Assay of Interphotoreceptor Retinoid-Binding Protein from the Eye

By Alice J. Adler, Gerald J. Chader, and Barbara Wiggert

Introduction

Interphotoreceptor retinoid-binding protein (IRBP) is an extracellular carrier of vitamin A and other lipids.[1-3] It occurs primarily in photosensi-

[1] A. J. Adler and K. J. Martin, Biochem. Biophys. Res. Commun. 108, 1601 (1982).
[2] G. I. Liou, C. D. B. Bridges, S.-L. Fong, R. A. Alvarez, and F. Gonzalez-Fernandez, Vision Res. 22, 1457 (1982).
[3] Y. L. Lai, B. Wiggert, Y. P. Liu, and G. J. Chader, Nature (London) 298, 848 (1982).

tive tissues (eye and pineal). In the vertebrate eye, IRBP is the major soluble component of the interphotoreceptor matrix (IPM, the narrow space between the neural retina and its supportive pigment epithelium). IRBP is synthesized and secreted by photoreceptor cells, and it is thought to participate in the transfer of retinoids between the retina and the pigment epithelium during the visual cycle. The major evidence for this role is that the major endogenous ligands of the IRBP are 11-*cis*-retinaldehyde in the dark and all-*trans*-retinol in the light.

Some properties of IRBP are relevant for the purification procedures presented in this chapter. First, IRBP is a mannose-containing glycoprotein and can, therefore, be separated from other proteins in the IPM by its interaction with concanavalin A (Con A). Second, IRBP is a large protein (the molecular weight of bovine IRBP is 133,000) and appears even larger (~250,000) by gel-sieving methods because of its elongated shape. Thus, it can easily be resolved from most other proteins by size-exclusion chromatography. Third, IRBP is quite acidic (isoelectric point ~4.5), making anion-exchange chromatography a useful preparative step. Finally, IRBP can easily be loaded with exogenous retinol, which can be monitored by absorbance at 330 nm, fluorescence, or radioactivity and constitutes a convenient marker for IRBP during purification.

Since 1982 many articles have been written about the ligands, physical properties, biosynthesis, gene structure, ontogeny, and possible role in diseases of IRBP. The interested reader can find studies from the laboratories of A. J. Adler, C. D. B. Bridges, G. J. Chader, and B. Wiggert, and J. C. Saari and A. H. Bunt-Milam. A previous article on the purification of IRBP has appeared in this series.[4]

Purification of IRBP

The customary starting point for isolation of IRBP is a crude preparation of IPM. The IPM can be obtained by vigorous washing of the retina for maximum yield of IRBP; there is no need to avoid any tissue damage. Even frozen, unwashed retinas may be used as the starting point, since much IRBP clings to the outer retinal surface; this type of preparation is particularly useful for large numbers of eyes. Purification of IRBP to homogeneity[5–9] consists of three chromatographic steps: Con A binding, anion exchange, and gel filtration. The second step can sometimes be

[4] S.-L. Fong, G. I. Liou, and C. D. B. Bridges, this series, Vol. 123, p. 102.
[5] A. J. Adler and C. D. Evans, *Invest. Ophthalmol. Visual Sci.* **26,** 273 (1985).
[6] A. J. Adler and C. D. Evans, *Biochim. Biophys. Acta* **761,** 217 (1983).
[7] T. M. Redmond, B. Wiggert, F. A. Robey, N. Y. Nguyen, M. S. Lewis, L. Lee, and G. J. Chader, *Biochemistry* **24,** 787 (1985).

omitted if the initial IPM preparation is fairly free of cellular contamination. In any case, the purity of IRBP should be assessed by sodium dodecyl sulfate (SDS)–polyacrylamide gel electrophoresis (PAGE) with silver staining; dye staining may not be sensitive enough to detect contaminating proteins.

If the aim of the study is to examine endogenous ligands of IRBP, Con A cannot be used in the preparation, because this treatment detaches retinoids from IRBP. In this case, a one-step partial fractionation with respect to size alone (size exclusion) is sufficient to produce a crude preparation containing IRBP as the only retinoid-binding protein, as all other known retinoid-binding proteins are much smaller.

Preparation of Interphotoreceptor Matrix from Vertebrate Eyes

The operational definition of IPM is taken to be the soluble material lying in the subretinal space (between the retina and the retinal pigment epithelium), material that can be collected by rinsing the formerly apposed surfaces of these two tissues, following separation, with isotonic buffer. In this procedure[5] all steps are carried out at 4°, and 1 mM phenylmethylsulfonyl fluoride (PMSF) is added to inhibit proteases. Eyes are trimmed to remove connective tissue and then are cut just posterior to the ora serrata. The anterior portion is discarded. The vitreous is removed carefully, without causing retinal or choroidal detachment. Final traces of vitreous are removed by washing the anterior retinal surface with phosphate-buffered saline (PBS), consisting of 0.14 M NaCl and 5 mM sodium phosphate (pH 7.2).

Retinas are excised by cutting at the optic nerve. They are pooled in PBS (2 ml/eye for bovine eyes, 1 ml for human, 0.2 ml for rat, for example), stirred at the slowest speed (20 rpm) on a magnetic stirrer for 30 min, and then filtered through plastic mesh (0.8 mm square pore size, Tetko Corp., Elmsford, NY) if there is a large mass of residual retina. The filtrate is saved, and the extraction procedure is repeated for 15 min (utilizing 1 ml buffer/bovine eye) on the residual retinas. The two retinal filtrates (washes) are combined. Meanwhile, material from the internal (apical) surface of the retinal pigment epithelium, still in the eyecup, is extracted by rinsing (gently pipetting) with PBS (e.g., 2 ml per bovine eye) for 30 sec. The IPM preparations from retina and epithelium are clarified by centrifugation (80 min at 20,000 g), combined, and concentrated with

[8] S.-L. Fong, G. I. Liou, R. A. Landers, R. A. Alvarez, and C. D. Bridges, *J. Biol. Chem.* **259**, 6534 (1984).

[9] J. C. Saari, D. C. Teller, J. W. Crabb, and L. Bredberg, *J. Biol. Chem.* **260**, 195 (1985).

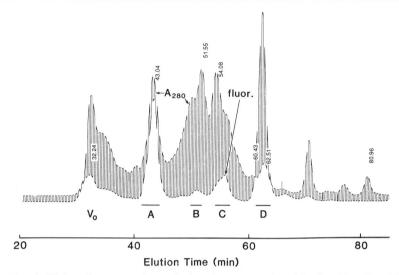

FIG. 1. High-performance, size-exclusion chromatography of bovine IPM. Two TSK columns (4000 SW and 3000 SW) are mounted in series. The effluent is monitored continuously by absorbance (full scale = 1 absorbance unit) and then (15 sec later) by fluorescence. Gel electrophoresis shows that peak A contains all the IRBP but no other retinoid-binding proteins.

macrosolute filters (Amicon B15 Minicon or Centricon, Danvers, MA). Typical amounts of protein collected (as determined by Lowry assays) are 7 mg per bovine eye, 2.5 mg for human, and 0.3 mg for rat. Approximate yields of IRBP in the crude IPM preparations for these species are 400 μg per bovine eye, 200 μg for human, 10 μg for rat. IPM may be stored at $-70°$.

Partial Purification of IRBP with Emphasis on Ligand Retention

Bovine IPM containing 10–15 mg protein in 250 μl of PBS can be fractionated by HPLC size-exclusion chromatography using two Beckman (Fullerton, CA) Spherogel columns in series, for better resolution. The instrument (Beckman System 344) has two detectors in tandem: first an absorbance detector (Model 160) set for 280 nm to monitor protein, then a fluorescence detector (Model 157) set for excitation at 305–395 nm and emission at 420–650 nm to visualize fluorescence of retinol bound to proteins. The solvent (PBS) flow rate is 0.5 ml/min.

Although four pooled peaks in Fig. 1 (A–D) display fluorescence, only peak A (of highest molecular weight) contains IRBP, as shown by SDS–PAGE. However, peak A is contaminated with about 20% of other proteins (e.g., subunit weights 63,000 and 44,000), and thus it cannot be used for studies where pure IRBP is required. Peaks C and D appear to contain cellular retinaldehyde-binding protein (CRALBP) and cellular retinol-binding protein (CRBP), respectively, probably present in the crude IPM preparation from tissue damage. The only retinoid-binding protein in peak A is IRBP; therefore, this one-step preparation is suitable for ligand-binding studies.

For example, one such study examined the effect of light adaptation on the endogenous retinoid ligands of bovine IRBP.[10] It was found that, for dark-adapted eyes, the ligands were (in nanomoles per bovine eye, which contains ~3 nmol IRBP) 0.11 11-*cis*-retinol, 0.09 all-*trans*-retinol, 0.14 11-*cis*-retinal, and 0.06 all-*trans*-retinal. For light-adapted eyes, the amount of all-*trans*-retinol was found to increase by a factor of 5 and that of 11-*cis*-retinal to decrease by a factor of 4.

Isolation of IRBP from Interphotoreceptor Matrix

Purification of IRBP from six bovine eyes is presented as an example. This procedure can be scaled down for smaller, or fewer, eyes. The first step is affinity chromatography. Bovine IPM containing approximately 40 mg total protein is stirred with 5 ml Con A–Sepharose beads (Pharmacia, Piscataway, NJ) in 20 ml of 0.14 M NaCl plus 10 mM Tris (pH 7.0) for 1 hr at room temperature. The supernatant is discarded, as are two batches of buffer used to wash the beads. The fraction containing IRBP is then liberated by stirring the beads in 20 ml of 50 mM α-methyl-D-mannoside (Sigma, St. Louis, MO) in the same buffer. The beads are removed by filtration. (Alternatively, the initial IPM/Con A–Sepharose mixture can be poured into a chromatography column for washing and for methylman-noside elution; this is faster if large volumes are required.) The yield of IRBP is approximately 300 μg per eye at this stage.[5,6] This partially puri-fied IRBP preparation may be stored at $-70°$. It is sometimes useful to incubate this preparation with tracer amounts of [3H]retinol to allow mon-itoring of the IRBP peak during subsequent steps in the purification.

After Con A treatment the preparation contains about 6% of an impu-rity at molecular weight 55,000 (as seen by PAGE). This can be removed by passage through a Sepharose CL-4B column (0.9 × 30 cm, conven-tional gel-filtration chromatography) or through a TSK 4000 SW column (size-exclusion HPLC) eluted with PBS and monitored by protein absor-bance at 280 nm. Fractions (1 ml each) of the principal peak (apparent M_r ~250,000), displaying at least half-maximum absorbance, are pooled.

[10] A. J. Adler and C. D. Evans, *in* "The Interphotoreceptor Matrix in Health and Disease" (C. D. Bridges and A. J. Adler, eds.), p. 65. Alan R. Liss, New York, 1985.

This preparation consists of essentially homogeneous IRBP; the yield is about 170 μg per bovine eye.

As a modification of this procedure, frozen, unwashed retinas can be used as the starting material instead of isolated IPM. For example, 300 frozen bovine retinas, obtained from Hormel, Inc. (Austin, MN), are thawed in, and then stirred in, 300 ml PBS. The resulting supernatant (a very crude IPM wash) is then treated with 40 ml Con A–Sepharose beads, as above. The preparation at this stage contains intracellular proteins from the retina and requires an anion-exchange step to ensure proper homogeneity of IRBP. The supernatant from methylmannoside treatment is applied to either a DEAE-cellulose column[4] for conventional anion-exchange chromatography or to a Pharmacia Mono Q column[7] for anion-exchange HPLC. In either case, elution is carried out with a 0–0.5 M NaCl gradient in 10 mM Tris buffer (pH 7.5) and is monitored at 280 nm. IRBP, identified by SDS–PAGE and [^{3}H]retinol binding,[11] elutes at approximately 0.33 M NaCl and is about 95% pure. Final purification to homogeneity, as assessed by SDS–PAGE with silver staining, is accomplished by size-exclusion HPLC, as above. If all three steps are used in the purification procedure, the yield of IRBP is approximately 100 μg per bovine eye.

Quantification of IRBP

The absorption coefficient of IRBP[12] (ε_{280} = 11.6 × 10^4 M^{-1} cm^{-1} or E_{280} for 1% = 8.7) can be used to quantify pure IRBP in fairly concentrated solutions. In large samples containing mixtures (but no other retinol-binding proteins), the amount of IRBP may be estimated qualitatively by incubating the solution with an excess of retinol, passing this mixture through a gel-sieving column, and measuring the amount of protein-bound retinol by absorbance, fluorescence, or (if radiolabeled retinol is used) scintillation counting.

For small amounts of IRBP, especially in mixtures such as tissue samples, the only accurate methods are immunochemical. For example, to examine the possible role of IRBP in retinal degeneration, it is important to quantify the protein in retinas from human donor eyes of patients with retinal degeneration and in animal models of retinal degeneration. A competitive enzyme-linked immunosorbent assay (ELISA)[13,14] has been

[11] B. Pfeffer, B. Wiggert, L. Lee, B. Zonnenberg, D. Newsome, and G. Chader, *J. Cell Physiol.* **117,** 333 (1983).

[12] A. J. Adler, C. D. Evans, and W. F. Stafford, *J. Biol. Chem.* **260,** 4850 (1985).

[13] A. Voller, A. Bartlett, and D. E. Bidwell, *J. Clin. Pathol.* **31,** 507 (1978).

[14] B. Wiggert, L. Lee, M. Rodrigues, H. Hess, T. M. Redmond, and G. J. Chader, *Invest. Ophthalmol. Visual Sci.* **27,** 1041 (1986).

found to be accurate in the microgram range using purified monkey IRBP and rabbit anti-monkey IRBP. This works well with human retinal samples of approximately 4 × 4 mm. Because IRBP is soluble and easily lost, these samples should consist of an intact sandwich of retina, pigment epithelium, and choroid. This method has not been found to be as successful in developmental studies using whole mouse eyes, where a high background interferes with the assay. In studies of IRBP in developing mouse eyes, a slot-blot assay[15,16] has been devised, which is accurate in the nanogram range using purified bovine IRBP and goat anti-bovine IRBP. One mouse eye is sufficient for analysis, although usually four or five mouse eyes should be pooled to avoid biological variation. Both of these methods should be validated, by means of another immunochemical method, for the species and tissue under investigation. In our hands, for example, these methods were confirmed by Western blotting and by immunohistochemistry, which, although qualitative, showed the same trends as ELISA and slot blots when various samples were compared.

Enzyme-Linked Immunosorbent Assay of IRBP

Primary antiserum against purified monkey IRBP is raised in rabbits. Tissue samples to be analyzed are homogenized in Tris buffer (10 mM Tris, 2 mM EDTA, 0.5 M NaCl, pH 7.5) and centrifuged at 100,000 g for 1 hr. All reagents should be reagent grade, and HPLC-grade water should be used throughout. Polystyrene (round bottomed) microtiter plates with 96 wells are coated with purified monkey IRBP (50 μl per well of a 1 μg/ml solution prepared in 66 mM NaHCO$_3$–34 mM Na$_2$CO$_3$ buffer, pH 9.5) at 4° for 16 hr. The plates are washed 3 times with 0.1% (v/v) Tween 20 in Dulbecco's phosphate-buffered saline (PBS) (200 μl per well). Blocking of nonspecific interactions is accomplished with 1% (w/v) bovine serum albumin (BSA) in PBS (100 μl per well, freshly prepared) for 30 min at ambient temperature.

Purified monkey IRBP is prepared in a series of dilutions, from 1 to 250 μg/ml in PBS.[14] Portions (25 μl) of either the sample (100,000 g supernatant) or dilutions of monkey IRBP are added to appropriate sample wells, followed by the addition of 25 μl of a 1 : 3750 dilution (in 1% BSA in PBS) of rabbit anti-monkey IRBP antiserum. Control wells contain only antibody or 1% BSA in PBS. The plates are wrapped with plastic wrap and incubated at 37° for 1 hr, followed by 3 washes as above.

β-Galactosidase anti-rabbit F(ab')$_2$ (Amersham, Arlington Heights,

---

[15] J. M. Gershoni and G. E. Palade, *Anal. Biochem.* **131**, 1 (1983).
[16] T. van Veen, P. Ekstrom, B. Wiggert, L. Lee, Y. Hirose, S. Sanyal, and G. J. Chader, *Exp. Eye Res.* **47**, 291 (1988).

IL) is diluted 1 : 250 in 0.1% Tween 20 in PBS containing 10 mM MgCl$_2$ and 1 mM 2-mercaptoethanol. Fifty microliters of this solution is added to each well, the plates are wrapped with plastic wrap, and incubation is carried out for 2 hr at 37°, followed by 3 washes as above. One hundred microliters of 3 mM o-nitrophenyl-β-D-galactopyranoside in PBS containing 10 mM MgCl$_2$ and 0.1 M 2-mercaptoethanol is added to each well, and the plates are incubated for 1 hr at 37°. A 50-μl portion of 1 M Na$_2$CO$_3$ is then added to each well, and the plates are read with a Titretek Multiscan (Flow Laboratories, McLean, VA) at 414 nm (see Fig. 2). Unknown samples can then be quantified relative to the monkey IRBP standards.

When purified human IRBP is substituted for monkey IRBP in coating the microtiter plates, essentially the same results are obtained. Thus, monkey IRBP can be used for direct quantification of IRBP in human tissue samples. This assay is accurate to about 1 μg IRBP.

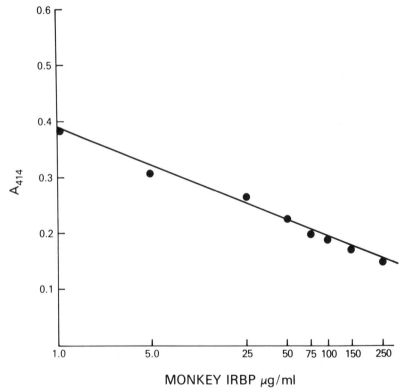

MONKEY IRBP μg/ml

FIG. 2. Typical competitive binding curve of the enzyme-linked immunosorbent assay (ELISA). Microtitration plates are coated with purified monkey IRBP at a concentration of 1 μg/ml. Rabbit antiserum to IRBP is used at a dilution of 1 : 7500. (Reproduced from Wiggert et $al.$[14])

Slot-Blot Quantification of IRBP

Primary antiserum against purified bovine IRBP is raised in goats. Tissue samples, such as whole mouse eyes or posterior segments, are homogenized in Tris buffer (composition given in the previous section), 60 μl per mouse eye, and centrifuged at 100,000 g for 60 min. Western blotting is performed on mouse samples of different postnatal ages, using a 1:150 dilution of goat anti-bovine IRBP. In this case, the antiserum reacts only with a single protein band of molecular weight 140,000, approximately the same M_r as purified bovine IRBP.

For slot-blot quantification of mouse IRBP relative to purified bovine IRBP standards,[16] an Immobilon polyvinylidene difluoride (PVDF) membrane (Millipore Corp., Bedford, MA) is first dipped in methanol (HPLC grade) for about 2 sec, then rinsed thoroughly with water (HPLC grade used throughout) followed by soaking in Tris-buffered saline (TBS, 0.9% NaCl in 20 mM Tris-HCl, pH 7.4). The membrane is then placed in a Microsample Filtration Manifold II (Schleicher and Schuell, Keene, NH) apparatus above 2 sheets of filter paper soaked with TBS, making certain that no air bubbles are trapped between the membrane and filter paper. A gentle vacuum is applied and TBS (~200 μl) is added to each sample slot to make certain that the flow rate is uniform in all slots. Purified bovine IRBP standards (0.5–75 ng in 200 μl of 0.01% BSA in TBS), blanks (1.0% BSA in TBS, 200 μl), and tissue samples (200 μl, diluted in 0.1% BSA in TBS) are added to the appropriate slots, and gentle vacuum is applied for 15 min to assure complete adsorption.

Following sample application, the membrane is removed from the blotting apparatus, and all subsequent steps are carried out in a covered plastic dish. Blocking of nonspecific interactions is accomplished by incubating the membrane with 100 ml of 5% (w/v) nonfat dried milk (Carnation) in TBS with gentle agitation at ambient temperature for 1 hr. The membrane is then rinsed with water followed by 3 washes with 0.1% BSA in TBS for 10 min with gentle agitation at ambient temperature. One hundred milliliters of goat anti-bovine IRBP (1:150 dilution in 1% BSA in TBS) is added to the membrane, followed by overnight incubation with gentle agitation at 4°. The membrane is rinsed with water, and then 3 washes are performed, as above, at ambient temperature. One hundred milliliters of affinity-purified, horseradish peroxidase (HRP)-conjugated rabbit anti-goat immunoglobulin G (IgG) (H + L) (Kirkegaard-Perry, Gaithersburg, MD) (1:1000 dilution in 1% BSA in TBS) is added, followed by incubation at 37° for 1 hr with gentle agitation. This is followed by a water rinse and 3 washes, as above, at ambient temperature.

In the final step, immediately before use, 60 mg of the color reagent 4-chloro-1-naphthol is dissolved in 20 ml methanol (at 4°), and 60 μl of the

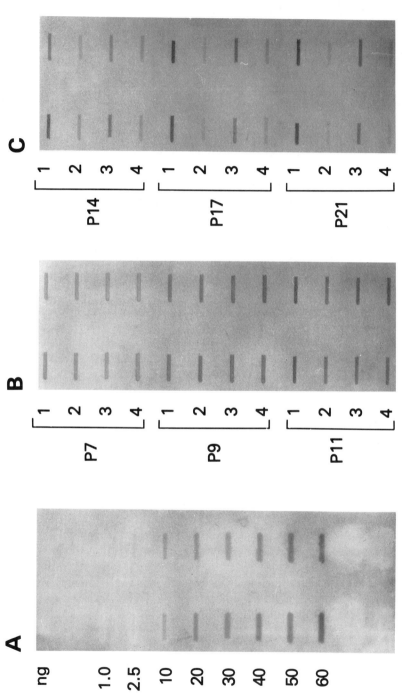

FIG. 3. Slot blots of soluble proteins from mouse eye. (A) Purified bovine IRBP, 1–60 ng; (B) mouse eye cytosol samples at postnatal day 7 (P7), P9, and P11; (C) mouse eye cytosol samples at P14, P17, and P21. 1, +/+, +/+; 2, rd/rd, +/+; 3, +/+, rds/rds; 4, rd/rd, rds/rds. Each sample is run in duplicate on the slot blots. Left-hand and right-hand columns represent one pair of the duplicates. (Reproduced from van Veen et al.[16]

substrate, 30% H_2O_2, is added (separately) to 100 ml TBS (at 4°). The color reagent and the substrate are then mixed together and added to the membrane. Color development is allowed to proceed for about 1.5 min followed by immediate rinsing with water to remove excess color reagent. The slot blots are photographed (see Fig. 3) and printed on Ilford paper, and the photographs are scanned using an LKB Ultra-Scan XL laser densitometer. Unknown samples are assayed and compared to bovine IRBP standards using a 2400 gel scan software package (LKB, Bromma, Sweden) and an IBM AT computer. As little as 2.5 ng IRBP can be accurately quantified in the mouse samples by this method.

Acknowledgments

This work was supported, in part, by National Institutes of Health Grant EY04368.

[22] Identification of Receptors for Retinoids as Members of the Steroid and Thyroid Hormone Receptor Family

By VINCENT GIGUÈRE and RONALD M. EVANS

Introduction

The identification of molecules and mechanisms that control cell fate during development of eukaryotic organisms is a central problem in modern molecular biology. One class of substances known to have dramatic effects on both cellular differentiation and proliferation are the retinoids. Until recently, the mechanism of action by which retinoids exert these effects was unknown. However, molecular studies of nuclear receptors for steroid and thyroid hormones have led to the identification of a superfamily of ligand-inducible regulatory proteins that includes receptors for retinoic acid.[1] This chapter describes the strategies devised to identify nuclear receptors for retinoic acid as members of the steroid and thyroid hormone receptor family and the potential application of these methods to the study of the role of retinoids in developmental processes, homeostasis, and disease.

[1] R. M. Evans, *Science* **240**, 899 (1988).

Molecular Cloning of Receptors for Retinoids

Background

The expression cloning of the human glucocorticoid receptor (hGR) not only provided the first completed structure of a steroid receptor but also revealed amino acid sequence identity with the oncogene *erbA*.[2,3] The relationship between hGR and *erbA* was later confirmed by the molecular cloning of specific receptors for estradiol, progesterone, and vitamin D_3 (reviewed in Ref. 1). The characterization of the *erbA* protoonco-gene product led to its identification as the thyroid hormone receptor.[4,5] Sequence comparison and mutational analysis of these proteins demonstrated structural features common to all steroid and thyroid hormone receptors. In particular, the receptors share a highly conserved cysteine-rich region encoding so-called zinc fingers which function as the DNA-binding domain.[6] The discovery of this common region led to the proposal that this segment might be diagnostic for related ligand-inducible nuclear transcription factors.

Strategies

Two variations of the idea proposed above have been used to identify novel receptors. In the first approach, a hybridization probe encoding a region of high homology between steroid hormone receptors is initially used to identify related sequences in the human genome by Southern blot analysis. A restriction enzyme fragment which is well resolved from other hybridizing bands is then chosen for direct genomic cloning. Restriction enzyme-digested genomic DNA is size-fractionated on an agarose gel, and the proper region is isolated for the construction of a partial genomic library. The library is then screened under conditions of low-stringency hybridization with the initial probe. Subsequently, the insert from a positive genomic clone is sequenced to confirm the cloning of a novel receptor gene. A hybridization probe derived from the genomic fragment is then used to screen cDNA libraries under high-stringency hybridization conditions. By taking advantage of this approach, the human mineralocorticoid receptor cDNA was isolated using a human glucocorticoid receptor hy-

[2] S. H. Hollenberg, C. Weinberger, E. S. Ong, G. Cerelli, A. Oro, R. Lebo, E. B. Thompson, M. G. Rosenfeld, and R. M. Evans, *Nature (London)* **318**, 635 (1985).

[3] C. Weinberger, S. H. Hollenberg, M. G. Rosenfeld, and R. M. Evans, *Nature (London)* **318**, 670 (1985).

[4] C. Weinberger, C. C. Thompson, E. S. Ong, R. Lebo, D. J. Gruol, and R. M. Evans, *Nature (London)* **324**, 641 (1986).

[5] J. Sap, A. Munoz, K. Damm, Y. Goldberg, J. Ghysdael, A. Leutz, H. Beug, and B. Venstrom, *Nature (London)* **324**, 635 (1986).

[6] R. M. Evans and S. H. Hollenberg, *Cell (Cambridge, Mass.)* **52**, 1 (1988).

bridization probe.[7] This method is particularly useful when the initial hybridization probe is derived from a receptor gene for which expression is ubiquitous, such as the glucocorticoid receptor.

A more direct approach to search for unrecognized hormone response systems is to systematically employ reduced stringency hybridization conditions to screen cDNA libraries made from mRNAs isolated from a variety of target tissues. The hybridization probe can either be a DNA fragment encoding the DNA-binding domain of a particular receptor (the most conserved region between receptors) or a consensus oligonucleotide corresponding to a consensus sequence encoding the DNA-binding domain of a number of steroid hormone receptors. Phage plaques or colonies giving positive signals are then rescreened under high-stringency conditions with cDNA probes derived from the DNA-binding domain of cloned receptors to eliminate clones of these receptors. The cDNA inserts of the remaining positive clones are then analyzed further by restriction enzyme mapping and DNA sequencing. For example, using the DNA-binding domain of the human estrogen receptor as a probe, two distinct cDNAs encoding cryptic steroid hormone receptors were identified.[8]

Eventually, it is a combination of both approaches that led to the discovery of the retinoic acid receptors. First, analysis of the integration site of a hepatitis B virus from a human hepatocellular carcinoma led to the fortuitous identification of a novel genomic sequence with striking similarity to the DNA-binding domain of the steroid hormone receptors. A hybridization probe derived from this sequence was used to isolate full-length cDNAs encoding novel members of the steroid hormone receptor family.[9,10] A second approach using a consensus oligonucleotide as a hybridization probe led to the isolation of a partial cDNA clone encoding a large portion of one of the novel proteins.[11] However, as explained in the next section, a strategy first had to be developed in order to identify the ligand associated with these novel receptors as retinoic acid.

Identification of the Specific Ligand

Finger Swap

As the ligands for novel gene products identified by low-stringency hybridization are unknown, a quick and sensitive assay had to be devel-

[7] J. L. Arriza, C. Weinberger, G. Cerelli, T. Glaser, B. L. Handelin, D. E. Housman, and R. M. Evans, *Science* 237, 268 (1987).
[8] V. Giguère, N. Yang, P. Segui, and R. M. Evans, *Nature (London)* 331, 91 (1988).
[9] V. Giguère, E. S. Ong, P. Segui, and R. M. Evans, *Nature (London)* 330, 624 (1987).
[10] H. de Thé, A. Marchio, P. Tiollais, and A. Dejean, *Nature (London)* 330, 667 (1987).
[11] M. Petkovitch, N. J. Brand, A. Krust and P. Chambon, *Nature (London)* 330, 444 (1987).

oped to reveal their identity. This strategy takes advantage of the modular structure of the steroid hormone receptors and the suggestion that functional domains may be interchangeable.[12] In the procedure illustrated in Fig. 1A, the DNA-binding domain of the novel receptor is replaced by the well-characterized DNA-binding domain of the human glucocorticoid receptor (the DNA-binding domain of any receptor for which the DNA recognition element has been elucidated can be used for the domain switch). To achieve the domain switch, the respective cDNAs must first be mutagenized *in vitro* to introduce common restriction enzyme sites on each border of the DNA-binding domains. The choice of the novel restriction enzyme sites is governed by two factors. First, for added convenience, it is preferable that these sites not be present in any of the two cDNAs or the expression vector. Second, one must try to minimize the changes resulting from the mutagenesis in the amino acid sequences encoded by the two cDNAs as drastic changes in amino acid sequence (a proline residue for a glycine residue, for example) could engender a nonfunctional hybrid receptor. Once the mutagenesis is completed, DNA fragments encoding the respective DNA-binding domain are interchanged to yield chimeric receptors bearing novel ligand specificity.

The assay system is established by transfecting recipient cells (CV-1 or HeLa cells are suitable) with the hybrid receptor cDNA and a glucocorticoid-responsive reporter gene. The chloramphenicol acetyltransferase (CAT) and the luciferase (LUC) gene linked to the mouse mammary tumor virus long-terminal repeat (LTR) (MMTV CAT or MMTV LUC) have been successfully used in this assay. Transfected cells are then systematically challenged with candidate ligands, and hybrid receptor activation is monitored by changes in reporter enzyme activity. As shown in Fig. 1B, retinoic acid elicits a dramatic increase in CAT activity induced by a hybrid receptor whereas other putative ligands do not induce a response.

High-Affinity Binding

To corroborate the identity of the novel gene product as the retinoic acid receptor (RAR), the binding properties of the expressed protein must also be evaluated. High levels of polypeptide expression from transfected RAR are essential to facilitate retinoic acid-binding experiments in transfected cells. Because plasmids containing the SV40 origin of replication can replicate to high copy numbers in COS cells and thus produce a high level of heterologous gene expression at 48–72 hr posttransfection,[13] an

---

[12] S. Green and P. Chambon, *Nature (London)* **327**, 75 (1987).
[13] Y. Gluzman, *Cell (Cambridge, Mass.)* **23**, 175 (1981).

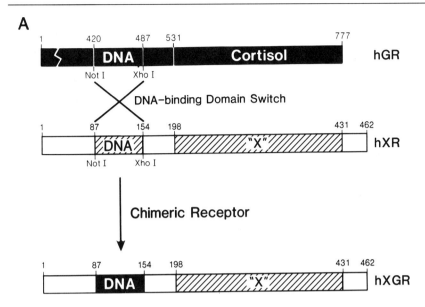

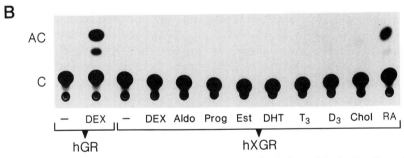

Fig. 1. Finger swap assay. (A) The modular structure of members of the family of steroid hormone receptors allows the exchange of functional domains between a candidate receptor and the glucocorticoid receptor. The resulting chimeric receptor (XGR) should stimulate a glucocorticoid-responsive reporter gene in the presence of the appropriate ligand. (B) Induction of CAT activity by a chimeric receptor in the presence of retinoic acid. The expression vector encoding the chimeric receptor was cotransfected with the MMTV CAT reporter plasmid in CV-1 cells which were cultured for 24 hr in the presence of candidate ligands. The ligands are dexamethasone (DEX), aldosterone (Aldo), progesterone (Prog), estradiol (Est), 5α-dihydrotestosterone (DHT), triiodothyroxine (T_3), vitamin D_3 (D_3), 25-hydroxycholesterol (Chol), and retinoic acid (RA).

expression vector containing the human RARα coding sequence, under the control of the promoter of the Rous sarcoma virus (RSV), and the SV40 origin of replication was constructed (see Fig. 2). Transfected COS-1 cells showed increased capacity to bind [^{3}H]retinoic acid which competes with unlabeled retinoic acid.[9] The use of this expression technique wil render possible, in the near future, the precise determination of the dissociation constants for the binding of [^{3}H]retinoic acid and a large number of analogs to human RARα and RARβ by Scatchard analysis.

Retinoic Acid Receptors

Anatomy

Using the hybrid receptor assay system described above, two distinct receptors for retinoic acid have been identified to date, referred to as RARα,[9,11] RAR,β,[14,15] and RARγ.[16,17] As described for other members of the steroid and thyroid hormone receptor family, the retinoic acid receptors are composed of a melange of regulatory domains in which ligand-binding, DNA-binding, and transactivation functions are part of an interactive cascade.[1] As expected, the three RAR proteins show a high degree of amino acid identity in the DNA- and hormone-binding domains (over 90%). The region located between the DNA- and hormone-binding domains, which probably functions as a hinge, shows moderate amino acid homology (between 60 and 70%). However, both the amino- and carboxy-terminal regions of the three receptors are totally unrelated in their amino acid sequence, which suggests that these domains may contribute to important functional differences between receptors.

Distribution

The existence of multiple RARs implies that a single ligand may have multiple receptors. Although the physiological significance of having multiple receptors for retinoic acid has yet to be assessed, one possibility is that they are expressed in tissue-specific fashion. Indeed, studies on the tissue distribution of the transcripts for RARα, RARβ, and RARγ have revealed marked differences in their spatial patterns of expression. The RARα gene generates two transcripts of 3.2 and 2.3 kilobases (kb) which

[14] N. J. Brand, M. Petkovitch, A. Krust, P. Chambon, H. de Thé, A. Marchio, P. Tiollais, and A. Dejean, *Nature (London)* **332,** 850 (1988).

[15] D. Benbrook, E. Lernhardt, and M. Pfahl, *Nature (London)* **333,** 669 (1988).

[16] V. Giguère, M. Shago, R. Zirngibl, P. Tate, J. Rossant, and S. Varmuza, *Mol. Cell. Biol.* **10,** 2335 (1990).

[17] A. Zelent, A. Krust, M. Petkovitch, P. Kastner, and P. Chambon, *Nature (London)* **336,** 714 (1989).

can be detected at low levels in all tissues examined[9,18] but were overexpressed in two specific areas of the brain, cerebellum and hippocampus,[9] and in the hematopoietic cell lines K562 and HL60.[18] The pattern of expression of the RARβ transcripts shows more variation. Two different transcripts of 3.0 and 2.5 kb can be detected in all tissues examined except the spinal cord and the liver where the minor transcript was undetectable.[18] However, high levels of expression of the RARβ transcripts were observed in the kidney, prostate gland, spinal cord, and cerebral cortex whereas somewhat lower levels were detected in the liver, spleen, uterus, ovary, breast, and testis.[15,18] No expression of the RARβ gene was detected in the K562 and HL60 cells. The different isoforms of the RARγ are expressed in the mouse embryo and preferentially in the adult skin.[16,17] The observation that the three RAR genes are differentially expressed suggests that their respective promoter and associated enhancer(s) respond to distinct metabolic and/or hormonal regulators. Finally, since retinoic acid has been shown to influence vertebrate limb pattern development and regeneration, the expression of the RARβ was investigated in the adult newt during regeneration of an amputated limb.[19] The presence of RARβ transcripts is detected at high but uniform levels in the mesenchymal regenerating blastema cells, indicating that retinoic acid might establish positional information through differential receptor activation of gene expression.

Assay for Transcriptional Regulation

To identify functional domains for hormone binding, DNA binding, and transactivation in steroid hormone receptors a screening assay that uses cultured cells transfected with two DNA expression vectors was developed (Fig. 2).[20] In the assay, an expression vector provides for the efficient production of receptor in cells that do not normally express the receptor gene while a reporter vector containing a LUC (or CAT) gene fused to a hormone-responsive promoter provides for the expression of the reporter enzyme. Applications of hormone or an experimental agonist will activate the luciferase gene, causing light to be emitted from the cell extracts. The level of light emitted is directly proportional to the effectiveness of the hormone–receptor complex in activating gene expression.

In the case of the retinoic acid receptors, a direct functional assay for transcriptional regulation by the retinoid–receptor complex was not pos-

[18] H. de Thé, A. Marchio, P. Tiollais, and A. Dejean, *EMBO J.* **8,** 429 (1989).
[19] V. Giguère, E. S. Ong, R. M. Evans, and C. Tabin, *Nature (London)* **337,** 566 (1989).
[20] V. Giguère, S. H. Hollenberg, M. G. Rosenfeld, and R. M. Evans, *Cell (Cambridge, Mass.)* **46,** 645 (1986).

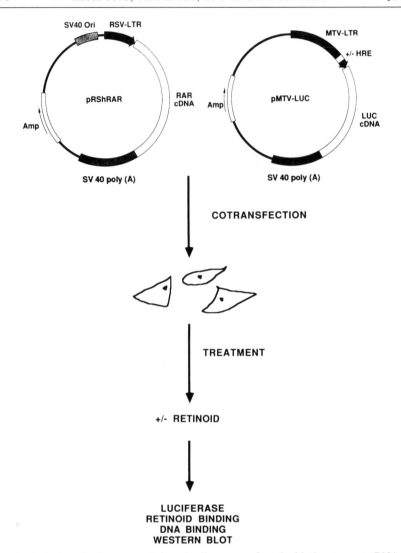

FIG. 2. Cotransfection assay. Cultured cells are transfected with the receptor cDNA in an expression vector (top left) and with a reporter plasmid containing a reporter gene linked to an appropriate hormone-responsive promoter (top right). To provide an internal control for transfection efficiency, a plasmid (not shown) containing the *Escherichia coli* β-galacto-sidase gene driven by the RSV LTR is cotransfected along with the expression and reporter plasmids. For the functional study of the RARs, the expression vector encodes a human RAR and the reporter plasmid the luciferase gene fused to the MMTV promoter in which the glucocorticoid-response elements (GREs) have been replaced by a palindromic TRE (see text for details). Transfected cells can be used to monitor induction of luciferase activity, retinoid binding, receptor binding to the DNA, and expression of RAR protein.

sible because the DNA sequences conferring retinoic acid responsiveness had not been identified. Once again, the homology in the amino acid sequence of the DNA-binding domains among various members of the steroid and thyroid hormone receptor family suggests a strategy to identify a retinoic acid-response element (RARE). Because the DNA-binding domains of the retinoic acid and thyroid hormone receptors are highly related, the possibility that the retinoic acid receptor could activate gene expression through a thyroid hormone-response element (TRE) was explored. Novel thyroid hormone-responsive promoters were first constructed by replacing the glucocorticoid-responsive elements present in the MMTV LTR with an oligonucleotide encoding a palindromic TRE (TREp). These promoters are then fused to the LUC gene to generate reporter plasmids such as ΔMTV–TREp–LUC. Using the cotransfection assay described above (Fig. 2), receptorless COS-1 cells are cotransfected with the reporter plasmid and the expression vector containing the RARα

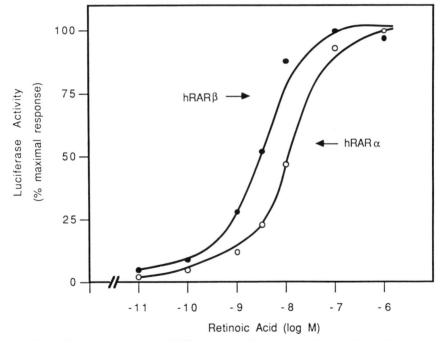

FIG. 3. Transcriptional assay. COS-1 cells transfected as described in Fig. 2 with expression vectors encoding either the human RARα or RARβ and the reporter plasmid ΔMTV–TREp–LUC were challenged with increasing concentrations of retinoic acid. The apparent ED_{50} values for retinoic acid in this particular experiment is 2 nM for RARβ and 10 nM for RARα.

or RARβ. The transfected cells are then treated with retinoids. As illustrated in Fig. 3, increasing concentrations of retinoic acid provoke a dose–response induction of LUC activity. This assay reveals that the concentration of retinoic acid leading to 50% of maximum LUC activity (ED$_{50}$) is significantly different between RARα and RARβ (2 and 10 nM, respectively). This system offers a rapid and sensitive assay to test the relative potency of a large number of natural and synthetic retinoids for their ability to activate receptor function. The availability of a reporter gene responsive to retinoic acid also provides a direct assay to test for the presence of, and the transcriptional response to, endogenous RAR(s) in a particular cell line. For example, endogenous RAR(s) present in F9 tetratocarcinoma cells can lead to high levels of LUC activity when transfected with the reporter plasmid ΔMTV–TREp–LUC and treated with retinoic acid.[21] Thus, TREp-containing vectors provide a simple and quantitative assay for retinoic acid responsiveness in cultured cells.

Conclusion

The discovery that the RARs belong to the superfamily of steroid and thyroid hormone receptors led us to propose that the interaction of retinoids with their intracellular receptors induces a cascade of regulatory events which results from the activation of specific sets of genes by the retinoid–receptor complex.[9] The identification of cDNAs encoding RARs that can be used to direct high-level expression of RAR protein together with the development of sensitive transcriptional assays opens new avenues of investigation of the pharmacology as well as the molecular mechanism of action of retinoids. In turn, these studies should lead to the development of better drugs for the treatment of a variety of diseases, including cancer.

Acknowledgments

We wish to thank Dr. M. Pfahl for the gift of the human RARβ cDNA clone. This research was supported by the Howard Hughes Medical Institute and grants from the National Institute of Health and the National Cancer Institute of Canada. V.G. is a Scholar of the Medical Research Council of Canada.

---

[21] K. Umesono, V. Giguère, C. K. Glass, M. G. Rosenfeld, and R. M. Evans, *Nature* (*London*) **336**, 262 (1988).

[23] Retinoic Acid Acylation: Retinoylation

By NORIKO TAKAHASHI and THEODORE R. BREITMAN

Introduction

Retinoic acid (RA) has numerous effects on many cell types *in vivo* and *in vitro*. One example is the RA-induced terminal differentiation of the human acute myeloid leukemia cell line HL60 to cells having many of the functional and morphological characteristics of mature granulocytes.[1] Retinoic acid induces the differentiation *in vitro* of cells from patients with acute promyelocytic leukemia[2] and, as a sole agent, induces complete remissions of patients with acute promyelocytic leukemia.[3]

A mechanism for the effects of RA on HL60 as well as on other cells has recently been provided by the discovery of two human nuclear retinoic acid receptors (hRARα and hRARβ)[4–6] which are members of the steroid/thyroid nuclear receptor multigene family. These two proteins have specific high-affinity binding sites for RA, and both of them are present in HL60[7] even though only the mRNA for hRARα is detected.[8] Many cell types responding to RA have a cellular RA-binding protein that is distinct from hRAR. However, not all target cells for RA contain cellular RA-binding protein, HL60 being one of the salient examples.[2] As an alternative explanation for the action of RA on these cells it was proposed[9] that RA, after activation in a CoA-mediated reaction, is covalently bound to proteins at sites that may compete with other modifications such as acylation or phosphorylation. Evidence for this reaction was recently reported.[10]

---

[1] T. R. Breitman, S. E. Selonick, and S. J. Collins, *Proc. Natl. Acad. Sci. U.S.A.* **77,** 2936 (1980).

[2] T. R. Breitman, S. J. Collins, and B. R. Keene, *Blood* **57,** 1000 (1981).

[3] M. E. Huang, Y. C. Ye, S. R. Chen, J. R. Chai, J. X. Lu, Z. Lin, L. J. Gu, and Z. Y. Wang, *Blood* **72,** 567 (1988).

[4] N. Brand, M. Petkovich, A. Krust, P. Chambon, H. de The, A. Marchio, P. Tiollais, and A. Dejean, *Nature (London)* **332,** 850 (1988).

[5] V. Giguere, E. S. Ong, P. Segui, and R. M. Evans, *Nature (London)* **330,** 624 (1987).

[6] M. Petkovich, N. J. Brand, A. Krust, and P. Chambon, *Nature (London)* **330,** 444 (1987).

[7] M. P. Gaub, Y. Lutz, E. Ruberte, M. Petkovich, N. Brand, and P. Chambon, *Proc. Natl. Acad. Sci. U.S.A.* **86,** 3089 (1989).

[8] H. de The, A. Marchio, P. Tiollais, and A. Dejean, *EMBO J.* **8,** 429 (1989).

[9] T. R. Breitman, B. R. Keene, and H. Hemmi, *Cancer Surv.* **2,** 263 (1983).

[10] N. Takahashi and T. R. Breitman, *J. Biol. Chem.* **264,** 5159 (1989).

Method for Detection of Retinoylated Protein(s)

Principle. Cells are incubated with radioactive RA and then extracted by the Bligh–Dyer procedure[11] using $CHCl_3$ and methanol to remove lipids including free RA. The delipidated pellet is the starting material for further analyses.

Reagents. RPMI 1640 nutrient medium (GIBCO, Grand Island, NY) supplemented with 10 mM 4-(2-hydroxyethyl)-1-piperazineethanesulfonic acid (HEPES) and 10% (v/v) fetal bovine serum (FBS) is used for cell growth. Bovine serum albumin (BSA), fatty acid-free, is from Sigma (St. Louis, MO), RA is from Sigma, and radioactive RA ([11,12-^{3}H(N)]RA, 40–60 Ci/mmol) is from Du Pont–New England Nuclear (Boston, MA). For isolating nuclei two mixtures are prepared, PCS and PCST. PCS contains 1 mM potassium phosphate (dilute 1 M potassium phosphate stock solution prepared by mixing 1 volume of 1 M monobasic potassium phosphate with 1 volume of 1 M dibasic potassium phosphate), 1 mM calcium chloride, 320 mM sucrose, 25 mM sodium bisulfite, 1 mM phenylmethylsulfonyl fluoride, 5 mM sodium acetate (use 1 M sodium acetate stock solution, pH 7–7.2), and 140 mM 2-mercaptoethanol. PCST is PCS containing 0.3% (v/v) Triton N-101 (Sigma). Sodium dodecyl sulfate (SDS) is obtained from Bio-Rad (Richmond, CA).

Cells. HL60 cells are maintained in suspension culture in RPMI 1640 medium supplemented with 10 mM HEPES and 10% (v/v) FBS (GIBCO). Cell cultures are incubated at 37° in a humidified atmosphere of 5% CO_2 in air and subcultured every week. Cell number is estimated on an electronic particle counter (Coulter Electronics Hialeah, FL), and viability is estimated by trypan blue dye exclusion.

Incorporation of Radioactive RA. HL60 cells growing exponentially are harvested by centrifugation and resuspended at a concentration of 5 × 10^5 cells/ml in medium consisting of RPMI 1640 supplemented with 10 mM HEPES and either 5% FBS or 5 μg insulin/ml and 5 μg transferrin/ml.[12] Unlabeled RA and radioactive RA are dissolved in absolute ethanol and diluted into the growth medium such that the final concentration of ethanol is no higher than 0.1%.

Cells are harvested by centrifugation (200 g, 5 min) and washed extensively with PBS containing 600 μg of BSA/ml. The cell pellet is then extracted by the Bligh–Dyer procedure with $CHCl_3/CH_3OH/H_2O$ (1 : 2 : 0.8)[11] and centrifuged at 10000 g for 5 min in a microcentrifuge. This extraction is repeated about 5 times. The delipidated pellet is then dried in

---

[11] E. G. Bligh and W. J. Dyer, *Can. J. Biochem. Physiol.* **37,** 911 (1959).
[12] T. R. Breitman, S. J. Collins, and B. R. Keene, *Exp. Cell Res.* **126,** 494 (1980).

a centrifugal vacuum device (Savant Instruments, Inc., Farmingdale, NY) and dissolved in various solutions depending on the test system.

Stability of Bound RA. The delipidated pellet prepared as described above is dissolved in 1% (w/v) SDS. The reaction mixture (100 μl), containing 40 mM Tris-HCl, pH 7.5, 2 mM ethylenediaminetetraacetic acid, and 0.4 mg proteinase K (Sigma), is incubated at 37° for 1 hr. The reaction mixture is chilled in an ice water bath, and BSA (final concentration 50–100 μg/ml) and trichloroacetic acid [TCA, final concentration of 10% (w/v)] is added. The mixture is centrifuged in a microcentrifuge at 10,000 g for 5 min. Radioactivity in the supernatant fraction and in the precipitate is determined in a liquid scintillation spectrometer.

For hydrolysis with NH$_2$OH, the dried delipidated pellet is dissolved in 1% (w/v) SDS and incubated under N$_2$ gas with 1 M NH$_2$OH–HCl (adjusted with NaOH to pH 8 or 10) at 23° for 4 hr. The reaction mixture is treated with BSA and TCA as described above. For hydrolysis with CH$_3$OH–KOH the dried delipidated pellet is incubated with 0.1 ml of 0.1 N KOH in CH$_3$OH for 2 hr at 20° under N$_2$ gas. The reaction mixture is centrifuged and the supernatant fraction is dried under N$_2$ gas. The residue is suspended in 300 μl of CHCl$_3$ and washed with four 300 μl portions of water. The washed CHCl$_3$ layer is dried by evaporation under a stream of N$_2$ gas. The residue is dissolved in a small volume of CHCl$_3$ and spotted on a Silica Gel G thin-layer plate (Baker Chemical Co., Phillipsburg, NJ) along with RA and methyl retinoate as unlabeled internal standards. Methyl retinoate is synthesized by esterification of RA with diazomethane

TABLE I
RELEASE OF ACID-SOLUBLE COUNTS BY
VARIOUS TREATMENTS[a]

Treatment	Acid-soluble disintegrations per minute (dpm) (% of total)
Proteinase K, 4 mg/ml, 37°, 1 hr	98
1 M Hydroxylamine, pH 8, 23°, 1 hr	90
CH$_3$OH/0.1 N KOH, 20°, 2 hr	80
CH$_3$OH/2 N HCl (80:20) 100°, 24 hr	74
0.1 N HCl, 37°, 15 min	23

[a] HL60 cells, labeled with 100 nM [^{3}H]RA for 36 hr, are extracted by the Bligh–Dyer procedure. The residue is dissolved in Tris buffer, pH 7.5, containing 1% SDS. There are about 15,000 dpm/ 1.5 × 10^6 cells. After treatment, the reaction mixture is adjusted to 10% TCA and centrifuged. Radioactivity is measured in both the supernatant fraction and the precipitate.

TABLE II
DISTRIBUTION OF RETINOYLATED
PROTEIN IN HL60 CELLS[a]

Fraction	Retinoylated protein (% total)
Low-speed pellet, nuclei	77
Low-speed supernatant	23

[a] HL60 cells (5×10^5/ml) are grown in the presence of 100 nM [^{3}H]RA for 24 hr. Cells are harvested by centrifugation, washed with PBS containing 600 μg BSA/ml, and fractionated. Each fraction is extracted by the Bligh–Dyer procedure and the residue dissolved in 1% SDS (w/v). Radioactivity is measured in a liquid scintillation counter.

using Diazald (Aldrich, Milwaukee, WI) according to the manufacturer's instructions. The plate is developed with benzene/methanol (9 : 1). Retinoic acid has an R_f of 0.35 and methyl retinoate has a R_f of 0.72 in this solvent system. Hydrolysis of the dried delipidated pellet with CH_3OH/2 N HCl (82 : 18) is at 95° for 24 hr under N_2 gas. The reaction mixture is chilled in an ice water bath and BSA (final concentration 50 to 100 μg/ml) and TCA [final concentration of 10% (w/v)] are added. The mixture is centrifuged in a microcentrifuge at 10,000 g for 5 min. Radioactivity in the supernatant fraction and in the precipitate is determined in a liquid scintillation spectrometer. Typical results are shown in Table I.

On silica gel thin-layer chromatography 89% of the radioactivity released by methanolic KOH hydrolysis migrates with methyl retinoate, and 11% migrates with RA. Based on these results RA appears to be bound to protein via an ester bond. The sensitivity to dilute alkali and NH_2OH is consistent with the formation of a retinoyl S-ester.[13]

[13] E. R. Stadtman, this series, Vol. 3, p. 228.

FIG. 1. Two-dimensional gel pattern of retinoylated protein from HL60 cells. HL60 cells (5×10^5/ml) were grown for 24 hr in the presence of 100 nM [^{3}H]RA. Cells (2×10^6) were harvested by centrifugation, washed with PBS containing BSA, and extracted by the Bligh–Dyer procedure. The residue, containing about 7×10^4 cpm, was dissolved in isoelectric focusing buffer and analyzed by two-dimensional electrophoresis and fluorography. (A) Coomassie blue-stained gel. (B) Fluorograph of the gel exposed to X-ray film for 10 days. Arrows indicate the position of the major radioactive protein. a, Actin (pI 5.3, M_r 45,000).

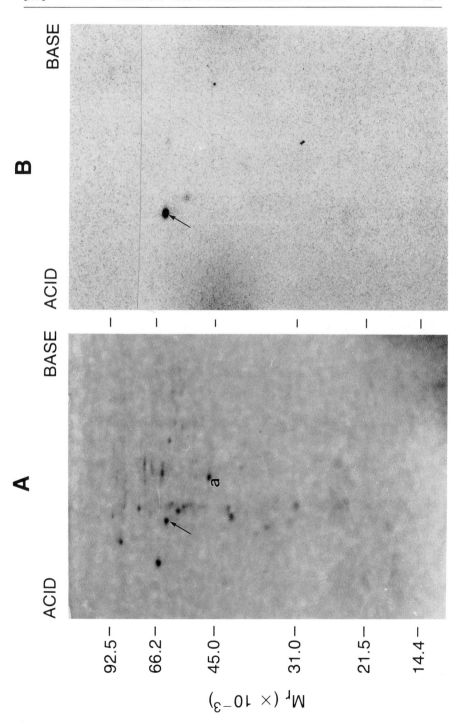

Subcellular Distribution of Retinoylated Protein. Cells incubated with RA are washed at least 5 times with PBS containing 600 μg of BSA/ml before fractionation. Nuclei are isolated at 4° by a detergent method.[14] A cell pellet (1–2 × 10⁶ cells) is washed 3 times with 1.5 ml of PCS each time. The cell pellet is suspended in 0.3 to 0.6 ml of PCST and transferred to a microcentrifuge tube. The cell suspension is agitated continuously and gently for about 10 min. The progress of the release of nuclei is monitored under a light microscope. The suspension is centrifuged at 200 g for about 1.5 min. The supernatant fraction is removed carefully, and the pellet is resuspended in 0.3 to 0.6 ml of PCST and treated as above. The second supernatant fraction is pooled with the first supernatant fraction and is the cytoplasmic/cell membrane fraction. As judged by phase microscopy, the nuclei in the pellet are clean without attached cytoplasmic tags or other debris. The yield of nuclei is about 80%. The nuclear and the supernatant fractions are extracted by the Bligh–Dyer procedure; the residue is dissolved in 5 mM Tris buffer, pH 7.5, containing 1% (w/v) SDS, and the radioactivity is measured on a liquid scintillation spectrometer. Typical results are shown in Table II.

Gel Electrophoresis. Two-dimensional gel electrophoresis is performed according to O'Farrell.[15] First-dimension isoelectric focusing gels contain 2% LKB (Bromma, Sweden) ampholytes (pH 3.5–10). Second-dimension gels are 12% acrylamide. Gels are fixed, stained with Coomassie blue R-250, and prepared for fluorography with ENTENSIFY (Du Pont–New England Nuclear) according to the manufacturer's instructions. The dried gels are exposed to Kodak XAR-5 film at less than −70°. Molecular weight markers are from Bio-Rad. An example is shown in Fig. 1.

[14] D. M. Berkowitz, T. Kakefuda, and M. B. Sporn, *J. Cell Biol.* **42**, 851 (1969).
[15] P. H. O'Farrell, *J. Biol. Chem.* **250**, 4007 (1975).

[24] Identification and Analysis of Retinoic Acid-Binding Proteins and Receptors from Nuclei of Mammalian Cells

By ANN K. DALY, CHRISTOPHER P. F. REDFERN, and BERNARD MARTIN

Introduction

The existence of specific binding sites for retinoic acid in nuclei was postulated several years ago, but it is only recently that nuclear proteins capable of specifically binding retinoic acid and distinct from cellular retinoic acid-binding protein (CRABP) have been identified.[1] Using salt extraction of DNase I-digested nuclei from F9 embryonal carcinoma cells which had been incubated in culture medium containing tritiated retinoic acid, we have shown that 20–40% of retinoic acid in the nucleus can be solubilized and is associated with a protein or proteins with a sedimentation constant of 4 S on sucrose density gradients.[1] Similar results have been obtained with HeLa human epithelial carcinoma cells, HL60 promyelocytic leukemia cells, and various melanoma cell lines, although the amount of nuclear binding activity varies between cell lines.[2] Our initial studies on nuclear retinoic acid-binding activity were carried out by incubating whole cells with tritium-labeled retinoic acid and analyzing nuclear fractions by sucrose density gradient centrifugation.[1] In subsequent studies on F9 cells, high-performance size-exclusion chromatography (HPSEC) was used to analyze the nuclear extracts.[3] This approach results in a considerable decrease in the time needed for each analysis, and, in addition, the increased sensitivity may make it possible to label nuclear retinoic acid-binding proteins by incubating nuclear extracts, rather than whole cells, with retinoic acid.

Incubating Cells with Labeled Retinoic Acid

Purification of all-trans-Retinoic Acid

all-*trans*-Retinoic acid photoisomerizes readily to give a mixture of stereoisomers with 13-*cis*- and all-*trans*-retinoic acid predominating. For

---

[1] A. K. Daly and C. P. F. Redfern, *Eur. J. Biochem.* **168,** 133 (1987).

[2] A. K. Daly, J. L. Rees, and C. P. F. Redfern, *Exp. Cell. Biol.* **57,** 339 (1989).

[3] M. Darmon, M. Rocher, M.-T. Cavey, B. Martin, T. Rabilloud, C. Delescluse, and B. Shroot, *Skin Pharmacol.* **1,** 161 (1988).

this reason there are two important precautions required for work with labeled retinoids: first, all manipulations of retinoic acid and retinoic acid-containing media should be done under darkroom conditions or in rooms equipped with yellow lights, and, second, the purity of the retinoic acid should be verified and the all-trans isomer repurified if necessary. We routinely check the purity of any new batch of all-*trans*-retinoic acid and after prolonged storage. If necessary, all-*trans*-retinoic acid can be repurified by reversed-phase high-performance liquid chromatography (HPLC). Methods for HPLC analysis and purification of retinoids are well established[4]; we use a Waters Novapak C_{18} reversed-phase Radial-Pak cartridge (4–5 μm particle size) and elute with a linear gradient of 75 : 25 (v/v) methanol/water increasing to 90 : 10 (v/v) methanol/water over 20 min at 2 ml/min. The solvents are buffered to pH 7.4 with 10 mM sodium acetate and the column effluent monitored at 343 nm. During purification of retinoic acid isomers it is important to avoid exposure of the isomer to light, including UV, and we therefore turn off the UV detector as the compound of interest is eluted from the column.

Cell Incubations

For experiments on F9 cells, we use cultures at subconfluent cell densities, typically three 75-cm^2 flasks, each containing 5 × 10^6 cells, per experiment. We have also used HeLa cells and some murine melanoma cell lines at densities of 10^7 cells per 75-cm^2 flask. HL60 cells are convenient to work with: the cells are grown in static suspension culture at densities of 0.3 × 10^6 to 0.9 × 10^6 cells/ml but may be pelleted by centrifugation and resuspended in a smaller volume for incubation with retinoic acid.

Incubation of cells with labeled retinoic acid is best done in serum-free medium. Retinoic acid binds with low affinity to serum albumin,[5] and this may reduce the concentration of free retinoic acid available for entry into the cells. For each 75-cm^2 flask of cells, the medium is withdrawn and replaced with 5 ml Dulbecco's modified Eagle's medium (DMEM) containing 20 nM all-*trans*-[11,12-^{3}H]retinoic acid at a specific activity of 15–50 Ci/mmol. In the presence of 10% (v/v) fetal calf serum, we find that it is necessary to increase the concentration of [^{3}H]retinoic acid to 0.2 μM to achieve a similar level of binding of retinoic acid to nuclear proteins. In our experiments, cells are incubated with labeled retinoic acid for 2–4 hr at 37°. It is important that the incubation with retinoic acid not be pro-

[4] A. M. McCormick, J. L. Napoli, and H. F. DeLuca, this series, Vol. 67, p. 220.
[5] D. S. Goodman, in "The Retinoids" (M. B. Sporn, A. B. Roberts, and D. S. Goodman, eds.), Vol. 2, p. 81. Academic Press, London and New York, 1984.

longed as all-*trans*-retinoic acid may be metabolized quite quickly. For example, with an initial retinoic acid concentration of 1.5×10^{-7} M in 35-mm dishes with 2 ml medium and approximately 10^6 cells, all-*trans*-retinoic acid has a half-life in the culture medium of the order of 8 hr.[6]

The specificity of binding of labeled retinoic acid to nuclear proteins can be tested by incubating the cells with the labeled retinoic acid in the presence of a 100-fold excess of unlabeled retinoic acid (see below).

Preparation of Nuclei

Many different methods have been developed for the isolation of cell nuclei.[7] For our studies on F9, S91, HeLa, and HL60 cells, nuclei are isolated by homogenization of 1.5×10^7 cells in a hypotonic buffer containing detergent and polyamines. After detachment from the tissue culture flask with 0.5 ml of 0.05% (w/v) trypsin, 0.02% (w/v) EDTA, an equal volume of fetal calf serum is added to neutralize protease activity, and the cells are pelleted by centrifugation at 2000 g for 5 min. The cells are then washed with phosphate-buffered saline (Dulbecco's formula, without calcium and magnesium, Flow Laboratories, McLean, VA) and resuspended in 10 ml of buffer A consisting of 15 mM KCl, 3.75 mM NaCl, 37.5 μM spermine, 125 μM spermidine, 0.5 mM EDTA, and 3.75 mM Tris-HCl, pH 7.4. The suspension is then centrifuged at 2000 g for 5 min and the cell pellet resuspended in 10 ml buffer A containing 0.1% digitonin; this and subsequent steps are carried out at 0–4°. Nuclei are released from the cells by homogenization with 15 strokes of a Dounce glass/glass homogenizer (Kontes, Vineland, NJ, pestle A) and pelleted by centrifugation at 1000 g for 5 min. The nuclei are then resuspended, homogenized, and recentrifuged twice more, washed once in buffer A without EDTA (buffer B), and finally resuspended in 0.5 ml buffer B. The purity of the nuclei should be assessed by phase-contrast microscopy. If whole cells, organelles other than nuclei, or other debris are present, the homogenization steps should be repeated.

Fractionation of Nuclei

A number of different protocols for the preparation of nuclear subfractions have been developed.[8] In general, these involve the extraction of

[6] C. P. F. Redfern and A. K. Daly, unpublished observations (1985).

[7] M. Muramatsu, *in* "Methods in Cell Physiology" (D. M. Prescott, ed.), Vol. 4, p. 195. Academic Press, London and New York, 1970.

[8] A. J. MacGillivray and G. D. Birnie (eds.), "Nuclear Structures: Isolation and Characterisation." Butterworth, London, 1986.

a b c d e f g h i j k

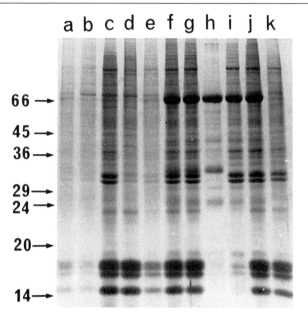

66 →
45 →
36 →
29 →
24 →
20 →
14 →

FIG. 1. Sodium dodecyl sulfate–polyacrylamide gel electrophoresis of proteins in nuclear extracts from F9 cells. The gel is 12.5% polyacrylamide and is stained with Coomassie blue. Lanes a to e are proteins remaining in the nuclear pellet, and lanes f to j are soluble proteins extracted into the supernatants after the following treatments: a and f, DNase I digestion followed by sonication; b and g, DNase I digestion and extraction with 0.6 M NaCl; c and h, DNase I digestion alone; d and i, extraction with 0.6 M NaCl alone; e and j, sonication and extraction with 0.6 M NaCl. A sample of unfractionated nuclei is in track k. Molecular weight markers are indicated at the arrows ($M_r \times 10^{-3}$). [Reprinted with permission from A. K. Daly and C. P. F. Redfern, *Eur. J. Biochem.* **168**, 133 (1987).]

nuclease-digested nuclei with high ionic strength buffers to solubilize the majority of nuclear proteins, leaving a residual insoluble structure, the nuclear matrix. The protein composition of the nuclear matrix may vary depending on the method used. In the protocol described below, the nuclear matrix lacks a residual internal nuclear structure and consists mainly of the nuclear lamin proteins.[9] Previous studies on steroid hormone receptors have used similar protocols for the solubilization of nuclear receptors,[10] although in some cases digestion steps have been omitted and only salt extraction has been used. For nuclear retinoic acid-binding proteins, considerably more material is solubilized if high-

[9] C. D. Lewis, J. S. Lebkovsky, A. K. Daly, and U. K. Laemmli, *J. Cell Sci. Suppl.* **1**, 102 (1984).
[10] P. S. Rennie, N. Bruchovsky, and H. Chang, *J. Biol. Chem.* **258**, 7623 (1983).

TABLE I

PERCENTAGE OF NUCLEAR TRITIUM LABEL RELEASED
BY DIFFERENT TREATMENTS[a]

Treatment	^{3}H released (% total)
Washing nuclei with buffer B	6.6 ± 3.9
DNase I digestion	9.3 ± 2.3
Sonication[b]	6.0 ± 0.4
DNase I digestion and extraction with 0.6 M NaCl	19.9 ± 2.4
Sonication and extraction with 0.6 M NaCl	36.3 ± 6.8
Extraction with 0.6 M NaCl alone	12.9 ± 0.9

[a] Reprinted with permission from A. K. Daly and C. P. F. Redfern, *Eur. J. Biochem.* **168,** 133 (1987).

[b] Sonication conditions are as described by A. K. Daly and C. P. F. Redfern, *Eur. J. Biochem.* **168,** 133 (1987).

salt extraction is combined with either sonication or DNase I digestion, and this also results in the solubilization of a considerably larger number of nuclear proteins (Fig. 1) than salt extraction alone.

In experiments with F9 cells, approximately 10% of the total cellular tritium label is accounted for by label associated with the nucleus.[1] Of this nuclear label, 20–40% can be released in soluble form by treatments which shear or digest the DNA, such as sonication or digestion with DNase I, followed by extraction with 0.6 M salt solutions (Table I). Although sonicating the nuclei solubilizes a greater proportion of nuclear-associated tritium label, digestion with DNase I gives lower backgrounds when the extracts are analyzed by sucrose density gradient centrifugation. The area of the 4 S peak obtained after sucrose density gradient centrifugation of nuclear extracts is similar when either sonication or DNase I digestion is used.

Nuclei (5–10 A_{260} units) are digested for 1 hr with DNase I (100 μg/ml, Pharmacia-LKB, Uppsala, Sweden) on ice in 0.5 ml buffer B containing 5 mM MgCl$_2$. It is important that the DNase I digestion be carried out at 0–4°: incubating isolated nuclei at 37° may result in the formation of insoluble complexes, and a number of nuclear proteins may become resistant to salt extraction.[11] After digestion, NaCl is added from a 4 M stock to a final concentration of 0.6 M, and the solution should become less turbid as the majority of nuclear proteins are solubilized.[12] If the

[11] T. D. Littlewood, D. C. Hancock, and G. I. Evan, *J. Cell Sci.* **88,** 65 (1987).

[12] Six-tenths molar NaCl is the minimum ionic strength that in our experience solubilizes the maximum amount of nuclear-associated tritium label.

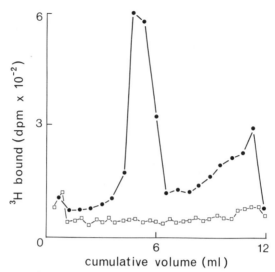

FIG. 2. Sucrose density gradient profile of DNase I/0.6 M NaCl extract of F9 cell nuclei, prepared after incubating the cells with 20 nM all-*trans*-[11,12-^{3}H]retinoic acid (15 Ci/mmol) in the presence (□) or absence (●) of a 100-fold excess of unlabeled retinoic acid. The gradients (5–20% sucrose) were prepared as described in the text and centrifuged for 60 hr at 200,000 g. The peak at 5–6 ml sediments at 4 S relative to IgG, hemoglobin, and myoglobin standards. [Reprinted with permission from A. K. Daly and C. P. F. Redfern, *Eur. J. Biochem.* **168,** 133 (1987).]

DNase I digestion has been insufficient, the solution will become viscous as salt is added, and a poor recovery of nuclear retinoic acid-binding proteins will be obtained. Finally, the solution is centrifuged at 15,000 g for 5 min in a microcentrifuge to pellet insoluble material, and the supernatant is retained for analysis by sucrose density gradient centrifugation or HPSEC. The supernatant may be stored frozen at −20° or below for later analysis.

Analysis of Nuclear Extracts

Analysis of DNase I/0.6 M NaCl extracts of nuclei from retinoic acid-treated cells has shown that labeled retinoic acid is bound to a protein or proteins with a sedimentation constant of approximately 4 S on sucrose density gradients (Fig. 2). The 4 S protein(s) is distinct from CRABP, which sediments at 2 S. Sucrose density gradient centrifugation is a time-consuming and lengthy process; as an alternative, high-performance size-exclusion chromatography can be used to effect a rapid size separation of proteins in nuclear extracts.[3]

Sucrose Density Gradient Centrifugation of DNase I-Digested Nuclear Extracts

We apply up to 0.5 ml of nuclear extract onto 5 to 20% linear sucrose density gradients which are centrifuged at 41,000 rpm (205,000 g) for 40–60 hr at 4° in a Sorvall TR-641 or Beckman SW41 swinging-bucket rotor. The gradients, final volume approximately 12 ml, are prepared using 0.6 M NaCl, 10 mM Tris-HCl, pH 7.4, as buffer. After centrifugation, the sucrose density gradients are fractionated using a suitable fractionation device or by piercing the bottom of the tube and collecting 0.3- to 0.5-ml fractions manually. As size standards, we run samples of sheep IgG (7 S), hemoglobin (4.5 S), and myoglobin (2 S) in parallel.

Analysis of Nuclear Extracts by High-Performance Size-Exclusion Chromatography

Ideally, a HPSEC column giving good separation of proteins in the range M_r 30,000 to 70,000 should be used. The 9.4 × 250 mm GF-250 column (Du Pont de Nemours, Paris, France) gives good results,[3] and the column may be calibrated using human albumin, egg albumin, and horse myoglobin (Sigma, St. Louis, MO) as standards. Nuclear extract (50 μl) is injected onto the column and eluted with 0.3 M potassium phosphate buffer, pH 7.8, at a flow rate of 1 ml/min. The effluent should be monitored at 280 nm and fractions collected for scintillation counting. Under these conditions, a peak of tritium label is eluted at a position corresponding to an M_r value of 45,000 (Fig. 3). This molecular weight is within the range expected for proteins sedimenting at 4 S on sucrose density gradients. The buffer concentration used for elution is important: 0.1 and 0.2 M potassium phosphate buffer give lower recovery of the M_r 45,000 peak, whereas increasing the buffer concentration beyond 0.3 to 0.4 M does not increase the amount of label eluted at this position.

Retinoid Binding Specificity and Further Characterization of Nuclear Retinoic Acid-Binding Proteins

For binding specificity experiments where intact, metabolically active cells are incubated with labeled all-*trans*-retinoic acid, it is important to identify the protein-bound tritium label extracted from nuclei. In the case of F9 cells incubated for 3 hr with 20 nM all-*trans*-retinoic acid, at least 90% of the tritium-label present in whole nuclei, the DNase I/salt extract, and the material sedimenting at 4 S (Fig. 4) is coeluted with all-*trans*-retinoic acid when analyzed by HPLC.[1] Some all-*trans*-retinoic acid also

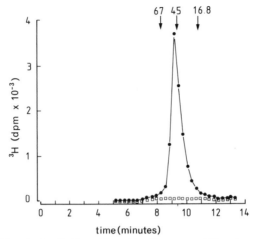

FIG. 3. HPSEC analysis of DNase I/0.6 M NaCl extract of F9 cell nuclei, prepared after incubating the cells with 20 nM all-*trans*-[11,12-^{3}H]retinoic acid (50 Ci/mmol) in the presence (□) or absence (●) of a 100-fold excess of unlabeled retinoic acid. The sample (50 μl) was applied and eluted from a 9.4 × 250 mm GF 250 column (Du Pont) with 0.3 M potassium phosphate, pH 7.8, as described in the text. The arrows mark the elution positions of molecular weight standards of 67,000, 45,000, and 16,800.

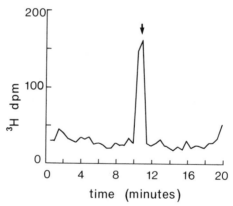

FIG. 4. Reversed-phase HPLC analysis of tritium label present in the 4 S tritium peak from sucrose density gradient centrifugation of nuclear extracts. One-third of each density gradient fraction was used for scintillation counting, and the remainders of each fraction comprising the 4 S peak were combined. Unlabeled all-*trans*-retinoic acid was added as internal standard. Methanol was added to 70% and the sample injected onto the HPLC column and analyzed as described in the text. The arrow marks the elution position of the all-*trans*-retinoic acid internal standard. [Reprinted with permission from A. K. Daly and C. P. F. Redfern, *Eur. J. Biochem.* **168**, 133 (1987).]

remains bound to the nuclear matrix; however, although this association is apparently specific,[1] the nature of the proteins involved are unknown. Incubating cells with tritiated retinoic acid in the presence of a 100-fold excess of unlabeled retinoic acid abolishes the 4 S/45,000 peak of bound label (Figs. 2 and 3). Binding of retinoic acid is thus specific, and this provides the basis of a competitive binding assay to compare binding affinities of different retinoids and retinoid analogs for the nuclear retinoic acid-binding activity.[3]

The solubility properties of the 4 S nuclear retinoic acid binding protein(s) have so far precluded further characterization of the protein. However, cDNA clones coding for three distinct retinoic acid receptors RARα, RARβ, and RAR-γ have recently been isolated.[13-17] The *in vitro* transcription/translation products of RAR cDNA clones have apparent molecular weights in the range 45,000–55,000.[2,14,15] The RAR-β protein, at least, cosediments with the nuclear retinoic acid-binding activity present in F9 cells, S91 cells, HeLa cells, and HL-60 cells,[2] and this nuclear binding activity may therefore represent the retinoic acid receptors. RARα and RARβ have been reported to differ in their affinities for retinoic acid,[16] and the expression of different combinations of RAR genes in particular cell types may cause problems in the interpretation of competitive binding assays. Furthermore, at least two transcripts each for RARα, RARβ, and RAR-γ[18,19] can be identified by Northern blotting and it is possible that alternative splicing may produce RAR isoforms with different retinoid- and DNA-binding specificities.

[13] M. Petkovich, N. Brand, A. Krust, and P. Chambon, *Nature (London)* **330,** 444 (1987).
[14] V. Giguere, E. Ong, P. Segui, and R. Evans, *Nature (London)* **330,** 624 (1987).
[15] D. Benbrook, E. Lernhardt, and M. Pfahl, *Nature (London)* **333,** 669 (1988).
[16] N. Brand, M. Petkovich, A. Krust, P. Chambon, H. de The, A. Marchio, P. Tiollais, and A. Dejean, *Nature (London)* **332,** 850 (1988).
[17] A. Krust, Ph. Kastner, M. Petkovich, A. Zelent, and P. Chambon, *Proc. Natl. Acad. Sci. U.S.A.* **86,** 5310 (1989).
[18] J. L. Rees, A. K. Daly, C. P. F. Redfern, *Biochem. J.* **259,** 917 (1988).
[19] J. L. Rees and C. P. F. Redfern, unpublished observations.

[25] Isolation and Binding Characteristics of Nuclear Retinoic Acid Receptors

By Anton M. Jetten, Joseph F. Grippo, and Clara Nervi

Introduction

Retinoids modulate cell proliferation and differentiation in a wide variety of cell systems *in vivo* as well as *in vitro*.[1-3] The molecular mechanisms by which retinoids regulate these processes are still largely unknown. It has been shown that retinoids act in many instances at nanomolar concentrations. Moreover, specific structural requirements are critical for the activity of retinoids. These observations suggest that the action of retinoids is mediated by specific, high-affinity receptors present in target cells. Recently, several specific nuclear retinoic acid receptor (RAR) proteins have been identified.[4-8] These receptors have a molecular weight of approximately 50,000 and are largely associated with the nucleus.[9] The nuclear retinoic acid receptors display a structural organization common to that of the steroid and thyroid hormone receptor family, with a DNA-binding domain, a ligand-binding domain, a "hinge" domain that connects the ligand- and DNA-binding domains, and an amino- and a carboxy-terminal domain. The RAR proteins exhibit an extensive sequence similarity, with the greatest homology occurring in the ligand-binding and DNA-binding domains. Retinoic acid receptors appear to act as ligand-responsive transcriptional factors and are likely to

[1] M. B. Sporn, A. B. Roberts, and D. S. Goodman (eds.), "The Retinoids," Vols. 1 and 2. New York, Academic Press, 1984.

[2] M. I. Sherman, "Retinoids and Cell Differentiation," CRC Press, Boca Raton, Florida, 1986.

[3] A. M. Jetten, in "Growth and Maturation Factors" (G. Guroff, ed.), Vol. 3, p. 251. Wiley, New York, 1985.

[4] M. Petkovich, N. J. Brand, A. Krust, and P. Chambon, *Nature (London)* **330,** 444 (1987).

[5] V. Giguere, E. S. Ong, P. Segui, and R. M. Evans, *Nature (London)* **330,** 624 (1987).

[6] N. Brand, M. Petkovich, A. Krust, P. Chambon, H. D. de Thé, A. Marchio, P. Tiollais, and A. Dejean, *Nature (London)* **332,** 850 (1988).

[7] D. Benbrook, E. Lernhardt, and M. Pfahl, *Nature (London)* **333,** 669 (1988).

[8] A. Krust, P. H. Kastner, M. Petkovich, A. Zelent, and P. Chambon, *Proc. Natl. Acad. Sci. U.S.A.* **86,** 5310 (1989).

[9] C. Nervi, J. Grippo, M. I. Sherman, M. A. George, and A. M. Jetten, *Proc. Natl. Acad. Sci. U.S.A.* **86,** 5854 (1989).

mediate many of the actions of retinoids on proliferation and differentiation.[10]

Assay of Nuclear Retinoic Acid Receptors

Materials

all-*trans*-[³H]Retinoic acid, 50 Ci/mmol (Du Pont–New England Nuclear, Boston, MA)

all-*trans*-Retinoic acid (Hoffmann-La Roche, Nutley, NJ)

Benzoic acid analog Ch55 (K. Shudo, Department of Pharmacology, University of Tokyo, Tokyo, Japan)[11,12]

Buffer A: 5 mM sodium phosphate (pH 7.4), 10 mM monothioglycerol, 10% (v/v) glycerol, 1 mM phenylmethylsulfonyl fluoride (PMSF), 10 U/ml aprotinin, 10 U/ml leupeptin

Buffer B: 10 mM Tris-HCl (pH 8.5), 10 mM monothioglycerol, 10% (v/v) glycerol, 1 mM PMSF, 10 U/ml aprotinin, 10 U/ml leupeptin, 0.8 M KCl

Buffer C: 10 mM Tris-HCl (pH 7.5), 10 mM monothioglycerol, 10% (v/v) glycerol, 50 mM NaCl, 1 mM PMSF, 10 U/ml aprotinin, 10 U/ml leupeptin

Superose 12 HR 10/30 size-exclusion HPLC column (Pharmacia, Piscataway, NJ)

Mono Q HR 5/5 anion-exchange HPLC column (Pharmacia)

Procedure

Preparation of Nuclear and Cytosolic Extracts.[9] Cells ($\sim$1–5 $\times$ 10⁸) from 10–15 confluent 150-cm² tissue culture dishes are rinsed twice with cold phosphate-buffered saline (PBS) containing 1 mM EDTA. Cells are removed from the dishes by standard trypsinization procedures and collected by centrifugation (5 min at 1000 g). Cells grown in suspension are collected directly by centrifugation. The cell pellet is washed twice with ice-cold PBS and once in ice-cold buffer A, then placed on ice. All the following procedures are carried out at 4°. The cells are resuspended in 3–5 ml buffer A and homogenized in a glass Dounce homogenizer (pestle

[10] A. M. Jetten, *in* "Mechanisms of Differentiation" (P. B. Fisher, ed.), CRC Press, Boca Raton, Florida, in press.

[11] H. Kagechika, E. Kawachi, Y. Hashimoto, and K. Shudo, *Chem. Pharm. Bull.* **32,** 4209 (1984).

[12] A. M. Jetten, K. Anderson, M. A. Deas, H. Kagechika, R. Lotan, J. I. Rearick, and K. Shudo, *Cancer Res.* **47,** 3523 (1987).

B) until more than 90% of the cells are disrupted. The number of strokes necessary for homogenization varies with the cell type; however, for most cell types 60–90 strokes appear to be sufficient. It is recommended that the formation of nuclei be monitored during homogenization. For this purpose an aliquot of the homogenate is diluted with buffer A and the percentage of nuclei determined in a hemacytometer via observation under a phase-contrast microscope.

The homogenate is centrifuged for 15 min at 1000 g. The supernatant is transferred to a polycarbonate tube and placed on ice. The pellet containing the nuclei is washed twice with 1 ml buffer A to reduce contamination by the cytosolic retinoic acid-binding protein (CRABP) present in the cell cytosol. The supernatants are combined and centrifuged for 30 min at 130,000 g using a 50Ti rotor (Beckman, Palo Alto, CA). The resulting supernatant is referred to as the cytosolic extract. The nuclear retinoic acid receptors associated with the nuclear pellet are solubilized by "salt extraction" in 3–6 ml of buffer B for 60 min. During this time the homogenate is resuspended every 10–15 min by up and down pipetting. The suspension is then transferred to a polycarbonate tube and centrifuged at 130,000 g in a 50Ti rotor. The resulting supernatant is referred to as the nuclear extract. For retinoic acid binding experiments, cytosolic and nuclear extracts can be used immediately, or they may be stored for several months at −70° without loss of retinoic acid-binding activity. The high-speed nuclear pellet can be used for the determination of the amount of DNA.

Retinoic Acid Binding Assay. Aliquots (3–5 μl) of all-*trans*-[³H]retinoic acid diluted in dimethyl sulfoxide (DMSO) are added to 1.5-ml Eppendorf tubes. Concentrations of [³H]retinoic acid in the range 1–10 nM have been found appropriate to measure nuclear retinoic acid-binding activity in a variety of cell types, including myeloblastic leukemia HL60, embryonal carcinoma PCC4.aza1R, and primary cultures of human epidermal and tracheobronchial epithelial cells. In parallel incubations, a 100- to 200-fold excess of unlabeled retinoic acid is included to determine nonspecific binding. After the addition of the cytosolic or nuclear extracts (0.3–0.5 ml) the samples are briefly mixed by gentle vortexing. Incubations of 3 hr at 4° are sufficient for experiments analyzing receptor properties, whereas longer incubations (12–18 hr) are required for quantitative estimations of the receptor–hormone complex, especially when low concentrations of labeled retinoic acid are used. At the end of the incubation period the unbound retinoic acid is removed by transferring samples to Eppendorf tubes containing dextran–charcoal pellets obtained after centrifugation of 50 μl of a charcoal–dextran suspension (3% acid-washed Norit A, 0.3% dextran C in 10 mM Tris-Cl, pH 7.4, and 0.02% sodium azide). The samples are placed on ice and mixed several times during the 10- to 15-min incubation period and then centrifuged for 15 min at 10,000 g to sediment the dextran-coated charcoal. The supernatant can then be analyzed for [^{3}H]retinoic acid-binding activity via sucrose gradient centrifugation, Superose 12 size-exclusion,[9] or Mono Q anion-exchange HPLC.

Sucrose Density Gradient Centrifugation. Linear 5–20% sucrose gradients in buffer A containing 0.4 M KCl are prepared in 12-ml polyallomer tubes using a gradient mixer and a peristaltic pump. After preparation, the gradients are kept at 4° until use. The [^{3}H]retinoic acid-labeled samples (400 μl) are layered on top of the 11.6-ml gradients and centrifuged at 4° in a swinging-bucket rotor (SW 41 Ti) for 63 hr at 41,000 rpm. After centrifugation fractions of 0.35 ml are collected by piercing the bottom of the tube using a Beckman fraction recovery system and removing the contents by a peristaltic pump. Fractions are collected directly into scintillation vials using a fraction collector. After the addition of a standard scintillation fluid, the radioactivity is measured in a liquid scintillation counter. Retinoic acid nuclear receptors are present in the nuclear extract and sediment as a single symmetrical peak between fractions 15 and 22, corresponding to an apparent molecular weight of 48,000–50,000.[9] This specific binding activity is present only in nuclear extracts and is not detectable in cytosolic extracts. CRABP (molecular weight 16,000) present in the cytosol sediments between fractions 20 and 28.

High-Performance Size-Exclusion Liquid Chromatography. HPLC analysis is performed using a Superose 12 HR 10/30 size-exclusion column. [^{3}H]Retinoic acid-labeled samples (0.4 ml) are centrifuged in an Eppendorf centrifuge at 10,000 g for 10 min just before they are injected into the HPLC. Buffer A containing 0.4 M KCl is used as eluent buffer. HPLC is performed at 4° using a flow rate of 0.5 ml/min. Fractions are collected at 1-min intervals directly into scintillation vials. After the addition of scintillation fluid, radioactivity is measured in a liquid scintillation counter. Alternatively, radioactivity can be determined via a radioflow detector (A250, Radiomatic Instruments, Tampa, FL). The nuclear retinoic acid receptors RARα, RARβ, and RARγ present in the nuclear extract elute at a retention time of 27 min (Figs. 1 and 2). This retention time corresponds to a molecular weight of 50,000.[9] Depending on whether the cells contain CRABP, the elution profile of the cytosolic fraction contains specific retinoic acid-binding activity eluting at a retention time of 32 min, and representing CRABP.

Analysis of Nuclear Receptor Activity via Mono Q Anion-Exchange HPLC. Before extracts are analyzed via anion-exchange HPLC, they are dialyzed at 4° against buffer C. Dialysis does not reduce the total retinoic

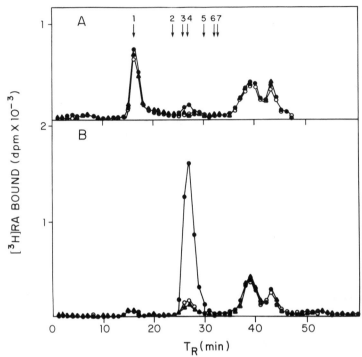

FIG. 1. Size-exclusion HPLC analysis of cytosolic and nuclear extracts prepared from myeloblastic leukemia HL60 cells. Cytosolic (A) and nuclear (B) extracts (300 μl) were incubated with 3 nM [³H]retinoic acid in the absence (●) or in the presence of a 200-fold excess of unlabeled retinoic acid (○) or benzoic acid analog Ch55 (▲) for 3 hr at 4°. Thereafter, the samples were treated with charcoal–dextran and fractionated over a Superose 12 HR 10/30 size-exclusion column using buffer A containing 0.4 M KCl as eluent. The flow rate was 0.5 ml/min. The column was calibrated using the following proteins as molecular weight markers: (1) Blue dextran (2,000,000); (2) phosphorylase b (97,000); (3) bovine serum albumin (66,000); (4) ovalbumin (45,000); (5) carbonate dehydratase (29,000); (6) murine CRABP (16,000); (7) cytochrome c (12,400). (From Ref. 9.)

acid-binding activity. The dialyzed extracts are incubated with [³H]retinoic acid, treated with dextran/charcoal as described above, and then applied to a Mono Q HR 5/5 HPLC column. Chromatography is carried out at 4° with a flow rate of 1 ml/min. Fractions of 1 min are collected. Proteins are eluted from the column using a noncontinuous NaCl gradient consisting of (1) 10 min buffer C; (2) 10 min linear gradient from 50 to 150 mM NaCl (final concentration); (3) 10 min 150 mM NaCl; (4) 20 min linear gradient from 150 to 350 mM NaCl. After the run the column is washed with buffer C containing 1 M NaCl (final concentration) and reequilibrated in buffer C. CRABP is eluted at 15 min (100 mM NaCl), and both RARα

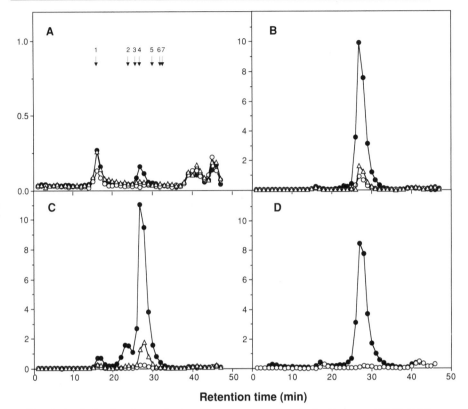

FIG. 2. Analysis of nuclear retinoic acid receptors, α, β, and γ by Superose HR 12 10/30 size-exclusion HPLC. Transient transfection of COS-1 cells either with no DNA (A; mock-transfected) or with an expression vector RARα0 [4] (B) RARβ0 [6] (C), or hRARγ0 [8] (D) containing the coding region of hRARγ, hRARβ, or hRARγ, respectively, was performed by the DEAE-dextran method [B. R. Cullen, K. Raymond, and G. Ju, *J. Virol.* **53**, 515(1985)]. Nuclear extracts were prepared and analyzed for specific [³H]retinoic acid-binding activity as described in the legend to Fig. 1. (●), [³H]Retinoic acid only; (○), [³H]retinoic acid and 200-fold excess unlabeled retinoic acid; (△), [³H]retinoic acid and 200-fold excess unlabeled Ch55. (From Ref. 9.)

and RARβ are eluted with 150 mM NaCl; however, RARβ is eluted at 22 min (Fig. 3), whereas RARα is eluted 2 min later.

Calculation of Results. Specific binding is determined as the difference between total and nonspecific binding in the relevant peak. The nuclear retinoic acid-binding activity can be calculated as the area under the peak using the trapezoidal rule. In the case radioactivity is determined via a Radiomatic A250 radioflow detector, which includes a personal computer for data processing, the elution profile is directly followed on

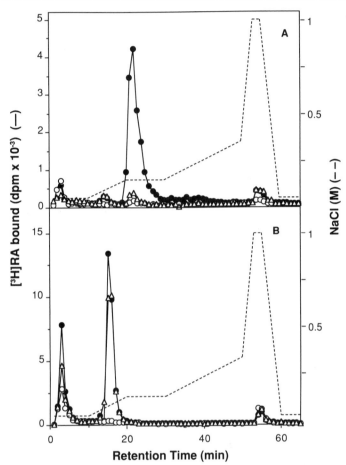

FIG. 3. Analysis of nuclear retinoic acid receptors and CRABP by Mono Q anion-exchange HPLC. Dialyzed nuclear extracts from COS-1 cells transfected with the expression vector RARβO (A) and cytosolic extracts from PCC4.aza1R (B) were prepared and labeled with [³H]retinoic acid. The samples were analyzed for specific [³H]retinoic acid-binding activity via Mono Q anion-exchange HPLC as described in the text. (●), [³H]Retinoic acid only; (○), [³H]retinoic acid and 200-fold excess of unlabeled retinoic acid; (△), [³H]retinoic acid and 200-molar excess of unlabeled Ch55. The dashed line represents the NaCl gradient.

the monitor and the total radioactivity in the peaks is automatically calculated via integration using the software provided by the manufacturer.

Comments

The addition of protease inhibitors, glycerol, and monothioglycerol in all buffers used during the preparation and analysis of nuclear and cytosolic extracts appears to be essential for assaying RAR proteins. The omission of these substances resulted in the complete elimination of, or greatly reduced, retinoic acid-binding activity.[9] Sodium molybdate, widely used in steroid receptor analyses at concentrations of 10–20 mM, does not modify retinoic acid receptor binding characteristics. Analysis of the nuclear extracts via Superose 12 size-exclusion HPLC at 25° instead of 4° reduces the total nuclear retinoic acid receptor binding activity by 20–50%; however, the binding of retinoic acid to CRABP is almost totally abolished. Elution of retinoic acid nuclear receptors from the Superose 12 HPLC column using buffer A containing KCl below 100 mM as eluent buffer reduces dramatically the specific retinoic acid-binding activity.

The nuclear retinoic acid receptors are distinguished from CRABP by molecular weight, cellular distribution, and retinoid-binding characteristics. The nuclear receptors RARα, RARβ, and RARγ have a molecular weight of approximately 50,000 and are largely (>95%) associated with the nucleus.[9] CRABP has a molecular weight of 16,000 and is generally found predominantly (>99%) in the cytosol.[13] However, in embryonal carcinoma PCC4.aza1R cells, which contain relatively high levels of CRABP, the nuclear extract contains about 10% of the total CRABP activity. This CRABP activity in the nuclear extract could be, at least in part, due to the contamination of the nuclear preparation with whole cells. The nuclear retinoic acid receptors can be separated from CRABP not only by size-exclusion chromatography (Fig. 2) but also via anion-exchange chromatography (Fig. 3).

Assay of [³H]retinoic acid-binding activity via size-exclusion HPLC can be used to perform saturation binding and Scatchard plot analyses in order to determine the binding affinity and number of receptors per cell.[9] The binding characteristics of the nuclear receptors are different from those of CRABP. The binding of retinoic acid to nuclear retinoic acid receptors exhibits a dissociation constant K_d of 0.5–1.5 nM. This binding affinity is about 10–50 times higher than that of CARBP. The nuclear receptors exhibit different binding specificities for certain retinoids than CRABP. For example, in contrast to CRABP, which does not bind biologically active, benzoic acid analogs of the Ch series, such as Ch55, the nuclear receptors do bind Ch55 with high affinity (Fig. 3).[9,11,13,14]

Acknowledgments

The authors thank Dr. P. Chambon for providing the expression vectors for hRARα, hRARβ, and hRARγ, and Dr. K. Shudo for the benzoic acid analogs of the Ch series.

[13] F. Chytil and D. E. Ong, Adv. Nutr. Res. 5, 13 (1983).
[14] Y. Hashimoto, H. Kagechika, E. Kawachi, and K. Shudo, Jpn. J. Cancer Res. 79, 473 (1988).

[26] Nuclear Retinoic Acid Receptors: Cloning, Analysis, and Function

By Magnus Pfahl, Maty Tzukerman, Xiao-Kun Zhang, Juergen M. Lehmann, Thomas Hermann, Ken N. Wills, and Gerhart Graupner

Introduction

The isolation and further characterization of nuclear receptors have contributed significantly to a better understanding of transcriptional regulation by steroid and thyroid hormones and, in particular, to that of the vitamin A derivative retinoic acid (RA). Three retinoic acid receptor (RAR) isoforms encoded by distinct genes have been cloned from mammalian species.[1-5] All of them share a similar domain structure with members of the steroid/thyroid receptor superfamily, and they are highly conserved within the regions D and E which correspond to the DNA- and ligand-binding/transcription activation domains, respectively (for a review, see Ref. 6).

The recent cDNA cloning of the RARδ from newt,[7] as well as the presence of different RNA splicing products of the closely related thyroid hormone receptors, suggests that the tissue-specific and developmental expression of RARs is even more complex. Owing to high sequence similarity and/or low template abundance, conventional techniques including Northern blotting, *in situ* hybridization, and RNase protection studies are not sensitive enough to resolve the question satisfactorily.

Cloning of Nuclear Retinoic Acid Receptors by Polymerase Chain Reaction

Whereas conventional cDNA cloning approaches are still time-consuming and require a multitude of skills, here we report an easy to follow

[1] D. Benbrook, E. Lernhardt, and M. Phahl, *Nature (London)* **333**, 669 (1988).
[2] M. Petkovich, N. J. Brand, A. Krust, and P. Chambon, *Nature (London)* **330**, 444 (1987).
[3] V. Giguere, E. S. Ong, P. Segui, and R. M. Evans, *Nature (London)* **330**, 624 (1987).
[4] N. Brand, M. Petkovich, A. Krust, P. Chambon, H. The, A. Marchio, P. Tiollais, and A. Dejean, *Nature (London)* **332**, 850 (1988).
[5] A. Zelent, A. Krust, M. Petkovich, P. Kastner, and P. Chambon, *Nature (London)* **339**, 714 (1989).
[6] R. M. Evans, *Science* **240**, 889 (1988).
[7] C. W. Ragsdale, Jr., M. Petkovich, P. B. Gates, P. Chambon, and J. P. Brockes, *Nature (London)* **341**, 654 (1989).

polymerase chain reaction (PCR) procedure for the isolation of RAR cDNA which can be accomplished within a few days. Amplification of RNA sequences via complementary DNA using the PCR is highly sensitive and also makes the cDNA available to address further questions.

Isolation of RNA. Total RNA is isolated by using the acid guanidinium isothiocyanate–phenol–chloroform extraction as described by Chomczynski and Sacchi.[8] Briefly, 10^7 cells are lysed in 1 ml of 4 M guanidinium isothiocyanate, 25 mM sodium citrate, pH 7, 0.5% sarcosyl, and 0.1 M 2-mercaptoethanol. Sequentially, 0.1 ml of 2 M sodium acetate, pH 4, 1 ml phenol, and 0.2 ml chloroform–isoamyl alcohol (49:1) are added. After thoroughly mixing and incubating for 10 min on ice, the suspension is centrifuged at 10,000 g for 20 min. The RNA is precipitated by adding 1 volume of 2-propanol.

Synthesis of cDNA. cDNA is synthesized from total RNA using Moloney murine leukemia virus (MoMLV) reverse transcriptase and following essentially the procedure as described by Gubler and Hoffman.[9] In modification to this, the cDNA synthesis is performed directly in the PCR buffer system (50 mM KCl, 10 mM Tris-HCl, pH 8.3, 1.5 mM MgCl$_2$, and 0.01% gelatin) using 2 μg of total RNA, 200 U MoMLV reverse transcriptase (BRL, Bethesda, MD), 200 μM of each dNTP, and 2 μg of oligo(dT)$_{12-18}$ primer in a final volume of 20 μl. Priming may also be accomplished by using either random hexamers or the downstream PCR primer. The reaction mixture is incubated for 1 hr at 42°. The reaction is heated for 10 min at 95° to denature the cDNA–RNA hybrids and to inactivate the reverse transcriptase.

Polymerase Chain Reaction Procedure. The PCR is performed in a biphasic amplification procedure, in order to circumvent the amplification of artifacts often occurring when the template DNA is present in low copy number.[10] In a first stage, only 1 pmol of primer is used to accomplish a limited amplification of the specific target cDNA and to reduce the generation of primer dimer artifacts.[10] In the second stage, the usual amount of the two amplification primers (100 pmol each) is added to preferentially amplify the cDNA of interest. The oligonucleotide primers are designed such that they have at least 22 specific annealing nucleotides and contain a naturally occurring or an artificially added restriction site which facilitates subcloning in later stages. For the RARs, the domains A, D, and F offer a reasonable divergence in the nucleotide composition to design the corresponding specific primers. In order to facilitate the amplification of long

[8] P. Chomczynski and N. Sacchi, *Anal. Biochem.* **162,** 156 (1987).
[9] U. Gubler and B. J. Hoffman, *Gene* **25,** 263 (1983).
[10] R. Watson, in "Amplifications" (Perkin-Elmer, Norwalk, CT), Issues 2 and 3, 1989.

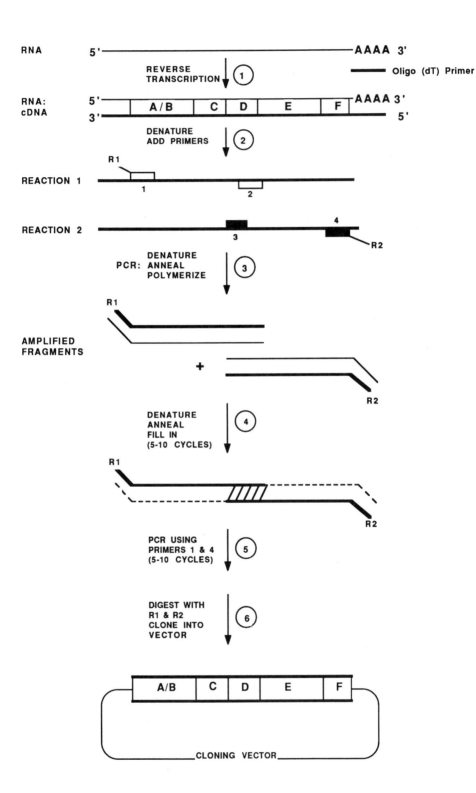

target sequences, appropriate primers were made to amplify the RAR sequences in two overlapping half-sites, that is, either the amino-terminal A, B, C, and D domains or the carboxy-terminal located D, E, and F domains (Fig. 1, Step 2, Reactions 1 and 2). Because the two half-sites are overlapping, the PCR technique can be applied to recombine these two amplified cDNA fragments into the complete RAR clone in a subsequent step (Fig. 1, Steps 4–6).

For the PCR reaction 4 μl of the reverse transcription reaction is directly used. All reactions are carried out in volumes of 50 μl PCR buffer, containing 400 μM of each nucleotide and 2.5 U of Taq polymerase (Perkin-Elmer Cetus, Norwalk, CT). For the first stage only 1 pmol of each upstream and downstream primer is added. Twenty cycles are performed with the usual profile of 50 sec denaturation at 94°, annealing at 54° for 2 min, and polymerization at 72° for 3 min using the Perkin-Elmer automated thermal cycler. In the second stage 100 pmol of each primer is added, and PCR is carried out for another 30 cycles.

In order to demonstrate how powerful this approach is, we show here the analysis of RAR isoforms that are expressed by the human pro-myeloblastic leukemia HL60 cell line, which can be differentiated into granulocytes by exposure to retinoic acid. This question has been discussed controversially in the literature recently.[11–12] As shown in Fig. 2, we were able to amplify cDNA from RARα as well as from RARγ. The significance of this finding is beyond the scope of this chapter and will be published elsewhere. One set of RARα primers and one set of primers for RARγ were used [see Fig. 2 (legend) for primers]. Specific DNA fragments are generated from HL60 cDNA with both sets of RAR primers. The RAR fragments produced were subsequently cloned into Bluescript (Fig. 2) and tested for specific restriction enzyme sites and by DNA sequencing (data not shown) to verify their identity.

In conclusion, we have described here a technique for analysis of mRNA as well as for cDNA cloning of RARs which is not only highly sensitive and easy to follow but very versatile. Although the PCR protocol given can be expected to amplify RAR cDNA from various sources, in

[11] Y. Hashimoto, M. Petkovich, M. P. Gaub, H. Kagechika, K. Shudo, and P. Chambon, *Mol. Endocrinol.* **3**, 1046 (1989).
[12] C. Nervi, J. F. Grippo, M. Sherman, M. D. George, and A. M. Jetten, *Proc. Natl. Acad. Sci. U.S.A.* **86**, 5854 (1989).

FIG. 1. Cloning of RAR cDNAs by the polymerase chain reaction. Schematic representation of the different steps for the cloning of RAR cDNAs by PCR. The details for the individual steps are described in the text.

1 2 3 4

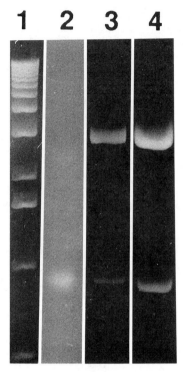

FIG. 2. cDNA cloning of the hormone-binding domains of RARα and RARγ by PCR. One-kilobase DNA ladder, BRL (lane 1); PCR amplification of the RARγ region, amino acids 176–454 (lane 2); primers were 5'-CCGGATCCCACCTGACAGCTATGAGCTG-3' and 5'-CCGAATTCCCCCACAACGGGGTAGGTCAGGG-3'. Subcloning of the RARγ fragment in Bluescript (lane 3); RARα fragment, amino acids 180–462, subcloned in Bluescript (lane 4); primers were 5'-GCTGGATCCTGACGCCGGAGGTG-3' and 5'-TTA-GAATTCACGGGGAGTGGGTGGCCGGGC-3'.

some cases it might be necessary to reoptimize time and temperature of the cycle profile and/or the MgCl₂ concentration, which has great influence on the Taq polymerase activity. It should also be pointed out that individual PCR clones can have nucleotide exchanges owing to the intrinsic error frequency of the Taq polymerase. Two analysis systems can be used to determine quickly the overall integrity of the PCR clone. One is the *in vitro* transcription/translation assay described below. With this system, erroneously introduced stop codons can be easily detected after the PCR fragments have been cloned into the Bluescript vector (Stratagene) or a comparable vector that allows *in vitro* RNA production. The second analysis system evaluates the functional integrity of the encoded receptor protein in the transient transfection system described below.

Analysis of Nuclear Receptor Domains by Hybrid Receptor
Construction: A Rapid Polymerase Chain Reaction
Procedure to Make Hybrid Proteins

The retinoic acid receptors, as well as other nuclear receptors, are made up of individual domains that separately contain DNA-binding and ligand-binding/transcriptional activation functions. Some of the more complex steroid hormone receptors also contain secondary activation domains.[13-14] Domains of individual receptors can be interchanged in hybrid receptors without loss of functions.[1,15] Such hybrid receptors have turned out to be of increasing value for the characterization of novel members of the steroid/thyroid receptor superfamily, for the functional analysis of receptor isoforms, and for the understanding of certain domain functions and receptor–receptor interactions (i.e., receptor heterodimer formation or oligomerization). Only in a very limited number of cases could functional hybrid receptors be designed utilizing existing restriction enzyme sites within the cDNA sequences. In most cases, new restriction sites needed to be created first by *in vitro* mutagenesis to allow the splicing of DNA sequences in a desired manner. Although a number of efficient *in vitro* mutagenesis systems have been developed and are commercially available, a considerable amount of DNA technology skills and time are required to obtain the desired hybrid receptor constructs in a convenient expression vector. To circumvent these lengthy procedures, we have used a PCR protocol that allows a rapid production of hybrid receptors or hybrid proteins. The procedure is outlined in Fig. 3. (The receptor cDNAs are cloned into Bluescript. This vector is convenient as it contains a large cloning box, and a number of specific primers for it are commercially available.)

Methods. The PCR procedure is carried out with 1 μg of receptor cDNA cloned in Bluescript vector and 1 μM of each primer in a 100-μl reaction volume containing 50 mM KCl, 10 mM Tris-HCl, pH 8.3, 1.5 mM MgCl$_2$, 0.01% (w/v) gelatin, dNTP mix (200 μM of each nucleotide), and 2.5 U of Taq polymerase. Five cycles of PCR are performed (denaturation for 1.5 min at 94°, annealing for 2 min at 37°, elongation for 2 min at 72°) using the Perkin-Elmer automated thermal cycler. Amplified DNA is separated and extracted from an agarose gel, digested with the appropriate restriction enzyme, and ligated into Bluescript vector.

We have used the procedure described here to construct a considerable number of hybrid receptors that all turned out to be functional. For

[13] S. M. Hollenberg, V. Giguere, P. Segui, and R. M. Evans, *Cell* **49**, 39 (1987).
[14] L. Tora, J. White, C. Brou, D. Tasset, N. Webster, E. Scheer, and P. Chambon, *Cell* **59**, 477 (1989).

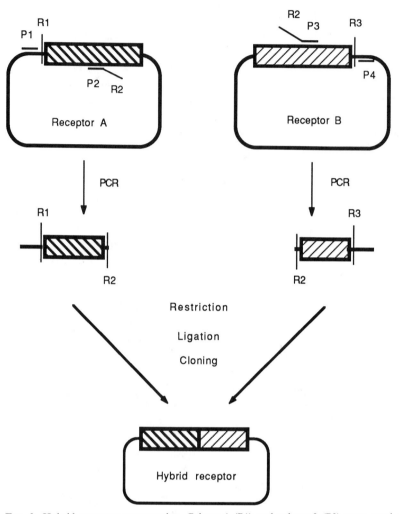

Fig. 3. Hybrid receptor construction. Primer 1 (P1) and primer 2 (P2) were used to amplify the desired portion of receptor A. Primer 3 (P3) and primer 4 (P4) were used to amplify the desired portion of receptor B by PCR. P1 and P4 are Bluescript primers. P2 and P3 were synthesized according to the DNA sequence in the hinge region of the receptor and contain at their 5' end a noncomplementary sequence that includes two random base pairs followed by a 6-base-pair restriction site. The restriction sites were used in a subsequent step to combine the fragments of the two different receptors and were designed such that the open reading frame of the coding sequence is maintained after ligation. The complete hybrid receptor was transcribed and translated *in vitro*, to monitor the intact open reading frame of the construct. The subsequent cloning into an eukaryotic expression vector allows further investigation of its functional properties.

all these receptors, we have chosen a common linkup site located 71 bases 5' of the beginning of the ligand-binding/caboxy-terminal domain. The functional analysis of one of these receptors, RA-ERII, is shown in Fig. 6B,C (also see Ref. 15).

RAR–DNA Interactions: *In Vitro* Systems for Retinoic Acid Receptor Synthesis and Protein–DNA Interactions

One of the major biological functions of nuclear retinoic acid receptors is to recognize specific DNA sequences. The study of this interaction, therefore, is important for understanding RAR function and allows us to determine specific DNA sequences recognized by RAR. To solve the problem of RAR availability in sufficient pure quantities we have synthesized RAR by transcription/translation *in vitro*. The *in vitro* synthesized protein binds specifically to a retinoic acid/thyroid hormone responsive element (TRE) described previously[15] in a gel retardation assay. We have shown elsewhere that the present *in vitro* system reflects *in vivo* functions of nuclear RARs.[16–18]

In Vitro Synthesis of Nuclear Retinoic Acid Receptors

To synthesize RAR protein *in vitro*, we take advantage of the presence of T7 or T3 RNA polymerase promoter sites in the Bluescript vector. The RAR construct is linearized with an appropriate restriction enzyme that cleaves "downstream" of the RAR insert. The linearized plasmid DNA is then treated with 200 μg/ml proteinase K at 37° for 30 min, followed by phenol/chloroform (1 : 1) extraction and ethanol precipitation. One microgram of template DNA is used for *in vitro* transcription in a 50-μl reaction mixture containing 40 mM Tris-HCl, pH 8.0, 8 mM MgCl$_2$, 2 mM spermidine, 30 mM dithiothreitol (DTT), 50 mM NaCl, 0.4 mM of each ribonucleotide, 1 unit RNase inhibitor (Stratagene), and 10 units of either T7 or T3 RNA polymerase. The transcription reaction is incubated at 37° for 30 min. To remove the DNA template, the reaction is diluted 10-fold with 40 mM Tris, pH 7.5, 6 mM MgCl$_2$, 10 mM NaCl, and incubated with 1 unit of

[15] G. Graupner, K. N. Wills, M. Tzukerman, X-K. Zhang, and M. Pfahl, *Nature (London)* **340**, 653 (1989).
[16] X-K. Zhang, K. N. Wills, G. Graupner, M. Tzukerman, T. Hermann, and M. Pfahl, submitted (1990).
[17] T. Hermann, X-K. Zhang, M. Tzukerman, K. N. Wills, G. Graupner, and M. Pfahl, submitted (1990).
[18] M. Tzukerman, X-K. Zhang, T. Hermann, G. Graupner, K. N. Wills, and M. Pfahl, submitted (1990).

RNase-free DNase I (BRL) at 37° for 15 min. The efficiency of transcription can be monitored by agarose gel electrophoresis.

RAR protein can be easily synthesized *in vitro* utilizing the rabbit reticulocyte translation system (Promega, Madison, WI). RNA template is incubated at 67° for 10 min and immediately cooled on ice to eliminate any secondary structure. Approximately 1 μg RNA is then mixed with 35 μl nuclease-treated reticulocyte lysate, amino acid mix (20 μM of each), and 1 unit of RNase inhibitor in a total volume of 50 μl. The translation reaction is carried out at 30° for 60 to 90 min. The synthesized receptor protein is then aliquoted and can be stored at $-70°$. Alternatively, 2.5 μl [^{35}S]methionine (>1200 Ci/mmol, New England Nuclear, Boston, MA) is included in the translation reaction. The labeled receptor protein can then

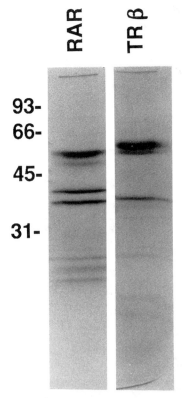

FIG. 4. SDS–PAGE analysis of *in vitro* synthesized RARβ/ε protein. *In vitro* translations were performed in the presence of [^{35}S]methionine.[1] Five microliters of the translation mixture of RARβ/ε or TRβ was loaded on a 10% reducing electrophoresis gel and the products visualized by autoradiography. Protein markers used were phosphorylase *b* (93K), bovine serum albumin (66K), ovalbumin (45K), and carbonated dehydratase (31K).

be analyzed by sodium dodecyl sulfate–polyacrylamide gel electrophoresis (SDS–PAGE) to determine the translation efficiency. Figure 4 shows that the predominant protein product synthesized is of the molecular weight of RARβ/ε^1 as predicted by the DNA sequence. A thyroid hormone receptor (TRβ) was also synthesized and is shown for comparison.

Gel Retardation Assay

We have used *in vitro* synthesized receptor proteins successfully to analyze protein–DNA interactions by gel retardation assays without any further purification.[15–18] In general, 5 μl of *in vitro* translated receptor protein is incubated with the ^{32}P-labeled specific DNA fragment in a reaction mixture containing 10 mM HEPES buffer, pH 7.9, 50 mM KCl, 1 mM

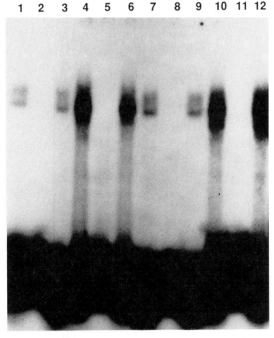

FIG. 5. Gel retardation assay of *in vitro* synthesized RAR protein. *In vitro* synthesized RARβ/ε was incubated with TRE monomer (lanes 1–3, 7–9) or TRE dimer (lanes 4–6, 10–12) either in the absence (lanes 1–6) or in the presence (lanes 7–12) of retinoic acid (10^{-7} M). The specific protein–DNA complex can be displaced by a 50-fold excess of unlabeled TRE (lanes 2, 5, 8, and 11) but not by a 50-fold excess of nonspecific DNA (lanes 3, 6, 9, and 12). The TRE used is a 16-bp perfect palindromic response element (TCAGGTCATGACCTGA) [C. K. Glass, J. M. Holloway, O. V. Devary, and M. G. Rosenfeld, *Cell* **54,** 313 (1988)] flanked by *Bgl*II adaptor sequences. When duplicated, two copies of the perfect palindrome are separated by 4 base pairs of the restriction site.

DTT, 2.5 mM MgCl, 10% glycerol, and 1 μg poly(dI–dC) at 25° for 20 min. The reaction mixture is then loaded on a 5% nondenaturing polyacrylamide gel containing 10 mM HEPES, pH 7.9, and 3.3 mM sodium acetate, pH 7.9. Electrophoresis is performed at 4°, 140 V (6.5 V cm^{-1}) for 3 hr with continuous circulation of the buffer (25 mM HEPES, pH 7.9, and 3.3 mM sodium acetate, pH 7.9). Figure 5 shows the interaction of *in vitro* synthesized RAR with a perfect palindrome TRE monomer and dimer. We routinely obtain more specific gel retardation with DNA fragments derived from plasmids compared to DNA fragments synthesized *in vitro*. RAR binds specifically to the TRE monomer and/or dimer. Binding competition can be achieved with a 50-fold excess of nonlabeled TRE but not with the same amount of nonspecific oligonucleotides. The receptor binds cooperatively to TRE since the binding of RAR to the TRE dimer is much stronger than that to the TRE monomer. Binding of the receptor to the TRE does not require the presence of ligand. However, addition of retinoic acid increases the binding affinity of RAR to the TRE slightly. In general, the DNA binding characteristics of RAR and other nuclear receptors (data not shown) reflect their *in vivo* function as previously discussed.[15–18]

Nuclear Retinoic Acid Receptor for Testing Retinoid Activities

RAR and other closely related nuclear retinoid receptors can be expected to mediate many, if not all, of the retinoid effects on gene transcription, although other retinoid-binding proteins may be important for different functions. The ability and capacity of retinoids to activate a particular nuclear receptor can now be easily assessed by a transient transfection assay.[1,15] In this assay, a eukaryotic expression vector that carries a RAR cDNA sequence is transfected into a tissue culture cell line (i.e., CV-1 cells) together with a reporter gene construct that can be induced by the activated RAR. We describe here an efficient procedure which allowed us to compare the gene activation capacity of a large number of individual synthetic retinoids. This system will be extremely useful for the evaluation of optimal, receptor-specific agonists and antagonists.

As an efficient expression vector we use the pECE vector that contains a SV40-derived promoter.[19] The reporter plasmid is derived from the pBL-CAT-2 construct[20] and contains a TRE dimer insert in the *Bam*HI site of the promoter element cloning box [see Fig. 5 (legend)]. This double

[19] L. Ellis, E. Clauser, D. O. Morgan, M. Edery, R. A. Roth, and W. J. Rutter, *Cell* **45,** 721 (1986).
[20] B. Luckow and G. Schütz, *Nucleic Acids Res.* **15,** 5490 (1987).

TRE has been shown to be an efficient retinoic acid responsive element.[15,16]

Transient Transfection

CV-1 cells are plated at 1.3 to 1.7 × 10⁶ per dish 16–24 hr prior to transfection in Dulbecco's modified Eagle's medium (DMEM) supplemented with 10% fetal calf serum (FCS). One to three hours before transfection, cells are fed with 10 ml DMEM supplemented with 10% charcoal-treated FCS. A modified calcium phosphate precipitation procedure is used for transient transfection.[21] One milliliter precipitate is prepared in the following way: 2 μg reporter plasmid, 0.1 to 1 μg receptor expression vector, 3 μg β-galactosidase control plasmid (pCH110, Pharmacia, Piscataway, NJ), and Bluescript plasmid to obtain a total amount of 20 μg DNA are dissolved in a total volume of 450 μl water. To this mixture 50 μl of 2.5 M CaCl₂ is added and mixed carefully. Five hundred microliters of BBS buffer, pH 6.95 [50 mM N,N-bis(2-hydroxyethyl)-2-aminoethanesulfonic acid, 250 mM NaCl, 1.5 mM Na₂HPO₄],[21] is added and mixed by inverting the tube repeatedly. The reaction mixtures are then incubated at room temperature for 10–15 min. During this time a very fine precipitate forms. One milliliter of this precipitate suspension is added per plate followed by incubation at 37° and 4% CO₂ for 14–18 hr. The medium is then exchanged with DMEM containing 10% charcoal-treated FCS with or without ligand. Incubation is continued for 24 hr at 37° and 6% CO₂. After that time, cells are harvested in HBS buffer, pH 7.1 (25 mM HEPES, 140 mM NaCl, 0.75 mM Na₂HPO₄), pelleted by centrifugation, and resuspended in 130 μl of 250 mM Tris-HCl (pH 7.8). The cells are lysed by 3 freeze/thaw cycles. Cellular debris is removed by centrifugation. The cell extract obtained can be used either directly for enzyme assays or may be stored at −20°.

β-Galactosidase Assay

The β-galactosidase assay is carried out in principle as described[22] but is modified to be run in microtiter plates. A reagent mix is freshly prepared by combining the following stock solutions (calculated per assay): 100 μl of 0.1 M sodium phosphate buffer, pH 7.3; 1.5 μl of 100× magnesium solution (0.1 M MgCl₂, 5 M 2-mercaptoethanol); and 33 μl ONPG solution (4 mg o-nitrophenyl-β-D-galactopyranoside per 1 ml sodium phosphate buffer). Fifteen microliters of extract is pipetted into each microliter well (using 15 μl of 250 mM Tris-HCl, pH 7.8, for background

[21] C. A. Chen and H. Okayama, *BioTechniques* **6**, 632 (1988).
[22] N. Rosenthal, this series, Vol. 152, p. 704.

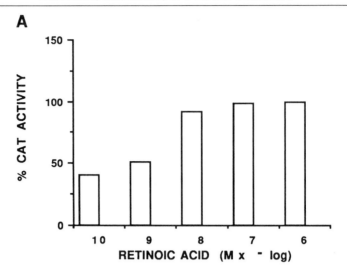

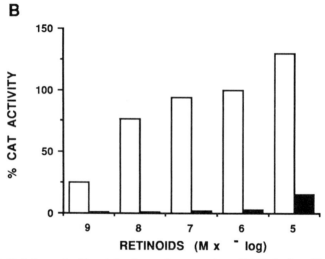

FIG. 6. Defining retinoid activity for nuclear receptors. (A) Retinoic acid-dependent induction of RARβ/ε activity. Increasing amounts of retinoic acid where added to tissue culture cells that were transfected with pECE-RARβ/ε and a TRE–CAT reporter gene.[15] CAT activity was quantitated by cutting out the separated reaction products from silica plates and measuring the amount of radioactivity in a scintillation counter. (B, C) Comparison of synthetic retinoids with retinoic acid. CV-1 cells were transfected with the hybrid receptor RAER-II[15] and an ERE–CAT reporter gene.[1] Cells were grown in the presence of various concentrations of retinoic acid (open bars) or synthetic retinoids (solid bars). Retinoid 1 shows low activation capacity (B); retinoid 2 shows high activation capacity (C). The modified CAT assay described in the text was used.

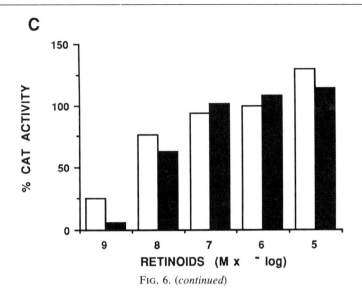

FIG. 6. (*continued*)

correction). After addition of 135 μl of the reagent mix to each well, the plate is incubated at 37° until a yellow color is visible (30–60 min). The reaction is stopped by adding 150 μl of 1 M Na$_2$CO$_3$. Absorbance is measured in a microtiter plate reader at 405 nm.

Chloramphenicol Acetyltransferase Assay

Chloramphenicol acetyltransferase (CAT) activity is determined by either the classic assay[23] or a modified assay using [³H]acetyl-coenzyme A as substrate.[24] The modified assay is more sensitive and less time-consuming. The assay is based on mixed-phase scintillation measurement where only the product, ³H-acetylated chloramphenicol, partitions out of the urea-containing aqueous phase into the nonpolar scintillation liquid. The substrate [³H]acetyl-CoA remains in the aqueous phase and is therefore not measured. For the reaction, 30 μl of the cell extract is combined with 20 μl of 250 mM Tris-HCl, pH 7.8, 20 μl chloramphenicol solution (5 mM chloramphenicol in 250 mM Tris-HCl, pH 7.8), and 20 μl [³H]acetyl-CoA solution (150 μM [³H]acetyl-CoA, 20 μCi/ml, in 75 μM HCl). For background activity determinations, the chloramphenicol solution is substituted by the Tris buffer. The reaction mixtures are incubated at 37° for 1 to 2 hr. The reaction is stopped by addition of 1 ml of 7 M urea. The

[23] C. M. Gorman, L. F. Moffat, and B. H. Howard, *Mol. Cell. Biol.* **2,** 1044 (1982).
[24] D. A. Nielsen, T.-C. Chang, and D. J. Shapiro, *Anal. Biochem.* **179,** 19 (1989).

mixture is transferred to a scintillation vial containing 4.5 ml scintillation cocktail (0.8% 2,5-diphenyloxazole in toluene). After shaking and phase separation, counts per minute (cpm) are determined. Counts per minute corrected for background and β-galactosidase activity represent the specific CAT activity.

In Fig. 6A the transient transfection results of a TRE–CAT reporter gene with RARβ/ε are shown. Induction of CAT activity is dependent on the retinoic acid concentration. In Fig. 6B,C, results obtained with the modified CAT assay[24] are shown using the RAER-II hybrid receptor[15] and an ERE–CAT reporter gene.[1] Retinoic acid is compared to two synthetic retinoids (obtained from M. Dawson), one of which shows a very low activity (Fig. 6B) whereas the other shows activity comparable to that of retinoic acid (Fig. 6C). The altered DNA binding specificity of the RAER-II hybrid receptor allows us to analyze its activity in various cell lines that also contain endogenous RAR.

The methods described here are rapid and sensitive procedures for the probing, cloning, and analysis of nuclear retinoic acid receptors and their ligands. The procedures can be easily adapted for the general analysis of other proteins.

[27] Radioimmunoassays for Retinol-Binding Protein, Cellular Retinol-Binding Protein, and Cellular Retinoic Acid-Binding Protein

By WILLIAM S. BLANER

Introduction

Serum retinol-binding protein (RBP) is responsible for the transport of retinol from retinoid stores in the liver to target tissues throughout the body. This protein has been extensively studied since its first isolation in 1968,[1] and a great deal is now known about its chemical structure and physiologic properties.[2,3] Much of the quantitative information regarding RBP levels in cells, tissues, and serum in both the normal and diseased states has come from studies employing radioimmunoassay (RIA) tech-

[1] M. Kanai, A. Raz, and D. S. Goodman, J. Clin. Invest. 47, 2504 (1968).
[2] D. S. Goodman, in "The Retinoids" (M. B. Sporn, A. B. Roberts, and D. S. Goodman, eds.), Vol. 2, p. 41. Academic Press, Orlando, Florida, 1984.
[3] W. S. Blaner, Endocr. Rev. 10, 308 (1989).

niques for the measurement of RBP. The first portion of this chapter describes procedures, which have been employed in many studies of RBP distribution and function, for the determination of RBP levels by RIA.

Within cells, retinol and retinoic acid are associated with specific intracellular retinoid-binding proteins, namely, cellular retinol-binding protein (CRBP) and cellular retinoic acid-binding protein (CRABP).[4] These two immunologically distinct proteins appear to be the most widely distributed of the intracellular retinoid-binding proteins, having been reported to be present in all rat tissues examined.[5] The greatest sensitivity and specificity for measuring either CRBP or CRABP levels, within tissues or cells, have been obtained through the use of RIA techniques. RIA procedures for the measurement of CRBP and CRABP levels are described in the second part of this chapter.

Radioimmunoassay for Retinol-Binding Proteins

Radioimmunoassay procedures for the determination of RBP levels provide sensitive, specific, and very reproducible methods for the measurement of RBP in serum,[6,7] tissues,[8] and isolated cells.[9] The RIA procedure is especially useful if RBP levels from a large number of samples are to be determined. The procedure described below was first reported by Smith et al.[10] for rat RBP; with appropriate substitutions of antisera, iodinated proteins, and standards, it can be used essentially as described for the measurement of human or rat RBP, transthyretin (TTR), and albumin.

Preparation of Antiserum

Purified RBP (human or rat) is required for preparation of monospecific polyvalent anti-RBP antisera. The procedures described in this volume for the purification of serum RBP[11] provide suitable RBP preparations for this purpose.

Anti-RBP antisera can be prepared readily in albino rabbits (2–3 kg body weight). A solution of purified RBP (either human or rat) is dis-

[4] F. Chytil and D. E. Ong, in "The Retinoids" (M. B. Sporn, A. B. Roberts, and D. S. Goodman, eds.), Vol. 2, p. 90. Academic Press, Orlando, Florida, 1984.

[5] M. Kato, W. S. Blaner, J. R. Mertz, K. Das, K. Kato, and D. S. Goodman, J. Biol. Chem. 260, 4832 (1985).

[6] F. R. Smith, A. Raz, and D. S. Goodman, J. Clin. Invest. 49, 1754 (1971).

[7] Y. Muto, J. E. Smith, P. O. Milch, and D. S. Goodman, J. Biol. Chem. 247, 2542 (1972).

[8] J. E. Smith, Y. Muto, and D. S. Goodman, J. Lipid Res. 16, 318 (1975).

[9] J. L. Dixon and D. S. Goodman, J. Cell. Physiol. 130, 7 (1987).

[10] J. E. Smith, D. D. Deen, Jr., D. Sklan, and D. S. Goodman, J. Lipid Res. 21, 229 (1980).

[11] W. S. Blaner and D. S. Goodman, this volume [19].

solved in 0.9% NaCl to a concentration of 1 mg/ml and emulsified in an equal volume of Freund's complete adjuvant (Difco Laboratories, Detroit, MI). Rabbits are immunized by intradermal injection, into approximately 15 sites on the back, of the RBP emulsion (1 mg RBP/rabbit). One month later, booster injections of RBP (1 mg/ml 0.9% NaCl) emulsified in an equal volume of Freund's incomplete adjuvant are given so that 250 μg of RBP is injected intradermally into approximately 10 sites over the back and 250 μg RBP is injected intramuscularly into the hind legs. The titer of anti-RBP antibodies reaches a maximum 10–14 days after this injection. The animal can be bled at this time to provide useful antiserum for the RIA or, alternatively, can be given a second booster injection of RBP to increase further the titer of anti-RBP antibodies. This second booster injection is given 1 month after the first and is given in a manner identical to the first booster injection. Anti-RBP antibody titer is highest 10–14 days after this second injection.

The rabbit anti-RBP antisera prepared in this manner should be distributed in small volumes in a series of plastic or glass vials or tubes and stored at −70°. Under these conditions, rabbit anti-RBP antiserum has been found to remain stable for over 15 years.

Iodination of Retinol-Binding Protein

Pure RBP is iodinated by the lactoperoxidase procedure essentially as described by Miyachi *et al.*[12] For this purpose, a stock solution containing 1 mg bovine milk lactoperoxidase (Sigma, St. Louis, MO) in 1 ml of 0.1 M sodium acetate buffer, pH 5.6, is prepared and stored at −20° prior to use. Immediately before use, it is necessary to prepare a 0.002% (v/v) H_2O_2 solution by first diluting 0.1 ml of 30% (v/v) H_2O_2 in 10 ml deionized water followed by further dilution of 0.1 ml of this newly made dilute H_2O_2 solution in 15 ml of deionized water.

The iodination of RBP is carried out by adding 10 μl of a stock solution of purified RBP (human or rat) containing 5 μg RBP in water, 25 μl of 0.4 M sodium acetate buffer, pH 5.6, and 1 mCi ^{125}I (New England Nuclear, Boston, MA) to a 10 × 75 mm glass tube and mixing gently. To this mixture, 10 μl of a lactoperoxidase solution, which is prepared immediately before use by diluting 0.1 ml of the stock lactoperoxidase solution in 10 ml of 0.1 M sodium acetate buffer, pH 5.6, is added, and the tube is again mixed gently. To start the enzymatic reaction, 10 μl of the 0.002% H_2O_2 solution is added, and the reaction tube is gently mixed and allowed to incubate at 37° for 30 min. The ^{125}I-labeled RBP is separated from

[12] Y. Miyachi, J. L. Vaitukatis, E. Nieschlag, and M. B. Lipsett, *J. Clin. Endocrinol. Metab.* **34**, 23 (1972).

nonprotein-bound ^{125}I by chromatography on a PD-10 column (Pharmacia LKB Biotechnology, Inc., Piscataway, NJ) in a buffer consisting of 70 mM Tris-HCl, pH 8.6, containing 70 mM NaCl and 1% bovine serum albumin (BSA).

The ^{125}I-labeled RBP is stored in 70 mM Tris-HCl buffer, pH 8.6, containing 70 mM NaCl and 5% BSA at $-20°$. Under these storage conditions, the ^{125}I-labeled RBP can be used in the RIA for up to 2 months.

Radioimmunoassay Procedure

Reagents. For the assay of RBP, five reagents are needed. These are RIA buffer, diluted ^{125}I-labeled RBP, diluted anti-RBP antiserum, a stock polyethylene glycol solution, and precipitation buffer.

RIA buffer: 50 mM Tris-HCl, pH 8.6, containing 1% BSA. This buffer should be filtered through Whatman #4 filter paper (Whatman International Ltd., Maidstone, UK), prior to use, to remove flocculent material. This solution can be stored at 4° for up to 2 weeks.

Diluted ^{125}I-labeled RBP: Within 2 hr of use, the ^{125}I-labeled RBP should be diluted in the RIA buffer so that approximately 20,000 counts per minute (cpm) of ^{125}I are present in 200 μl of RIA buffer.

Diluted anti-RBP antiserum: Within 2 hr of use, the anti-RBP antiserum should be diluted so that the final concentration of antiserum present in the RIA will be sufficient to bind 60% of the ^{125}I-labeled RBP present in the assay tube in the absence of added unlabeled RBP. [For rabbit anti-RBP antiserum collected after two booster injections, this will normally be a 1/10,000–1/20,000 (v/v) final (in the assay tube) dilution of the antiserum.]

Polyethylene glycol stock solution: 260 g polyethylene glycol 8000 (average molecular weight 7000–9000) (J. T. Baker Chemical Co., Phillipsburg, NJ) is added to 310 ml deionized water and allowed to dissolve overnight at room temperature to give a 50% (w/v) solution. This solution can be stored for several months at 4°.

Precipitation solution: Prepared immediately before needed, 0.2 g bovine γ-globulin (Schwarz/Mann Biotech, Cleveland, OH) is dissolved in 12 ml deionized water and added to 56 ml RIA buffer. To this γ-globulin-containing RIA buffer is added 32 ml of the polyethylene glycol stock solution. The solution should become lactescent on addition of the polyethylene glycol to the γ-globulin-containing RIA buffer. As described, the procedure provides 100 ml of precipitation solution; however, the procedure can be scaled up as needed.

Procedure. The reagents for the RBP assay are added to 12 × 75 mm polystyrene conical tubes in the following order:

1. 100 μl of sample to be assayed (either a standard solution of RBP or a diluted serum sample or diluted tissue homogenate)
2. 200 μl of the diluted ^{125}I-labeled RBP (~20,000 cpm)
3. 200 μl of the diluted anti-RBP antiserum [final antiserum dilution 1/10,000–1/20,000 (v/v)]

The assay tubes are mixed well with a vortex mixer and incubated at 4° for 3 days. For precipitation of the immunoglobulins, 1.5 ml of freshly prepared ice-cold precipitation buffer is added to each assay tube and thoroughly mixed with a vortex mixer. The assay tubes are centrifuged at 4000 g for 30 min at 4°. The resulting clear supernatant is removed by aspiration, and the pelleted ^{125}I present in each tube is determined.

Standard solutions containing known quantities of purified RBP are prepared in RIA buffer. The concentration of the pure RBP is determined spectrophotometrically using an $E_{1\,\text{cm}}^{1\%}$ of 19.4.[1] For the assay of serum (or plasma) RBP levels, samples are diluted in RIA buffer. The diluted sera can be stored for at least 1 week at −20°. Tissue levels of RBP can be measured in homogenates prepared from the tissues. Tissues are homogenized in 4 volumes (v/w) of RIA buffer containing 1% Triton X-100 with a Polytron homogenizer (Brinkmann Instruments, Westbury, NY) at setting 5 for 30 sec. The homogenate is centrifuged at 8000 g for 30 min at 4°, and the resulting supernatants are used directly for the radioimmunoassay of RBP. Similarly, levels of RBP in isolated cells can be obtained by homogenizing a cell pellet of approximately 10^7 cells in 1.0 ml RIA buffer containing 1% Triton X-100 with a Polytron homogenizer, at setting 5 for 30 sec. The cell homogenates are centrifuged at 8000 g for 30 min at 4°, and the supernatants are removed for RBP assay. If further dilution of the tissue or cell supernatants is necessary, the dilutions should be made in RIA buffer containing 1% Triton X-100.

Discussion

The procedure described above provides a highly reproducible method for measuring RBP levels in serum (or plasma), tissues, and cells. For the measurement of human serum RBP levels the within- and between-assay coefficients of variation are, respectively, 6.8 and 8.4%.[13] The assay for RBP is very sensitive and is able to measure accurately RBP levels as low as 3 ng per tube. Figure 1 shows a standard displacement curve for puri-

[13] G. D. Friedman, W. S. Blaner, D. S. Goodman, J. H. Vogelman, J. L. Brind, R. Hoover, B. H. Fireman, and N. Orentreich, *Am. J. Epidemiol.* **123,** 781 (1986).

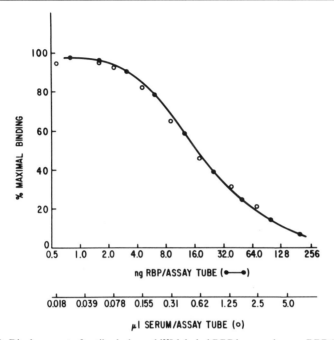

FIG. 1. Displacement of antibody-bound ^{125}I-labeled RBP by pure human RBP (●) and by serial dilutions of an individual human serum specimen (○).

fied human RBP and diluted human serum. The useful range of this assay is between approximately 3 and 50 ng RBP per assay tube.

It has not been possible to use polyvalent rabbit anti-rat RBP antiserum to measure human RBP levels, nor has it been possible to use polyvalent rabbit anti-human RBP antiserum to measure rat RBP levels. Antiserum directed against rat RBP, however, has been used successfully to measure relative levels of RBP in mouse plasma.[14]

As a word of caution, the most critical factor for ensuring success in carrying out the RBP RIA is maintaining the pH of all solutions at 8.6. If this pH is not maintained, the immunoglobulins are not properly precipitated. This can be most directly seen with the precipitation buffer, which will not become lactescent on addition of the polyethylene glycol solution.

It is now possible to purchase purified human RBP from Calbiochem Corp. (San Diego, CA) and Chemicon International, Inc. (Temecula, CA) and polyclonal anti-RBP from Chemicon and Accurate Chemical and Sci-

[14] A. Brouwer, W. S. Blaner, A. Kukler, and K. J. van den Berg, *Chem.–Biol. Interact.* **68,** 203 (1988).

entific Corp. (Westbury, NY). Although it seems likely that these commercially available reagents will be useful for the RIA of RBP, this has not been reported in the literature.

Radioimmunoassays for Cellular Retinoid-Binding Protein and Cellular Retinoic Acid-Binding Protein

Radioimmunoassays for CRBP have been reported by four laboratories,[5,15–18] whereas two radioimmunoassays for CRABP[5,16] have been described in the literature. Unfortunately, the tissue CRBP and/or CRABP levels reported by different laboratories have not always been in agreement. The basis for the differences in the levels observed between laboratories is not fully clear; however, methodological differences in how the antibodies against CRBP or CRABP were produced and the presence or absence of detergents in the radioimmunoassays may account for some of the lack of agreement between laboratories.

The procedures described below have been employed in a variety of studies exploring the cellular and tissue distributions of CRBP and CRABP.[5,19–22] The procedures for the two RIAs are identical except for the concentration of antibodies used in each assay.

Preparation of Antisera

CRBP and CRABP must be homogeneously pure for the preparation of polyvalent antisera according to the procedures described below. For the procedures described below, CRBP and CRABP were purified from rat testis homogenates by protocols described elsewhere[23]; however, the purification procedures described previously[24] provide comparably purified CRBP and CRABP for use as immunogens.

[15] N. Adachi, J. E. Smith, and D. S. Goodman, *J. Biol. Chem.* **256,** 9471 (1981).
[16] D. E. Ong and F. Chytil, *J. Biol. Chem.* **257,** 13385 (1982).
[17] G. Fex and G. Johannesson, *Cancer Res.* **44,** 3029 (1984).
[18] U. Eriksson, K. Das, C. Busch, H. Nordlinder, L. Rask, J. Sundelin, J. Sällström, and P. A. Peterson, *J. Biol. Chem.* **259,** 13464 (1984).
[19] W. S. Blaner, K. Das, J. R. Mertz, S. R. Das, and D. S. Goodman, *J. Lipid Res.* **27,** 1084 (1986).
[20] W. S. Blaner, H. J. F. Hendriks, A. Brouwer, A. M. de Leeuw, D. L. Knook, and D. S. Goodman, *J. Lipid Res.* **26,** 1241 (1985).
[21] M. Kato, K. Kato, W. S. Blaner, B. S. Chertow, and D. S. Goodman, *Proc. Natl. Acad. Sci. U.S.A.* **82,** 2488 (1985).
[22] W. S. Blaner, M. Galdieri, and D. S. Goodman, *Biol. Reprod.* **36,** 130 (1987).
[23] M. Kato, K. Kato, and D. S. Goodman, *J. Cell Biol.* **98,** 1696 (1984).
[24] F. Chytil and D. E. Ong, *in* "The Retinoids" (M. B. Sporn, A. B. Roberts, and D. S. Goodman, eds.), Vol. 2, p. 89. Academic Press, Orlando, Florida, 1984.

Preparation of Anti-CRBP. Antiserum against CRBP is prepared by immunizing a female white turkey (weighing ~8 kg) with 1 ml of a solution of rat testis CRBP (1 mg/ml) in 0.9% NaCl emulsified with an equal volume of Freund's complete adjuvant (Difco). The immunizations are given intramuscularly into multiple sites in the breast muscle. One month after the initial injections, a series of multiple small booster injections consisting of a total of 0.2 mg CRBP emulsified with Freund's incomplete adjuvant is given at the same location as the original injections. A second series of multiple small booster injections, identical in content and site to the first series of booster injections, is given 1 month after the initial booster injections. Antibody titers for CRBP reach a maximum 6 to 9 days after the second booster injection, and blood is collected from the animal during this time interval for use in the RIA for CRBP.

The IgG fraction of the turkey plasma is obtained by column chromatography using the procedure of Saif and Dohms.[25] Purified monospecific IgG is then obtained by immunosorbent affinity chromatography of the whole turkey IgG fraction with CRBP linked to Sepharose 4B (Pharmacia LKB Biotechnology).[23] For use in the CRBP RIA, 1 mg of purified turkey IgG against CRBP is mixed with 49 mg preimmune turkey IgG in 5.0 ml of 0.9% NaCl and stored at −80° until use.

Preparation of Anti-CRABP. For use as an immunogen, purified CRABP is conjugated with poly(L-lysine). The conjugation is accomplished by adding 300 µg of succinylated poly(L-lysine) (Sigma) and 500 µg of 1-ethyl-3-(3-dimethylaminopropyl)carbodiimide hydrochloride (Sigma) to a solution of CRABP (1 mg/2.7 ml water) and allowing the reaction to proceed overnight at room temperature. The reaction mixture is dialyzed overnight against 0.9% NaCl prior to use in immunization. A portion of the solution of CRABP conjugated to the poly(L-lysine) equivalent to 340 µg CRABP is emulsified with 1.2 ml of Freund's complete adjuvant and with 120 µg *Mycobacterium butyricum* (Difco) and injected intramuscularly into multiple sites of the breast of a female white turkey (weighing ~8 kg). An intramuscular booster injection consisting of a portion of the solution of CRABP conjugated to poly(L-lysine) (containing 200 µg CRABP) emulsified with 0.6 ml of Freund's incomplete adjuvant is given 1 month after the initial immunization. The titer of antibodies against CRABP reaches a maximum 5 to 7 days after the booster injection, and blood should be taken at this time for use in the CRABP RIA.

Purified monospecific IgG against CRABP is obtained by first purifying whole turkey IgG[25] and subjecting this fraction to immunosorbent affinity chromatography on CRABP linked to Sepharose 4B.[23] One milli-

[25] Y. M. Saif and J. E. Dohms, *Avian Dis.* **20,** 79 (1976).

gram of purified turkey antibodies against CRABP is mixed with 49 mg preimmune turkey IgG in 5.0 ml of 0.9% NaCl and stored at −80° for use in the RIA for CRABP.

Preparation of Anti-Turkey IgG. Antibodies for the immunoprecipitation of turkey IgG in the RIAs are prepared in albino rabbits (∼3 kg body weight). Turkey IgG purified from preimmune turkey plasma[25] is dissolved in 0.9% NaCl, to a concentration of 1 mg/ml, and emulsified in an equal volume of Freund's complete adjuvant. The turkey IgG emulsion is injected intradermally (1 mg/rabbit) over the shoulders of the rabbit in about 15 separate sites. After 1 month, intradermal booster injections of turkey IgG (0.7 mg/rabbit), emulsified in Freund's incomplete adjuvant, are given in multiple sites over the midback. The highest titer of antibodies against turkey IgG is present 6 weeks after the booster injections. The whole antiserum taken at this time is stored at −80° and used without further purification in the RIAs for CRBP and CRABP.

Iodination of CRBP and CRABP

Both CRBP and CRABP are acylated with ^{125}I-labeled Bolton–Hunter reagent (4000 Ci/mmol) (New England Nuclear) according to the manufacturer's instructions. Briefly, 10 μg of CRBP or CRABP, in 5 μl of 0.1 M borate buffer, pH 8.5, is added to the dried iodinated ester, and the reaction mixture is continuously mixed for 20 min at 0° and then allowed to stand for 40 min at 0°. Following this incubation, the reaction is quenched by adding 100 μl of 0.5 M glycine in 0.1 M borate buffer, pH 8.5, and incubating for 60 min at 0°. The acylated proteins are separated from nonprotein-bound ^{125}I by chromatography on PD-10 columns (Pharmacia LKB Biotechnology) using 10 mM sodium phosphate, pH 7.2, containing 150 mM NaCl and 1% BSA. Following this procedure, the specific activities of the resulting acylated CRBP and CRABP preparations normally are in the range of 20 to 40 mCi/mg.

Radioimmunoassay Procedures

The radioimmunoassay procedures for CRBP and CRABP are identical. All reagents are dissolved or diluted in a buffer (RIA buffer) consisting of 50 mM imidazole-HCl, pH 7.4, containing 150 mM NaCl, 0.03% BSA, 0.1% thimerosal (Sigma), 0.01% leupeptin (Peninsula Laboratories, San Carlos, CA), and 1% Triton X-100. All procedures are carried out on ice or at 4°. Reagents are added to polystyrene conical tubes (12 × 75 mm) and incubated in the following sequence:

1. 100 μl of sample to be assayed (either a standard solution of CRBP or CRABP, or a diluted tissue or cellular cytosol)

2. 50 μl of diluted anti-CRBP IgG solution (containing 0.04 μg purified specific IgG and 1.96 μg of preimmune IgG) or diluted anti-CRABP IgG solution (containing 0.02 μg of purified specific IgG and 0.98 μg of preimmune IgG)
3. 300 μl of RIA buffer; incubate for 3–4 hr
4. 50 μl of ^{125}I-labeled CRBP or ^{125}I-labeled CRABP (~20,000 cpm ^{125}I); incubate 20–24 hr at 4° with gentle agitation
5. 25 μl of rabbit antisera against turkey IgG; incubate 12–18 hr at 4° with gentle agitation.

Samples are centrifuged at 4000 g for 30 min at 4°, the supernatants are removed by aspiration, and the precipitates are assayed for [125]I.

The standard solutions of CRBP or CRABP are prepared from purified CRBP or CRABP diluted in RIA buffer. The concentrations of pure CRBP and CRABP are estimated spectrophotometrically using an $E_{1\,cm}^{1\%}$ of 14.[26] For both the CRBP RIA and the CRABP RIA, the working ranges of the assays are between 1 and 16 ng of CRBP or CRABP per assay tube.

Tissue cytosol preparations are prepared from perfused (with ice-cold 0.9% NaCl) freshly dissected tissues. The dissected tissues are minced, and 500 mg of the tissue mince is added to 3.0 ml RIA buffer and homogenized with a Polytron homogenizer (Brinkmann Instruments) at setting 5 for 30 sec. The tissue homogenates are centrifuged at 105,000 g for 1 hr, and the resulting clear supernatant is diluted appropriately with RIA buffer for assay. Cell pellets containing approximately 10^7 cells are homogenized in 1.0 ml RIA buffer with a Polytron homogenizer on setting 5 for 30 sec. Cytosols from the cell homogenates are prepared by centrifugation of the homogenates at 105,000 g for 1 hr.

Discussion

The procedures described above provide sensitive, specific, and reproducible methods for measuring CRBP and CRABP levels in tissues or isolated cells. Levels of CRBP and CRABP determined with these two RIA procedures, for some tissues and cell types, are given in Table I. For the CRBP RIA, the values for the within-assay and for the between-assay variability are estimated to be 2.7 and 4.8%, respectively.[5] The lower detection limit of the CRBP RIA is 1 ng CRBP/assay tube; hence, it is possible to measure accurately tissue levels as low as 0.07 μg CRBP/g tissue or cellular levels as low as 1 ng CRBP/10^6 cells. The within-assay and between-assay variabilities for the RIA of CRABP are estimated to be

[26] A. C. Ross, Y. I. Takahashi, and D. S. Goodman, *J. Biol. Chem.* **253**, 6591 (1978).

TABLE I
LEVELS OF CRBP AND CRABP IN VARIOUS RAT
TISSUES AND CELLS

Tissue or cell	CRBP[a]	CRABP[a]
Liver[b]	40.0 ± 9.7	1.1 ± 0.2
Small intestine[b]	10.3 ± 1.4	0.8 ± 0.2
Brain[b]	4.4 ± 0.5	1.2 ± 0.3
Skin[b]	3.5 ± 0.6	12.6 ± 2.3
Fat[b]	1.2 ± 0.3	0.2 ± 0.1
Testis[b]	17.8 ± 4.8	9.5 ± 1.1
Seminal vesicle[b]	2.9 ± 0.7	30.4 ± 7.9
Hepatic parenchymal cells[c]	470 ± 238	5.6 ± 3.9
Hepatic stellate cells[c]	236 ± 89	8.7 ± 5.7

[a] Tissue levels of CRBP and CRABP are given as micrograms CRBP or CRABP per gram wet weight of the tissue. Cellular levels of CRBP and CRABP are given as nanograms CRBP or CRABP per 10^6 cells. All values are given as the mean ± 1 S.D.

[b] Levels of CRBP and CRABP are from Ref. 5.

[c] Levels of CRBP and CRABP are from Ref. 20. When these levels are expressed on the basis of cellular protein, the hepatic parenchymal cell levels of CRBP and CRABP are 243 ± 124 ng CRBP/mg protein and 2.9 ± 3.3 ng CRABP/mg protein; hepatic stellate cell levels of CRBP and CRABP are 1256 ± 477 ng CRBP/mg protein and 44 ± 30 ng CRABP/mg protein.

4.1 and 5.5%, respectively.[5] Lower detection limits for the CRABP RIA are 0.07 μg CRABP/g tissue and 1 ng CRABP/10^6 cells.

In carrying out the assays for CRBP and CRABP, it is important to adhere to the protocols described above. The times and temperatures of incubations are important for the sensitivities (affecting the working ranges) of each of the RIAs. The preparations of ^{125}I-labeled CRBP or ^{125}I-labeled CRABP should have specific activities between 20 and 40 mCi/mmol. If CRBP or CRAPB are either more or less heavily labeled with ^{125}I, the standard displacement curves are found to shift to lower sensitivity. The ^{125}I-labeled CRBP and ^{125}I-labeled CRABP can be stored in 1% BSA at −20° for up to 1 month. Tissue or cell cytosol preparations can be stored overnight at 4° or for up to 14 days at −80°. Storage of cytosol preparations for longer than 14 days at −80° results in a decrease in the immunoreactive CRBP and CRABP contents of the cytosol preparations.

It is necessary to titer each new batch of second antibody (rabbit anti-turkey IgG) employed in the assays, to ensure that this antibody is not present in limiting concentrations. For eight different rabbit anti-turkey IgG antisera examined by this author, 25 μl of each antisera provided a large excess of anti-turkey IgG antisera; however, this need not always be true.

Acknowledgments

The author would like to thank Drs. DeWitt S. Goodman and John E. Smith for their help and support. This work was supported by National Institutes of Health Grants DK 05968 and HL 21006.

[28] X-Ray Crystallographic Studies on Retinol-Binding Proteins

By MARCIA NEWCOMER and T. ALWYN JONES

Introduction

Structural information for proteins which interact with retinoids, in their transport or metabolism, or which mediate retinoid function(s) is critical to our understanding of the role(s) of natural or synthetic retinoids in physiology and medicine. X-Ray crystallographic studies of this class of proteins can provide the details we need to understand retinoid recognition at the molecular level. The steps involved in a structural determination can be briefly summarized as follows. (1) The protein must be isolated in relatively large quantity (>10 mg) with a very high degree of purity. (2) Protein crystals suitable for X-ray diffraction data collection are prepared. (3) X-Ray diffraction data are collected. (4) The "phase problem" is solved by an appropriate method. (5) An electron density map is calculated. (6) A model for the protein structure is built into the electron density. (7) The model is refined. Once a high-resolution structure is determined, the structures of the protein complexed with ligand or ligand analogs are relatively straightforward to determine.

The three-dimensional structure of the retinol-binding protein (RBP) from human serum has been determined by the process briefly outlined above.[1] For information with respect to the standard methods in protein structure determination employed, the reader is referred to Volumes 114

---

[1] M. E. Newcomer, T. A. Jones, J. Åquist, J. Sundelin, U. Eriksson, L. Rask, and P. A. Peterson, *EMBO J.* **3,** 1451 (1984).

and 115 of this series. Some of the biochemical details of the process, relevant only to the solution of the structure of RBP, are described in this chapter. Now that these methods are available, we can probe the structural aspects of the specificity of the retinol-binding protein for its ligand as well as for its carrier protein transthyretin (TTR). The three-dimensional structure of transthyretin, which is found complexed with the retinol-binding protein in human serum, has also been determined.[2]

Protein Purification

In general, when preparing protein for crystallization one attempts to employ only the more "gentle" methods in protein purification. For example, lyophilization is best avoided. Only reagents of the highest purity should be used, as trace contaminants may adversely affect crystallization or reproducibility.

The retinol-binding protein used to prepare the crystals for our X-ray crystallographic structure determination was prepared from the urine of patients suffering from chronic cadmium poisoning. In such patients, RBP is a major urinary protein and can be purified by the method of Peterson and Berggård.[3] The protein isolated from urine is largely holo and capable of binding its carrier protein transthyretin. In the preparation, the concentrated urinary proteins are chromatographed on a Sephadex G-100 column in 20 mM Tris-HCl at pH 8.0. Further purification is achieved on DEAE-Sephadex (20 mM Tris-HCl, pH 7.2, NaCl gradient) and Sephadex G-200. The final step in the preparation involves binding of the RBP to a transthyretin affinity column and elution by a low ionic strength buffer.

We have purified RBP from serum by a method reported by Shingleton *et al.*[4] As a starting material we use outdated human plasma. From 1.6 liters of plasma we obtain 30 mg RBP with a A_{330}/A_{280} ratio normally greater than 0.97. RBP fully saturated with retinol has an A_{330}/A_{280} ratio of 1.03.[5] In the purification, serum is applied to a DE-52 (Whatman) column in 50 mM phosphate buffer, 50 mM NaCl (pH 7.5), and RBP elutes in the middle one-third of a gradient 50 to 500 mM in NaCl. Subsequently, the protein is chromatographed on a Sephadex G-75 (1 mM Tris–acetate) after exhaustive diafiltration into the low ionic strength buffer to promote RBP–TTR dissociation. This column step must generally be repeated to

[2] C. C. F. Blake, M. J. Geisow, S. J. Oatley, B. Rérat, and C. Rérat, *J. Mol. Biol.* **121,** 339 (1978).
[3] P. A. Peterson and I. Breggård, *J. Biol. Chem.* **246,** 25 (1971).
[4] J. L. Shingleton, M. K. Skinner, and D. E. Ong, *Biochemistry* **28,** 9647 (1989).
[5] Y. Muto, Y. Shidoji, and Y. Kanda, this series, Vol. 110, p. 840.

adequately separate RBP from TTR. Further purification is achieved on a Mono Q (Pharmacia, Piscataway, NJ) column (10 mM Tris, pH 8, 0 to 500 mM NaCl).

Both the protein isolated from serum and that isolated from urine are able to bind retinol and the carrier protein transthyretin. However, RBP isolated from either source is heterogeneous with respect to charge, and the multiple forms can be readily resolved on native gels (Pharmacia Phast System, 8–25% gradient gels, native buffer strips). Furthermore, we have observed that during storage in the cold, the protein isolated from serum changes in the relative amounts of the charged species. Most notably, the major band becomes that of a more negatively charged species rather than the more positively charged species. Such a change in charge could be due to deamidation of a glutamine or asparagine residue. We are able to resolve these species on the Pharmacia Mono Q column under the conditions described above.

Crystal Preparation

Normally one must scan numerous variations in precipitant, precipitant concentration, pH, buffer, and ionic strength, among other variables, to determine the conditions which promote the growth of single crystals. The most efficient way to do this is with hanging-drop vapor-diffusion experiments. In a typical experiment a 5- to 10-μl drop of protein in a solution of precipitant such as ammonium sulfate or polyethylene glycol (PEG), at a concentration lower than one which immediately precipitates the protein, is suspended above a well containing a solution of higher precipitant or salt concentration. In the sealed chamber the water slowly diffuses out of the protein drop, allowing a gradual increase in the concentrations of precipitant and protein. If set up in a multiwell tissue culture dish or a sealed petri dish, the drops can easily be scanned for crystal growth under a microscope.

RBP crystallizes in fine needles from ammonium sulfate.[6] With the addition of $CuSO_4$ to the ammonium sulfate large hexagonal plates can be grown.[7] The crystals, however, are disordered. Unable to improve the quality of these crystals, we switched to RBP isolated from rabbit serum and obtained diffraction-quality crystals from 6% polyethylene glycol 6000, 1 mM $CdCl_2$, 10 mM Tris, pH 8.0, 75 mM NaCl. These crystals of rabbit serum RBP were subsequently used to seed experiments of human

[6] H. Haupt and K. Heide, *Blut* **24,** 94 (1972).
[7] M. E. Newcomer, A. Liljas, J. Sundelin, J. Rask, and P. A. Peterson, *J. Biol. Chem.* **259,** 5230 (1984).

serum RBP set up in equivalent conditions. We used the "microseeding" technique: a crystal is thoroughly crushed in a small volume of stabilizing solution and the solution is spun briefly in a tabletop centrifuge. With a glass fiber one touches the surface of the centrifuged stabilizing solution and then the drop set up for crystallization. Only seeded drops of human serum RBP, set up as described below, ever produced crystals. Once human serum RBP gave crystals, further seeding was done with these crystals.

Crystallization of RBP isolated from urine is effected by adding 1 volume of RBP (12 mg/ml, 10 mM Tris, pH 8.0, 150 mM NaCl) to 1 volume of 6% PEG 6000, 2 mM CdCl$_2$. Drops (10–20 μl) of the mixture are applied to a siliconized glass surface, and the glass plate is inverted and suspended over a reservoir of 0.3 M KCl. The interface between the well and the plate is sealed with a generous application of vacuum grease. After a few days the glass plate is removed and the drops are seeded. Crystals grow over the course of 2 to 4 weeks.

Once large (>0.5 mm) crystals have been grown they can be mounted in glass capillary tubes. RBP crystals are elongated along one axis, allowing one to readily align the crystal along the length of the capillary. Enough liquid is left on the crystal such that the drop in which it is lying is about halfway up the crystal when viewed from the side. Excess liquid must be removed from the capillary, and then a dampened wick of filter paper is inserted. The capillary is sealed with a low-melting point wax. The crystal is now ready for X-ray data collection.

RBP crystallizes in the space group $P2_12_12_1$ with one molecule per asymmetric unit. The cell dimensions are $a = 45.7$, $b = 48.7$, and $c = 76.5$ Å. It is the b axis which lies along the length of the crystal.

RBP isolated from human serum was crystallized by Ottonello et al.[8] from 4.5 M NaCl. Their crystals are space group $R3$, and the asymmetric unit contains two molecules. Our RBP purified from serum does not readily crystallize under our PEG/CdCl$_2$ conditions but under conditions similar to those described by Ottonello et al. gives $R3$ crystals. However, if we resolve our RBP (isolated from serum) into its two major forms which differ in charge, and use only the more negatively charged of the two forms, crystals can be obtained from PEG 6000/CdCl$_2$ with the same morphology as those we obtained from the urine protein. Part of the problem in reproducing the $P2_12_12_1$ PEG-grown crystals may be due to the variability between commercially available batches of polyethylene glycol, as it has been reported that different batches of PEG vary consider-

[8] S. Ottonello, G. Maraini, M. Mammi, H. Monaco, P. Spadon, and G. Zanotti, J. Mol. Biol. **163**, 679 (1983).

ably with respect to trace contaminants.[9] It is prudent to use only material prepared in a single batch or purify the PEG on an ion-exchange resin to standardize the procedure.

Diffraction Data Collection

The number of data points (reflections) to be measured is determined by (1) the volume of the unit cell, (2) the number of asymmetric units within the unit cell, and (3) the resolution of the diffraction data. For 2.0-Å resolution data for the retinol-binding protein there are roughly 11,000 unique reflections. Generally, the crystal size and quality for RBP are good, and data collection can be done on a standard four-circle diffractometer. Since it is unlikely that a single crystal will survive the X-ray hours necessary to measure 11,000 reflections, the data are measured in shells of increasing resolution which contain about 2,000 reflections each. A standard group of 100 reflections is collected for each crystal to allow for scaling between crystals.

Phase Determination

The solution of the X-ray structure requires more information than that obtained by simply measuring the intensity of the diffracted waves. One must also determine the phase of these waves. One method of phase determination, termed multiple isomorphous replacement (MIR), requires that one prepare crystals which have a heavy atom, for example, Hg or Au, specifically bound to the protein. It is of great importance that the derivatized crystals remain isomorphous with the native protein crystals. Isomorphism is possible because protein molecules are not closely packed in the crystal lattice. On the contrary, protein–protein contacts are minimal in the crystal and are usually limited to a small number of salt bridges between charged groups on the surfaces of neighboring molecules. About 50% of the crystal volume is channels of buffer. To scan for heavy atom derivatives, crystals are soaked in a stabilization solution which contains, in addition to the precipitant, buffer, and any other additives, millimolar amounts of heavy atom compounds.

RBP crystals were soaked in a variety of heavy atom solutions in order to obtain derivatives. However, the crystals were unstable in most of the solutions tried. The heavy atom compounds used for the solution of the structure of RBP were K_2HgI_4 and $KAu(CN)_2$, in concentrations of 3 and 5 mM, respectively. In the latter case it was critical to remove the crystal

---

[9] F. Jurnak, *J. Cryst. Growth* **76**, 577 (1986).

from the heavy atom soak solution after 5 hr and mount immediately in a glass capillary.

Ligand Analog Studies

Once the structure of the native protein has been determined, structures of the protein complexed with ligand analogs can be readily determined. The ligand analog–protein complexes are prepared and drops set up for crystallizations under conditions equivalent to those for the native protein. Native crystals can be used to seed drops of protein–analog experiments. The diffraction data for the crystal are measured, and a difference electron density map can be calculated.

[29] *In Situ* Hybridization of Retinoid-Binding Protein Messenger RNA

By DIANNE ROBERT SOPRANO and DEWITT S. GOODMAN

Introduction

In situ hybridization is a very powerful technique to localize specific mRNA molecules directly within a cell. This technique provides a combination of molecular and morphological information about individual cells in a tissue, in contrast to the information that can be obtained from the analysis of a pool of RNA isolated from a tissue or organ. Because it combines both molecular and morphological analyses, *in situ* hybridization is a valuable technique in several areas of biomedical research, including, for example, developmental biology, cell biology, genetics, and pathology.

Since the first reports of the detection of mRNA by *in situ* hybridization,[1,2] great technical advances have been made in the area of molecular genetics that have greatly increased the power and potential of this methodology. This has resulted in the application of *in situ* hybridization techniques by many investigators to study the localization of specific mRNA molecules within individual cells of a large number of tissues or populations of cells in tissue culture.

[1] P. R. Harrison, D. Conkie, J. Paul, and K. Jones, *FEBS Lett.* **32,** 109 (1973).
[2] P. R. Harrison, D. Conkie, N. Affara, and J. Paul, *J. Cell Biol.* **63,** 302 (1974).

We have used the technique of *in situ* hybridization to localize retinol-binding protein (RBP) mRNA and transthyretin (TTR) mRNA in fetuses of a wide range of gestational ages,[3] brain,[4] kidney,[5] adipose tissue,[5] and F9 teratocarcinoma cells.[6] We are also conducting similar studies on the localization of the specific mRNAs for cellular retinol- and retinoic acid-binding proteins (CRBP and CRABP) in fetal and adult rat tissues (D. S. Goodman *et al.*, unpublished studies). In all tissues examined to data we have found very specific localization of each transcript within cell types.

The technique of *in situ* hybridization can be divided into four parts: (1) preparation of tissue sections; (2) probe preparation; (3) hybridization of tissue sections; and (4) autoradiographic analysis and staining of tissue sections. We have organized this chapter into these four parts for clarity of presentation.

Materials and Solutions

The following solutions and materials should be made or purchased before performing each step in the *in situ* hybridization protocol.

Preparation of Tissue Sections

Isotonic saline

Paraformaldehyde solution: 4% paraformaldehyde prepared in 0.1 M sodium phosphate, pH 7.4

Phosphate-buffered saline: 10 mM potassium phosphate, pH 7.4, 0.145 M NaCl (PBS; sterile)

Sucrose solution: 15% sucrose prepared in sterile PBS

OCT embedding compound (Miles Laboratory, Naperville, IL)

Isopentane

Liquid nitrogen

Polylysine-coated slides: poly(L-lysine) (Sigma, St. Louis, MO) dissolved at a concentration of 50 μg/ml in 10 mM Tris, pH 8.0; slides are immersed in this solution for 15 min and dried overnight in a dust-free environment

[3] A. Makover, D. R. Soprano, M. L. Wyatt, and D. S. Goodman, *Differentiation* **40**, 17 (1989).

[4] M. Kato, D. R. Soprano, A. Makover, J. Herbert, and D. S. Goodman, *Differentiation* **31**, 228 (1986).

[5] A. Makover, D. R. Soprano, M. L. Wyatt, and D. S. Goodman, *J. Lipid Res.* **30**, 171 (1989).

[6] D. R. Soprano, K. J. Soprano, M. L. Wyatt, and D. S. Goodman, *J. Biol. Chem.* **263**, 17897 (1988).

Preparation of Probe

[35]S-Labeled UTP (specific activity 800 Ci/mmol, New England Nuclear, Boston, MA) or [^{32}P]UTP (specific activity 3000 Ci/mmol, New England Nuclear)

10 mM stock solutions of each of UTP, ATP, CTP, and GTP prepared in sterile water, pH adjusted to 7.5 and stored in small aliquots at −70°

Transcription buffer, 5× concentrated: 200 mM Tris-HCl, pH 7.5, 30 mM MgCl$_2$, 10 mM spermidine; autoclaved and stored at room temperature

0.4 mM dithiothreitol (DTT) prepared in sterile water and stored at −70°

RNasin: RNase inhibitor (Promega Biotec, Madison, WI)

SP6 RNA polymerase (Promega Biotec) or T7 RNA polymerase (Promega Biotec)

RQ1 DNase: RNase-free DNase (Promega Biotec)

DNase buffer: 400 mM Tris, pH 8.0, 100 mM NaCl, 60 mM MgCl$_2$; autoclave before use and store at 20°

50% phenol (saturated with 1 M Tris buffer, pH 8) and 50% 25:1 chloroform–isoamyl alcohol (phenol/chloroform)

3 M potassium acetate, pH 5.0; autoclave before use

Hybridization of Tissue Sections

Proteinase K solution: 1 μg/ml proteinase K dissolved in 20 mM Tris-HCl, pH 7.5, and 2 mM CaCl$_2$

Triethanolamine solution: 0.1 M triethanolamine, pH 8.0; prepared fresh each day in sterile water and pH adjusted with glacial acetic acid

Acetic anhydride solution: 0.25% acetic anhydride prepared fresh each day in 0.1 M triethanolamine, pH 8.0

20× SSC: 3 M NaCl, 300 mM sodium citrate, pH 7.5

2× box buffer: 20 mM Tris-HCl, pH 7.5, and 1.2 M NaCl; autoclave before use

Formamide (Fluka Chemical Co., Buchs, Switzerland)

Prehybridization buffer: 50% formamide, 600 mM NaCl, 10 mM Tris-HCl, pH 7.5, 0.02% Ficoll, 0.02% polyvinylpyrrolidone, 0.1% bovine serum albumin, 1 mM EDTA, 0.5 μg/ml denatured salmon sperm DNA, and 50 μg/ml yeast transfer RNA; prehybridization buffer is prepared by mixing appropriate amounts of the following stock solutions, and diluting with deionized or distilled water (autoclaved), to yield the final concentrations above: formamide, 5 M NaCl (autoclaved), 1 M Tris-HCl, pH 7.5 (autoclaved), 6% Ficoll (filter sterilized), 6% polyvinylpyrrolidone (filter sterilized), 6% bovine serum albumin (filter sterilized), 250 mM EDTA (autoclaved), 10 mg/ml salmon sperm DNA (autoclaved), and 50 mg/ml yeast transfer RNA (phenol/chloroform extracted 2 times and ethanol precipitated 2 times)

Hybridization buffer: same as the prehybridization buffer with the addition of 10% dextran sulfate (Sigma)

1 M DTT: prepared in sterile water and stored at $-70°$

RNase A and T solution: 20 μg/ml RNase A and 2.2 μg/ml RNase T1 dissolved in 10 mM Tris, pH 8.0, and 0.5 M NaCl

Autoradiographic Analysis and Staining

NTB-2 nuclear track emulsion (Kodak Chemical Co., Rochester, NY)

D-19 developer mixed per manufacturer's directions (Kodak)

Rapid fixer mixed per manufacturer's directions (Kodak)

Hematoxylin stain, Gill's formulation #2 (Fisher Scientific Co., Fairlawn, NJ)

Eosin: 1% prepared in 100% ethanol

0.1 M sodium bicarbonate

Ethanol

Xylene

Permawash (Fisher)

Experimental Procedures

Preparation of Tissue Sections

The method of choice for the preparation of tissue sections to be used for *in situ* hybridization of retinoid-binding protein transcripts depends on the tissue of interest. We have successfully employed the *in situ* hybridization technique to localize specific mRNAs for RBP, CRBP, CRABP, and TTR by utilizing tissues which were perfused and fixed *in situ* or tissues fixed after removal from the animal. The method and time of fixation which result in optimal *in situ* hybridization signals are rather tissue specific and must be optimized by the investigator. We provide some illustrative details for rat tissues, which we have investigated extensively; however, the reader should bear in mind that these descriptions should be employed mainly as guidelines to begin new studies.

For studies of adult rat tissues, the animals are first anesthetized by a suitable method; we generally use an injection of ketamine hydrochloride (80–100 mg/kg body weight) containing 4% acepromazine. After the rat has been anesthetized, the chest is opened and the systemic circulation is rinsed by cardiac perfusion with approximately 20 ml cold isotonic saline. This results in the removal of the majority of blood in the animal. When the perfusate appears colorless to light pink, a catheter is passed into the ascending aorta, and approximately 300 ml of paraformaldehyde solution is perfused through the animal. For small animals (e.g., mice or newborn rats) the amount of paraformaldehyde used in the perfusion can vary from 3 to 300 ml depending on the size of the animal.

For some tissues fixation is now complete (e.g., brain or newborn rat kidneys). However, for other tissues, such as the adult kidney, it is necessary to fix the tissue further by immersing the kidney in the paraformaldehyde solution for 1 to 12 hr at 4°. When rat embryos are fixed, the pregnant dam is anesthetized and the embryos are removed *in utero* (at 7 to 11 days of gestation) or dissected from the uterus and extraembryonic membranes (at 13 to 20 days of gestation). The embryos (whether *in utero* or dissected out from the uterus) are fixed in the paraformaldehyde solution at 4° for 1 to 6 hr (time also depends on the size of the embryo).

Following fixation, all tissues are immersed in the sucrose solution for 1 to 3 hr, again depending on the size of the tissue, at 4°. The tissue is next embedded in OCT embedding compound and frozen in isopentane cooled in liquid nitrogen. The tissue embedded in OCT compound can be stored for an extended period of time at −70°. When ready, serial sections of 5- to 10-μm thickness are cut on a cryostat and mounted on polylysine-coated slides. The slides are immediately frozen and stored at −70° in slide boxes containing desiccant capsules. The slides can be stored for a long period of time (at least 6 months) until ready for hybridization.

Preparation of Probe

The preparation of a probe suitable for *in situ* localization of retinoid-binding protein transcripts involves consideration of both the method of preparing the probe and the radioisotope which will be utilized. First, it is necessary to obtain or prepare a specific cDNA encoding the protein of interest (see below). Probes for *in situ* hybridization can then be prepared by nick translation of the cDNA,[7] by oligo labeling of DNA,[8] or by the *in vitro* synthesis of single-stranded antisense cRNA[9] utilizing the cDNA as a template. We found greater sensitivity and a reduction of background

[7] P. W. J. Rigby, M. Dieckmann, C. Rhodes, and P. Berg, *J. Mol. Biol.* **113**, 237 (1977).
[8] A. Feinberg and B. Vogelstein, *Anal. Biochem.* **132**, 6 (1983).
[9] D. A. Melton, P. A. Krieg, M. R. Rebagliati, T. Maniatis, K. Zinn, and M. R. Green, *Nucleic Acids Res.* **12**, 7035 (1984).

hybridization when single-stranded antisense cRNA probes for retinoid-binding protein transcripts were employed.

With regard to labeling, probes (whether they are nick-translated cDNA or *in vitro* synthesized cRNA) can be prepared utilizing ^{3}H-, ^{35}S-, or ^{32}P-labeled nucleotides. The choice of the radioisotope to be utilized in the probe preparation depends largely on the experimental question to be addressed. We generally prepare ^{32}P-labeled probes only in our initial experiments to assess whether we are obtaining good, specific hybridization signals. Slides hybridized with ^{32}P-labeled probes can be exposed directly to X-ray film for a few hours or overnight to obtain quick results; however, because of the large energy emitted by ^{32}P, such slides cannot be used in autoradiographic analysis of tissue sections to determine cellular localization of the hybridization pattern. After we have demonstrated that the tissue sections and the probes are yielding good hybridization signals, we prepare ^{35}S-labeled probes; these allow autoradiographic analysis of the cellular localization of the hybridization signal, and, hence, of the specific transcript, generally within a 4- to 6-week exposure time. If additional information is desired concerning the subcellular localization of a retinoid-binding protein transcript, ^{3}H-labeled probes can be prepared; however, these are generally of low specific activity and require long exposure times (up to 6 months) for the signal to be clearly visualized.

For the preparation of single-stranded antisense strand cRNA probes for retinoid-binding protein transcript localization, it is first necessary to obtain the appropriate cDNA cloned into a vector which contains an RNA polymerase promoter. There are several excellent molecular biology techniques manuals available which describe in detail methodology for the cloning of cDNAs and the isolation of plasmid DNA.[10–12] We have used the pGem series of vectors obtained from Promega Biotec for our studies with retinoid-binding protein transcripts. These vectors contain both the SP6 and T7 RNA polymerase promoters on opposite strands of the plasmid, which allows the synthesis of cRNA in both orientations (sense and antisense strand) from the same plasmid construct. Once the appropriate constructs have been prepared, plasmid DNA is linearized downstream from the promoter site to be used in the synthesis of the cRNA, using a convenient restriction site and restriction enzyme, generally in the multicloning region. This ensures that the probe contains sequences only from

[10] S. L. Berger and A. R. Kimmel (eds.), this series, Vol. 152.

[11] T. Maniatis, E. F. Fritsch, and J. Sambrook, "Molecular Cloning: A Laboratory Manual." Cold Spring Harbor Laboratory, New York, Cold Spring Harbor, 1982.

[12] F. M. Ausubel, R. Brent, R. E. Kingston, D. D. Moore, J. A. Smith, J. G. Seidman, and K. Struhl, "Current Protocols in Molecular Biology." Wiley, New York, 1987.

the promoter to the point of linearization and does not contain large amounts of plasmid sequences. A portion of the linearized DNA is electrophoresed on a 1% agarose gel to determine that it has been completely linearized, while the remainder is extracted with phenol/chloroform 3 times and precipitated with ethanol. The DNA is resuspended in sterile water at a concentration of 0.5 $\mu g/\mu l$.

Before transcribing cRNA it should be noted that it is imperative that all reagents, tubes, and pipette tips are RNase-free and that gloves be worn at all time to eliminate contamination with RNases present on the hands. We generally double-autoclave all glassware, tubes, pipette tips, and reagents to ensure that they are RNase-free and aliquot all reagents into small volumes to be used only once and then discarded.

The 10-μl transcription reaction mixture contains 1 μg of linearized plasmid DNA, 100 μCi ^{35}S-labeled UTP or 50 μCi ^{32}P-labeled UTP, 10 μM UTP, 500 μM each of ATP, CTP, and GTP, 10 μM DDT, 1× transcription buffer (40 mM Tris, pH 7.5, 6 mM MgCl$_2$, 2 mM spermidine), 15 units RNasin, and 10–15 units of either SP6 RNA polymerase or T7 RNA polymerase. Alternatively, an entire transcription kit containing all the necessary reagents, buffers, and enzymes except the linearized plasmid DNA and the radioactive nucleotide can be purchased from Promega Biotec. The reaction mixture is incubated for 1 hr at 40°. The mixture is then diluted 10-fold with DNase buffer and digested with 3 units of RNase-free DNase (RQ1 DNase) at 37° for 20 min to remove the plasmid DNA template. The newly prepared cRNA probe is then extracted with phenol/chloroform and precipitated with ethanol using 0.1 volume of a 3 M potassium acetate, pH 5.0, solution and 2 volumes of ethanol. After centrifugation the probe is resuspended in RNase-free water and stored at $-70°$ until used. We generally use the ^{32}P-labeled probes within a few days after synthesis; however, the ^{35}S-labeled probes can be stored for up to 1 month. The specific activity of the probes generally ranges from 1 to 2 × 10^8 cpm/μg for ^{35}S-labeled probes and 5 to 6 × 10^8 cpm/μg for ^{32}P-labeled probes.

After preparation and before use, the probes are assessed by electrophoresis of a small aliquot on a mini 5% polyacrylamide/urea gel; the gel is dried and exposed to X-ray film to demonstrate that the probe is a single band migrating at the appropriate size location for a cRNA of the expected size.

Hybridization of Tissue Sections

The hybridization procedures that we have used for the *in situ* localization of TTR, RBP, CRBP, and CRABP transcripts are a modification of the procedures described by Wilcox, Gee, and Roberts[13] and Fremeau *et al.*[14] The slides are removed from the freezer ($-70°$) and immediately covered with approximately 50 μl of proteinase K solution for 10 min. Slides are then transferred to dipping slide racks and dipped in sequence in the following solutions: (1) 5 min in sterile water, (2) 5 min in triethanolamine solution, (3) 10 min in acetic anhydride solution, (4) 5 min in sterile 2× SSC, and (5) 5 min in sterile 2× SSC. After the second 2× SSC wash, the SSC is removed from the slides by gently blotting around the tissue with a Kimwipe. At this point you should have prepared an RNase-free box (airtight plastic utility boxes work well) which is used *only* for this purpose. This box is prepared by placing a pad of Whatman #3 paper into the box and then saturating the Whatman #3 paper with sterile 2× box buffer diluted just before use with an equal volume of formamide. Slides are placed in the box on top of the saturated Whatman #3 paper and overlayed with approximately 50 μl of prehybridization buffer per tissue section. The slides in the box are incubated for 2 hr at 53–55°.

After this prehybridization step, the prehybridization buffer is removed by blotting the sides of the tissue sections with a Kimwipe. At this time, a solution of the appropriate probe in the hybridization buffer should be prepared by adding approximately 2500 cpm/μl (final concentration) of cRNA probe to the hybridization solution. If both the antisense and the sense (control) probes are being studied at the same time, check to be sure that the number of counts per minute in each of the hybridization solutions are very similar. The hybridization solution containing the probe is heated at 68° for 10 min and then placed on ice. Ten microliters of a 1 M DDT solution per milliliter of hybridization buffer is added at this time if you are using ^{35}S-labeled probes. Tissue sections are overlayed with approximately 50 μl of the hybridization buffer containing the probe, kept in the same utility boxes saturated with the box buffer plus formamide, and incubated at 53–55° overnight.

After overnight hybridization, the slides are washed to remove unhybridized probe. Slides are removed from the utility boxes, placed on slide racks, and washed for 30 min in sterile 2× SSC at room temperature. Slides are then removed from the slide rack, blotted dry with a Kimwipe, and placed in another airtight utility box (again a box that is used only for this purpose). Approximately 50 μl of RNase A and T solution is overlayed on each tissue section, and the box containing the slides is incubated at 37° for 45 min. This treatment digests any unhybridized single-

---

[13] J. N. Wilcox, C. E. Gee, and J. L. Roberts, this series, Vol. 124, p. 510.
[14] J. L. Fremeau, Jr., J. R. Lundblad, D. B. Pritchett, J. N. Wilcox, and J. L. Roberts, *Science* **234**, 1265 (1986).

stranded probe and does not digest any double-stranded probe/transcript hybrids. Following the RNase digestion the slides are placed back on the slide racks and washed as follows: (1) Wash 2 times for 60 min in a large excess volume of 2× SSC at 53–55°. (2) Wash for 3 hr in a large excess volume of 0.1× SSC containing 0.5% sodium pyrophosphate and 14 mM 2-mercaptoethanol (2-mercaptoethanol is used with only ^{35}S-labeled probes) at 53–55°. (3) Following the 3-hr wash, the wash solution plus slides are allowed to cool to room temperature overnight.

The next day the tissue sections are dehydrated through 50, 70, and 90% (v/v) ethanol containing 0.3 M ammonium acetate for 2 min each at room temperature. Slides are vacuum dried in a desiccator for 30 min. At this point the slides can be directly exposed to X-ray film in a cassette, or they can be dipped in emulsion for autoradiographic analysis.

Autoradiographic Analysis and Staining

All steps must be performed in a dark room. The NTB-2 nuclear track emulsion is heated in a water bath to 42° and then aliquoted into 20-ml portions in 50-ml tubes. These tubes can be wrapped in aluminum foil and stored in the dark. The 20-ml aliquots are remelted by heating to 42° when needed and each diluted with 20 ml of 0.3 M ammonium acetate. This solution should be mixed well gently to avoid air bubbles. The solution is poured into a small vessel and kept in a 42° water bath. Slides are dipped for approximately 2 sec, blotted on a paper towel, and allowed to hang until dry; drying may take as long as 2 to 3 hr. After thorough drying the slides are placed in black slide boxes along with 1 or 2 desiccating caplets. The slide box is carefully sealed and stored in the refrigerator at 4° until ready for development. The time of exposure of the hybridized tissue section to the emulsion (before development) depends on the transcript and the tissue. We generally expose liver sections probed with RBP cRNA for approximately 1 week, whereas kidney sections probed with RBP cRNA can require exposures of 4 to 6 weeks.

When the slides are ready to be developed they are dipped in sequence in the following solutions in the dark room: (1) D-19 developer for 3.5 min,

---

FIG. 1. Localization of RBP mRNA in the adult rat kidney. (a, c) Bright-field micrographs of kidney transverse sections demonstrating the anatomical regions under study [outer stripe of medulla (OSM), delineated between dashed lines]. (b, d) Dark-field micrographs of adjacent kidney sections hybridized to a ^{35}S-labeled RBP cRNA antisense probe. Magnification: a and b, ×83; c and d, ×207. All sections are Giemsa stained. OSM, Outer stripe of medulla; ISM, inner stripe of medulla; Cx, cortex. (From Makover et al.[5])

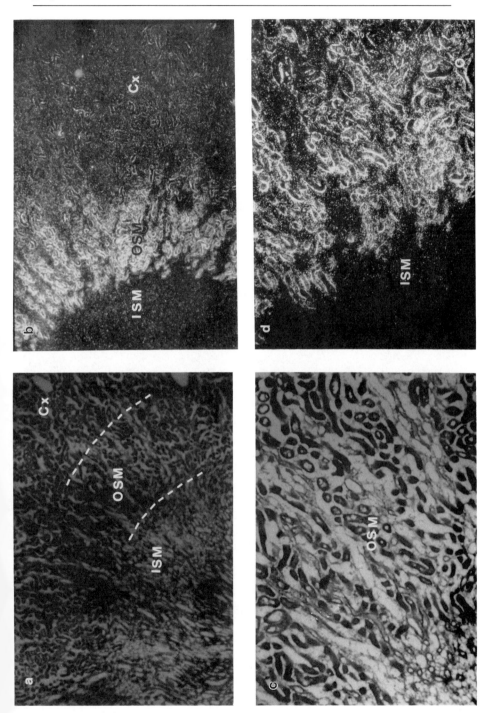

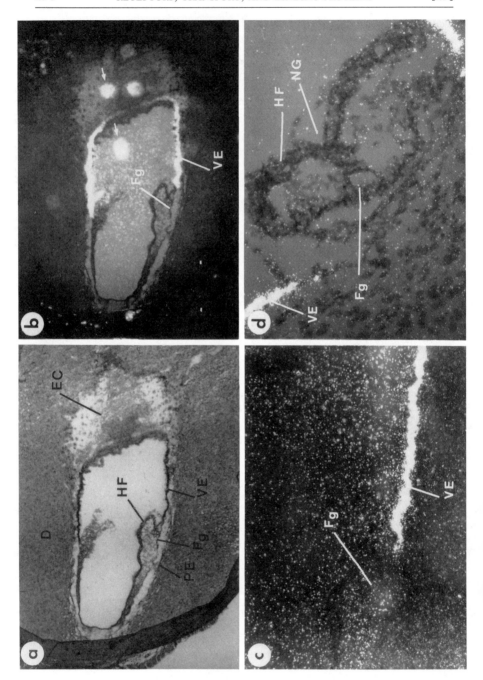

(2) water for 30 sec, (3) rapid fixer for 5 min, (4) running wash of tap water for 5 min, (5) Permawash for 5 min, and (6) tap water, in which the slides are kept until ready for staining. The tissue sections on the slides are stained with hematoxylin/eosin by placing the slides on slide racks and treating the slides in sequence as follows: (1) rinse in deionized water, (2) dip in hematoxylin stain for 4–20 sec, (3) rinse in running tap water for 2 min, (4) dip for 1 min in deionized water, (5) dip 3 times in 0.1 M sodium bicarbonate for 2 to 5 min each, (6) rinse in running tap water for 5 min, (7) dip for 2 min in deionized water, (8) dip in 50% ethanol for 2 min, (9) dip in 70% ethanol (v/v) for 2 min, (10) dip in 90% ethanol (v/v) for 2 min, (11) dip in 100% ethanol (v/v) for 2 min, (12) dip in eosin stain for 5 min, (13) dip 3 times in 100% ethanol (v/v) for 2 min each, (14) dip in 50% xylene/ 50% ethanol (v/v) for 2 min, (15) dip 2 times in xylene for 5 min each, and (16) place a coverslip over the tissue section with permount.

Tissue sections can be examined and photographed (after careful cleaning of the slides to remove emulsion on the back side of the slide) using either dark-field or bright-field illumination. Under bright-field illumination the individual grains indicating the localization of the transcript of interest appear black, whereas under dark-field illumination they appear white. Generally it is much easier to observe grains using dark-field illumination; however, under these conditions the tissue morphology may be difficult to visualize. We have had good success using dark-field illumination with a NTB 10 blue filter. This allows good localization of the grain pattern along with visualization of some histological details.

Illustrative Results

Representative *in situ* hybridization micrographs are shown in Figs. 1 and 2. Figure 1 demonstrates the localization of RBP mRNA in the kidney of adult Sprague-Dawley rats. Sections of kidney were hybridized with antisense strand ^{35}S-labeled RBP cRNA. An intense hybridization signal was found to be specifically localized in one particular anatomical region of the kidney (Fig. 1b,d), the outer stripe of the medulla. The results

---

FIG. 2. Localization of RBP mRNA in the 9-day rat conceptus. (a) Bright-field micrograph demonstrating the anatomy of the conceptus. (b–d) Dark-field micrographs. All sections were hybridized to a RBP ^{35}S-labeled cRNA antisense probe and Giemsa stained. (a–c) Longitudinal sections. (d) Transverse section through the head fold. Magnification: a and b, ×88; c, ×219; d, ×438. The large white dots (arrowheads) in (b) are dark-field staining artifacts. D, Decidua; HF, head fold; Fg, foregut; PE, parietal endoderm; VE, visceral endoderm; EC, ectoplacetal cone; NG, neural groove. (From Makover *et al.*[3])

demonstrate that RBP mRNA is specifically localized in the S_3 segment of the proximal tubules. Only background hybridization was observed in the kidney cortex and the inner stripe of the medulla. In addition, analysis with a sense strand ^{35}S-labeled RBP cRNA probe demonstrated only background hybridization signal (data not shown). This work is described and discussed in detail in Ref. 5.

Immunohistochemical studies have shown that immunoreactive RBP is localized in the proximal convoluted tubular cells of the renal cortex.[15] It is curious that RBP mRNA in the kidney is localized in an anatomical region (the S_3 segment of the proximal tubules) different from that of immunoreactive RBP. In contrast to RBP mRNA, CRBP mRNA has been localized, by *in situ* hybridization analysis, in the proximal convoluted tubular cells of the renal cortex (N. Rajan and D. S. Goodman, unpublished). Both immunoreactive CRBP and RBP are localized in these cells.[14] RBP mRNA has also been demonstrated, by *in situ* hybridization, in rat perinephric fat tissue.[5] We have hypothesized that RBP synthesized in extrahepatic tissue may function in the recycling of retinol back to the liver or to other target tissues.

Figure 2 illustrates the localization of RBP mRNA within developing rat fetuses. Sections of 9-day rat conceptuses were hybridized with ^{35}S-labeled RBP cRNA. Specific hybridization was found to be localized only in the visceral endoderm; only background, nonspecific localization was observed in the remainder of the conceptus and uterus. The specific hybridization signals for RBP mRNA could be seen clearly as black grains in the bright-field micrograph (Fig. 2a) and as white grains in the dark-field micrographs (Fig. 2b–d). Detailed data on the patterns and cellular sites of expression of the genes for RBP and TTR during the early to mid stages of rat embryogenesis are provided in Ref. 3.

Similar studies on the expression and localization of CRBP mRNA and of CRABP mRNA during rat embryogenesis have also been carried out in our laboratories at Columbia University (D. S. Goodman *et al.*, unpublished studies). These studies have demonstrated specific and evolving patterns of localization of both of these transcripts in the developing nervous system. The results suggest that the cellular retinoid-binding proteins and their ligands play important roles in the development of the nervous system.

In summary, our studies utilizing *in situ* hybridization techniques coupled with Northern blot analysis and immunohistochemistry have allowed us to localize specific retinoid-binding protein mRNAs in individual cells within a number of organs and tissues. This information can provide the

[15] M. Kato, K. Kato, and D. S. Goodman, *J. Cell Biol.* **98,** 1696 (1984).

foundation for hypotheses and future research concerning the physiological and biochemical roles of each of the retinoid-binding proteins.

Acknowledgments

The research reported and referred to in this chapter was supported by National Institutes of Health Grants HL21006 and DK05968.

[30] Gel Electrophoresis of Cellular Retinoic Acid-Binding Protein, Cellular Retinol-Binding Protein, and Serum Retinol-Binding Protein

By Georges Siegenthaler

Introduction

Cellular retinoid-binding proteins are involved in the mechanism of action of retinoids.[1,2] However, their exact role is not fully understood. The elucidation of retinoid-binding proteins requires the availability of a rapid and sensitive method endowed with a high power of resolution for the detection, discrimination, and characterization of these proteins.

Several different retinoid-binding proteins can be detected in a target tissue for vitamin A. The cellular retinoic acid- and retinol-binding proteins (CRABP, CRBP), which are specific for retinoic acid and retinol, respectively, are among the best known.[1-3] The transport and storage of retinoids may also involve nonspecific binding proteins such as fatty acid-binding proteins (FABP).[4,5] Unfortunately, the molecular weights of CRABP, CRBP, and FABP are very close to each other, around 15,000, a fact which renders the gel sieving technique ineffective, in spite of the various specificity tests one can perform with different ligands.

[1] F. Chytil and D. E. Ong, *in* "The Retinoids" (M. D. Sporn, A. B. Roberts, and D. S. Goodman, eds.), Vol. 2, p. 89. Academic Press, Orlando, Florida, 1984.
[2] M. Mader, D. E. Ong, D. Summerball, and F. Chytil, *Nature (London)* **335**, 733 (1988).
[3] P. N. McDonald and D. E. Ong, *J. Biol. Chem.* **262**, 10550 (1987).
[4] B. P. Sani, R. D. Allen, C. M. Moorer, and B. W. McGee, *Biochem. Biophys. Res. Commun.* **147**, 25 (1987).
[5] F. Fukai, T. Kase, T. Shidotani, T. Nagai, and T. Katayama, *Biochem. Biophys. Res. Commun.* **147**, 899 (1987).

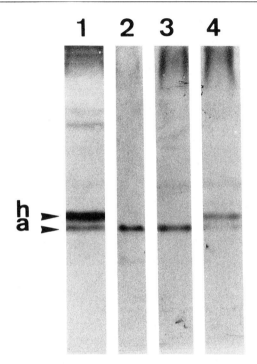

FIG. 1. Electrophoresis–immunoblotting analysis with goat anti-human RBP serum of normal serum, chorion, and epithelial extracts of human oral mucosa. Lane 1, Holo- (h) and apo-RBP (a) from human serum; lanes 2 and 3, apo-RBP of chorion and epithelial extracts; lane 4, reconstitution of holo-RBP from apo-RBP of epithelial extracts with 10 μM of retinol. (From Ref. 8 with permission of the publisher.)

Serum retinol-binding protein (RBP), which is supposed to supply retinol from the blood vessels to target cells, might also be an important contaminant in tissue protein extracts[6-8] (Fig. 1) or in the cytosol of cultured hepatic cells. It has been reported that these cells are able to synthesize RBP in addition to CRBP and CRABP.[9] The study of these binding proteins necessitates specific techniques other than gel sieving, charcoal-dextran separation, or sucrose density gradient centrifugation. In the serum, holo-RBP (the retinol–RBP complex) is found to represent 98% of the total RBP (holo- and free or apo-RBP).[10] Holo-RBP is entirely bound

[6] G. Siegenthaler and J.-H. Saurat, J. Invest. Dermatol. 88, 403 (1987).
[7] G. Siegenthaler and J.-H. Saurat, Eur. J. Biochem. 166, 209 (1987).
[8] G. Siegenthaler, J. Samson, J.-P. Bernard, G. Fiore-Donno, and J.-H. Saurat, J. Oral Pathol. 17, 106 (1987).
[9] J. L. Dixon and D. S. Goodman, J. Cell. Physiol. 130, 14 (1987).
[10] G. Siegenthaler and J.-H. Saurat, Biochem. Biophys. Res. Commun. 143, 418 (1987).

to transthyretin (TTR), formerly named prealbumin, forming a complex with a molecular weight of 76,000. The tissue level of TTR is lower than the serum level and may not be sufficient to bind RBP completely, so that some RBP could remain free. Moreover, it has been shown that RBP binds retinoic acid (RA) as well as retinol *in vitro*.[7–10] The resulting RA–RBP complex does not appear to possess strong binding properties toward TTR because of important structural alterations of the protein core. For these reasons, RA–RBP will remain free in protein extracts. It should be mentioned that the RA–RBP complex migrates faster on polyacrylamide gel electrophoresis (PAGE) than holo-RBP and apo-RBP and just before the CRABP peak.[10]

The mode of action of RBP involves plasma membrane receptors for RBP, CRBP, and esterifying and hydrolyzing enzymes for retinol and its esters.[11] Studies of such a model also necessitate specific radiobinding techniques which allow one to measure the levels of holo-RBP and CRBP, whereas an estimation of apo-RBP is possible with PAGE immunoblotting techniques. Sedimentation analysis on sucrose density gradients[12] and gel sieving chromatography[13,14] connected to or independent of high-performance liquid chromatography (HPLC) methods[15,16] are currently used for the study of retinoid-binding proteins. However, these methods are extremely time-consuming, and only a limited number of samples can be analyzed simultaneously. In order to diminish the background radioactivity, one has to treat the samples with dextran-coated charcoal after incubation with [³H]retinoids, to remove excess free labeled ligand. If the samples are analyzed by HPLC, previous filtration of the samples through 0.45-μm filters is indispensable, to avoid blocking of the column by unsedimented charcoal particules. Moreover, during analysis with the above technique, the sample is diluted about one-tenth, a fact which decreases considerably the resolution of the various radioactive peaks.

In this chapter, we describe procedures for the detection, characterization, and measurement by polyacrylamide gel electrophoresis of CRABP, CRBP, RBP and other retinoid-binding proteins. The results of

[11] S. Ottonello, S. Petrucc, and G. Maraini, *J. Biol. Chem.* **262**, 3975 (1987).

[12] D. E. Ong and F. Chytil, *Proc. Natl. Acad. Sci. U.S.A.* **73**, 3976 (1976).

[13] G. Siegenthaler, J.-H. Saurat, R. Hotz, M. Camenzind, and Y. Mérot, *J. Invest. Dermatol.* **86**, 42 (1986).

[14] G. Siegenthaler, J.-H. Saurat, and M. Ponec, *Exp. Cell. Res.* **178**, 114 (1988).

[15] H. E. Shubeita, M. D. Patel, and A. M. McCormick, *Arch. Biochem. Biophys.* **247**, 280 (1986).

[16] E. A. Allegretto, M. A. Kelly, C. A. Donaldson, N. Levine, J. W. Pike, and M. R. Haussler, *Biochem. Biophys. Res. Commun.* **116**, 75 (1983).

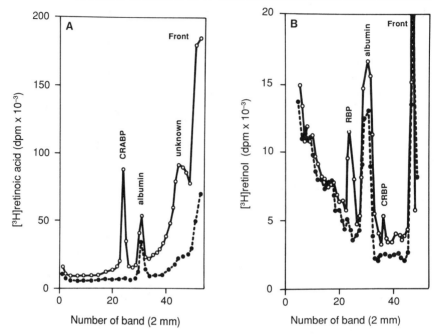

FIG. 2. Electrophoresis analysis of labeled retinoid-binding proteins from human skin. (A) Epidermis extract incubated with [³H]retinoic acid in the absence (○) or presence (●) of excess retinoic acid. (B) Dermis extract incubated with [³H]retinol in the absence (○) or presence (●) of retinol. (From Ref. 7 with permission of the publisher.)

typical experiments for tissue protein extract or cytosol are presented in Figs. 2 and 3.

Separation of Retinoid-Binding Proteins by Electrophoresis

Principle. The protein extracts are incubated with the various labeled retinoids before being subjected to vertical slab PAGE under nondenaturing conditions. The proteins are then separated according to their different electrophoretic mobilities in the gel, as a consequence of their net charge and molecular weight. After electrophoresis, the gel is divided into lanes and cut into 2-mm bands which are counted for the determination of the radioactivity profile. The technique of PAGE allows the separation of CRABP, CRBP, RBP and other retinoid-binding proteins in one run. Moreover, 10 different samples can be analyzed in the same gel.

During gel electrophoresis, there is practically no dissociation of the ligands from the binding proteins, whereas the complexes of retinoid-binding proteins with other proteins dissociate readily. For example, the

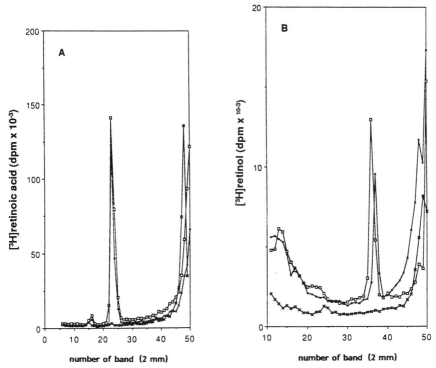

FIG. 3. Electrophoresis analysis of CRABP and CRBP from cultured differentiated human keratinocytes. (A) Cytosol incubated with [³H]retinoic acid in the absence (□) or presence of excess retinoic acid (X) or retinol (●). (B) Cytosol incubated with [³H]retinol in the absence (□) or presence of excess retinol (X) or retinoic acid (■). (From Ref. 14 with permission of the publisher.)

holo-RBP–TTR complex is totally dissociated into holo-RBP and TTR. This phenomenon might explain the higher level of CRABP found with the PAGE technique by comparison with gel sieving, which reveals the presence of high molecular weight retinoid acid-binding proteins besides CRABP.[8,14] It is not necessary to remove excess labeled retinoid, which either migrates with the front of the gel in a sharp band (retinoic acid) or does not migrate at all (retinol). Suspensions of membrane extracts can also be analyzed directly on PAGE without further separation.

Protein Transfer, Immunoblotting, and Autoradiography

Proteins separated by PAGE are electrophoretically transferred onto a nitrocellulose sheet. Binding proteins can be identified by specific antibodies. The transfer and the binding of proteins to nitrocellulose do not

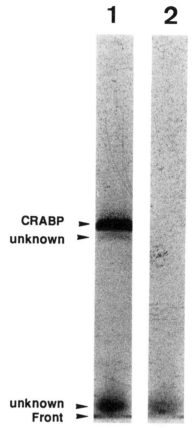

FIG. 4. Direct autoradiography of a nitrocellulose sheet. A human epidermal extract incubated with [³H]retinoic acid was separated on a polyacrylamide gel and transferred to a nitrocellulose sheet. The nitrocellulose sheet was subjected to autoradiography for 1 week with Hyperfilm. In lane 1, the CRABP band and unknown retinoic acid-binding proteins can be detected. Lane 2 shows the disappearance of the radioactive bands in the presence of an excess of unlabeled retinoic acid.

constitute denaturing conditions for the retinoid-binding proteins, as indicated by the fluorescence of the holo-RBP band when the nitrocellulose sheet is exposed to ultraviolet light (360 nm). This is also confirmed by the profiles obtained when protein extracts incubated with labeled retinoids are separated on PAGE and transferred onto nitrocellulose sheets, before being subjected to autoradiography with X-ray film (see Fig. 4). Low amounts of retinoid-binding protein can be detected, depending on the duration of exposure time. This procedure increases the resolution, as the gel does not have to be cut into bands. The background is lower, free retinoic acid being removed upon transfer.

Tissue and Serum Preparation

Skin is one of the most difficult tissues with respect to extraction of proteins. The technique described below has been developed for skin and may also be used for other tissues. Approximately 20 mg of lyophilized human skin is homogenized in an ice bath, in 800 μl of ice-cold extraction buffer [50 mM Tris-HCl, 25 mM NaCl, 2.5 mM EDTA, 1 mM dithiothreitol (DTT), pH 7.5], with three 30-sec strokes of a Polytron PT7 tissue homogenizer (Kinematica, Luzern, Switzerland), at full speed. Supernatants are obtained by centrifugation at 100,000 g at 4° for 60 min, distributed in 100-μl aliquots, and frozen at $-20°$ or used immediately. Fresh or frozen serum with or without delipidation (to remove retinol from RBP) is used as the RBP standard in the PAGE–immunoblotting technique.

Incubation with Retinoids

All manipulations are done under dim or yellow light. The supernatants containing 200–300 μg protein in 100 μl are incubated with 600 nM (saturating conditions for skin extract) of all-*trans*-[11,12-³H]retinoic acid (50 Ci/mmol, Du Pont, Paris, France) or all-*trans*-[11,12-³H]retinol (50 Ci/mmol, Du Pont). For this purpose, alcoholic solutions of retinoids containing 50 μg/ml of butylated hydroxytoluene as antioxidant are first deposited in glass microtubes, and the solvent is evaporated with a stream of N_2. The supernatants are then added, mixed with the pipette, and incubated at 4° for 16 hr. For binding studies with ligands of higher affinity, an incubation period of 1 hr at 0° is sufficient. For binding specificity or competition studies, a 200-fold excess of unlabeled ligand is added, in the same manner as for radioactive compounds. At the end of the incubation period an aliquot of the supernatant (90 μl) is subjected to PAGE.

Polyacrylamide Gel Electrophoresis

Stock Solutions

Acrylamide, 45%: dissolve 56.25 g of acrylamide and 1.5 g N,N'-methylenebisacrylamide in 125 ml distilled water and filter the solution

Tris-HCl buffers: 1 M, pH 6.8, and 1 M, pH 8.8

Concentrated (10×) reservoir buffer: dissolve 30.28 g Tris (0.25 M) and 146.38 g glycine in a total volume of 1000 ml of distilled water (final pH 8.8); dilute one-tenth before use

Slab Gel Preparation. Quantities are indicated for an LKB-Pharmacia (Uppsala, Sweden) vertical electrophoresis system with two gels (14 cm × 16 cm × 1.5 mm). Each gel contains 10 wells. For the resolving gel, mix

9.6 ml of Tris-HCl buffer, pH 8.8, 14.9 ml water, 5.2 ml 45% acrylamide, and 12 μl N,N,N',N'-tetramethylethylenediamine (TEMED). Degas and add 300 μl of a freshly prepared solution of sodium persulfate (30 mg/ml). For the 7.5% stacking gel, mix 1.25 ml of Tris buffer, pH 6.8, 7.35 ml water, 0.64 ml 45% acrylamide, and 5 μl TEMED. Degas and add 200 μl of sodium persulfate solution.

Sample Application. Add 20 μl of a solution containing 0.05% bromphenol blue and 25% glycerol to each sample, in order to facilitate its application into the well.

Determination of Radioactivity. After electrophoresis, the gel is divided into lanes and cut into 2-mm bands with a homemade cutter system made of razor blades separated with 2-mm spacers. The bands are collected and treated overnight with 400 μl Protosol (Du Pont) in 5-ml scintillation vials before 4 ml of Pico-fluor (Packard) is added. The radioactivity is then determined in a β-counter with counting efficiency of 47%. Typical migration profiles are represented in Figs. 2 and 3.

Protein Transfer

Proteins separated by PAGE are electrophoretically transferred to nitrocellulose filters at 4° in a 0.25 mM sodium phosphate solution at pH 6.5 for 2.5 hr under the electrical conditions recommended by the manufacturers.

Autoradiography of Nitrocellulose Sheets

Once the transfer is finished the nitrocellulose sheet is completely dried with a blower-type dryer. For optimal sensitivity and resolution, the nitrocellulose sheet is placed in close contact with a Hyperfilm (Amersham) in an X-ray cassette. It is important that the side of the film bearing the emulsion is in contact with the nitrocellulose sheet surface, where the proteins have been transferred (see results in Fig. 4). Under these conditions an exposure of 1 week is sufficient for a radioactive band of approximately 5000 dpm.

Immunoblotting of Retinoid Binding Protein

After the transfer, the nitrocellulose sheet is treated in a blocking solution for 2 hr [5 mM sodium phosphate, 0.13 mM NaCl, pH 7.2 (PBS), containing 3% defatted powdered milk], prior to a 2-hr incubation at room temperature in PBS containing 0.5% bovine serum albumin, 0.2% Tween 20, and an appropriate amount of goat antiserum directed against human RBP. The filter is then washed 3 times with the incubation solution and further incubated for 1 hr at room temperature with horseradish peroxidase-labeled rabbit anti-goat IgG Fab fragment (Cappel, Cochranville, PA). The filter is then washed 3 times in Tris buffer (100 mM Tris-HCl, pH 7.4) to remove excess second antibody. The bound enzyme is visualized by incubating the sheet in the above Tris buffer containing 0.5 mg/ml diaminobenzidine and 0.03% H_2O_2. After the bands appear, the reaction is stopped, and the background coloration is partially removed by replacing the Tris buffer containing the substrates with a solution of 3% acetic acid in water (see Fig. 1).

Conclusion

The electrophoresis technique enables one to study in detail the extra- and intracellular retinoid-binding proteins in human normal skin and in skin diseases affecting keratinization (for review, see Ref. 17). Moreover, this method was used with success to show that terminal differentiation in cultured human keratinocytes was associated with increased levels of CRABP.[14]

Acknowledgments

This work was supported in part by the Swiss National Science Foundation, Grant 3,874,088. I thank Dr. A. Capponi for rewriting the manuscript. I gratefully acknowledge Raymonde Hotz and Evelyne Leemans for excellent technical assistance and Sylvie Deschamps for typing the manuscript.

[17] G. Siegenthaler and J.-H. Saurat, *Pharmacol. Ther.* **40**, 45 (1989).

[31] Cellular Retinoic Acid-Binding Protein from Neonatal Rat Skin: Purification and Analysis

By CHRISTOPHER P. F. REDFERN and ANN K. DALY

Introduction

Cellular retinoic acid-binding protein (CRABP) is a member of a family of low molecular weight, hydrophobic ligand-binding proteins which includes cellular retinol-binding protein, myelin protein P2, and fatty acid-binding proteins.[1] Skin is an important target organ for retinoic acid, but the function of CRABP and the role of retinoic acid in skin development are unknown. A knowledge of the properties of CRABP and its affinity for various retinoids is important for elucidating the biological role of CRABP and for designing new synthetic retinoids. CRABP is abundant in neonatal rat skin, and we have developed a method for the purification of CRABP from this tissue.[2]

Preparation of Homogenates from Skin

Because of its mechanical and barrier properties, skin is not an easy tissue to homogenize. The dermis is rich in cross-linked collagen fibers, and the differentiated cells of the epidermis are hard to break open owing to the high degree of keratinization and the formation of cornified envelopes. Tissue homogenization methods used for tissues such as liver are therefore not suitable. For our studies on CRABP from rat skin[2] we have used tissue from neonatal rats 1–3 days old: at this stage hair has not grown and the skin is easier to homogenize than skin from adult rats. For a typical purification, we use the trunk skin from 400 neonatal rats. The tissue (~300 g wet weight), stored initially at $-70°$, is cooled in liquid nitrogen and ground to a fine powder using a pulverizing mill. The powder is then suspended in 2 volumes of 10 mM Tris-HCl, pH 7.4, 1 mM EDTA, 0.1 mM phenylmethylsulfonyl fluoride (PMSF), and homogenized using a Polytron or a blender.

CRABP may be purified from dermis and epidermis separately. A variety of mechanical, chemical, and enzymatic methods for separating dermis and epidermis have been described.[3] For our studies, we separate the dermis and epidermis of rat skin by incubating the skin, dermis side down, for 30 min at 37° in phosphate-buffered saline (PBS, Dulbecco's formula, Flow Laboratories, Rickmansworth, UK) containing 0.25% (w/v) trypsin. The skin is then transferred to 10% (v/v) fetal calf serum in either tissue culture medium or PBS and the epidermis peeled off with forceps. The separated tissues should be washed in PBS before homogenization, and as a precaution we include aprotinin (40 kIU/ml) and soybean trypsin inhibitor (10 μg/ml) in the homogenization buffer.

[1] J. Sundelin, S. R. Das, U. Eriksson, L. Rask, and P. A. Peterson, J. Biol. Chem. 260, 6494 (1985).
[2] A. K. Daly and C. P. F. Redfern, Biochim. Biophys. Acta 965, 118 (1988).
[3] C. J. Skerrow and D. Skerrow, in "Methods in Skin Research" (D. Skerrow and C. J. Skerrow, eds.), p. 609. Wiley, Chichester, 1985.

Assay for Cellular Retinoic Acid-Binding Protein during Purification

The elution of CRABP from chromatography columns during purification can be conveniently monitored by adding all-*trans*-[11,12-^{3}H]retinoic acid to the tissue extract. In addition, we also assay for CRABP activity as retinoic acid will dissociate from CRABP during purification (dissociation rate constant −0.0035 min^{-1} at 0°) and it is useful to compare the chromatographic properties of CRABP holoprotein and apoprotein.

The assay for CRABP activity involves the incubation of tritium-labeled all-*trans*-retinoic acid with aliquots of each column fraction followed by the separation and determination of free and protein-bound radioactivity. Duplicate 100-µl samples of column fractions are incubated with 20 pmol all-*trans*-[^{3}H]retinoic acid for 1–2 hr at 4° in the presence or absence of a 200-fold excess of unlabeled retinoic acid. To separate free and bound retinoic acid we routinely use one of two methods, depending on the stage of purification. In addition to the two methods given below, we have also used sucrose density gradient centrifugation[2] to separate free and protein-bound retinoic acid. However, this method is inconvenient because of the lengthy centrifugation runs required and limited sample capacity.

During the initial stages of CRABP purification, the total protein concentration of column fractions is high, and in the CRABP assay unbound [^{3}H]retinoic acid can be removed effectively by absorption to dextran-coated charcoal (Method 1). The protein concentration of fractions eluted during the final chromatographic steps is low (<100 µg/ml), and to assay these samples for CRABP we increase the sample protein concentration by adding hemoglobin (final concentration 1 mg/ml) to minimize losses of CRABP during the assay and use gel filtration (Method 2) to separate free and protein-bound [^{3}H]retinoic acid. Retinoic acid may dissociate rapidly from CRABP (estimated[2] dissociation rate constants are −0.149 min^{-1} at 23° and −0.0035 min^{-1} at 0°), and it is important that the separation of free and bound retinoic acid should be carried out as quickly as possible at 0–4°.

Method 1: Separation of Free and Bound [^{3}H]Retinoic Acid Using Dextran-Coated Charcoal. After incubating aliquots of column fractions with [^{3}H]retinoic acid, an equal volume of a suspension of 1–5% (w/v) activated charcoal in 0.5% (w/v) dextran T-500, 0.1% (w/v) gelatin, 10% (w/v) glycerol, 0.02% (w/v) sodium azide, 1.5 mM EDTA, and 10 mM Tris-HCl, pH 7.4, is added and the tubes incubated for 5 min at 4° with occasional mixing. The dextran-coated charcoal is then removed by centrifugation for 3 min at 10,000 g and 4°, and a 100-µl aliquot taken for scintillation counting.

Method 2: Separation of Free and Bound [³H]Retinoic Acid by Gel Filtration. In this method, free and bound [³H]retinoic acid are separated using 5-cm columns of BioGel P-10, bed volume approximately 1 ml, equilibrated with 0.1 M NaCl, 1 mM EDTA, 10 mM Tris-HCl, pH 7.4. CRABP is eluted in the void volume, and fractions (6 drops or ~100 μl) are collected for scintillation counting.

Purification of Cellular Retinoic Acid-Binding Protein

CRABP is separated from other skin proteins in four chromatographic steps on the basis of charge (two steps), molecular size, and hydrophobicity. All stages in the purification are performed at 4°, and the initial steps, involving acid precipitation, ion-exchange chromatography, and gel-filtration chromatography are based on the scheme devised by Ong and Chytil[4] for CRABP from rat testis.

Step 1. Acid Precipitation and Ion-Exchange Chromatography

The crude tissue homogenate is centrifuged at 50,000 g for 1 hr to remove debris, and the supernatant is titrated to pH 5.25 with glacial acetic acid. The precipitate is removed by centrifugation for 10 min at 50,000 g. all-*Trans*-[³H]Retinoic acid (100 pmol) can be added to the supernatant at this stage, or to CRABP eluted from the subsequent gel-filtration column, to monitor the elution of CRABP holoprotein. The supernatant is then dialyzed overnight against 5 liters of 10 mM Tris–acetate, pH 8.3, 0.1 mM PMSF, and, after centrifugation to remove protein precipitating under the low-salt conditions, passed through a DEAE-Sephacel (Pharmacia-LKB, Uppsala, Sweden) column (bed volume 100 ml), equilibrated with 10 mM Tris–acetate, pH 8.3. The column is washed with 10 mM Tris–acetate, pH 8.3, until the eluate has an OD_{280} less than 0.1, and CRABP is eluted with a 500- or 1000-ml linear gradient of 0 to 0.2 M NaCl in 10 mM Tris–acetate, pH 8.3, at a flow rate of 100 ml/hr. Using a 1000-ml salt gradient, CRABP is eluted at volumes between 450 and 650 ml (Fig. 1). Fractions containing CRABP are pooled and concentrated to 10 ml by ultrafiltration (Amicon ultrafiltration cell with XM5 membrane).

Step 2. Gel Filtration

CRABP eluted from the DEAE-Sephacel column is then gel filtered through Sephadex G-75 (80-cm column, bed volume 250 ml), equilibrated with 0.2 M NaCl, 5 mM Tris-HCl, pH 7.4, 0.02% (w/v) sodium azide.

[4] D. E. Ong and F. Chytil, *J. Biol. Chem.* **253**, 4551 (1978).

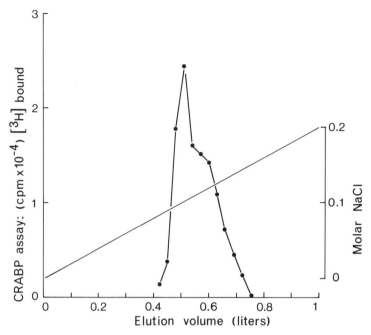

FIG. 1. Elution of CRABP from DEAE-Sephacel using a 1-liter linear gradient of 0 to 0.2 M NaCl (sloping line, right ordinate). Duplicate 100-μl aliquots from every third 10-ml fraction were assayed for CRABP activity, using dextran-coated charcoal to separate free and protein-bound retinoic acid. CRABP activity (left ordinate) is expressed as counts per minute (cpm) specifically bound per 100-μl sample.

CRABP is eluted at an apparent molecular weight, relative to RNase, chymotrypsinogen, and bovine serum albumin standards, of 10,500–13,700. Peak fractions containing CRABP are pooled for separation by chromatofocusing.

Step 3. Chromatofocusing

During the chromatofocusing and subsequent hydrophobic interaction chromatography steps, it is important both to monitor the elution of [³H]retinoic acid-labeled CRABP and to assay column fractions for CRABP activity because apo- and holo-CRABP are eluted from the columns at slightly different positions (Figs. 2 and 3). To monitor the elution of CRABP holoprotein, all-*trans*-[³H]retinoic acid (100 pmol) can be added, if necessary, to the CRABP fractions pooled from the gel-filtration step.

For chromatofocusing, CRABP is dialyzed overnight against 2 liters of

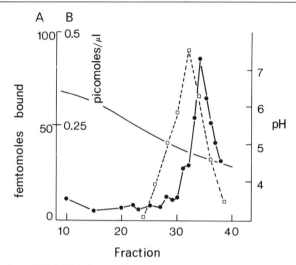

FIG. 2. Elution of CRABP holoprotein and apoprotein from a PBE 94 chromatofocusing column. Elution profiles of holoprotein (●) and apoprotein (□) in relation to the pH of the eluent (right ordinate) were determined by measuring the amount of tritium label (left ordinate, A, expressed as femtomoles of retinoic acid bound per 100-μl sample) and by assaying the retinoic acid-binding activity (left ordinate, B, expressed as picomoles retinoic acid/μl/ fraction) in each fraction using the BioGel P-10 column assay. [Reprinted, with permission, from A. K. Daly and C. P. F. Redfern, "Purification and properties of cellular retinoic acid binding protein from neonatal rat skin," *Biochim. Biophys. Acta* **965**, 118 (1988).]

25 mM histidine, pH 6.25, and loaded onto a 20- to 30-cm PBE 94 (Pharmacia-LKB) chromatofocusing column (bed volume 15–25 ml), equilibrated in 25 mM histidine, pH 6.25. CRABP is then eluted with Polybuffer 74 (Pharmacia-LKB), pH 4, diluted 1 : 8 with distilled water, at a flow rate of 20 ml/hr. Under these conditions, holo- and apo-CRABP are eluted at a buffer pH within the range 4.5–5.1; apo-CRABP is eluted slightly ahead of holo-CRABP (Fig. 2).

Step 4. Hydrophobic Interaction Chromatography

Peak fractions containing apo- and holo-CRABP eluted from the chromatofocusing step are pooled and solid ammonium sulfate added to 30% saturation. Insoluble material is removed by centrifugation at 14,000 *g* for 20 min, and the supernatant is passed through a Phenyl-Sepharose CL-4B (Pharmacia-LKB) column (bed volume 4 ml) equilibrated with 30% saturated ammonium sulfate, 10 mM Tris-HCl, pH 7.4, 1 mM EDTA. CRABP is eluted with a 50- or 60-ml linear gradient of ammonium sulfate from 30 to 0% saturation at a flow rate of 10 ml/hr. CRABP holoprotein is eluted

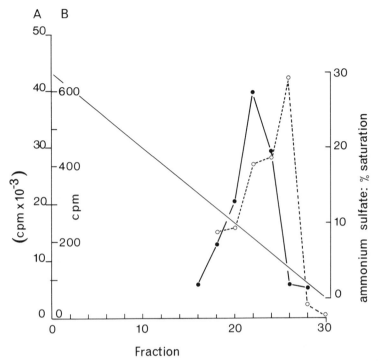

FIG. 3. Hydrophobic interaction chromatography (Phenyl-Sepharose) of CRABP holo-protein and apoprotein. CRABP was eluted with a linear gradient of ammonium sulfate from 0 to 30% of saturation (right ordinate). CRABP apoprotein was assayed using 100-μl aliquots of each 2-ml fraction and expressed as counts per minute (cpm) specifically bound [^{3}H]reti-noic acid per sample (○, left ordinate, A). The elution of CRABP holoprotein was monitored by the scintillation counting of 100-μl aliquots of each 2-ml fraction (●, left ordinate, B, expressed as cpm).

slightly ahead of the apoprotein (Fig. 3) so it is important to monitor the elution of both forms of CRABP. Fractions containing CRABP, which is eluted toward the end of the ammonium sulfate gradient at about 2.5% of saturation (Fig. 3), are concentrated and desalted by ultrafiltration. For concentrating small volumes we use Centricon-10 microconcentrator units supplied by Amicon (Stonehouse, Gloucestershire, UK).

General Comments

CRABP can be stored at $-20°$ in 50% glycerol and is stable for at least 18 months. The purity of the product should be assessed by sodium dode-cyl sulfate–polyacrylamide gel electrophoresis; in our hands, CRABP

TABLE I
PURIFICATION OF CRABP FROM NEONATAL RAT SKIN[a,b]

Purification step	Extract volume (ml)	Protein (mg)	CRABP specific activity (pmol/mg protein)	Recovery (%)	Purification (-fold)
First supernatant	250	820	12.7	—	—
Second supernatant	250	690	13.8	91.7	1.09
DEAE-Sephacel	8.5	88	80.3	67.9	6.34
Sephadex G-75	45	19.6	244.7	46.2	19.32
Chromatofocusing	35	2.9	1517	42.3	119.8
Phenyl-Sepharose	25	0.24	14,720	33.8	1162.1

[a] Reprinted, with permission, from A. K. Daly and C. P. F. Redfern, "Purification and properties of cellular retinoic acid binding protein from neonatal rat skin," *Biochim. Biophys. Acta* **965,** 118 (1988).

[b] CRABP levels at different stages in purification were assayed using either sucrose density gradient centrifugation or BioGel P-10 columns (chromatofocusing and hydrophobic interaction chromatography steps) to separate free and protein-bound retinoic acid.

migrates with an apparent molecular weight of 14,800.[2] Purification steps are given in Table I.

Using this purification scheme, we have obtained a yield of 250 μg CRABP from approximately 300 g (wet weight) of rat skin tissue. With respect to molecular weight, charge, and hydrophobicity, CRABP from rat testis and 13-day-old embryonic chick skin has similar properties to CRABP from neonatal rat skin. However, using this purification scheme we have not obtained the same degree of purity of rat testis CRABP as has been obtained with neonatal rat skin CRABP. For tissues other than skin additional purification steps may therefore be necessary. The chromatography results suggest that chromatofocusing followed by hydrophobic interaction chromatography may be used for the preparative separation of CRABP apoprotein and holoprotein.

[32] Tissue Distribution of Cellular Retinol-Binding Protein
and Cellular Retinoic Acid-Binding Protein:
Use of Monospecific Antibodies for Immunohistochemistry
and cRNA for *in Situ* Localization of mRNA

By Christer Busch, Puspha Sakena, Keiko Funa,
Hans Nordlinder, and Ulf Eriksson

Introduction

Knowledge concerning the tissue and cellular expression of cellular retinol-binding protein (CRBP) and cellular retinoic acid-binding protein (CRABP) may be important in order to understand the physiology of vitamin A. Information on the localization of both proteins is of particular interest to identify target cells, transport routes, and storage systems for the vitamin. By comparing the cellular expression of CRBP and CRABP in the context of the recently identified group of nuclear receptors for retinoic acid,[1-4] knowledge of the detailed function of the intracellular proteins may also be generated.

The most widely used technique to monitor expression of CRBP and CRABP in tissues and cells has been based on the binding of their respective radiolabeled ligands in tissue or cell extracts.[5] Following separation of free versus bound ligand, the amounts of the proteins can be estimated. The validity of the results is largely dependent on the presence of endogenous retinoids and whether other proteins with overlapping ligand-binding specificies and biochemical characteristics similar to those of CRBP and CRABP are present in the extracts. The ability of several fatty acid-binding proteins, which display some features similar to the two retinoid-binding proteins, to bind vitamin A compounds may thus largely influence the measurements. The ligand-binding techniques require tissue homogenates and, therefore, do not yield information regarding the cellular localization of the two retinoid-binding proteins unless highly purified cell populations are analyzed. The use of immunohistochemical techniques for the detection of CRBP and CRABP in tissue sections, however, has been only partly successful, and this may relate to two basic problems. First,

[1] D. Benbrook, E. Lernhardt, and M. Pfahl, *Nature (London)* **333,** 669 (1988).
[2] N. J. Brand, M. Petkovich, A. Krust, P. Chambon, H. de The, A. Marchio, P. Tiollais, and A. Dejean, *Nature (London)* **332,** 850 (1988).
[3] V. Giguere, E. S. Ong, P. Segui, and R. M. Evans, *Nature (London)* **330,** 624 (1988).
[4] M. Petkovich, N. J. Brand, A. Krust, and P. Chambon, *Nature (London)* **330,** 444 (1987).
[5] M. M. Bashor, D. O. Toft, and F. Chytil, *Proc. Natl. Acad. Sci. U.S.A.* **70,** 3483 (1973).

because of their highly evolutionarily conserved nature, CRBP and CRABP are poor immunogens, and it is difficult to raise specific antibodies. Second, both retinoid-binding proteins are structurally very similar to a number of abundant fatty acid-binding proteins, making it difficult to ascertain that polyclonal antibodies recognize exclusively the appropriate antigen. The drastic differences in expression and tissue localization of both proteins that have been reported in the literature may reflect these facts.[6-10] Both problems can be overcome, and in this chapter we discuss the generation, characterization, and use of monospecific antipeptide antibodies to CRBP and CRABP as well as the cellular localization of the corresponding mRNAs using in situ hybridization with ^{35}S-labeled cRNA probes.

Immunohistochemical Techniques

Generation and Characterization of Monospecific Antipeptide Antibodies to CRBP and CRABP

The poor immunogenicity of intact CRBP and CRABP and the potential problems with polyclonal antibodies that may cross-react with a number of closely related and highly abundant fatty acid-binding proteins warranted the production of antibodies with a given specificity. Given the fact that the primary structures for most of the identified members belonging to this group of proteins have been determined, we decided to raise antibodies to synthetic peptides corresponding to discrete and unique peptide segments derived from the two retinoid-binding proteins. Apart from providing an unlimited supply of well-defined antigens, the use of peptides as immunogens offers several advantages. One is that antibodies can be produced to highly evolutionarily conserved proteins that are nonimmunogenic or poorly immunogenic when using the intact protein as the immunogen.[11]

The selection of peptide segments was mainly based on two criteria. First, the amino acid sequences should differ as much as possible from the

[6] D. E. Ong, J. A. Crow, and F. Chytil, *J. Biol. Chem.* **257**, 13385 (1983).
[7] M. Kato, W. S. Blaner, J. R. Mertz, K. Das, K. Kato, and D. S. Goodman, *J. Biol. Chem.* **260**, 4832 (1985).
[8] W. S. Blaner, K. Das, J. R. Mertz, S. R. Das, and D. S. Goodman, *J. Nutr.* **27**, 1084 (1986).
[9] U. Eriksson, K. Das, C. Busch, H. Nordlinder, L. Rask, J. Sundelin, J. Sallstrom, and P. A. Peterson, *J. Biol. Chem.* **259**, 13464 (1984).
[10] U. Eriksson, E. Hansson, H. Nordlinder, C. Busch, J. Sundelin, and P. A. Peterson, *J. Cell. Physiol.* **133**, 482 (1987).
[11] H. E. Schmitz, H. Atassi, and M. Z. Atassi, *Immunol. Commun.* **12**, 161 (1983).

TABLE I
AMINO ACID SEQUENCES OF THE CRABP AND CRBP PEPTIDES
USED FOR IMMUNIZATION AND OF CORRESPONDING PEPTIDE SEGMENTS
OF RELATED PROTEINS[a]

Protein	Peptide 1	Peptide 2
CRABP	GEGFEEETV - -DGRKC	CTQTLLEGDGPKT
CRBP	GKEFEEDLTGIDDRKC	CVQK - - - GEKEGR
CRBP II	GVEFDEHTKGIDGRNV	CVQK - - - GEKENR
P2	GQEFEETTA - -DNRKT	QVQK - - - WNGNET
aP2	GVEFDEITA - -DDRKV	QVQK - - - WDGKST
Heart FABP	GVEDDEVTA - -DDRKV	HVQK - - - WDGQET
Intestinal FABP	GVDFAYSLA - -DGTEL	GKFK - RVDNGKEL
Liver FABP	GEECELETM - -TGEKV	TTFK - - - - - - -GI
MGDI	GVEFDETTA - -DDRKV	QVQK - - - WNGQET

[a] Gaps in the amino acid sequences have been introduced in order to optimize the alignments.

corresponding regions in the related proteins. Second, the peptide segments should be hydrophilic and, consequently, likely to be exposed on the surface of the proteins. The latter question was addressed by analyzing the amino acid sequences using a computer program to locate possible antigenic sites.[12] Several regions in both proteins fulfill both criteria, and we selected residues 68–81 and 95–107 of bovine CRABP[13] and the corresponding regions in rat CRBP.[14] The four peptides selected as the immunogens are listed in Table I. For comparison, Table I also lists the corresponding peptide segments of CRBP II,[15] of the five fatty acid-binding proteins (FABP),[16–20] and of the closely related mammary gland-derived growth inhibitor (MGDI).[21] During the initial part of this work the se-

[12] H. P. Wood and K. R. Hopps, *Proc. Natl. Acad. Sci. U.S.A.* **78,** 3824 (1981).
[13] J. Sundelin, S. Das, U. Eriksson, L. Rask, and P. A. Peterson, *J. Biol. Chem.* **260,** 6494 (1985).
[14] J. Sundelin, H. Anundi, L. Tragardh, U. Eriksson, P. Lind, H. Ronne, P. A. Peterson, and L. Rask, *J. Biol. Chem.* **260,** 6488 (1985).
[15] E. Li, L. A. Demmer, D. A. Sweefser, D. E. Ong, and J. I. Gordon, *Proc. Natl. Acad. Sci. U.S.A.* **83,** 5779 (1986).
[16] K. Kitamura, M. Suzuki, A. Suzuki, and K. Uyemura, *FEBS Lett.* **115,** 27030 (1980).
[17] D. A. Bernlohr, C. W. Anjus, M. D. Lane, M. A. Bolanowski, and T. J. Kelly, Jr., *Proc. Natl. Acad. Sci. U.S.A.* **81,** 5468 (1984).
[18] J. C. Sacchetti, B. Said, J. Schultz, and J. I. Gordon, *J. Biol. Chem.* **261,** 8218 (1986).
[19] D. H. Alpers, A. W. Strauss, R. K. Ockner, N. M. Bass, and J. I. Gordon, *Proc. Natl. Acad. Sci. U.S.A.* **81,** 313 (1984).
[20] K. Takahashi, S. Odani, and R. Ono, *FEBS Lett.* **115,** 27 (1982).
[21] F.-D. Bohmer, R. Kraft, A. Otto, C. Wernstedt, U. Hellman, A. Kurtz, T. Muller, K. Rohde, G. Etzold, W. Lehmann, P. Langen, C.-H. Heldin, and R. Grosse, *J. Biol. Chem.* **262,** 15137 (1987).

quence of CRBP II was not available. On publication, however, it became evident that peptide 2 of CRBP is almost identical to the corresponding segment in CRBP II. Because antibodies raised to this peptide most likely will not discriminate between the two CRBP forms we focus on the antipeptide antibodies raised to peptide 1 of CRBP and to the two peptides derived from CRABP.

To enhance their immunogenicity during the immunization procedure, all peptides were separately conjugated via the cysteine residues to keyhole limpet hemocyanin (KLH) using succimidyl-3-(2-pyridyldithio)propionate (SPDP, Pharmacia, Uppsala, Sweden). When injected subcutaneously into rabbits, all peptide conjugates generated antisera with high titers to the peptides that served as the antigen, as revealed by an ELISA technique.[10] As expected, the titers to the native proteins were significantly lower.

To enrich for antibodies recognizing the native antigens, immunopurified immunoglobulin (Ig) fractions from all antipeptide antisera were generated and subsequently analyzed. Although some peptides used as antigens displayed significant similarities in their primary structures, for example, peptide 1 of CRBP and CRABP, no cross-reacting antibodies could be detected when analyzing intact CRBP and CRABP by the ELISA procedure. When analyzing the reactivity of the isolated Ig fractions using the peptides that served as the antigens, we detected some cross-reacting antibodies. However, the reactivity to the homologous peptide was always at least 3 orders of magnitude stronger. The difference in reactivity may reflect the more flexible nature of the free peptide compared to the corresponding peptide segment in the intact protein. The binding characteristics of the immunopurified Ig fractions to peptide 1 of CRBP and CRABP are summarized in Fig. 1. In addition, the Ig fraction to peptide 2 of CRABP specifically recognized CRABP using this technique (not shown). The abilities of the immunopurified Ig fractions to specifically recognize the appropriate antigens were further proved in experiments employing the myelin protein P2. This protein is highly homologous to CRBP and CRABP (see Table I) yet no cross-reactivity could be recorded to the Ig fractions raised to peptide 1 of CRBP and CRABP, respectively. Combined, these results suggested that the generated Ig fractions recognized the appropriate antigens with a high degree of specificity.

To demonstrate the usefulness of the isolated Ig fractions in immunohistochemistry it appeared essential to carry out immunoblotting analyses. The ability of the two Ig fractions to CRABP to specifically recognize CRABP in tissue homogenates has previously been demonstrated.[10] As illustrated in Fig. 2, the Ig fractions to peptide 1 of CRBP displayed the appropriate specificity, and only a stained band corresponding to the

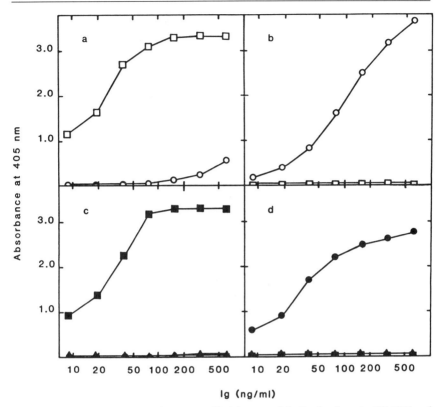

FIG. 1. Reactivities of the immunopurified immunoglobulin fractions to CRABP and CRBP using an ELISA procedure. Fixed amounts of peptide 1 of CRBP (□), native CRBP (■), peptide 1 of CRABP (○), native CRABP (●), and native myelin protein P2 (▲) were separately coated onto microtiter plates. The reactivities of the immunopurified IgG fractions derived from the antipeptide antisera to peptide 1 of CRBP (a, c) and to peptide 1 of CRABP (b, d) were determined by serial dilution. Bound antibodies were visualized using goat anti-rabbit Ig labeled with alkaline phosphatase. The enzyme substrate was p-nitrophenyl phosphate, and formation of p-nitrophenol was determined after 30 min in a Titretek microtiter plate reader at 405 nm.

migration of CRBP could be visualized in a rat liver homogenate separated by sodium dodecyl sulfate–polyacrylamide gel electrophoresis (SDS–PAGE). Thus, the isolated Ig fractions displayed the desired specificities and could be used to localize the corresponding antigens in tissue sections.

To localize CRBP and CRABP we have employed the avidin–biotin complex (ABC) technique,[22] the detailed protocol for which is outlined in

[22] F. M. Hsu, L. Rainer, and H. Fanger, *J. Histochem. Cytochem.* **29,** 577 (1981).

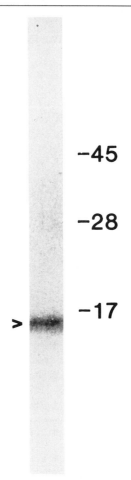

FIG. 2. Immunoblotting analysis using the immunopurified immunoglobulin fraction to peptide 1 of CRBP. Rat liver cytosol (200 μg total protein) was subjected to SDS–PAGE using a linear 15% gel, and the separated proteins were electrophoretically transferred to a nitrocellulose filter. Nonspecific protein binding was blocked by incubating the filter in 5% (w/v) nonfat dry milk dissolved in 20 mM Tris, pH 7.5/150 mM NaCl/0.1% (v/v) Tween 20 for 2 hr at room temperature. Immunopurified Ig to peptide 1 of CRBP (10 μg/ml) was added and incubated for 2 hr at room temperature. The filter was washed several times, and bound antibodies were visualized using alkaline phosphatase-labeled goat anti-rabbit Ig. Nitroblue tetrazolium and 5-bromo-4-chloro-3-indolyl phosphate were used as substrates. The migration of prestained marker proteins (lysozyme, 17K; trypsin inhibitor, 28K; carboanhydrase, 45K) are indicated on the right-hand side.

TABLE II
IMMUNOHISTOCHEMICAL STAINING TECHNIQUE

A. Blocking of endogenous peroxidase activity
  1. 0.3% hydrogen peroxidase in phosphate-buffered saline (PBS), 30 min
  2. Rinse in PBS, 10 min
B. Incubation with primary antibody
  1. Saturate nonspecific protein binding by incubation with 1% bovine serum albumin in PBS, 10 min
  2. Rinse in PBS, 10 min
  3. Apply the primary antibody (25 μl) diluted in 1% bovine serum albumin in PBS (normally between 5 and 50 μg of Ig/ml, titration is necessary) at room temperature, 30 min
  4. Rinse in PBS, 10 min
C. Incubation with secondary antibody
  1. Apply biotinylated goat anti-rabbit Ig (Vectastain) appropriately diluted in 1% bovine serum albumin in PBS, 30 min.
  2. Rinse in PBS, 10 min
D. Avidin–biotin–peroxidase complex (ABC)
  1. Avidin–biotinylated horseradish peroxidase complex (Vectastain) diluted 1/200 in PBS at room temperature, 30 min
  2. Rinse in PBS, 10 min
E. Dissolve the peroxidase substrate, 10 mg of 3-amino-9-carbazole, in 6 ml of dimethyl sulfoxide and add to 50 ml 20 mM sodium acetate buffer, pH 5.4; immediately before application add 4 μl of 30% hydrogen peroxide and incubate the sections at room temperature for 15 min
F. Immerse in tap water, 15 min
G. Counterstain in Mayer's hematoxylin, 5 min
H. Immerse in tap water, 10 min
 I. Mount the sections using glycerin–gelatin

Table II. The antipeptide Ig fractions are compatible with most fixation techniques. Routinely we use either formalin fixation or Bouin's fixative. Representative rat liver sections stained with preimmune Ig and Ig to peptide 1 of CRBP are shown in Fig. 3a,b, respectively. Using the monospecific antipeptide antibodies, we have established the cellular localization of CRBP and CRABP in a number of rat and human tissues, and some of our results are presented in Table III. Details regarding the localization are available elsewhere.

In Situ Hybridization Techniques

To identify cells expressing CRBP and CRABP we have also employed *in situ* hybridization techniques for the cellular localization of the respective mRNAs. This approach is not subject to the same difficulties and limitations as the immunohistochemical localization of the two pro-

TABLE III
TISSUE DISTRIBUTION OF CRBP AND CRABP IN A NUMBER OF RAT AND HUMAN
TISSUES AS REVEALED BY IMMUNOCYTOCHEMICAL LOCALIZATION
USING MONOSPECIFIC ANTIBODIES

| | | Staining intensity[a] | |
| | | CRBP | CRABP |
Tissue	Localization	I (rat/human)	I (rat/human)
Liver	Hepatocytes	+++/+++	+/+
	Stellate cells	+++++/+++++	−/−
	Kupffer cells	−/−	−/−
	Endothelial cells	−/−	−/−
	Bile duct epithelium	+/+++	+/+
Kidney	Glomeruli	−/−	−/−
	Proximal tubular cells	+++/+++	++/++
	Distal tubular cells	++/++	+++/+++
	Collecting duct cells	+/+	+/+
	Urothelium		
	Surface epithelium	+++/+++	++/+++
	Lower layers	+/+	+/+
Respiratory tract	Alveolar septal cells	++/++	−/−
	Respiratory epithelium	++/++	+/+
	Mucous gland epithelium	+/+	+/+
Testis	Leydig cells	++/++++	+/++
	Sertoli cells	+++/+++	+++/+++
	Germinal cells		
	Spermatogonia	+++/+++	++/ne[b]
	Spermatocytes	+++/+++	−/ne

[a] The staining intensity in the different cell types was scored visually and categorized as follows: (+++++) intensive, (++++) strong, (+++) moderate, (++) weak, (+) slightly above background, and (−) not detected.
[b] ne, Not examined.

teins, although it is more technically demanding. The stringent conditions used for hybridization and autoradiographic detection minimize the risk of cross-hybridization between mRNAs encoding the group of proteins related to CRBP and CRABP. Thus, we feel that *in situ* hybridization complements the immunohistochemical localization. In the following section we describe the *in situ* hybridization of the two mRNAs encoding the

FIG. 3. Localization of CRBP in rat liver tissue sections by immunohistochemical techniques (a,b) and the corresponding mRNA (c,d) using *in situ* hybridization with a radiolabeled CRBP cRNA probe. The background staining using preimmune Ig (a) and the nonspecific hybridization using a sense CRBP cRNA probe (c) demonstrate the specificities of the two techniques employed. The bars indicate 50 μm (a,b) and 17 μm (c,d), respectively.

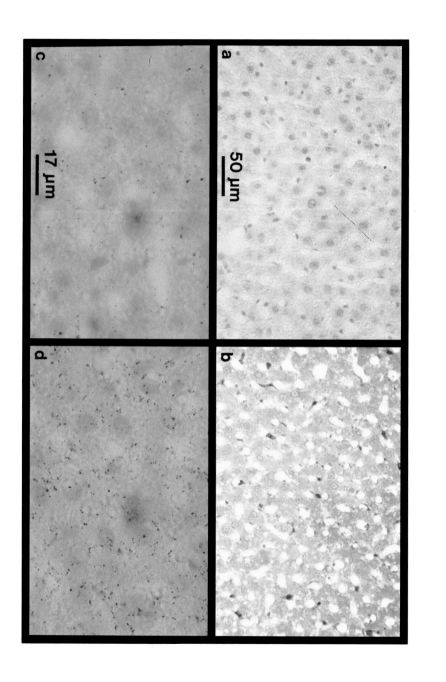

two retinoid-binding proteins using radiolabeled cRNA transcribed by *in vitro* transcription systems.[23,24]

Slide Preparation. To ensure retention of tissue sections, slides are washed with detergent (Extran, Merck, Darmstadt, FRG), rinsed with tap water, cleaned with 70% ethanol, and air-dried thoroughly. The slides are subsequently immersed for a few hours in 2% γ-aminopropyltriethoxysilane (Sigma, St. Louis, MO) in water at 60°. Dried slides are kept in dust-free boxes before activation (usually no longer than 1 month). Before sectioning of tissues, the slides are activated by immersing in 10% glutaraldehyde in phosphate-buffered saline (PBS) for 30 min. Finally, the slides are rinsed in water and dried.

Tissue Preparation. Formalin-fixed, paraffin-embedded human or rat tissues are used in most of our experiments. Rat tissues are obtained from animals perfused with 4% formalin, proceeded by rinsing with PBS containing 1.5% nuclease-free bovine serum albumin. In other experiments, freshly obtained tissues are immediately fixed in 4% paraformaldehyde in PBS. To avoid RNase contamination during sectioning, the microtome is cleaned in 0.1% diethyl pyrocarbonate (an RNase inhibitor) in water. The tissues are sectioned (5 μm) and collected on the siliconized slides. The slides are kept overnight at 37°.

Preparation of RNA Probes. The 580-base pair (bp) *Hin*dIII/*Bam*HI fragment of human CRBP cDNA[23] and the 420-bp *Hin*dIII/*Bam*HI fragment of bovine CRABP cDNA[24] are subcloned into *in vitro* transcription vectors (pGEM3, Promega Biotec, Madison, WI). The two plasmids are linearized with *Hin*dIII and transcribed using the T7 promoter to obtain the antisense RNA probe. The sense CRBP RNA probe, used as a background control, is generated from a *Bam*HI-digested plasmid using the SP6 promoter. ^{35}S-Labeled RNA is generated using a kit (Promega Biotec) essentially according to the supplier. Typically, a final specific activity of $0.5-1.0 \times 10^9$ dpm/μg of RNA is obtained.

In Situ Hybridization. *In situ* hybridization is performed as described elsewhere.[25] Briefly, the paraffin-embedded sections are deparaffinized by immersing in xylene, then graded ethanols, and digested with 0.1 mg/ml of proteinase K (Sigma) at 37° for 15 min before the hybridization. The sections are then rinsed in 2× SSC (1× SSC is 0.15 M NaCl and 15 mM sodium citrate, pH 7.2) and incubated at room temperature in 0.25% acetic anhydride in 0.1 M triethanolamine. Following an incubation in 0.1

[23] V. Colantuoni, R. Cortese, M. H. L. Nilsson, J. Lundvall, C.-O. Båvik, U. Eriksson, P. A. Peterson, and J. Sundelin, *Biochem. Biophys. Res. Commun.* **130,** 431 (1985).

[24] M. H. L. Nilsson, N. K. Spurr, P. Sakena, C. Busch, H. Nordlinder, P. A. Peterson, L. Rask, and J. Sundelin, *Eur. J. Biochem.* **173,** 45 (1988).

[25] K. Funa, L. Steinholtz, E. Nou, and J. Berg, *Am. J. Clin. Pathol.* **88,** 216 (1987).

M Tris-HCl and 0.1 M glycine for 30 min, the slides are rinsed in 2× SSC and immersed in 50% formamide/2× SSC at 50°. Subsequently, the labeled cRNA probe (1 × 10^6 cpm/section) in 50% formamide/2× SSC/10 mM dithiothreitol (DTT) containing 20 μg of sheared herring sperm DNA, 20 μg of *Escherichia coli* tRNA, and 40 μg of nuclease-free bovine serum albumin is applied in a total volume of 10–30 μl/section. The section is covered with coverglass and incubated in a humidified chamber at 50° for 3 hr. To remove nonhybridized RNA, the slides are rinsed in 4 changes of 50% formamide, 2× SSC at 50°, treated with RNase A, and finally rinsed in 2× SSC. Following dehydration in graded ethanols, the sections are autoradiographed with NTB 2 nuclear track emulsion (Eastman Kodak, Rochester, NY) diluted 1 : 1 with distilled water. After exposure for 4 days at 4°, the slides are developed in Dektol developer (Eastman Kodak) at 15° for 5 min and air-dried. Sections are counterstained with Mayer's hematoxylin.

To ascertain that specific hybridization has occurred, we feel that a number of control hybridizations are necessary. Routinely we use the following controls: (1) positive and negative tissue sections to check the quality of the probes; (2) a positive tissue section treated with RNase prior to hybridization to confirm that the probe hybridizes with cellular RNA; (3) checking all tissues for the presence of cellular RNA by hybridization with a cRNA probe to human β-actin. Examples of representative *in situ* hybridizations of CRBP in rat liver using sense and antisense cRNA probes are given in Fig. 3c,d.

Conclusion

Using the reagents described in this chapter we have found that both CRBP and CRABP are expressed in many functionally and structurally different cells and tissues. A theme which seems to be repeated in some tissues where a functional differentiation can be observed is that a gradient of the two proteins exists with respect to apparent concentrations. Thus, more differentiated cells express higher levels of CRBP and CRABP. Attempts to explain the mode of action of vitamin A should include consideration of these facts regarding the localization of the two proteins.

Acknowledgments

This work was supported by grants from the Swedish National Cancer Foundation. The support from Drs. Per A. Peterson, Johan Sundelin, and Magnus Nilsson is greatly acknowledged.

[33] Retinoids Bound to Interstitial Retinol-Binding Protein during the Visual Cycle

By ZHENG-SHI LIN, SHAO-LING FONG, and C. D. B. BRIDGES

Introduction

Interstitial retinol-binding protein (IRBP) is a large glycoprotein found in the interphotoreceptor matrix that fills the space between the cells of the retina and the subjacent pigment epithelium. On sodium dodecyl sulfate (SDS)–polyacrylamide gels, the apparent molecular weight of vertebrate IRBP is 134,200, except in the teleosts, where it is about one-half this value.[1] IRBP is believed to have an important function in the visual cycle in that it appears to be involved in the transport of retinoids between the cells of the retina and the pigment epithelium.[2,3] For example, IRBP prepared from cattle eyes has been found to have 11-*cis*-retinal, 11-*cis*-retinol, and all-*trans*-retinol bound to it.[2–6] Furthermore, when the rhodopsin in excised bovine eyes is bleached, the amount of all-*trans*-retinol bound to IRBP increases by a factor of 5 to 7.

Cattle, however, are not suitable experimental animals for physiological studies on light and dark adaptation. In contrast, the visual cycle of frogs has been intensively studied,[7] and, more recently, the retinoids bound to IRBP during the visual cycle in intact, living frogs have been investigated.[8] The results of this investigation have provided further support for the postulated role of IRBP in the visual cycle. They suggest that when rhodopsin is bleached, IRBP transports all-*trans*-retinol from the retina to the pigment epithelium and that when rhodopsin regenerates it utilizes 11-*cis*-retinal delivered to the rod outer segments by IRBP. Lin

[1] C. D. B. Bridges, G. I. Liou, R. A. Alvarez, R. A. Landers, A. M. Landry, Jr., and S.-L. Fong, *J. Exp. Zool.* **239**, 335 (1986).

[2] S.-L. Fong, G. I. Liou, R. A. Alvarez, and C. D. B. Bridges, *J. Biol. Chem.* **259**, 6534 (1984).

[3] J. C. Saari, D. C. Teller, J. W. Crabb, and L. Bredberg, *J. Biol. Chem.* **260**, 195 (1985).

[4] C. D. B. Bridges, R. A. Alvarez, S.-L. Fong, F. Gonzalez-Fernandez, D. M. K. Lam, and G. I. Liou, *Vision Res.* **24**, 1581 (1984).

[5] A. J. Adler, and C. D. Evans, *in* "The Interphotoreceptor Matrix in Health and Disease" (C. D. Bridges and A. J. Adler, eds.), p. 65. Academic Press, New York, 1985.

[6] A. H. Bunt-Milam, J. C. Saari, and D. L. Bredberg, *in* "The Interphotoreceptor Matrix in Health and Disease" (C. D. Bridges and A. J. Adler, eds.), p. 151. Academic Press, New York, 1985.

[7] C. D. B. Bridges, *Exp. Eye Res.* **22**, 435 (1976).

[8] Z.-S. Lin, S.-L. Fong, and C. D. B. Bridges, *Vision Res.* **29**, 1699 (1989).

et al.[8] showed that frog IRBP carried all-*trans*-retinol, 11-*cis*-retinol, and 11-*cis*-retinal as endogenous ligands. The amount of all-*trans*-retinol bound per mole of IRBP increased when the frogs were initially exposed to strong bleaching lights, then declined as bleaching continued. The 11-*cis*-retinal bound to IRBP diminished in the light and increased in the dark. Concurrently, however, no significant changes were found in the amount of 11-*cis*-retinol.

The methods summarized in this chapter are based in part on those reported by Lin *et al.*,[8] and we describe procedures used for preparing IRBP, together with its endogenous retinoids, rapidly and efficiently from interphotoreceptor matrix obtained from light- and dark-adapted frogs.

Light and Dark Adaptation

Strong light adaptation is used in these types of experiments.[7] In order to start from standard conditions, frogs are usually dark adapted overnight. The dark-adapted frogs are then placed in a white tank containing a few inches of water for periods up to 2 hr, typically under the illumination of four 75-W incandescent bulbs at a distance of 1 m. Dark adaptation is studied by placing these light-adapted frogs into darkness.

Interphotoreceptor Matrix

All procedures are carried out under dim red light, either in a cold room or on ice. At various times, groups of light- or dark-adapted frogs (usually six in number) are decapitated and the heads and bodies pithed. The eyes are enucleated, the cornea cut away, and the lens removed. To prevent loss of interphotoreceptor matrix and to minimize cell disruption (particularly in light-adapted frogs), no attempt should be made to detach the retina from the underlying pigment epithelium and choroid. The combined retina, pigment epithelium, and choroid are gently removed with curved iris forceps and gently stirred in 1 ml of ice-cold Tris-buffered saline (10 mM Tris-HCl, 150 mM NaCl, pH 7.5). This procedure leaches out most of the interphotoreceptor matrix. The tissues are centrifuged at low speed (1200–1900 rpm for 10–20 min; HS-4 rotor or equivalent), then the supernatant is centrifuged at high speed (1 hr at 100,000 g; 40,000 rpm, AH-650 rotor). The supernatant ("interphotoreceptor matrix") contains 60–85% of the IRBP in the eye. The pellets from the two centrifuging operations are combined and homogenized in 2 ml Tris-buffered saline and centrifuged at high speed (see above). The supernatant contains cytosol and a small proportion of the interphotoreceptor matrix containing the remaining IRBP. The pellet contains rhodopsin.

Rhodopsin is determined by homogenizing the pellet in 2 ml of 100 mM hydroxylamine in 50 mM Tris-HCl (pH 7.5) and extracting a 0.5-ml portion for 30 min on ice with an equal volume of 2% (w/v) L-1690 sucrose ester detergent, or equivalent.[9,10] The mixture is centrifuged, and the absorbance spectrum of the supernatant extract is measured before and after complete bleaching with white light (5–10 min). The rhodopsin concentration is then determined from the absorbance loss at 500 nm using an extinction coefficient for frog rhodopsin of 42,000 M^{-1} cm^{-1} (Ref. 11).

Separation of Interstitial Retinol-Binding Protein from Cellular Retinaldehyde-Binding Protein and Cellular Retinol-Binding Protein

Both cellular retinaldehyde-binding protein (CRAlBP) and cellular retinol-binding protein (CRBP) are present in the cytosol of the retina and pigment epithelium.[12-14] In spite of the care taken to minimize cell disruption, frog interphotoreceptor matrix preparations always contain CRA1BP and CRBP, which are likely sources of 11-*cis*-retinol, 11-*cis*-retinal, and all-*trans*-retinol.[15] The strategy is to obtain IRBP free of these retinoid-binding proteins as rapidly as possible and in a minimum volume. Additionally, the procedure used should not cause significant loss of bound retinoids. Since IRBP loses nearly all of its endogenous retinoids when purified on a concanavalin A affinity column, it is necessary to use an ion-exchange resin. This has the advantage that IRBP can be rapidly eluted at comparatively high concentration in a small volume, thus facilitating subsequent extraction of retinoids and reducing the losses of IRBP that characteristically occur in dilute solutions of the protein.

Under dim red light and in the cold, the interphotoreceptor matrix samples (diluted to a NaCl concentration of 125 mM) are applied to 2-ml columns of DE-52 (Whatman, Maidstone, England) packed in Pasteur pipettes and equilibrated with 125 mM NaCl in 10 mM Tris-HCl (pH 7.5). On DE-52 columns, CRBP from bovine and frog eyes elutes close to this NaCl concentration (S.-L. Fong and C.D.B. Bridges, unpublished observations).[2] Elution is initially with about 28 ml of the same buffer at a flow rate of 0.3 ml/min. The IRBP is then rapidly eluted with 300 mM NaCl in

[9] K. Nashima, M. Mitsudo, and Y. Kito, *Biochim. Biophys. Acta* **579,** 155 (1979).

[10] S.-L. Fong, A. T. C. Tsin, C. D. B. Bridges, and G. I. Liou, this series, Vol. 81, p. 133.

[11] C. D. B. Bridges, *Vision Res.* **11,** 841 (1971).

[12] G. W. Stubbs, J. C. Saari, and S. Futterman, *J. Biol. Chem.* **254,** 8529 (1979).

[13] J. C. Saari, S. Futterman, and L. Bredberg, *J. Biol. Chem.* **253,** 6432 (1978).

[14] S. Futterman, J. C. Saari, and S. Blair, *J. Biol. Chem.* **252,** 3267 (1977).

[15] J. C. Saari, L. Bredberg, and G. G. Garwin, *J. Biol. Chem.* **257,** 13329 (1982).

10 mM Tris-HCl, pH 7.5. Most of the IRBP emerges in the second or third 1-ml fractions, which are monitored at 280 nm. To verify that these fractions are not contaminated with CRAlBP (which is eluted in the first 28 ml), portions are dot-blotted in duplicate on two nitrocellulose sheets. One sheet is probed[16] with rabbit anti-bovine CRAlBP antibodies (provided to the authors by Dr. J. C. Saari, University of Washington, Seattle) and the other with rabbit anti-frog IRBP antibodies prepared by the authors.

Extraction of Retinoids from Interstitial Retinol-Binding Protein and Analysis by High-Performance Liquid Chromatography

The extraction and analysis of retinoids is carried out in dim red light and under nitrogen. A 0.2-ml aliquot from a 1-ml DE-52 column fraction is retained for gel electrophoresis and quantitation of IRBP (see below). The remaining 0.8 ml is mixed with an equal volume of chilled ethanol containing 0.2 μg/ml d-α-tocopherol as an antioxidant. The mixture is extracted with two 0.5-ml volumes of n-hexane. The hexane extract is evaporated to dryness under a stream of nitrogen and dissolved in 100 μl n-hexane containing 5% dioxane. The procedure in the authors' laboratory is to inject a 10- to 30-μl volume with a Waters (Milford, MA) WISP automatic injector on to a Supelcosil 3-μm column (4.6 $\times$ 150 mm) and to elute with 5% dioxane in n-hexane at a flow rate of 0.7 ml/min. The absorbance is monitored with a Kratos (Ramsey, NJ) detector at 325 nm. This wavelength is optimal for retinol, and, if desired, the detector can be shifted to longer wavelengths more appropriate for retinal. At regular intervals (preferably before each sample is run), standard mixtures containing known amounts of authentic 11-, 13- and all-*trans*-retinals and -retinols must be injected to permit accurate quantitation.

Typical chromatograms of retinoids extracted from IRBP are illustrated in Fig. 1. In the sample from dark-adapted frogs (Fig. 1A), the major retinoids present are 11-*cis*-retinal and 11-*cis*-retinol. In this example, there are also minor peaks due to all-*trans*-retinal and all-*trans*-retinol. In other instances, all-*trans*-retinol is hardly detectable, showing that IRBP-containing fractions do not contain significant amounts of CRBP, which has all-*trans*-retinol as its only endogenous ligand.[15] However, the all-*trans*-retinol peak becomes prominent when the IRBP has been obtained from light-adapted frogs, as shown in Fig. 1B. Concomitantly, there is reduction in the height of the 11-*cis*-retinal peak. The relatively large peak preceding 11-*cis*-retinal in Fig. 1B is not a retinoid

16 F. Gonzalez-Fernandez, S.-L. Fong, G. I. Liou, and C. D. B. Bridges, *Invest. Ophthalmol. Visual Sci.* **26,** 1381 (1985).

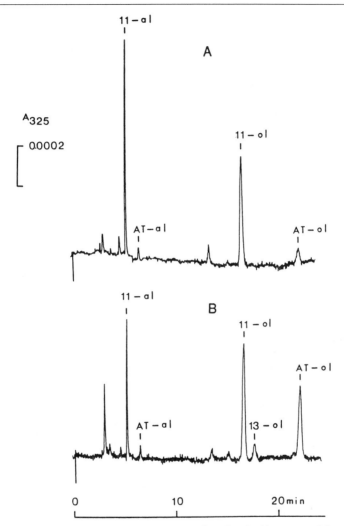

FIG. 1. High-performance liquid chromatography of retinoids extracted from DE-52-processed IRBP obtained from interphotoreceptor matrix prepared from dark-adapted and light-adapted frog eyes. (A) Dark adapted for 24 hr; (B) light adapted for 1.0 hr. Peak identification: 11-al, 11-*cis*-retinal; AT-al, all-*trans*-retinal; 11-ol, 11-*cis*-retinol; 13-ol, 13-*cis*-retinol; AT-ol, all-*trans*-retinol. Quantities of retinoids injected amounted to 20–40 pmol. Mobile phase, 5% dioxane in *n*-hexane; flow rate, 0.7 ml/min; detection wavelength, 325 nm; column, Supelcosil 3 μm (4.6 × 150 mm). Downward deflection represents the injection point.

but has not been identified: the small 13-*cis*-retinol peak in this chromatogram is an artifact of the extraction process.

Determination of Interstitial Retinol-Binding Protein

The method of Ball[17] has been found to be very satisfactory. Measured volumes of IRBP-containing column fractions are first electrophoresed in SDS–polyacrylamide gels (10%).[18] As standards, various levels of IRBP ranging from 0.5 to 10 μg are run in the same gel. The IRBP should be prepared and purified according to Fong *et al.*[19] The gel is stained overnight in 300 ml 0.05% Serva Blue R (Serva, Heidelberg, FRG) in 25% (v/v) 2-propanol and 10% (v/v) glacial acetic acid. It is destained twice by diffusion with the same volume of 10% (v/v) methanol in 10% (v/v) glacial acetic acid. The regions of the gel containing stained IRBP and nearby areas serving as controls are cut out with a razor or scalpel blade and the stain quantitatively extracted over a 24-hr period at 37° with 3% SDS in 50% (v/v) 2-propanol. The amount of stain extracted, determined from its absorbance at 595 nm, is linearly proportional to the amount of IRBP loaded on the gel, provided this does not exceed 10 μg.

[17] E. H. Ball, *Anal. Biochem.* **155**, 23 (1986).
[18] U. K. Laemmli, *Nature (London)* **227**, 680 (1970).
[19] S.-L. Fong, G. I. Liou, and C. D. B. Bridges, this series, Vol. 123, p. 102.

[34] Purification of Cellular Retinoic Acid-Binding Protein from Human Placenta

By MICHIMASA KATO, MASATAKA OKUNO, and YASUTOSHI MUTO

Introduction

Although human placentas are sufficiently abundant for medical studies, they are not always suitable for purification studies. In most purification studies, researchers are not interested in the major components of the placenta, namely, blood and connective tissue. The former gives a large volume of homogenate, and the latter makes homogenization more difficult. In addition, the human placenta is a very poor source of cellular retinoic acid-binding protein (CRABP). Only 320 μg of CRABP was isolated from 2 kg of placentas in our previous studies,[1] thus making the method not highly recommended.

This chapter deals mainly with the initial treatment of placentas, an approach that provides a process improvement for large-scale preparation of relatively small molecules. The molecular weight of bovine CRABP has been calculated to be 15,460 from its amino acid sequence.[2] A higher value (16,300) was estimated in a study of CRABP from bovine eyes.[3]

Assay Methods

Human CRABP and cellular retinol-binding protein (CRBP) are not separated from each other in the early purification steps including size-exclusion procedures. Endogenous retinol, bound to CRBP, is readily detected by its prominent fluorescence (excitation 350 nm, emission 475 nm) during the course of purification. Thus, the monitoring of fluorescence serves to mark the presence of both binding proteins in early steps. In later steps, radioactivity of the bound exogenous [^{3}H]retinoic acid and weak fluorescence serve to identify CRABP. If high-performance size-exclusion chromatography is available, a quantitation of retinoic acid-binding activity will be helpful to estimate the concentration of CRABP in each sample. In our previous study, an aliquot (1 ml) of sample was incubated overnight with 80 nM [^{3}H]retinoic acid in the presence or absence of a 200-fold molar excess of cold retinoic acid, and was then analyzed by a high-performance liquid chromatography (HPLC) system.[1] In the above preparation, [^{3}H]retinoic acid (Amersham Japan, Tokyo) was diluted with cold retinoic acid to obtain 1 Ci/mmol in ethanol.

Purification Procedures

Fractionation procedures outlined below are carried out at 4°. The columns and eluted samples are covered with aluminum foil to avoid exposure of the bound retinoids to white light. Table I presents the sequence of six fractionation steps of our previous study[1] including the following: (a) eight column runs of gel filtration on Sephadex G-50 (10 × 125 cm, Pharmacia Fine Chemicals, Uppsala, Sweden); (b) rechromatography of combined fractions from (a); (c) ion-exchange chromatography at pH 6.4 on DEAE-cellulose DE-52 (3.2 × 40 cm, Whatman, Maidstone, UK); (d) ion-exchange chromatography at pH 8.4 on DEAE-

[1] M. Okuno, M. Kato, H. Moriwaki, M. Kanai, and Y. Muto, *Biochim. Biophys. Acta* **923**, 116 (1987).
[2] J. Sundelin, S. R. Das, U. Eriksson, and P. A. Peterson, *J. Biol. Chem.* **260**, 6494 (1985).
[3] J. C. Saari, S. Futterman, and L. Bredberg, *J. Biol. Chem.* **253**, 6432 (1978).

TABLE I
PURIFICATION OF CRABP FROM HUMAN PLACENTA

Procedure	Protein (mg)	CRABP (mg)	CRBP (mg)	Purification (-fold)		Recovery (%)	
				CRABP	CRBP	CRABP	CRBP
Soluble acetone powder extract	181,000	2.40	8.72	—	—	—	—
Sephadex G-50 (I)	6,000	2.06	7.12	26	25	86	82
Sephadex G-50 (II)	975	1.59	6.73	123	143	66	77
DEAE-Cellulose, pH 6.4	111	1.09	5.01	740	940	45	57
DEAE-Cellulose, pH 8.4	22.4	0.46	4.04	1,550	3,740	19	46
Sephadex G-50 (III)	9.9	0.41	3.59	3,120	7,530	17	41
SP-Sephadex, pH 4.9	0.32 (CRABP) 2.7 (CRBP)	0.32	2.70	75,400	20,700	13	31

cellulose DE-52 (1.5 × 15 cm); (e) gel filtration on Sephadex G-50 (2 × 95 cm); (f) ion-exchange chromatography at pH 4.9 on SP-Sephadex C-50 (1.5 × 30 cm, Pharmacia). The most time-consuming steps in the above study are the extraction procedure and the initial cumbersome steps of the gel filtration. Many more placentas can be handled by a new extraction method planned to reduce the bulk of the homogenate. This new protocol described below will be helpful in large-scale purifications.

Step 1. Extraction

Frozen placentas (below −60°) are crushed using a hammer. Resulting pieces are then homogenized in a Waring blendor (not by a Polytron-type homogenizer) in 2 tissue volumes (v/w) of 10 mM Tris-HCl buffer, pH 7.6, containing 4 mM EDTA, 0.05% NaN$_3$, and 0.01% Foy (gabexate mesylate; Ono Pharmaceutical Co. Ltd., Osaka, Japan). After centrifugation at 20,000 g, to the supernatant is added a 1/10 original tissue weight of CDR (Cell Debris Remover; Whatman), preequilibrated with homogenizing buffer. After 1 hr of stirring at 4°, the suspension is centrifuged at 20,000 g.

The resulting supernatant is filtered through a 1/50 original tissue weight of CDR sandwiched between two sheets of filter paper (No. 1, Whatman) on a Büchner funnel. The filtrate is then concentrated using the hollow fiber system from Amicon (Danvers, MA) with an HIP 10-20 cartridge to obtain a 1/5 original tissue volume (v/w) of placentas. The concentrated solution is lyophilized to dryness, then homogenized with one-half the original tissue volume (v/w) of ice-cold acetone. The homogenate is filtered on a Büchner funnel through No. 2 filter paper (Whatman). The insoluble material on filter paper is blow-dried. The dried powder is rehomogenized with one-half the original tissue volume (v/w) of homogenizing buffer. After centrifugation at 20,000 g, the resulting supernant is filtered through a 1/100 original tissue volume (v/w) of CDR as described above. The filtrate is then ultrafiltered using a XM300 Diaflo membrane (Amicon) under a weak (0.1 kg/cm^2) pressure of nitrogen gas in an Amicon cell (2 liters). If the thin channel system (Amicon) is used, the efficiency of ultrafiltration will be improved. After extensive ultrafiltration, the resulting filtrate is concentrated by a YM5 Diaflo membrane (Amicon) to obtain a 1/100 volume (v/w) of original placentas. Although the molecular weights of the proteins in the above preparation vary widely, only a trace amount of albumin (66,000) passes through the XM300 Diaflo membrane. Fortunately, however, CRABP is readily filtered through the XM300 membrane.[4]

---

[4] M. Kato, M. Okuno, S. Nishiwaki, and Y. Muto, unpublished.

Step 2. Gel Filtration

Size-exclusion chromatography as an initial fractionation step is essential to obtain the best recovery of CRABP, but it is not the first choice in the analysis of endogenous retinoid bound to CRABP, because the elution profile of apo-CRABP is slightly different from that of holo-CRABP under conditions of conventional ion-exchange chromatography. Moreover, the charging of a sample with exogenous retinoic acid has a undesirable effect on the recovery of CRABP from the anion-exchange column. We recommend that the charging of apo-CRABP with retinoic acid should be done before size-exclusion chromatography.

For example, 50 ml of extract prepared from 5 kg of placentas is incubated with 50 μl of an ethanolic solution of [^{3}H]retinoic acid (3.3 mM solution, 100 mCi/mmol) with stirring overnight at 4°. For the above sample, a 5 × 100 cm column of Sephadex G-50 is adequate to process the whole sample at once. The column is equilibrated with 50 mM Tris-HCl buffer, pH 7.6, containing 200 mM NaCl, 4 mM EDTA, and 0.05% NaN$_3$ operating at a flow rate of 40 ml/hr. Fractions (10 ml each) are collected and analyzed for fluorescence and radioactivity. When [^{3}H]retinoic acid is not available, the fluorescence of each fraction serves to mark the presence of CRABP together with CRBP. The sharp peak (fluorescence or radioactivity) for retinoids bound to the binding proteins should be distinguished from the other peaks for fluorescence (or radioactivity) eluting in the void volume and in the included volume of the column. The fractions comprising the fluorescence peaks of the binding proteins are separately combined and concentrated.

Step 3. Anion-Exchange Chromatography

In our previous study, two column runs with different buffer systems were employed for the middle phase of purification using Whatman DE-52 cellulose.[1] As shown in Table I, both chromatographies produced adequate recovery of CRBP. However, the anion-exchange chromatography at pH 8.4 yielded an obviously low recovery of CRABP. Taking the improvement in the extraction procedure into consideration, we now propose that a single column run of anion-exchange chromatography is sufficient to process a sample, because impurities are effectively reduced in Steps 1 and 2.

The concentrated solution containing CRABP and CRBP from Step 2 is simultaneously desalted and equilibrated on a small column of Sephadex G-25 previously washed with 20 mM imidazole–acetate buffer, pH 6.4. The resulting solution is then applied to a column (2.5 × 20 cm for 5 kg of placentas) of Whatman DE-52 cellulose equilibrated with the above buffer at a flow rate of 40 ml/hr. After washing with 2 column volumes of buffer, a linear gradient of imidazole–acetate buffer, pH 6.4, from 20 to 200 mM (total gradient volume 1 liter) is used to elute the binding proteins. Fractions (10 ml each) are collected and analyzed for fluorescence and absorbance at 280 nm.

When [^{3}H]retinoic acid is not available as the marker for CRABP, analysis of fluorescence should be done very carefully. In anion-exchange chromatography, CRABP is eluted slightly faster than CRBP. Thus, the fluorescence of retinoic acid, bound to CRABP, forms a gently sloping shoulder on the prominent peak of retinol complexed with CRBP. If [^{3}H]retinoic acid is available, CRABP is readily located as a sharp peak of radioactivity. Generally, the fractions comprising the entire peak of fluorescence are pooled for the concurrent purification of CRBP. Figure 1 represents a previous elution profile from anion-exchange chromatography at pH 6.4[1] conducted similarly to the present protocol. Treading on the heels of the most prominent fluorescence peak (CRABP and CRBP), two peaks for fluorescence and radioactivity of [^{3}H]retinoic acid are observed. The fluorescence and radioactivity are attributed to retinol and [^{3}H]retinoic acid bound to retinol-binding protein (RBP). However,

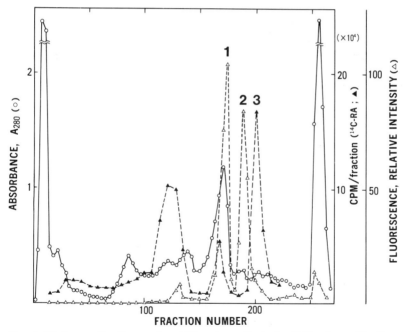

FIG. 1. Elution profile of a sample gel-filtered on a Sephadex G-50 column. The fluorescence at peak 1 is due to retinoic acid and retinol bound to CRABP and CRBP. The fluorescence at peak 2 and radioactivity at peak 3 are due to retinol and [^{3}H]retinoic acid, respectively, bound to RBP.

CRABP and/or CRBP is easily distinguished from RBP because of its unique absorption spectrum.[3]

Step 4. Second Gel Filtration

An additional gel filtration before the final preparation of CRABP yields a significant increase in purification (Table I). A relatively small column (2 × 100 cm, for example) can be used. Fully charging apo-CRABP with retinoic acid should be performed at this step. The concentrated sample from Step 3 is charged with cold retinoic acid (final concentration, 10 μM) and then chromatographed at a flow rate of 2 ml/cm²/hr or less. Fractions (3 ml) are assayed for protein absorbance at 280 nm and fluorescence. The fractions comprising a single prominent peak of fluorescence contain both CRABP and CRBP.

Step 5. Cation-Exchange Chromatography

Fractions corresponding to binding proteins are pooled from a Sephadex G-50 column, concentrated, and desalted on a small column of Sephadex G-25 preequilibrated with 10 mM sodium acetate buffer, pH 4.9. The resulting sample is then applied to a SP-Sephadex C-50 column (Pharmacia, 1.5 × 30 cm for 5 kg of placentas), prepared and equilibrated using the same buffer, and eluted with a gradient of 10 to 130 mM sodium acetate buffer, pH 4.9 (total gradient volume 200 ml). Fractions (3 ml) are collected at a flow rate of 8 ml/hr and assayed for protein absorbance at 280 nm, retinoid absorbance at 350 nm, and fluorescence. This final procedure results in the complete separation of CRABP from CRBP and in the isolation of each of these proteins as purified homogeneous entities. The fractions containing binding proteins are readily identified by either fluorescence or absorbance at 350 nm because other proteins (impurities) show no characteristics of the bound retinoids. Elution of CRABP is preceded by that of CRBP.

Assessment of Purity

Conventional polyacrylamide gel electrophoresis in 0.1% sodium dodecyl sulfate is performed according to the discontinuous system of Laemmli,[5] using a 15% separating gel in 375 mM Tris-HCl buffer, pH 8.8, and a 5% stacking gel in 125 mM Tris-HCl buffer, pH 6.8. If analytical isoelectric focusing is to be performed, apo-CRABP in the sample should be charged with retinoic acid prior to electrophoresis.[1]

[5] U. K. Laemmli, *Nature (London)* **227,** 680 (1970).

[35] Differentiation of Embryonal Carcinoma Cells in Response to Retinoids

By PATRICIO ABARZÚA and MICHAEL I. SHERMAN

Introduction

Embryonal carcinoma (EC) cells, the stem cells of teratocarcinomas, offer an attractive experimental model for studying differentiation because they resemble early embryonic stem cells.[1] Embryonal carcinoma cells were initially isolated from primary cultures of cells derived from teratocarcinomas, and from these many clonal cell lines have been developed. The propensity for differentiation of cells from these lines can be assessed *in vivo;* such studies have indicated a broad spectrum of differentiation potential among the lines.[2] When so-called pluripotent EC cells (e.g., PCC3, PC-13) are injected subcutaneously into syngeneic mice, they give rise to teratocarcinomas which typically contain many differentiated cell types that can represent all three primary germ layers. At the other extreme, so-called nullipotent EC cells (e.g., Nulli-SCC1, F9) show little or no histological evidence of differentiation. In fact, this classification is artificial insofar as there is probably a continuum of potential for differentiation in tumors if one considers large numbers of EC lines.

In tissue culture, the ability of EC cells to differentiate also varies markedly. Cells from most EC lines remain undifferentiated or differentiate at low frequency when growing exponentially. Some lines differentiate readily on reaching confluence. When EC cells are prevented from attaching to the substratum, they form multicellular aggregates, and under these conditions cells from some of the lines undergo differentiation.[3] A variety of chemicals have been shown to induce EC cell differentiation; retinoids are among the most potent chemical inducers of EC cell differentiation, with effective concentrations typically in the nanomolar range. Even EC cells such as Nulli-SCC1 and F9, which differentiate poorly in tumors, do so readily in culture in response to retinoids.[4,5] Because the

[1] G. R. Martin, *Science* **209,** 768 (1980).

[2] G. R. Martin, *in* "Teratocarcinoma Stem Cells" (L. M. Silver, G. R. Martin, and S. Strickland, eds.), p. 690. Cold Spring Harbor Laboratory, Cold Spring Harbor, New York, 1983.

[3] M. I. Sherman, *in* "Teratomas and Differentiation" (M. I. Sherman and D. Solter, eds.), p. 189. Academic Press, New York, 1975.

[4] S. Strickland and V. Mahdavi, *Cell (Cambridge, Mass.)* **15,** 393 (1978).

[5] A. M. Jetten, M. E. R. Jetten, and M. I. Sherman, *Exp. Cell Res.* **124,** 381 (1979).

mechanisms by which retinoids induce differentiation of EC cells might be relevant to their biochemical and biological effects on a large variety of cell types, and because differentiation is easily monitored in EC cells, this cell system is useful for studying retinoid action. We describe below techniques for the culture of EC cells from established lines and for evaluating the differentiation of these cells in response to retinoids.

Culture of Embryonal Carcinoma Cells

Although human EC cells have been cultured,[2] as a rule they are less responsive to retinoids than are murine EC cells. Murine EC-like cells derived from embryos *in vitro* are also available[6,7]; however, culture conditions for these cells can be more demanding than those required for conventional EC cell lines, and it is likely that the response to retinoids is the same. The methods described here are generally applicable to conventional murine EC lines. Many such lines are documented.[2] As mentioned above, the cells show widely differing propensities for differentiation in tumor form; however, cells from most EC lines differentiate readily in response to all-*trans*-retinoic acid (RA) (except for mutant or variant lines selected for lack of retinoid responsiveness). Cells with a high propensity for differentiation, and which can give rise to a variety of cell types in tumors or in culture, are particularly useful for studying patterns of differentiation. As a rule, such cultures are less likely to contain foci of cells refractory to retinoid action (see below). On the other hand, monitoring differentiation in these cultures can be complicated by the different phenotypes of the cells produced. Therefore, in experiments to evaluate the potencies of various retinoids, EC cell lines with more restricted differentiation profiles are generally easier to assess.

The culture of murine EC cells can be carried out in any laboratory equipped with standard tissue culture equipment (laminar flow hood, inverted microscope, humidified CO_2 incubator, etc.). Because of their rapid growth rate (doubling time 10–14 hr), EC cells can be readily expanded for many experimental procedures involving protein and nucleic acid isolation. Embryonal carcinoma cells are not contact-inhibited; they pack tightly in monolayer cultures, giving very high final densities. The cells are also capable of growing clonally in semisolid medium.

[6] G. R. Martin and L. F. Lock, *in* "Teratocarcinoma Stem Cells" (L. M. Silver, G. R. Martin, and S. Strickland, eds.), p. 635. Cold Spring Harbor Laboratory, Cold Spring Harbor, New York, 1983.

[7] E. J. Robertson, M. H. Kaufman, A. Bradley, and M. J. Evans, *in* "Teratocarcinoma Stem Cells" (L. M. Silver, G. R. Martin, and S. Strickland, eds.), p. 647. Cold Spring Harbor Laboratory, Cold Spring Harbor, New York, 1983.

Materials

Dulbecco's modified Eagle's medium (DMEM), 1 g/liter glucose (Gibco, Grand Island, NY)

Fetal calf serum (FCS), heat-inactivated at 55° for 20 min and stored at −20° (Gibco)

Ca^{2+}- and Mg^{2+}-free phosphate-buffered saline (PBS) (Gibco)

0.1% Gelatin (swine skin, Type I; Sigma, St. Louis, MO) dissolved in double-distilled water by mild heating; while still warm, the solution is sterile-filtered through a 0.45-μm Nalgene tissue culture-grade disposable filter unit, then stored at 4°

Trypsin–EDTA solution (0.05% Trypsin, 0.53 mM EDTA) (Gibco), dispensed into aliquots and stored frozen at −20°

Penicillin–streptomycin solution (10,000 U/ml penicillin G, 10 mg/ml streptomycin), stored frozen at −20° (Gibco)

Kanamycin solution (10 mg/ml), stored frozen at −20° (Gibco)

Retinoids (e.g., RA) are dissolved in ethanol (final concentration ∼10 mM); RA is stored as a stock solution for a maximum of 2 weeks in the dark at −70°, whereas other retinoids might be less soluble in ethanol than RA and/or more stable when stored at −70°

Methods

Since many retinoids are light-sensitive, it is desirable to carry out all tissue culture operations in a room with amber lighting. Embryonal carcinoma cells are grown in DMEM supplemented with fresh glutamine (4 mM) and 10% FCS (S-DMEM) at 37° in a humidified atmosphere of 5% CO_2 in air. The use of antibiotics is optional. Working concentrations of antibiotics are 100 U/ml penicillin, 100 μg/ml streptomycin, and/or 100 μg/ml kanamycin. Although EC cells grow well in the low concentrations of glucose present in S-DMEM, higher concentrations (e.g., 4.5 g/liter) will result in accelerated growth. The drawback of using the high glucose formulation is that the medium becomes acidic very quickly, and this can lead to death of the cultures, particularly when they reach high densities. Thus, if large numbers of cells are required, the cultures can be grown in high glucose, but the medium will have to be changed at least every other day and the cells will have to be passaged on a regular basis, for example, every second or third day at a split ratio of 1 : 5 to 1 : 20.

For subculturing, cells are rinsed once with PBS and incubated for 2–3 min at room temperature with enough trypsin–EDTA solution to cover the cells (normally 0.5 ml/25 cm² flask). Some EC cell lines are more difficult to remove from the substratum and require longer incubation or incubation at 37°. In some instances gentle shaking of the tissue culture vessel will help to detach the cells from the plastic. However, EC cells are very sensitive to trypsin–EDTA and will lyse if treated too vigorously or for too long. To ensure good recovery of viable cells, trypsin digestion can be monitored under an inverted microscope. Trypsin is inactivated by adding prewarmed S-DMEM. A single-cell suspension is usually obtained by gently pipetting the mixture several times. Embryonal carcinoma cells from some lines tend to be dislodged from the substratum as small aggregates rather than single cells. For routine subculturing this does not present a problem. However, if cell counting is to be performed, the potential for experimental error is too great to be ignored. We have found that by gently passing the suspension through a 19-gauge needle, with care to minimize bubbling, greater than 98% of the aggregates can be dispersed into single cells with little or no noticeable effect on cell viability. The cells should be allowed to settle on, and attach to, the substratum for at least 6 hr before any additions are made to the medium. Some EC cells (e.g., F9, Nulli-SCC1) adhere very poorly to tissue culture vessels. A gelatin coating can overcome this problem: at least 10 min prior to seeding the cells, enough 0.1% gelatin solution is added to the vessel to completely cover the surface. After standing at room temperature, the gelatin solution is aspirated and the vessel rinsed once with PBS before medium is added. When aggregate formation is desired, cells are plated in bacteriological grade dishes.[3]

Induction of Differentiation

To induce differentiation of cells in monolayers or aggregates with RA, a 10 mM solution in ethanol is diluted at least 100-fold in S-DMEM, and appropriate aliquots are added to cells to give final concentrations between 1 nM and 1 μM. For RA and most other retinoids, lower concentrations do not produce a noticeable effect, and higher concentrations can be toxic. (If serum-free medium is used, the retinoid concentration will have to be lowered dramatically or toxicity will result.) The final ethanol concentration should be less than 0.5%. For long-term cultures, medium containing RA is replaced at least every 48 hr. Since EC cells can metabolize RA and certain other retinoids very effectively (see Ref. 8), in experiments where continuous presence of retinoid is critical, RA should be added every 24 hr. Although not an inducer of differentiation by itself, 1 mM dibutyryl cyclic adenosine 3',5'-monophosphate (dbcAMP) can be added to enhance the differentiation-inducing potency of retinoids on at least some EC cells.[4]

---

[8] M. L. Gubler and M. I. Sherman, this volume [60].

Comments

There are several factors that the investigator should consider when designing an experimental protocol with EC cells. For example, after retinoids are added to EC cells, cell proliferation will be dramatically curtailed beyond 24 hr as differentiation occurs. Some cell death is also likely. Therefore, if differentiation is extensive or complete at the end of the experiment, there might only be two or three population doublings from the time of retinoid addition. Another important consideration is the existence in most EC cell cultures of a small proportion of cells which will be at least transiently refractory to retinoid-induced differentiation. (This phenomenon is particularly marked in cell lines such as Nulli-SCC1 or F9 which generally differentiate less readily than pluripotent lines, but the extent of retinoid refractoriness should be evaluated empirically whenever a cell line is used for the first time.) The result is the emergence with time of colonies of EC-like cells against a background of flat and enlarged differentiated derivatives.[9] This is most evident when cells are induced to differentiate at high density. Thus, with some EC lines, if cultures are treated with retinoids when they are relatively dense, cells which are refractory to retinoid action can eventually overgrow differentiated cells, and this could lead to inappropriate interpretations of experimental results. It is, therefore, desirable in long-term experiments to plate cells at low density and, if large amounts of cells are required, either to use larger culture dishes (e.g., Nunc 245 × 245 × 20 mm plates; Inter Med, Roskilde, Denmark) or to increase the number of dishes per experiment. This need not necessarily require significant scaling up of subsequent biochemical procedures: for example, total RNA can be extracted by the guanidinium thiocyanate procedure from as many as 10 large Nunc plates by merely doubling the amount of solution routinely used for a single plate (see below), which is then transferred sequentially from plate to plate.

Assessment of Embryonal Carcinoma Cell Differentiation

It is not within the scope of this chapter to review all of the alterations that accompany differentiation of EC cells. Instead, we concentrate on some consistent changes exhibited by EC cells that have been widely used as convenient indicators of differentiation. Generally, these can be divided into disappearance of properties characteristic of undifferentiated EC cells (EC cell markers) and appearance of properties shared by many or most of the cell types that result from retinoid-induced differentiation of EC cells *in vitro* (differentiation markers).

[9] M. I. Sherman, K. I. Matthaei, and J. Schindler, *Ann. N.Y. Acad. Sci.* **359,** 192 (1981).

Embryonal Carcinoma Cell Markers

Morphology. Although there are differences in the morphology of EC cells from various lines and a trained investigator can distinguish cells from different lines, EC cells are commonly small with large nucleus/cytoplasm ratios and prominent nucleoli. The cells typically (although not always) grow in tightly packed clusters, and because they fail to produce an abundant extracellular matrix, cellular boundaries within clusters are generally indistinct. When EC cells differentiate, they can assume a variety of phenotypes, but in general the cells are larger and flatter with reduced nucleus/cytoplasm ratios and more discretely defined cell boundaries. Morphology is a convenient way to monitor differentiation qualitatively as an experiment progresses. However, there is a great danger in using morphological change as a sine qua non for differentiation as retinoids and other chemical inducers can alter morphology without inducing overt cell differentiation.[10] *Note:* Conclusions about differentiation of EC cells should never be based exclusively on cell morphology. In fact, it is prudent to use two or more markers when evaluating the differentiation status of cells in culture.

Growth Properties. Two other features of EC cells that are consistently and dramatically reduced or extinguished as a result of differentiation are colony-forming efficiency (CFE) in liquid culture[11] and anchorage-independent growth in semisolid medium.[12] Such reductions might not necessarily be indicative of differentiation: cytotoxic agents would obviously give a similar result. However, this eventuality can be eliminated by toxicity testing of the retinoid at effective concentrations against EC cells at nonclonal densities in liquid medium. With this caveat, these tests can be reliably used not only to signal cell differentiation, but also to indicate the time at which cells become irreversibly committed to differentiate (i.e., they go on to generate differentiated cells even after the inducer is removed).[11] In fact, a convenient way to assay for these properties is in the same way that commitment studies are carried out: cells treated with retinoid in S-DMEM for at least 48 hr (a period adequate for commitment to differentiation to occur in the great majority of cells[11]) are washed extensively with PBS to remove retinoid and treated with trypsin–EDTA as described above. Fresh S-DMEM is added and the cells are counted. Cells are plated in liquid or semisolid medium (the latter contains a 1 : 1 mixture of 2× S-DMEM and 1.92% methylcellulose) at

---

[10] M. I. Sherman, M. A. Eglitis, and R. Thomas, *J. Embryol. Exp. Morphol.* **93**, 179 (1986).

[11] M. I. Sherman, M. L. Gubler, U. Barkai, M. I. Harper, G. Coppola, and J. Yuan, *Ciba Found. Symp.* **113**, 42 (1985).

[12] S.-Y. Wang and L. Gudas, *J. Biol. Chem.* **259**, 5899 (1984).

densities low enough to allow accurate counting of colonies (e.g., 200 or fewer colonies/60-mm-diameter culture dish). It is desirable to plate triplicate dishes for each retinoid-treated sample.

Cultures should be incubated for at least 7 days in liquid medium and for as many as 14 days in semisolid medium. At the end of the culture period, colonies are counted after fixing and staining with 0.2% crystal violet (liquid culture) or 0.05% iodonitrotetrazolium (methylcellulose). An inverse relationship is to be expected between number of colonies and EC cell differentiation.

SSEA-1. One of the most useful antigenic markers for following differentiation of EC cells is the surface antigen SSEA-1 (stage-specific embryonic antigen-1).[13] This surface marker is expressed on most or all EC cells but is rapidly lost when the cells differentiate, regardless of the inducer added or direction of differentiation. Expression of this marker can be easily assessed by indirect immunofluorescence using a mouse monoclonal antibody developed by Solter and Knowles.[13] Standard protocols for indirect immunofluorescence work reasonably well with most EC cells. However, cells from some EC lines do not attach firmly to glass coverslips and are lost during washing. With such cells the procedure can be carried out in gelatin-coated culture dishes (gelatin coating of glass coverslips does not markedly improve EC cell adhesion). For the assay, retinoid-treated cells are rinsed twice with PBS and fixed for 10 min with cold methanol or paraformaldehyde. After two rinses with PBS, a small circle (0.5–1 cm in diameter) is etched in the center of the dish with a scalpel. The area immediately outside the circle is dried with filter paper. Immunostaining is then performed by incubating the cells inside the etched circle with 50 to 70 μl antibody, appropriately diluted. After washing and exposure to fluorescein-labeled antiimmunoglobulin, a drop of glycerol/PBS (9:1, v/v) is placed in the circle and covered with a coverslip for microscopic analysis.

Viral Gene Expression. Another common property shared by EC cells is that they are nonpermissive to expression of viral gene products following infection with papovaviruses and type-C retroviruses.[14] Although the mechanisms by which EC cells restrict viral gene expression are not yet fully understood, it is known that following differentiation the block to virus expression is removed. This attribute has been exploited to evaluate the differentiation status of cultured EC cells. Treated or untreated EC cells are infected at high multiplicity with either polyoma or SV40 virus. At appropriate times (see below), cells are assayed for T-antigen expres-

[13] D. Solter and B. B. Knowles, *Proc. Natl. Acad. Sci. U.S.A.* **75,** 5565 (1978).
[14] S. Astigiano, M. I. Sherman, and P. Abarzua, *Environ. Health Perspect.* **80,** 25 (1989).

sion by indirect immunofluorescence. The procedure is essentially as described for SSEA-1 antigen analysis. Positive cells are identified by strong and specific nuclear (but not nucleolar) staining. The results can be quantitated by analyzing several randomly chosen fields and scoring positive nuclei among a total of, for example, 1000 cells. Maximum expression occurs 3 to 4 days after infection, but significant immunostaining can be detected as early as 24 hr postinfection. Infected mouse 3T3 cells and uninfected EC cells should be used as positive and negative controls, respectively, in these experiments. More than 70% of SV40-infected 3T3 cells should exhibit T-antigen staining. Infected but untreated EC cells should be uniformly negative, and, even in fully differentiated cultures, the fraction of T-antigen-positive cells will be much smaller than that observed for 3T3 fibroblasts.

Differentiation Markers

The fact that EC cells can differentiate into a large variety of cell types implies that a variety of markers can be used to assess and monitor EC cell differentiation. The choice of markers to monitor, therefore, will depend in large part on the cell line being analyzed and the type of inducer.

Plasminogen Activator

Plasminogen activator was one of the first criteria used to monitor differentiation of EC cells.[15] Very low levels of plasminogen activator are produced and secreted by EC cells. Synthesis and secretion of this enzyme are enhanced by as much as two orders of magnitude upon differentiation of EC cells into endodermlike cells, particularly parietal endoderm (e.g., F9 cells treated with RA plus dbcAMP or micromolar levels of RA alone). Even when cells differentiate to other phenotypes there appears to be a transient (within 48 hr) but significant increase in plasminogen activator secretion. For these reasons this differentiation marker is useful for following retinoid-induced differentiation in early stages. Plasminogen activator is also a very sensitive parameter since elevated secretion of this enzyme by only a small percentage of cells in the culture can be readily detected.

A quantitative assay has been used to measure secretion of plasminogen activator by EC cells.[16] This assay measures plasminogen-dependent fibrinolysis of ^{125}I-labeled fibrinogen. Cells are cultured and induced to

[15] M. I. Sherman, S. Strickland, and E. Reich, *Cancer Res.* **36**, 4208 (1976).
[16] S. Strickland, E. Reich, and M. I. Sherman, *Cell (Cambridge, Mass.)* **9**, 231 (1976).

differentiate as described above. Twenty-four hours before a time point is to be taken, medium is removed, cells are rinsed thoroughly with PBS, and S-DMEM medium containing heat-inactivated, plasminogen-depleted serum[17] is added. After incubation, conditioned medium is collected and stored at −20° and cells are counted. For the assay, a mixture of ^{125}I-labeled fibrinogen and unlabeled fibrinogen (30 μg, ~4000 cpm/μg) in a 1 : 11 mixture of PBS/water is added to each well of 24-well plates and allowed to dry at 37°. Dried plates can be stored for several days at room temperature.

Two hours prior to assay, 0.3 ml of a 10% solution of plasminogen-depleted serum in 0.1 M Tris-HCl, pH 7.4, is added to each well. After a 2-hr incubation at 37°, each well is washed twice with 0.1 M Tris-HCl, pH 8.1. For measuring plasminogen activator secretion, 20 μl conditioned medium and 5 μl plasminogen (10 μg) in 225 μl of 0.1 M Tris-HCl, pH 8.1, are added to each well and incubated for 2 hr at 37°. Reaction mixtures are collected, each well is washed with 250 μl Tris buffer, and the wash solution is pooled with the reaction mixture. Samples are then counted in a γ counter to determine the release of [125]I. Samples should be run in triplicate. Two controls should always be included: in one, wells are incubated with 25 μl trypsin–EDTA solution to determine the total number of ^{125}I counts per minute that can be released with protease; a second control consists of medium previously incubated without cells (also for 24 hr at 37°), to determine background levels of soluble [125]I. When characterizing a new cell line, it is also prudent to include control samples lacking plasminogen to ensure that all fibrinolysis is plasmin-dependent. For quantitation, the solubilized counts in the control medium samples are subtracted from experimental values, and the results are normalized for trypsin-degradable ^{125}I-labeled fibrin and expressed on a cell number basis. Plasminogen activator can also be qualitatively determined *in situ* by the fibrin–agar assay as outlined by Sherman *et al.*[15]

Structural Proteins

There are several cell surface and cytoskeletal proteins for which antibodies are available. These proteins can be easily detected by indirect immunostaining. Fibronectin is synthesized by EC cells but is not retained in association with the cell surface. On cell differentiation into parietal endoderm, fibronectin can accumulate as part of the basement membrane.[18] It also becomes surface-associated when fibroblastlike cells

---

[17] L. Ossowski, J. C. Unkeless, A. Tobia, J. P. Quigley, D. B. Rifkin, and E. Reich, *J. Exp. Med.* **137**, 112 (1973).

[18] J. Wolfe, V. Mautner, B. Hogan, and R. Tilly, *Exp. Cell Res.* **118**, 63 (1979).

are formed. Increased synthesis and extracellular matrix deposition of three other basement membrane proteins, namely, collagen IV, laminin, and entactin, can be detected when EC cells differentiate into endo-dermlike cells.[19,20] On the other hand, α-fetoprotein antibodies are useful for staining visceral endodermlike cells.[21] These latter markers might be of more limited utility, depending on patterns of differentiation of the EC lines under study.

Cytoskeletal proteins have also been employed to evaluate EC cell differentiation, although such markers are perhaps more useful as secondary indicators. Keratin intermediate filaments are detected by immunofluorescence when EC cells differentiate into epithelial cells.[22] Monoclonal antibodies TROMA-1 and TROMA-3 react with different intermediate filament components present in gut and fetal liver as well as mesoderm and endoderm-derived epithelia in adult animals. Both fail to react with EC cells, but they detect an ordered intermediate filament network present in differentiated cells: TROMA-1 reacts strongly with both visceral and parietal endoderm cells, whereas TROMA-3 is more specific, reacting only with parietal endoderm.[23] Two other intermediate filament components not expressed in EC cells, Endo A and Endo B, are readily detectable in trophoblast, visceral endoderm, parietal endoderm, and liver cells but not in myoblasts, neuroblastoma cells, or keratinocytes.[24] Differentiation of EC cells into glial-like cells and myoblastlike cells is accompanied by neurofilament and desmin expression, respectively.[25]

mRNA Analyses

cDNA probe technology can be utilized to evaluate EC cell differentiation. These probes can often be used to study differentiation at the molecular level before measurable morphological or biochemical changes occur. Several cDNA clones for genes either up- or downregulated have been reported, some representing structural proteins or enzymes previously known to be regulated, as described above (e.g., collagen IV,

[19] A. Grover and E. D. Adamson, J. Biol. Chem. 260, 12252 (1985).

[20] A. R. Cooper, A. Taylor, and B. L. M. Hogan, Dev. Biol. 99, 510 (1983).

[21] B. L. M. Hogan, A. Taylor, and E. Adamson, Nature (London) 291, 235 (1981).

[22] D. Paulin, H. Jakob, F. Jacob, K. Weber, and M. Osborn, Differentiation 22, 90 (1982).

[23] R. Kemler, P. Brulet, M. T. Schnebeley, J. Gaillard, and F. Jacob, J. Embryol. Exp. Morphol. 64, 45 (1981).

[24] A. Grover, R. Oshima, and E. D. Adamson, J. Cell Biol. 96, 1690 (1983).

[25] E. M. V. Jones-Villeneuve, M. W. McBurney, K. A. Rogers, and V. I. Kalnins, J. Cell Biol. 94, 253 (1982).

laminin, entactin, α-fetoprotein),[26–28] others encoding novel proteins of unknown function.[27,29]

For RNA isolation from EC cells, we have found a slightly modified version of the procedure described by Chomczynski and Sacchi[30] to be particularly useful and convenient. Because of its simplicity, it allows simultaneous processing of a large number of samples, as would be required, for example, in time-course experiments. Embryonal carcinoma cells are plated as described above (preferentially in culture dishes rather than flasks). At the end of the incubation period, cells are washed with PBS and lysed *in situ* by the addition of 2 ml/100-mm dish of guanidine thiocyanate solution (4 M guanidine thiocyanate, 25 mM sodium citrate, pH 7.0, 0.5% sarcosyl, 0.1 M 2-mercaptoethanol). The lysate is then transferred with the help of a rubber policeman either to a siliconized and baked Corex 15-ml glass tube or to a disposable 14-ml polypropylene tube (Falcon 2059).

The RNA is extracted by adding 0.2 ml of 2 M sodium acetate, pH 4.0, 2 ml water-saturated phenol, and 0.4 ml chloroform. After vortexing for 10 sec, the suspension is cooled on ice for 15 min and centrifuged at 8000 rpm (e.g., Sorval SS-34 rotor) for 20 min at 4°. The aqueous phase containing the RNA is transferred to a new tube, mixed with 1 ml ice-cold 2-propanol, and incubated at $-20°$ to precipitate the RNA. Care should be taken to leave behind the interphase, where DNA is present. The RNA is pelleted by centrifugation, resuspended in 0.3 ml guanidine thiocyanate solution, transferred to a 1.5-ml microcentrifuge tube and precipitated again with 1 volume ice-cold 2-propanol or 2 volumes ice-cold ethanol. The RNA pellet is collected by centrifugation for 10 min in a microcentrifuge, washed once with 70% ice-cold ethanol, vacuum-dried, and dissolved in 50–100 μl water or 0.5% sodium dodecyl sulfate. If undissolved material is present at this point, the solution is centrifuged for 5 min in a microcentrifuge to pellet it, and the RNA can be further purified by ethanol precipitation from 0.3 M sodium acetate. Yields from a 100-mm dish vary from 100 to 200 μg of total RNA depending on cell density. The OD_{260}/OD_{280} ratio in 10 mM Tris-HCl, pH 7.5, is normally not less than 2. RNA purified by this procedure is undegraded and stable for many months when stored at $-20°$ and can be used for cDNA synthesis,

[26] S.-Y. Wang and L. Gudas, *Proc. Natl. Acad. Sci. U.S.A.* **80**, 5880 (1983).
[27] S.-Y. Wang, G. J. LaRosa, and L. J. Gudas, *Dev. Biol.* **107**, 75 (1985).
[28] P. R. Young and S. M. Tilghman, *Mol. Cell. Biol.* **4**, 898 (1984).
[29] R. A. Levine, G. J. La Rosa, and L. J. Gudas, *Mol. Cell. Biol.* **4**, 2142 (1984).
[30] P. Chomczynski and N. Sacchi, *Anal. Biochem.* **162**, 156 (1987).

Northern blots, poly(A)$^+$, selection and S1 protection assays (see, e.g., Ref. 31).

[31] J. F. Grippo and M. I. Sherman, this series, Vol. 190 [16].

[36] Parasite Retinoid-Binding Proteins

By Brahma P. Sani

Introduction

The major theory to explain the molecular mechanism of retinoid action in the control of epithelial differentiation and growth centers on the well-known cellular binding proteins for retinol and retinoic acid, CRBP and CRABP, respectively,[1-4] and on the recently discovered nuclear receptors for retinoic acid (RARs).[5-7] The role of retinoids in parasites is largely unknown although the occurrence and uptake of retinol and retinoic acid, as well as the formation of retinol from β-carotene by parasitic helminths, have been described by several investigators. Growth of some developmental stages of helminth parasites may depend on host levels of retinol.[8,9] Sturchler et al.[10] showed that the retinol concentration in adult *Onchocerca volvulus* was 8 times higher than that of the surrounding host tissues. Although the exact function of retinoids in parasites is not clear, it may be assumed, by analogy to host tissues, that they control certain vital biological functions of the parasites, such as differentiation, growth, and reproduction. In our efforts to study the interactions of retinoids with parasitic components, we discovered and partially characterized specific

[1] M. M. Bashor, D. O. Toft, and F. Chytil, *Proc. Natl. Acad. Sci. U.S.A.* **70**, 3483 (1973).
[2] B. P. Sani and D. L. Hill, *Biochem. Biophys. Res. Commun.* **61**, 1276 (1974).
[3] D. E. Ong and F. Chytil, *J. Biol. Chem.* **250**, 6113 (1975).
[4] B. P. Sani and D. L. Hill, *Cancer Res.* **36**, 409 (1976).
[5] M. Petkovich, N. J. Brand, A. Krust, and P. Chambon, *Nature (London)* **330**, 444 (1987).
[6] V. Giguere, E. S. Ong, P. Sequi, and R. M. Evans, *Nature (London)* **330**, 624 (1987).
[7] N. Brand, M. Petkovich, A. Krust, P. Chambon, H. Dethe, A. Marchio, P. Tiollais, and A. Dejean, *Nature (London)* **332**, 850 (1988).
[8] D. Mahalanabis, K. N. Jalan, T. K. Maitra, and S. R. Agarwal *Am. J. Clin. Nutr.* **29**, 1372 (1976).
[9] D. M. Storey, *Z. Parasitenkd.* **67**, 309 (1982).
[10] D. Sturchler, B. Holzer, A. Hanck, and A. Oegremont, *Acta Trop.* **40**, 261 (1983).

METHODS IN ENZYMOLOGY, VOL. 189

parasite retinol-binding protein (PRBP) and parasite retinoic acid-binding protein (PRABP) from several filarial parasites.[11–13] Some of the physicochemical characteristics of the parasite binding proteins are distinct from those of their host-tissue counterparts. In particular, the protein–ligand interactions in host tissues were sensitive to organomercury compounds[14] whereas in parasitic tissues they were insensitive to the mercurials. Another major difference was that ivermectin, a potent and widely used antiparasitic drug, competed efficiently with retinol-binding sites on PRBP, but not with the host CRBP.[13]

Isolation of Parasite Retinoid-Binding Proteins

Lyophilized samples of adult filarial parasites (*Onchocerca volvulus, Onchocerca gibsoni, Dipetalonema viteae, Brugia pahangi,* and *Dirofilaria immitis*) are homogenized in 10 mM sodium phosphate, pH 7.2, containing 0.15 M NaCl (PBS), and centrifuged at 100,000 g for 90 min at 4°. The supernatants are diluted to a protein concentration of 12 mg/ml. Cytosolic extracts from homogenized rat tissues and 12- to 13-day-old chick embryo skin are also prepared.[2,4]

Molecular Size Profiles and Specificity of Parasite Retinoid-Binding Proteins

Soluble extracts (60 mg protein) from *D. immitis* are incubated with either [³H]retinol (3 nmol) or [³H]retinoic acid (2.5 nmol) for 1 hr at room temperature and 3 hr at 4°, then filtered through a column (2.5 × 10 cm) of Sephadex G-100 equilibrated with PBS. Both protein and radioactivity distributions of the elution profile are determined. Samples (5 mg each) of bovine serum albumin (M_r 68,000), ovalbumin (44,000), β-lactoglobulin (36,000), myoglobin (17,800), and cytochrome c (12,400) are run through the same column as standards of M_r determination. Figure 1 presents the chromatographic profiles of radioactivity distribution and absorbance at 280 nm. The second radioactive peak from the left, which represents the PRBP–[³H]retinol complex, corresponds to M_r 19,000. In a separate run on a Sephadex G-100 column under similar conditions, the [³H]retinoic acid–PRABP complex is also retarded to a position corresponding to M_r 19,000.

[11] B. P. Sani and J. C. W. Comley, *Tropenmed. Parasitol.* **36,** 20 (1985).
[12] B. P. Sani, A. Vaid, J. C. W. Comley, and J. A. Montgomery, *Biochem. J.* **232,** 577 (1985).
[13] B. P. Sani and A. Vaid, *Biochem. J.* **246,** 929 (1988).
[14] B. P. Sani, S. M. Condon, and C. K. Banerjee, *Biochim. Biophys. Acta* **624,** 226 (1980).

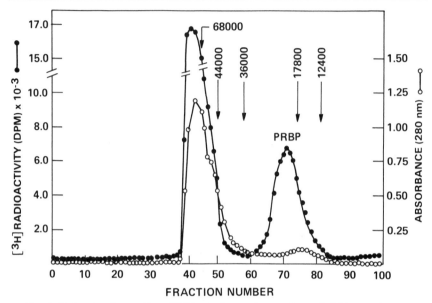

FIG. 1. Molecular size profiles of [³H]retinol–protein complexes. Soluble protein extracts from *D. immitis* (60 mg protein) after incubation with [³H]retinol (2.5 nmol) were passed through a Sephadex G-100 column (2.5 × 100 cm) equilibrated with PBS, and fractions (5.5 ml each) were collected. Protein standards, bovine serum albumin (M_r 68,000), ovalbumin (44,000), β-lactoglobulin (36,000), myoglobin (17,800), and cytochrome *c* (12,400), were run separately through the same column under identical conditions.

The fractions corresponding to [³H]retinol–PRBP and [³H]retinoic acid–PRABP from the above gel filtration step are concentrated by lyophilization to 1 mg protein/ml. Portions (0.5 mg protein) of the protein–[³H]retinoid complexes are centrifuged at 180,000 *g* for 18 hr at 4° through 5–20% (w/v) sucrose gradients with bovine serum albumin and ovalbumin as external standards.[4] In the radioactivity profiles obtained after fractionation of the sucrose gradients, the [³H]retinol–PRBP and [³H]retinoic acid–PRABP peaks correspond to a calculated $s_{20,w}$ value of 2.0 S. Similar 2 S parasite binding protein peaks may also be observed after incubation of the parasite extracts (2 mg protein) with either [³H]retinol (300 pmol) or [³H]retinoic acid (300 pmol) followed by sucrose density gradient sedimentation. The sedimentation values of 2 S for PRBP and PRABP are similar to those reported to CRBP and CRABP from host tissues.[1-4] The M_r values of 19,000 for both PRBP and PRABP, however, are higher than the reported M_r values of 14,000–17,800 for CRBP and CRABP as determined under similar conditions. Ligand specificity of both PRBP and PRABP is assessed with sucrose density gradient experiments in which

TABLE I
Parasite Retinol- and Retinoic
Acid-Binding Proteins[a]

| | [³H]Retinoid bound (pmol/mg protein) | |
Parasite	Retinol (PRBP)	Retinoic acid (PRABP)
Dirofilaria immitis	3.1	5.2
Dipetalonema viteae	2.8	5.4
Brugia pahangi	2.7	4.9
Onchocerca volvulus	3.0	5.6
Onchocerca gibsoni	3.1	5.8

[a] The binding proteins were examined by sucrose density gradient analysis; [³H]retinol and [³H]retinoic acid specifically bound to their respective binding protein peaks were measured.

the radioligand binding is challenged with 200-fold molar excesses of unlabeled retinol and retinoic acid. The radioactivity profiles of the sucrose gradients reveal that unlabeled retinol almost completely eliminates the radioactive 2 S PRBP peak, whereas a similar excess of retinoic acid does not cause any inhibition of binding of [³H]retinol. Likewise, in the case of radioactive PRABP peak, the unlabeled retinol shows virtually no competition.

A comparative study on the quantitative amounts of PRBP and PRABP in filarial parasites produced the results presented in Table I. Levels of ligand bound to the extractable parasite proteins (amounts/mg) were assumed to be directly related to the amounts of the binding protein. All of the five parasites analyzed contained between 2.7 and 3.1 pmol of retinoic acid bound/mg of protein. These levels were similar to the CRABP levels in mammalian and avian tissues.[11,12] The levels of PRBP in these parasites, however, ranged between 4.8 and 5.8 pmol/mg (Table I), which is generally much higher than the CRBP levels in rat and chick embryo tissues. Such an increase in the PRBP levels in parasites is in agreement with the observation by Sturchler et al.[10] that O. volvulus contains higher amounts of retinol as compared to host tissues.

Mercurial Sensitivity of Retinoid Binding to Binding Proteins

We have earlier demonstrated that retinol and retinoic acid binding to CRBP and CRABP are sensitive to mercurials and that the inhibition is

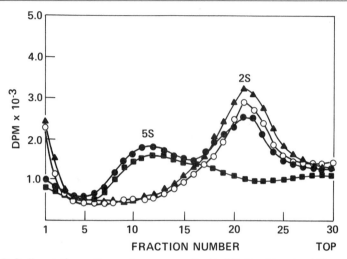

FIG. 2. Sedimentation patterns of sucrose gradients of *D. immitis* extract (2 mg protein) and rat testis extract (2 mg protein) plus [³H]retinoic acid (300 pmol) in the presence or absence of CMPS. ● rat testis plus [³H]retinoic acid; ■ rat testis plus [³H]retinoic acid and 1 mM CMPS; ○, *D. immitis* plus [³H]retinoic acid; ▲, *D. immitis* plus [³H]retinoic acid and 2 mM CMPS.

reversible in the presence of dithiothreitol.[4,14] We examined whether such mercurial-sensitive binding sites are also present in PRBP and PRABP. Figure 2 illustrates the effect of 1–2 mM *p*-chloromercuriphenylsulfonic acid (CMPS) on the binding of PRABP and CRABP. The binding of retinoic acid to testis CRABP was completely abolished in the presence of 1 mM CMPS. The PRABP–ligand interactions in *D. immitis* preparations, however, were totally insensitive up to 2 mM CMPS. In fact, 2 mM CMPS slightly stimulated the binding of [³H]retinoic acid to PRABP. We also tested four other parasites, *O. volvulus, O. gibsoni, D. viteae,* and *B. pahangi,* for mercurial sensitivity; none of them were sensitive. In a similar manner, a systematic examination of the sensitivity of the [³H]retinol–PRBP interaction to CMPS revealed that, in contrast to avian and mammalian CRBP, the parasitic binding protein expressed a complete lack of mercurial sensitivity. These observations establish that fundamental differences exist between the host and parasites with regard to their mode of ligand–protein interactions.

Retinol–Ivermectin Cross-Competition for Parasite Retinol-Binding
Protein Binding

Because retinoids may be involved in the control of vital biological functions of the parasites, and because PRBP and PRABP could be the

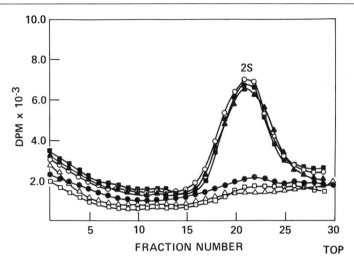

FIG. 3. Sucrose density gradient sedimentation profiles of *D. immitis* extract (1.5 mg protein) and rat testis cytosol (2.0 mg protein) plus [³H]retinol (300 pmol) in the presence or absence of 200-fold molar excesses of retinol or ivermectin. ○, *D. immitis* plus [³H]retinol; ●, *D. immitis* plus [³H]retinol and retinol; △, *D. immitis* plus [³H]retinol and ivermectin; ■, rat testis plus [³H]retinol; □, rat testis plus [³H]retinol and retinol; ▲, rat testis plus [³H]retinol and ivermectin.

mediators of such functions, it was important to examine whether any of the known antiparasitic drugs exert their action by nullifying the functional role of the binding proteins through their interactions with them. The avermectins are a family of natural compounds that are extremely active against nematode and arthropod parasites.[15] Several members of the avermectin family possess different degrees of antiparasite activity, with ivermectin (22,23-dihydroavermectin B_{1a}) being the most potent.[15,16]

We examined the binding affinities of avermectins for PRBP and PRABP, as distinct from their binding affinities for CRBP and CRABP. Figure 3 illustrates the sucrose density gradient sedimentation profiles of [³H]retinol in the presence or absence of 200-fold molar excesses of retinol or ivermectin after incubation with soluble protein extracts from *D. immitis* and rat testis. Ivermectin showed 100% inhibition of [³H]retinol binding to PRBP, whereas the drug was without any effect on the binding of [³H]retinol to CRBP from rat testes. Retinol itself showed 100% inhibition of binding of [³H]retinol to either PRBP or CRBP. The results thus

[15] M. H. Fisher and M. Mrozik, *in* "Macrolide Antibiotics" (S. Omura, ed.), p. 553. Academic Press, New York, 1984.

[16] W. C. Campbell, M. H. Fisher, E. O. Stapley, G. Albers-Schonberg, and T. A. Jacob, *Science* **221**, 823 (1983).

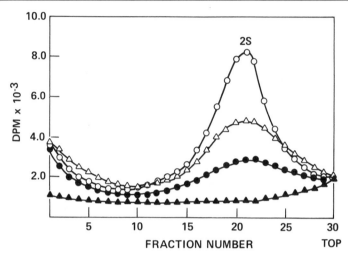

FIG. 4. Sedimentation patterns on sucrose gradients of *D. immitis* extract (1.5 mg protein) and rat testis cytosol (2 mg protein) plus [³H]ivermectin (200 pmol) in the presence or absence of 200-fold molar excesses of retinol or ivermectin. ○, *D. immitis* plus [³H]ivermectin; ●, *D. immitis* plus [³H]ivermectin and ivermectin; △, *D. immitis* plus [³H]ivermectin and retinol; ▲, rat testis plus [³H]ivermectin.

clearly indicate that ivermectin distinguishes between retinol-binding sites on the parasite and the host-tissue binding protein. Ivermectin also showed 100% inhibition of [³H]retinol binding to PRBP from *O. volvulus*, *O. gibsoni,* and *B. pahangi*. The drug, however, failed to show any significant inhibition of retinol binding to CRBP from any of the rat tissues, such as kidney, liver, eye, small intestine, colon, and testes. Although ivermectin abolished [³H]retinol binding in filarial parasites, the drug was virtually without competition with [³H]retinoic acid for the binding sites on PRABP or CRABP. This indicates that ivermectin may not interfere with any possible parasitic functions involving the retinoic acid–PRABP complex.

Although ivermectin inhibits the binding of [³H]retinol to PRBP, we had no evidence to show that the retinol-binding sites were in fact occupied by ivermectin. Figure 4 represents the sucrose gradient sedimentation pattern of *D. immitis* and rat testis extracts after incubation with [22,23-³H]ivermectin (10 Ci/mmol) in the presence or absence of 200-fold molar excesses of the test compounds. *Dirofilaria immitis* extracts display a 2 S radioactive peak attributable to ivermectin-binding protein. Competition with unlabeled ivermectin abolishes this peak almost completely, indicating the specific nature of binding. Competition with a similar molar excess of retinol, however, diminished the peak only to about 50%. This

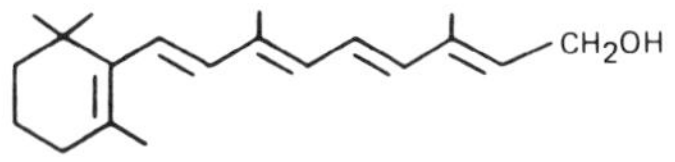

all-trans-Retinol

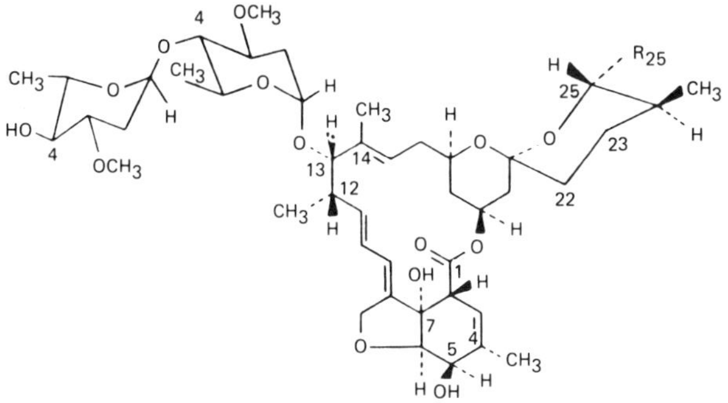

22, 23 — Dihydroavermectin B₁ (Ivermectin)

a series: $R_{25} = CH(CH_3)C_2H_5$
b series: $R_{25} = CH(CH_3)_2$

FIG. 5. Chemical structures of retinol and ivermectin.

suggests that ivermectin has a higher affinity than retinol for the retinol-binding sites on PRBP, and it may explain why ivermectin displays anti-parasitic activity at extremely low concentrations. An important observation was that rat testis cytosol did not contain an [³H]ivermectin-binding protein (Fig. 4), indicating that ivermectin has no affinity for either CRBP/CRABP or any other proteins in host tissues.

Our data indicate that both retinol and ivermectin bind to the 2 S PRBP extracted from the filarial parasites and that there exists a cross-competition of the drug for the retinol binding site and of retinol for the ivermectin binding site. A similar cross-competition is absent in the host tissues. Since ivermectin and retinol have dissimilar chemical structures (Fig. 5), the competition between the two compounds for the PRBP binding site is intriguing. A common hydrophobic moiety in their structures may be responsible for the recognition of the binding site in PRBP, but it does not recognize such a site in CRBP. As parasites apparently lack visual function, the retinol concentrated in them may be utilized in the control of vitamin A functions other than vision. We do not yet know whether ivermectin blocks any such functions, nor do we know that pharmacological levels of ivermectin are operative by such a mechanism. In susceptible parasites, ivermectin acts as a potent stimulant of γ-aminobutyric acid (GABA)-mediated Cl^- conductance and interferes with GABA-mediated signal transmission from interneurons to excitatory motor neurons with the result of paralysis as well as suppression of reproductive processes.[15–17] Although retinol has not been implicated in GABA-mediated functions, it is involved in reproduction. Thus, the cross-competition between retinol and ivermectin for PRBP binding sites may be important in the understanding of the mechanism of action of vitamin A in parasites and of avermectins in antifilarial chemotherapy.

Acknowledgments

This work was supported by U.S. Public Health Service Grant CA 34968 and World Health Organization Grant 83001.

[17] I. S. Kass, C. C. Wang, J. P. Walrond, and A. O. W. Stretton, *Proc. Natl. Acad. Sci. U.S.A.* **71,** 6211 (1980).

[37] Purification of Cellular Retinoic Acid-Binding Proteins Types I and II from Neonatal Rat Pups

By JOHN STUART BAILEY and CHI-HUNG SIU

Introduction

Retinoic acid can have a profound influence on the state of differentiation of many different cell types. Different concentrations of retinoic acid can elicit distinctly different effects.[1,2] This raises the possibility that specific cellular effects of retinoic acid may be mediated by distinct proteins which have different affinities for retinoic acid. Neonatal rat tissue has been described as a rich source of retinoid-binding proteins.[3] In addition to the major species of cellular retinoic acid-binding protein (CRABP), termed CRABP I hereafter, we have discovered a novel cellular binding protein for retinoic acid, termed CRABP II, in neonatal rats.[4] CRABP II shares many similarities with the more abundant CRABP I.

[1] M. K. Edwards, J. F. Harris, and M. W. McBurney, *Mol. Cell. Biol.* **3,** 2280 (1983).
[2] A. M. Jetten, *in* "Modulation and Mediation of Cancer by Vitamins" (F. L. Meyskens and K. N. Prasad, eds.), p. 177. Karger, Basel, 1983.
[3] D. E. Ong, *J. Biol. Chem.* **259,** 1476 (1984).
[4] J. S. Bailey and C-H. Siu, *J. Biol. Chem.* **263,** 9326 (1988).

However, it has a lower affinity for retinoic acid. The observation that CRABP I and CRABP II have different affinities for the ligand retinoic acid suggests they may mediate unique aspects of retinoic acid action. The purification of CRABP I and CRABP II from neonatal rat pups is quite similar. For this reason the purification of CRABP I is discussed concurrently with the purification of CRABP II. Differences in the protocols are indicated. In addition, this procedure can also be used for the purification of cellular retinol-binding protein (CRBP I) from rat pups and CRABP I and II from mouse pups.[5]

Purification Procedure

Step 1: Preparation of Tissue Extract. The initial steps of the purification protocol are modified from the purification procedure for CRABP from rat testis.[6] Unless otherwise stated, all procedures with the exception of chromatography are performed at 4°. Approximately 100 neonatal rats (700–800 g) are homogenized in 1.6 liters of 20 mM Tris HCl, pH 7.2, containing 0.25 mg/ml of phenylmethylsulfonyl fluoride. Homogenization is carried out in a Waring blendor at 15-sec intervals for a total of 2 min. Insoluble debris is removed by centrifugation at 14,000 g for 15 min. The supernatant is collected and titrated to pH 5.1 with glacial acetic acid, and the sample is again centrifuged at 14,000 g for 15 min to remove precipitated material. The supernatant is collected, and 1/10 volume of CM-BioGel (Bio-Rad, Richmond, CA) equilibrated with 10 mM sodium acetate, pH 5.1, is added to the extract. After gentle mixing for 30 min, the CM-BioGel is removed by sedimentation. The supernatant is collected and titrated to pH 7.5 with 10 N NaOH. The volume of the extract at this stage is generally about 2 liters. The sample can be concentrated to a total volume of 100 ml in approximately 2 hr using a hollow fiber concentration cell (A/G Technology, Needham, MA). The amount of total protein recovered after this step is approximately 14.5 g.

Step 2: Gel Filtration on Sephadex G-75. In order to facilitate the detection of CRABP I and CRABP II, [^{3}H]retinoic acid (48–50 Ci/mmol, New England Nuclear, Boston, MA) is added to the extract prior to gel filtration. Generally, a single addition of 1 μCi of retinoic acid is sufficient to detect CRABP I and CRABP II in all subsequent chromatographic steps. The concentrated extract (100 ml) is loaded onto a Sephadex G-75 column (5.0 × 90 cm) equilibrated with 50 mM Tris-HCl, pH 7.5, containing 0.2 M NaCl. The column is eluted at a flow rate of 2 ml/min with the

[5] J. S. Bailey and C-H. Siu, *Biochem. Cell Biol.* **66,** 750 (1988).
[6] D. E. Ong and F. Chytil, *J. Biol. Chem.* **253,** 4551 (1978).

same buffer. Under these conditions, three distinct [³H]retinoic acid peaks elute from the column. The first peak eluted with the void volume is due to nonspecific binding. The second peak centering at an approximate volume of 1100 ml contains CRABP I and CRABP II as well as CRBP I and CRBP II. The third ³H peak elutes with the included volume of the column and is likely due to unbound [³H]retinoic acid. Fractions containing CRABP I and CRABP II are collected and concentrated to a final volume of 50 ml by ultrafiltration. The sample can be safely stored at −20° at this stage or immediately processed for further purification.

Step 3: Anion-Exchange Chromatography on DEAE-Cellulose. The concentrated sample obtained from the Sephadex G-75 column is dialyzed extensively against 10 mM Tris–acetate buffer, pH 8.3. After exhaustive dialysis, the sample is loaded onto a DEAE-cellulose column (2.5 × 20 cm, Whatman DE-52) equilibrated with 10 mM Tris–acetate, pH 8.3, at a flow rate of 1 ml/min. The column is washed sequentially with 50 ml each of 10 mM Tris–acetate, 50 mM Tris–acetate, and 100 mM Tris–acetate buffers, all at pH 8.3. The protein which remains bound to the column is eluted at 1 ml/min with a 400-ml linear gradient from 0.1 to 0.3 M Tris–acetate, pH 8.0, and 5-ml fractions are collected. At this step, CRABP I and CRABP II are effectively separated. Typical results for the elution of CRABP I and CRABP II from DEAE-cellulose are shown in Fig. 1. CRABP II elutes as the first radioactive peak, which is usually centered at fraction 21 after the beginning of the gradient elution. The second radioactive peak is CRABP I and is usually centered at fraction 38 of the gradient profile. The size of the CRABP II radioactive peak generally represents 5% of that in the CRABP I peak. The two radioactive peaks are pooled separately.

Step 4: Chromatofocusing Step (for Purification of CRABP I). The major contaminating protein present in the CRABP I sample after the DEAE ion-exchange chromatography is a protein doublet which migrates with an apparent molecular weight of 22,000 in sodium dodecyl sulfate (SDS)–polyacrylamide gels. The majority of this contaminant is removed after the chromatofocusing step. Samples containing CRABP I obtained from the DEAE-cellulose step are dialyzed against 20 mM histidine-HCl, pH 5.8. The sample is then loaded onto a chromatofocusing column (1.2 × 16 cm; PBE 9-4 resin obtained from Pharmacia, Piscataway, NJ) equilibrated with 20 mM histidine-HCl, pH 5.8. The column is eluted at a flow rate of 0.25 ml/min with 160 ml of Polybuffer 7-4, pH 4.0, (Pharmacia). Under these conditions, a linear pH gradient is generated between pH 5.8 and 4.2. CRABP I elutes as a single radioactivity peak at a pH of approximately 4.75. The major protein peak elutes between pH 4.9 and 5.1.

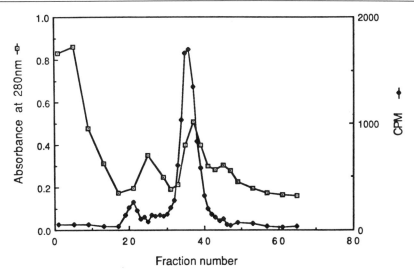

FIG. 1. Separation of CRABP I and CRABP II by anion-exchange chromatography on DEAE-cellulose. The fractions containing CRABP I and CRABP II obtained from the Sephadex G-75 column were pooled and concentrated prior to dialysis with 10 mM Tris–acetate, pH 8.0. The column was loaded and washed as described in the text. CRABP I and CRABP II were eluted with a linear gradient from 0.1 to 0.3 M Tris–acetate, pH 8.0. Fractions (5 ml each) were collected, and 100 μl was used for the determination of radioactivity. CRABP I eluted as the major radioactivity peak centered at fraction 38, whereas CRABP II eluted as a minor radioactivity peak centered at position 21. [From J. S. Bailey and C.-H. Siu, *J. Biol. Chem.* **263**, 9327 (1988).]

Step 5: Anion-Exchange High-Performance Liquid Chromatography. The conditions for the purification of CRABP I and CRABP II by HPLC are identical. The elution positions for both proteins can be determined by monitoring the absorbance of the column eluant at 350 nm, which detects the bound retinoid in the protein–ligand complex. Although CRABP I has sufficient bound retinoic acid, it is necessary to add an excess of unlabeled retinoic acid to CRABP II prior to the HPLC steps. Unlabeled retinoic acid should be added to the sample to a final concentration of 1 μM prior to dialysis. Samples containing either CRABP I or CRABP II are dialyzed exhaustively against 20 mM Tris-HCl, pH 7.2. An aliquot of approximately 1 mg of protein is injected onto a TSK DEAE 5PW column (7.5 × 75 mm) equilibrated with 20 mM Tris-HCl, pH 7.2 (Buffer A). The column is eluted at a flow rate of 1 ml/min with 0.5 M NaCl in 20 mM Tris-HCl, pH 7.2 (buffer B), using the following gradient profile: 0–5 min, 0% B; 5–20 min, linear to 50% B; 20–25 min, 50% B; 25–30 min, linear to 0% B. Under these conditions, CRABP II elutes as a single symmetrical peak with a retention time of 14.2 min, whereas CRABP I elutes with a retention time of 14.7 min.

Samples containing CRABP I or CRABP II are collected and diluted with an equal volume of HPLC-grade water. An aliquot of up to 1 mg protein is injected onto the same DEAE 5PW column equilibrated with 20 mM sodium acetate, pH 5.8. The column is eluted at a flow rate of 1 ml/min with 0.5 M sodium acetate, pH 5.8 (buffer B), and the gradient profile outlined above. Under these conditions, CRABP II elutes with a retention time of 14.5 min, whereas CRABP I elutes with a retention time of 14.2 min. Typical results for the HPLC purification of CRABP II are shown in Fig. 2. Homogeneity is usually obtained for both proteins after this step. Generally the A_{350}/A_{280} ratio for both CRABP I and CRABP II is 1.4 after this step. The purity of CRABP I and CRABP II can be assessed by SDS–PAGE[7] in a 15% polyacrylamide gel followed by silver staining.[8] CRABP II and CRABP I run with apparent M_r values of 15,000 and 14,600, respectively.

Ligand-Binding Assay for Simultaneous Quantification of CRABP I and CRABP II

Chromatographic analysis on a TSK DEAE 5PW column can be used to quantify the levels of CRABP I and CRABP II in crude tissue extracts. Because accurate quantification of both proteins is dependent on the saturable binding with [3H]retinoic acid, endogenous retinoic acid is generally first destroyed by exposing the tissue extracts to long-wavelength ultraviolet light for 4 hr prior to the assay procedure.[9]

Step 1: Preparation of Tissue Extract. Tissue samples are homogenized in a Dounce homogenizer in 2–3 volumes of 10 mM Tris–acetate, pH 8.0. Particulate material is removed by centrifugation at 14,000 g for 15 min. The tissue samples are then titrated to pH 5.1 with 1 M acetic acid, and the precipitated material is removed by centrifugation at 14,000 g for 15 min. The supernatant is then titrated to a pH between 7.5 and 8.0 with 1 M NaOH to obtain a cell-free extract.

Step 2: Ligand Binding and Chromatographic Analysis. The tissue samples are adjusted with 10 mM Tris–acetate, pH 8.0, to a total protein concentration of less than 5 mg/ml. One microcurie of [3H]retinoic acid (48 Ci/mmol, New England Nuclear) is added to 200 μl of tissue extract. To control samples, a 250-fold molar excess of unlabeled retinoic acid is

[7] V. K. Laemmli, *Nature* (*London*) **227**, 680 (1970).
[8] J. H. Morrissey, *Anal. Biochem.* **117**, 307 (1981).
[9] D. Ong and F. Chytil, this series, Vol. 67, p. 288.

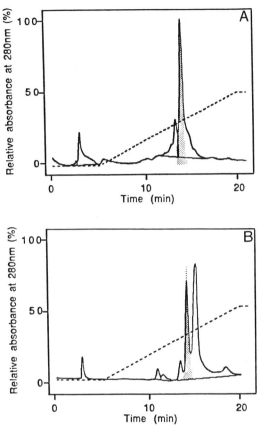

FIG. 2. Purification of CRABP II by anion-exchange HPLC. (A) Aliquots of 1 mg protein containing CRABP II were injected on a TSK DEAE 5PW column equilibrated with 20 mM Tris-HCl, pH 7.2. The column was eluted with the gradient profile shown (buffer B: 0.5 M NaCl in 20 mM Tris-HCl, pH 7.2). (B) Samples collected from the first HPLC step were diluted with an equal volume of water and injected onto the same DEAE 5PW column equilibrated with 20 mM sodium acetate, pH 5.8. The column was eluted with the gradient profile shown (buffer B: 0.5 M sodium acetate, pH 5.8). The flow rate was 1 ml/min, and the column eluate was monitored for absorbance at either 280 or 350 nm (shaded). [From J. S. Bailey and C.-H. Siu, *J. Biol. Chem.* **263**, 9328 (1988).]

added just prior to the addition of labeled ligand. Incubation is carried out overnight at 4° in the dark. After incubation, 50 μl of a charcoal–dextran suspension[10] is added to each sample to remove unbound ligand. After a 15-min incubation with gentle mixing, the charcoal–dextran is removed by centrifugation at 13,000 g for 5 min. A sample of 200 μl is injected onto

[10] C. A. Strott, this series, Vol. 36, p. 34.

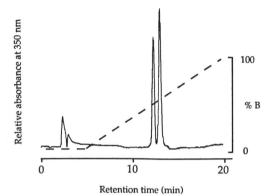

FIG. 3. Separation of CRABP I and CRABP II by anion-exchange HPLC. CRABP I and CRABP II (40 μg each) were injected onto a TSK DEAE 5PW column equilibrated with 20 mM Tris–acetate, pH 8.0. The column was eluted at 1 ml/min with a linear gradient to 0.5 M Tris–acetate, pH 8.0, using the gradient profile shown. The eluate was monitored for absorbance at 350 nm.

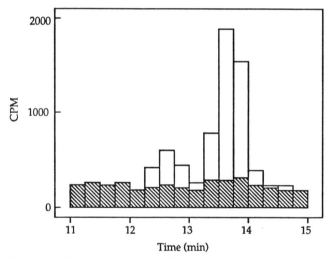

FIG. 4. Separation of two distinct retinoic acid-binding activities in a neonatal rat extract. A neonatal rat extract (500 μg) incubated with [³H]RA in the presence (hatched bars) or absence (open bars) of a 250-fold excess of unlabeled retinoic acid was injected onto a DEAE 5PW column and eluted as described in the legend to Fig. 3. Fractions (0.25 ml) were collected beginning at 11 min for determination of radioactivity.

a DEAE 5PW column equilibrated with 20 mM Tris–acetate, pH 8.0 (buffer A). The column is eluted with 0.5 M Tris–acetate, pH 8.0 (buffer B) using the following gradient profile: 0–5 min, 0% B; 5–20 min, linear to 100% B; 20–25 min, 100% B; 25–30 min, linear to 0% B. Under these conditions, purified CRABP II and CRABP I elute with retention times of 12.6 and 13.8 min, respectively (Fig. 3). For tissue extracts, 0.25-ml samples are collected at 15-sec intervals beginning at time 11 min. Two distinct [3H]retinoic acid peaks elute at 12.75 and 13.75 min, respectively. Neither peak is present in samples which have been incubated with an excess of the unlabeled ligand. The amount of CRABP I and CRABP II present in a sample is determined by the amount of specific [3H]retinoic acid recovered minus the amount recovered in the same fraction of the control sample. Figure 4 shows the HPLC separation of CRABP I and CRABP II present in a neonatal rat extract.

Comments. One common problem with this assay is loss of adequate resolution between CRABP I and CRABP II. This can occur for one of two reasons. First, injecting an amount in excess of 1 mg of protein onto the column can lead to tailing of protein peaks and loss of resolution. Second, if the column has deteriorated owing to excessive accumulation of adsorbed material, a loss of resolution can occur. This is most noticeable after prolonged column use. We have found that daily washing of the column with 0.2 M NaOH followed by 20% methanol is a suitable precaution against rapid accumulation of adsorbed material and subsequent column deterioration.

[38] Purification of Adipocyte Lipid-Binding Protein from Human and Murine Cells

By Valerie Matarese, Melissa K. Buelt, Laurie L. Chinander, and David A. Bernlohr

Introduction

The adipocyte lipid-binding protein (ALBP) is a cytosolic polypeptide of approximately 15 kDa expressed in adipose tissue and cultured adipogenic cell lines.[1,2] The protein has been purified from human adipose

[1] D. A. Bernlohr, T. L. Doering, T. J. Kelly, Jr., and M. D. Lane, *Biochem. Biophys. Res. Commun.* **132**, 850 (1985).
[2] K. M. Zezulak and H. Green, *Mol. Cell. Biol.* **5**, 419 (1985).

tissue[3] and murine 3T3-LI adipocytes.[4] ALBP is one member of a multigene family of proteins which have affinity for fatty acids, retinoic acid, or retinol. Both human and murine ALBP bind long-chain fatty acids and retinoic acid but have no measurable affinity for retinol. ALBP may be important for both fatty acid and retinoic acid utilization in the adipocyte. In particular, this protein may facilitate the cellular utilization of retinoic acid or fatty acids by delivering the ligand obtained at the plasma membrane to the appropriate microsomal esterification system.

Murine ALBP has been shown to be a substrate, both *in situ*[5] and *in vitro*,[6] for the insulin receptor tyrosine kinase. The possibility that covalent phosphorylation of ALBP may affect ligand trafficking has heightened interest in defining the role this protein plays in retinoic acid or fatty acid metabolism. Hence, purification of large amounts of homogeneous ALBP is essential.

Two primary experimental procedures allow for the preparation of ALBP from either human or murine adipose cells. ALBP possesses two reduced sulfhydryl groups,[3,4] one at the amino terminus and one buried within the ligand-binding domain.[7] The readily accessible amino-terminal sulfhydryl group makes chromatography on activated-thiol Sepharose 4B an effective purification step. More importantly, isolation of ALBP necessitates the handling of large volumes of crude extracts containing dissolved lipid in sufficient quantity to make conventional chromatographic procedures unfeasible. Aqueously dispersed lipid is easily removed by chromatography on Lipidex 1000 resin. The removal of endogenous lipid is absolutely required in order to purify the protein and to perform *in vitro* binding assays.

Human Adipocyte Lipid-Binding Protein

The procedure described here details the isolation of ALBP from human subcutaneous adipose tissue. Adipose tissue is obtained during routine cosmetic surgery (abdominoplasty, "tummy tuck") and consists of intact cells. Frozen adipose tissue, typically 100 g, is freed of connective tissue, minced into buffer A [10 mM $NaPO_4$, pH 7.0, 20 mM NaCl, 1 mM EDTA, 0.1 mM phenylmethylsulfonyl fluoride (PMSF); 4 ml/g tissue] and

[3] C. Baxa, R. S. Sha, M. K. Buelt, V. Matarese, A. Smith, L. L. Chinander, K. L. Boundy, and D. A. Bernlohr, *Biochemistry* **28**, 8683 (1989).

[4] V. Matarese and D. Bernlohr, *J. Biol. Chem.* **263**, 14544 (1988).

[5] R. C. Hresko, M. Bernier, R. D. Hoffman, J. R. Flores-Riveros, K. Liao, D. M. Laird, and M. D. Lane, *Proc. Natl. Acad. Sci. U.S.A.* **85**, 8835 (1988).

[6] L. L. Chinander and D. A. Bernlohr, *J. Biol. Chem.* **264**, 19564 (1989).

[7] M. K. Buelt and D. A. Bernlohr, *Biochemistry* in press (1990).

homogenized at 4° with six 30-sec bursts using a Polytron homogenizer. Following centrifugation of the extract (5000 g, 20 min) to remove cellular debris, the solubilized proteins are collected by careful removal of the aqueous layer under the fat cake and further clarified by filtration through 6–8 layers of cheesecloth. The bulk of the dissolved lipid is then removed by passage of the extract at room temperature through Lipidex 1000 (Packard) equilibrated in Buffer A. The resin (30–50 ml) is packed into a 60-ml disposable syringe, and the extract is allowed to pass slowly through the column. The progress of the delipidation can be readily monitored by the change in resin color from white to yellow. In this fashion a reasonably clear, slightly colored extract is obtained. The resin can be regenerated with successive washes of water, methanol, butanol, methanol, and finally water without significant loss of lipid binding capacity.[3,4]

The protein extract is titrated to pH 5.0 with glacial acetic acid, incubated 4–12 hr at 4°, and an acid-soluble fraction obtained by centrifugation at 7000 g for 60 min. The acid-soluble proteins are concentrated by ultrafiltration (Amicon YM5 Danvers, MA), centrifuged to remove residual debris, and applied at room temperature to a 3 × 40 cm Sephadex G-100 column equilibrated in buffer B (50 mM sodium acetate, pH 4.85). The low molecular weight proteins (as determined by electrophoresis analysis) are pooled and exhaustively delipidated in batch by incubation with Lipidex resin at 37° for 3 to 4 hr. The low molecular weight protein–Lipidex slurry is filtered through a 12-ml syringe fitted with siliconized glass wool, and a clear, colorless protein fraction is obtained. The sample is concentrated by ultrafiltration as before and titrated to pH 8.0 with 1.0 M Tris base. One to two milliliters of activated-thiol Sepharose 4B (Pharmacia, Piscataway, NJ) in degassed buffer C (10 mM Tris-HCl, pH 8.0, 100 mM NaCl, 1 mM EDTA) is added to the sample and incubated with gentle agitation at 4° for 48 to 96 hr. The reaction is transferred into a 12-ml syringe fitted with Whatman No. 1 paper, washed with excess buffer C, and covalently bound proteins are eluted by the inclusion of 25 mM dithiothreitol (DTT) in buffer C. The eluted proteins are dialyzed once against buffer C containing 1 mM DTT and several times against buffer C. To remove all traces of contaminating proteins, the preparation is passed at 25° through a 2-ml column of DEAE-Sephadex equilibrated in buffer C lacking EDTA. Human ALBP passes through this resin whereas all contaminating proteins bind tightly. Homogeneous human ALBP is stored in buffer C at 4° for several weeks without any observed loss in ligand binding capacity.

The purity of the preparations is evaluated by denaturing electrophoresis, amino acid analysis, and direct chemical sequencing. The amino terminus of human ALBP is blocked to sequencing, presumably via an

N^α-acetyl group as is murine ALBP (V. Matarese and D. A. Bernlohr, unpublished observations), so amino acid composition is frequently the method of choice for analysis.

Murine Adipocyte Lipid-Binding Protein

The purification protocol for murine ALBP isolated from cultured 3T3-L1 adipocytes is essentially that described by Matarese and Bernlohr.[4] Monolayers of cultured adipocytes are incubated at room temperature for 15 min with buffer A containing 50 μg/ml digitonin to permeabilize the plasma membrane. Cytosolic proteins released from the monolayers are recovered and clarified of cellular debris by centrifugation at 16,000 g for 30 min at 4°. The extract is dialyzed against buffer B, and an acid-soluble extract is obtained by centrifugation as before. The sample is delipidated by chromatography through Lipidex 1000, concentrated by ultrafiltration (Amicon YM5), and applied to a 5 × 90 cm column of Sephadex G-75 equilibrated in buffer B (all steps are performed at room temperature). Murine ALBP is identified in the low molecular weight fraction of the eluate by denaturing electrophoresis, concentrated by ultrafiltration, and delipidated at 37° as described earlier for human ALBP. The delipidated extract is applied at room temperature to a CM-Sephadex column equilibrated in buffer B. The column is washed with buffer B, and murine ALBP is eluted by the inclusion of 0.5 M NaCl in buffer B. The CM-Sephadex eluate is titrated to pH 8.0 with 1.0 M Tris base and incubated at 4° with activated-thiol Sepharose 4B at approximately a 10-fold molar excess of resin sulfhydryls to protein sulfhydryls. After 48–96 hr the protein–resin slurry is poured into a column and washed with buffer C, then homogeneous murine ALBP is eluted with 25 mM DTT in buffer C. The purified protein is dialyzed once against buffer C containing 1 mM DTT and exhaustively against buffer C alone.

Retinoid Binding by Adipocyte Lipid-Binding Protein

Two assays have been utilized to assess the retinoid binding capacity of various ALBP preparations. It must be stressed that binding assays yield equivocal results if protein samples are used prior to delipidation at 37°. Lipids, either endogenous or fortuitous by virtue of making an extract from fat cells, contaminate the preparations until exhaustive high-temperature delipidation is performed.

Several reports describe retinol binding by either cellular retinol-binding protein (CRBP) I or CRBP II using fluorescent spectroscopy.[8,9] In

[8] D. E. Ong and F. Chytil, *J. Biol. Chem.* **253,** 828 (1978).
[9] D. E. Ong, *J. Biol. Chem.* **259,** 1476 (1984).

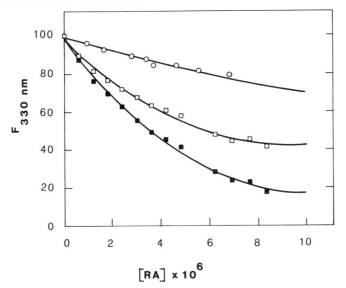

FIG. 1. Fluorescence titration of human ALBP by retinoic acid. Human ALBP (1.1 μM) in 10 mM Tris-HCl, pH 8.0, 100 mM NaCl was titrated with 0.5–8.5 μM retinoic acid. In the assay, stock retinoic acid (10 mM in ethanol) was sequentially added to human ALBP to obtain the indicated concentrations, and the tryptophan emission at 330 nm was determined (285 nm excitation). The corrected fluorescence intensity (□) was determined by subtraction of the fluorescence of N-acetyl-L-tryptophanamide (○) arising from inner filter effects from the observed total values (■).

contrast, fluorometric titrations of ALBP with a variety of retinoids did not yield any alteration in ligand fluorescence (L. L. Chinander and D. A. Bernlohr, unpublished observations). However, titration of human and murine ALBP with retinoic acid did result in an 80% decrease in the intrinsic protein fluorescence. Correction of this total fluorescence change for inner filter effects using N-acetyltryptophanamide[10] yielded a saturable binding isotherm (Fig. 1). The addition of retinol had no effect on the intrinsic tryptophan fluorescence of ALBP.[3] The concentration of retinoic acid which gave a half-maximal change in fluorescence was about 2 μM. This value may not, however, be equivalent to the dissociation constant. Cogan et $al.$[10] have pointed out that retinoic acid can exist in any of several micromolecular states, indicating that the free ligand concentration is not formally equivalent to the total ligand minus free ligand concentration. Hence, we conclude that the $K_{0.5}$ value for retinoic acid binding to ALBP is 2 μM.

[10] V. Cogan, M. Kopelman, S. Mokaday, and M. Shinitzky, $Eur. J. Biochem.$ **65,** 71 (1976).

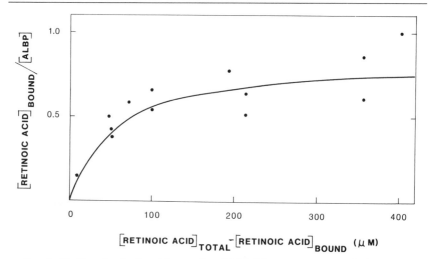

FIG. 2. Binding of retinoic acid to murine ALBP. Liposomes composed of 4 mM phosphatidylcholine, 1.3 mM cholesterol, and varying concentrations of [³H]retinoic acid in 10 mM Tris-HCl, pH 7.4, 100 mM NaCl were incubated for 60 min at 25° with 1 $\mu$$M$ murine ALBP. The liposomes were pelleted by centrifugation at 100,000 g for 15 min at 4°, and the protein-bound ligand was determined by measuring the radioactivity remaining in the supernatant relative to that in a buffer blank. [Adapted from V. Matarese and D. Bernlohr, *J. Biol. Chem.* **263**, 14544 (1988).]

The second assay we have utilized to measure retinoic acid binding to ALBP preparations is an adaptation of the liposome delivery assay of Brecher *et al.*[11] developed for fatty acid binding. Inclusion of [11,12-³H]retinoic acid or [15-³H]retinol into liposomes containing phosphatidylcholine and cholesterol allows for the development of an isotopic retinoid binding assay. ALBP is incubated with liposomes containing various concentrations of retinoic acid for 30 to 60 min at room temperature. The liposomes are pelleted by centrifugation at 400,000 g for 15 min at 4°, and the radioactivity in the supernatant, compared to that in a buffer blank, is determined. Typical results are shown in Fig. 2.

Retinoic acid, but not retinol, saturably and stoichiometrically bound to murine ALBP. The $K_{0.5}$ value was about 60 $\mu$$M$, quite different than the 2 $\mu$$M$ value obtained for fluorometric titration of human ALBP with retinoic acid. This discrepancy points out that $K_{0.5}$ values are not dissociation constants and cannot be compared between assays using different methodologies, but can be compared only within an assay. In the case of the

[11] P. Brecher, R. Saouaf, J. M. Sugarman, D. Eisenberg, and K. LaRosa, *J. Biol. Chem.* **259**, 13395 (1984).

liposome assay several dispersed forms of retinoic acid are present, and we are unable to determine the true molar free ligand concentration.

Comment

The purification procedures described result in the isolation of homogeneous ALBP devoid of any bound ligand. Chromatography at 37° on the lipophilic resin Lipidex 1000 removed any bound lipids and rendered the ALBP and other proteins amenable to purification. The failure to remove dissolved lipids resulted in protein preparations intractable to purification. In contrast, other retinoid-binding proteins[8,9] copurify with a bound ligand. It is unclear at this time if one should consider ALBP as a retinoid-binding protein, a fatty acid-binding protein, or both. Hence, we have named this protein simply a "lipid-binding protein." The availability of homogeneous ALBP and several assays to test its binding characteristics will allow for a detailed analysis of this protein.

[39] Binding of 11-*cis*-Retinaldehyde to Cellular Retinaldehyde-Binding Protein from Pigment Epithelium

By Maria A. Livrea *and* Luisa Tesoriere

Introduction

Found in the retina and retinal pigment epithelium (RPE) cells, the cellular retinaldehyde binding protein (CRALBP) is the only cellular retinoid-binding protein known to carry 11-*cis*-retinaldehyde as an endogenous retinoid.[1-3] Its unique localization to tissues in the eye suggests that this protein may be an operative component of the visual cycle, probably acting as a substrate carrier for the stereospecific dehydrogenase of the retinal pigment epithelium.[4] This potential function led us to investigate its binding properties and affinity. The procedure to analyze the binding of 11-*cis*-retinaldehyde to the CRALBP from bovine RPE is described below.

[1] S. Futterman, J. C. Saari, and S. Blair, *J. Biol. Chem.* **252**, 3267 (1977).
[2] G. W. Stubbs, J. C. Saari, and S. Futterman, *J. Biol. Chem.* **254**, 8529 (1979).
[3] J. C. Saari, L. Bredberg, and G. G. Garwin, *J. Biol. Chem.* **257**, 13329 (1982).
[4] J. C. Saari, and D. L. Bredberg, *Biochim. Biophys. Acta* **716**, 266 (1982).

11-cis-Retinaldehyde Binding Assay

All operations are carried out under dim red illumination (ruby red bulb, 15 W) to preserve light-sensitive material. all-*trans*-[³H]Retinol (2.5 Ci/mmol, New England Nuclear, Boston, MA) is the source of the tritiated retinaldehyde required for the binding assay. Enzyme oxidation[1] is used to produce all-*trans*-[³H]retinaldehyde. 11-*cis*-[³H]Retinaldehyde prepared by photoisomerization of all-*trans*-[³H]retinaldehyde is purified and quantified by high-performance liquid chromatography (HPLC), using a Waters (Milford, MA) μPorasil column (0.4 × 30 cm) and 2% diethyl ether in petroleum ether as the eluent[5] at a flow rate of 5 ml/min and a monitoring wavelength of 365 nm. The same procedure is applied to obtain 11-*cis*- from all-*trans*-retinaldehyde (Sigma, St. Louis, MO).

Methods for the preparation of CRALBP from retina have been described in this series,[6,7] and Saari and Bredberg have reported an improved procedure to purify CRALBP from both bovine retina and retinal pigment epithelium.[8] Cultured RPE cells are not recommended as a source of CRALBP, as evidence indicates that this protein is rapidly depleted in culture.[9]

The binding assay was originally developed utilizing a CRALBP preparation, whose purity degree was estimated at 75%, obtained from the pigment epithelium cytosol of 40–50 fresh bovine eyes according to the described method[7] with minor modifications.[10] The binding analysis reported here utilizes purified CRALBP. The amount of protein is determined by the method of Bradford,[11] with 5 μg of γ-globulin as a standard (micromethod).

Apoprotein is prepared by removing the endogenous ligand by exposure of CRALBP (8 μg/ml) to light from a 100-W incandescent lamp in an ice bath for 2 min. For the binding assay 2.5 μl of ethanol solution of 11-*cis*[³H]retinaldehyde is added to 5 μg of apoprotein in a total volume of 1.2 ml of TRIS-HCl pH 7.5, to reach a final concentration of 1.4–88 × 10^{-8} M (5,800–50,000 cpm) and incubated at 25° for 6 hr. This incubation time is chosen in order to assure equilibrium even at the lowest of the retinoid concentrations.

[5] J. P. Rothmans and A. Kropf, *Vision Res.* **15**, 1301 (1975).

[6] J. C. Saari, S. Futterman, G. W. Stubbs, and L. Bredberg, this series, Vol. 67, p. 296.

[7] J. C. Saari, this series, Vol. 81, p. 819.

[8] J. C. Saari and L. Bredberg, *Exp. Eye Res.* **46**, 569 (1988).

[9] C. D. B. Bridges, M. S. Oka, S.-L. Fong, G. I. Liou, and R. A. Alvarez, *Neurochem. Int.* **8**, 527 (1986).

[10] M. A. Livrea, A. Bongiorno, L. Tesoriere, C. Nicotra, and a. Bono, *Experientia* **43**, 582 (1987).

[11] M. Bradford, *Anal. Biochem.* **72**, 248 (1976).

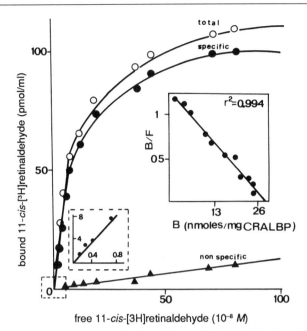

FIG. 1. Saturation curve for the binding of 11-*cis*-[³H]retinaldehyde to CRALBP. The dashed square shows the specific binding curve relative to 11-*cis*-retinaldehyde concentration between 0.5 and 1.4 × 10⁻⁸ *M*. The inset shows the Scatchard plot derived from the specific binding data. *B* represents the amount of specifically bound 11-*cis*-[³H]retinaldehyde and *F* the concentration of free 11-*cis*-[³H]retinaldehyde.

At the same time, samples are made containing the same concentration of labeled 11-*cis*-retinaldehyde in the presence of an excess of unlabeled ligand. To check binding at the very low 11-*cis*-[³H]retinaldehyde concentrations (0.5–1.4 × 10⁻⁸ *M*) a preparation of 11-*cis*-retinaldehyde, whose specific radioactivity is increased 5 times, is used under all other assay conditions described. Separation of bound from free retinaldehyde is achieved in accordance with Bashor.[12,13] Aliquots of the incubation mixture (4 μg protein) are applied to a column (0.8 × 10 cm) of Sephadex G-25 (Fine), equilibrated and then eluted with 50 m*M* Tris-HCl, pH 7.5, at a flow rate of 25 ml/hr. Fractions (0.5 ml) are collected in counting vials and the radioactivity determined after addition of 5 ml of liquid scintillation mixture. The radioactivity measured in the void volume of 8 ml is considered to be due to the bound 11-*cis*-retinaldehyde. Nonspecific bind-

[12] M. M. Bashor, D. O. Toft, and F. Chytil, *Proc. Natl. Acad. Sci. U.S.A.* **70**, 3483 (1973).
[13] M. M. Bashor and F. Chytil, *Biochim. Biophys. Acta* **411**, 87 (1975).

TABLE I
BINDING PROPERTIES FOR PHYSIOLOGICAL LIGANDS OF ISOLATED CELLULAR RETINOID-BINDING PROTEINS

Binding protein	Source	Substrate	K_d(M)	B_{max} (nmol/mg)	Assay method
CRBP	Rat liver[a]	all-trans-Retinol	1.6×10^{-8}	—	Fluorometric titration[b]
	Bovine adrenal glands[c]	all-trans-Retinol	5.5×10^{-7}	8.0	Cold acetone filtration[d]
	Cultured mouse skin cells[e]	all-trans-Retinol	2.5×10^{-8}	—	Dextran–charcoal[f]
CRBP II	E. coli-derived rat CRBP II[g]	all-trans-Retinol	1.0×10^{-8}	—	Fluorometric titration[b]
CRABP	Bovine adrenal glands[c]	all-trans-Retinoic acid	6.9×10^{-8}	2.2	Cold acetone filtration[d]
	Rat testis[h]	all-trans-Retinoic acid	4.2×10^{-9}	—	Fluorometric titration[b]
	Mouse skin[e]	all-trans-Retinoic acid	3.0×10^{-8}	—	Dextran–charcoal[f]
	Rat skin[i]	all-trans-Retinoic acid	8.0×10^{-9}	—	Dextran–charcoal[i]
CRABP II	Rat pups[j]	all-trans-Retinoic acid	6.5×10^{-8}	—	Fluorescence titration[b]
CRALBP	Bovine retinal pigment epithelium[k]	11-cis-Retinaldehyde	9.0×10^{-8}	29.8	Gel filtration[l]

a D. E. Ong and F. Chytil, J. Biol. Chem. 253, 828 (1978).
b U. Cogan, M. Kopelman, S. Mokady, and M. Shinitzky, Eur. J. Biochem. 65, 71 (1976).
c M. Ninomiya, M. Suganuma, N. S. Paik, Y. Muto, and H. Fujiki, FEBS Lett. 233, 255 (1988).
d C. L. Ashendel and R. K. Boutwell, Biochim. Biophys. Res. Commun. 99, 543 (1981).
e M. R. Haussler, C. A. Donaldson, M. A. Kelly, D. J. Mangelsdorf, G. T. Bowden, W. J. Meinke, F. L. Meyskens, and N. Sidell, Biochim. Biophys. Acta 803, 54 (1984).
f S. Dokoh, J. W. Pike, J. S. Chandler, J. M. Mancini, and M. R. Haussler, Anal. Biochem. 116, 211 (1981).
g E. Li, B. Locke, N. C. Yang, D. Ong, and J. Gordon, J. Biol. Chem. 262, 13773 (1987).
h D. E. Ong and F. Chytil, J. Biol. Chem. 253, 4551 (1978).
i A. K. Daly and C. P. F. Redfern, Biochim. Biophys. Acta 965, 118 (1988).
j J. S. Bailey and C. H. Siu, J. Biol. Chem. 263, 9326 (1988).
k Data from M. A. Livrea, A. Bongiorno, L. Tesoriere, C. Nicotra, and A. Bono, Experientia 43, 582 (1987).
l M. M. Bashor and F. Chytil, Biochim. Biophys. Acta 411, 87 (1975).

ing, defined as that binding resistant to a 200-fold excess of nonradioactive 11-*cis*-retinaldehyde, is subtracted from total binding to obtain specific binding in all experiments. All values must be corrected by a factor of 0.98 because of the contribution of radioactivity which elutes in the void volume when 11-*cis*-[³H]retinaldehyde is applied to the gel column with no protein sample present.

The protein-binding sites approach saturation at $88 \times 10^{-8} M$ 11-*cis*-retinaldehyde (Fig. 1). Inspection of the binding curve at the lowest 11-*cis*-retinaldehyde concentrations reveals linearity of the binding so that cooperative site interactions could be excluded even at very low concentrations of ligands.

Analysis of Kinetic Data

Equilibrium binding data are evaluated by the Scatchard plot,[14] using a least-squares regression analysis, and by the Hill plot, carried out by fitting the Hill equation directly to the experimental data, using the nonlinear regression method reported by Atkins[15] that guarantees to find the minimum for the sum of squares of residuals. Both methods are indicative of a single class of binding sites and concur to calculate an apparent equilibrium constant (K_d) of $0.9 \times 10^{-7} M$ and a maximum binding capacity (B_{max}) of 29.8 nmol/mg protein.

Binding Properties of Retinoid-binding Proteins

Table I summarizes the binding properties reported for some of the cellular retinoid-binding proteins from different sources.

[14] G. Scatchard, *Ann. N.Y. Acad. Sci.* **51**, 660 (1949).
[15] G. L. Atkins, *Eur. J. Biochem.* **33**, 175 (1973).

[40] Interactions of Retinoids with Phospholipid Membranes: Optical Spectroscopy

By WILLIAM STILLWELL and STEPHEN R. WASSALL

Introduction

The amphipathic structure of retinoids indicates that these compounds should partition into the lipid bilayer portion of membranes, and perhaps, in part, it is there they perform their physiological function. Many years ago vitamin A was shown to affect the structure and permeability of several biological membranes including those of erythrocytes,[1] lysosomes,[2] and mitochondria.[3] Retinoids have also been shown to affect phospholipid packing in monolayers,[4,5] microviscosities of erythrocyte ghost[6] and fibroblast[7] membranes, acyl chain ordering and dynamics in phospholipid model membranes,[8,9] and lipid bilayer permeability to anions, cations, and neutral solutes.[9,10] These experiments have generally demonstrated a large difference in the ability to perturb membranes between the two most common retinoid forms not involved in the visual process, retinol and retinoic acid.

Here we discuss two simple optical techniques to monitor the effect of retinoids on bilayer permeability to water, a neutral solute (erythritol), and an anion (carboxyfluorescein). From these studies we conclude that retinoic acid affects bilayer permeability to a much larger extent than does retinol. We also describe how optical techniques can be used to deduce information about the effect of retinoids on lipid phase transitions.

Lipid Vesicles

Three types of lipid vesicles are used, multilamellar vesicles (MLV) for the water permeability and erythritol measurements, large unilamellar vesicles (LUV) to monitor carboxyfluorescein (CF) release, and small (sonicated) unilamellar vesicles (SUV) for the lipid vesicle turbidity studies. For the experiments presented here, vesicles are made either from synthetic DMPC (dimyristoylphosphatidylcholine) or DPPC (dipalmitoylphosphatidylcholine) or from egg PC (unsaturated phosphatidylcholine extracted from egg yolks). These lipids may be purchased from either Sigma (St. Louis, MO) or Avanti Polar Lipids (Birmingham, AL); all-*trans*-retinoids are available from Aldrich (Milwaukee, WI).

MLV are prepared by the original method of Bangham[11] at a tempera-

[1] J. T. Dingle and J. A. Lucy, *Biochem. J.* **84**, 611 (1962).
[2] J. T. Dingle and J. A. Lucy, *Biol. Rev.* **40**, 422 (1965).
[3] C. R. Seward, G. Vaughan, and E. L. Hove, *J. Biol. Chem.* **241**, 1229 (1966).
[4] A. D. Bangham, J. T. Dingle, and J. A. Lucy, *Biochem. J.* **90**, 133 (1964).
[5] O. A. Roels and D. O. Shah, *J. Colloid Interface Sci.* **29**, 279 (1969).
[6] R. G. Meeks, D. Zaharevitz, and R. F. Chen, *Arch. Biochem. Biophys.* **207**, 141 (1981).
[7] A. M. Jetten, R. G. Meeks, and L. M. de Luca, *Ann. N.Y. Acad. Sci.* **359**, 398 (1981).
[8] S. R. Wassall and W. Stillwell, *Bull. Magn. Reson.* **9**, 85 (1987).
[9] S. R. Wassall, T. M. Phelps, M. R. Albrecht, C. A. Langsford, and W. Stillwell, *Biochim. Biophys. Acta* **939**, 393 (198).
[10] W. Stillwell, M. Ricketts, H. Hudson, and S. Nahmias, *Biochim. Biophys. Acta* **688**, 653 (1982).
[11] A. D. Bangham, *in* "Progress in Biophysics and Molecular Biology" (J. V. A. Butler and D. Noble, eds.), Vol. 18, p. 29. Pergamon, New York, 1968.

ture in excess of 10° above the gel to liquid crystalline phase transition temperature (DMPC, 23.6°; DPPC, 41.3°; egg PC, −15°). Phospholipids and retinoids are codissolved in chloroform in a pear-shaped flask, then the chloroform is removed under nitrogen followed by overnight vacuum pumping. (Alternatively, retinoids can be added to preformed vesicles by incubating the aqueous vesicles with small volumes of concentrated retinoid dissolved in ethanol.) Stock solutions of MLV containing 13.1 mM phospholipid with either 0, 1, or 3 membrane mol% retinol or retinoic acid are made by vortexing the warm 80 mM glucose, 2 mM histidine, 2 mM Tris-HCl, pH 7.4, buffer until all of the lipid comes off the flask walls, resulting in a very milky MLV suspension.

LUV are used to follow CF permeability. MLV composed of either egg PC or 80 mol% egg PC/20 mol% retinoid are first made in 100 mM CF, 25 mM phosphate, pH 7.0. The MLV dispersion is passed 500 times through a Liposor Liposome Cylinder (Lidex Technologies, Bedford, MA) and then sequentially extruded through 1.0 and 0.2 μm Nucleopore (Pleasanton, CA) filters.[12] Nonsequestered CF is removed from the resultant LUV dispersion on a Sephadex G-50 column.

For the vesicle turbidity experiments, SUV are made from the MLV by sonicating for about 10 min with a 30-sec, 50% duty cycle at level 7 of a Heat Systems W-220F Cell Disruptor. Sonication is stopped after the milky MLV become translucent with a slight bluish sheen, characteristic of SUV. A brief, 5-min, centrifugation with a tabletop centrifuge is sufficient to remove nonhydrated lipids from the MLV and titanium particles from the sonicated SUV.

Water Permeability Measurements

Bangham[13] demonstrated that MLV behave as almost perfect osmometers, swelling and shrinking in response to osmotic gradients. When MLV made in a high osmotic strength buffer are rapidly mixed into a buffer of lower ionic strength, they swell. The initial decrease in absorbance is proportional to the vesicle volume change and when normalized for the initial absorbance yields the osmotic permeability, $d(1/A)/dt\%$.[14] In our experiments, approximately 100 μl of the MLV made in high osmotic strength buffer is rapidly mixed into 2.0 ml of the low ionic strength swelling buffer (2.0 mM histidine, 2.0 mM Tris-HCl, pH 7.4). Both solu-

[12] M. J. Hope, M. B. Bally, G. Webb, and P. R. Cullis, *Biochim. Biophys. Acta* **812,** 55 (1985).
[13] A. D. Bangham, J. de Gier, and G. D. Greville, *Chem. Phys. Lipids* **1,** 225 (1967).
[14] M. C. Blok, L. L. M. Van Deenen, and J. de Gier, *Biochim. Biophys. Acta* **433,** 1 (1976).

tions have been equilibrated for about 5 min at the appropriate experimental temperature. The initial rapid MLV absorbance change associated with swelling is followed at 450 nm on a Beckman (Palo Alto, CA) DU-8 Computing Spectrophotometer. Cuvette temperatures are carefully controlled at within 0.1° from 10 to 50°, the practical range for the Beckman temperature controller.

Figure 1A shows the effect of 1 mol% and Fig. 1B of 3 mol% retinol and retinoic acid on water permeability in DMPC MLV. Several pieces of important information can be obtained from these curves. The general conclusion is that retinoids enhance permeability and that the enhancement is greater for retinoic acid than retinol. When incorporated at the 1 mol% level, neither retinoid substantially alters permeability of the gel state bilayers ($T \leq 25.7°$) whereas both significantly increase permeability in the liquid crystalline state ($T \geq 25.7°$). When retinoid is incorporated at 3 mol%, qualitatively the same behavior is exhibited with substantially higher rates of permeability.

From the dependence on temperature, activation energies E_a for the osmotically driven permeability of water across the membrane can be determined using the Arrhenius relationship

$$\ln P = -(E_a/RT) + \ln P_0 \qquad (1)$$

where P is the rate of permeability, E_a the activation energy, R the gas constant, T the temperature in degrees Kelvin, and P_0 a constant. Plots of $\ln P$ versus $10^3/T$ yield straight lines in the liquid crystalline phase, and activation energies may be calculated from the slope. Incorporation of 1% retinol increases the activation energy from 3.6 kcal/mol for pure DMPC to 6.3 kcal/mol, and with 1% retinoic acid the increase is even greater, to 7.6 kcal/mol. This trend appears somewhat surprising in view of the enhancement in permeability associated with incorporation of retinoid. The implication is that an intrinsically faster transport process with higher activation energy dominates permeability in the presence of retinol or retinoic acid. For instance, the formation of membrane channels would greatly enhance permeability but could possess a higher activation energy.

The data presented in Fig. 1 also demonstrate that a large increase in permeability occurs during the gel to liquid crystalline phase transition. In fact, this can be used to determine the transition temperature for synthetic phospholipids whose phase transitions fall within the temperature control range of the spectrophotometer. Table I compares the effect of retinol and retinoic acid on the temperature and approximate width of the transition as calculated from osmotic permeability measurements for bilayers composed of DMPC. The onset temperature is decreased and the width in-

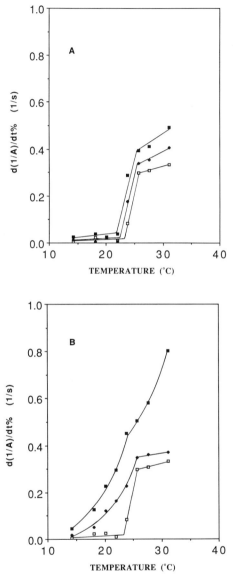

FIG. 1. Water permeability [expressed as the MLV swelling rate $d(1/A)/dt\%$] across bilayers made of either 100 mol% DMPC or 99 mol% DMPC/1 mol% retinoid (A) or 97 mol% DMPC/3 mol% retinoid (B): □, no retinoid; ◆, retinol; ■, retinoic acid.

TABLE I
EFFECT OF RETINOL AND RETINOIC ACID ON PHASE-TRANSITION
TEMPERATURE (T_c) AND WIDTH OF TRANSITION
FOR DMPC LIPID VESICLES[a]

Retinoid	Concentration (mol%)	Change in transition temperature (°)	Transition width (°)
Retinol	0.0	0.0	2.5
	1.0	−1.5	3
	3.0	−5.2	7.8
Retinoic acid	0.0	0.0	2.5
	1.0	−1.6	4.3
	3.0	−6.0	8.5

[a] As determined by osmotic water permeability measurements on MLVs. The published T_c value for DMPC is 23.6°.

creased by both retinoids, with retinoic acid exhibiting a larger effect than retinol. At 3 mol% incorporation of retinoic acid, the effect is so large that the phase transition is almost obliterated.

Our studies illustrate that the osmotic swelling approach offers a simple method of measuring membrane permeability and monitoring lipid phase transitions using equipment available to almost any laboratory. However, the method has certain limitations. The curves obtained from osmotic measurements are variable, and the slope of the initial absorbance change is not always easy to estimate. As a result we have routinely taken the average of between 6 and 10 determinations for each sample. Variability is greatly diminished by removing with centrifugation the large nonhydrated lipid particles often found in MLV before measuring swelling rates. Water permeability can be easily measured for relatively impermeable, saturated synthetic lipid membranes. In contrast, unsaturated lipids or lipids extracted from natural sources which contain a large amount of unsaturations are usually too permeable to measure water movement without use of stop-flow equipment. Nevertheless, vesicle swelling rates for unsaturated lipids can be decreased enough to be followed with a chart recorder by (1) lowering the temperature, (2) adding agents such as cholesterol[15] which decrease permeability in the liquid crystal state, or (3) measuring osmotic swelling following the movement of a permeant nonelectrolyte across the bilayers.

[15] M. C. Blok, L. L. M. Van Deenen, and J. de Gier, *Biochim. Biophys. Acta* **464,** 509 (1977).

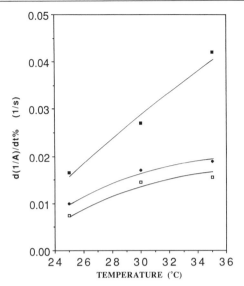

FIG. 2. Effect of incorporation of retinol and retinoic acid into egg PC MLVs on the osmotic swelling velocities [$d(1/A)/dt\%$] in isotonic erythritol: □, 0 mol% retinoid; ◆, 20 mol% retinol; ■, 20 mol% retinoic acid.

Erythritol Permeability

Although glucose is almost impermeable to lipid bilayers during the time course of the experiments reported here (0–30 sec), there is a graded permeability to other neutral solutes (e.g., glycerol > urea > erythritol).[13] If MLV are made in 40 mM glucose and rapidly mixed into 40 mM erythritol, the initial vesicle swelling rate is limited by the movement of erythritol into the vesicles. Hence, the swelling rates are considerably decreased. By this method it is possible to measure the relative permeability rates for even very unsaturated acyl chain bilayers. Retinoid-induced permeability to erythritol is presented in Fig. 2 for MLV made from unsaturated phosphatidylcholine extracted from egg yolks. Although the permeability rates are much less than for the DMPC bilayers (Fig. 1), retinoic acid is again shown to increase permeability more than retinol.

Carboxyfluorescein Permeability

At 100 mM the fluorescent anion CF self-quenches,[16] so that LUV prepared with CF in the intravesicular aqueous volume are initially poorly

[16] J. N. Weinstein, S. Yoshikami, P. Henkart, R. Blumenthal, and W. A. Hagins, *Science* **195**, 489 (1977).

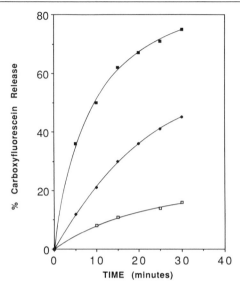

FIG. 3. Leakage of CF from egg PC LUV containing retinol or retinoic acid: □, 0 mol% retinoid; ◆, 20 mol% retinol; ■, 20 mol% retinoic acid. Results are expressed as the percentage of initially sequestered CF leaking as a function of time.

fluorescent. The effect of retinoids on leakage of the anion from vesicles is followed by observing the increase in fluorescence intensity (excitation at 470 nm, emission at 515 nm) as a function of time. Total trapped CF is ascertained by release in Triton X-100,[17] to enable estimation of the amount of CF leaked as a percentage of initially sequestered CF. Figure 3 clearly indicates that, consistent with the osmotic permeability experiments on nonelectrolytes, retinoic acid enhances permeability of the anion much more than does retinol.

Lipid Vesicle Turbidity Measurements

By DSC (differential scanning calorimetry), retinoids have been shown to remove the pretransition and to decrease the temperature and increase the width of the main gel to liquid crystalline phase transition in DPPC bilayers.[18] A similar effect on the phase transition for DMPC vesicles was measured by the osmotic swelling experiments reported above. An alternative method of measuring phase transition temperatures is pro-

[17] J. N. Weinstein, R. Blumenthal, S. O. Sharrow, and P. A. Henkart, *Biochim. Biophys. Acta* **509**, 272 (1978).
[18] S. R. Wassall, W. Stillwell, and M. Zeldin, *Proc. Indiana Acad. Sci.* **95**, 149 (1986).

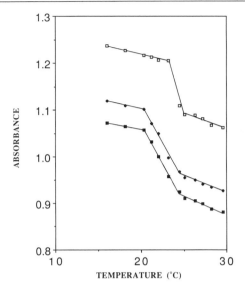

FIG. 4. Turbidity (absorbance at 450 nm) of SUV made from DMPC bilayers containing either 0 (□), 1 (◆), or 3 (■) mol% retinol. The membranes were cooled from 30°.

vided by observations of lipid turbidity. This easily measured parameter has been shown to abruptly increase on cooling SUV from the liquid crystal to the gel state. Specifically, Blok et al.[14] demonstrated that the transition temperature for DMPC or DPPC vesicles determined from optical absorbance measurements (turbidity) correlated well with those obtained by calorimetry. They also showed that low levels of a membrane contaminant (4% egg phosphatidic acid) caused a decrease in the measured transition temperature. Therefore, lipid vesicle turbidity was also employed as a method to monitor retinoid induced changes in phase behavior.

Absorbances at 450 nm for DMPC and DPPC vesicles with 0, 1, and 3 mol% retinol or retinoic acid were obtained. The curves for retinol–DMPC vesicles are shown in Fig. 4. Slightly different amounts of vesicles were used for each of the three retinol concentrations tested to offset the curves and allow for easier presentation of the transitions. They clearly confirm that introduction of the retinoid broadens the phase transition and reduces the onset temperature. Similar curves were recorded for retinoic acid–DMPC, retinol–DPPC, and retinoic acid–DPPC vesicles. The effects on phase behavior seen for retinoids by the other techniques are thus supported. However, although no new information is obtained by this method, it is emphasized that the turbidity method does have its virtues.

The technique requires much less specialized instrumentation than DSC and is much faster than the osmotic swelling method.

Additional Considerations

Although the methods presented here have the advantages of being simple and involving equipment generally available to most laboratories, they are not without shortcomings. All lipid vesicles are not osmotically active. SUV, though of well-defined size, are osmotically insensitive,[19] and it is controversial whether even the better defined large unilamellar vesicles (LUV) can behave as osmometers.[20] Osmotic permeability and vesicle turbidity are not the best methods to monitor phase behavior. The optical methods are a poor substitute for the direct DSC approach, which can provide transition enthalpies, and they are not as sensitive or accurate as NMR (nuclear magnetic resonance), ESR (electron spin resonance), or fluorescence polarization. Osmotic permeability and turbidity are also limited to the temperature range available to the spectrophotometer. The methods cannot be used for unsaturated acyl chain phospholipids which have phase transition temperatures below the freezing point of water. An additional possible problem concerns the instability of phosphatidylethanolamine-containing vesicles in the presence of retinal (owing to Schiff base formation).

Conclusions

Retinoids can be incorporated into phospholipid bilayers and, at relatively low concentrations (1 and 3 membrane mol%), can affect the basic membrane properties of permeability and lipid phase transition, with retinoic acid exerting a larger perturbing effect than retinol. The results obtained from the optical techniques presented here are consistent with those obtained from different methods. Retinoic acid to a larger extent than retinol has been shown to alter the following: membrane permeability to K^+, I^-, and glucose[10]; electrical conductivity of planar bimolecular lipid membranes[10]; membrane acyl chain order[8,9]; and oxidative phosphorylation of coupled mitochondria.[21]

[19] S. M. Johnson and N. Buttress, Biochim. Biophys. Acta 307, 20 (1973).
[20] E. Hantz, A. Cao, J. Escaig, and E. Taillondier, Biochim. Biophys. Acta 862, 379 (1986).
[21] W. Stillwell and S. Nahmias, Biochem. Int. 6, 385 (1983).

[41] Interactions of Retinoids with Phospholipid Membranes: Electron Spin Resonance

By STEPHEN R. WASSALL and WILLIAM STILLWELL

Introduction

Retinoids are essential for the maintenance of health.[1] Their amphiphilic nature suggests that they will locate within the phospholipid bilayer of membranes, which may constitute a site of action. In addition, the toxicity of high doses of retinoids poses a serious problem to their claimed efficacy in treating a number of cancers.[2] Membrane disruption may be a cause. Several studies have demonstrated that retinoids affect the properties of membranes. Enhanced permeability to nonelectrolytes and ions on incorporation of retinoids into phospholipid membranes has been extensively reported.[3-5] Calorimetric and optical methods have shown that all-*trans*-retinol, retinal, and retinoic acid broaden the gel to liquid crystalline phase transition and depress its onset temperature in PC (phosphatidylcholine) bilayers.[4,6] Reduced microviscosities in rat erythrocytes and mouse fibroblasts caused by retinoids have been measured by fluorescence polarization,[7,8] whereas magnetic resonance techniques have detected retinoid-associated changes in molecular ordering and dynamics within phospholipid model membranes.[9-13] Further work, however, is required to establish a clear picture.

In this chapter, the application of ESR (electron spin resonance) spin label techniques as a method of assay for the interactions of retinoids with

[1] L. M. DeLuca and S. S. Shapiro (eds.), *Ann. N.Y. Acad. Sci.* **359** (1981).
[2] D. F. Birt, *Proc. Soc. Exp. Biol. Med.* **183**, 311 (1986).
[3] W. Stillwell and M. Ricketts, *Biochem. Biophys. Res. Commun.* **97**, 148 (1980).
[4] W. Stillwell, M. Ricketts, H. Hudson, and S. Nahmias, *Biochim. Biophys. Acta* **688**, 653 (1982).
[5] W. Stillwell and L. Bryant, *Biochim. Biophys. Acta* **731**, 483 (1983).
[6] S. R. Wassall, W. Stillwell, and M. Zeldin, *Proc. Indiana Acad. Sci.* **95**, 149 (1986).
[7] R. G. Meeks, D. Zaharevitz, and R. F. Chen, *Arch. Biochem. Biophys.* **207**, 141 (1981).
[8] A. M. Jetten, R. G. Meeks, and L. M. DeLuca, *Ann. N.Y. Acad. Sci.* **359**, 398 (1981).
[9] S. P. Verma, H. Schneider, and I. C. P. Smith, *Arch. Biochem. Biophys.* **162**, 48 (1974).
[10] S. R. Wassall and W. Stillwell, *Bull. Magn. Reson.* **9**, 85 (1987).
[11] C. A. Langsford, M. R. Albrecht, T. M. Phelps, W. Stillwell, and S. R. Wassall, *Biophys. J.* **51**, 239a (1987).
[12] S. R. Wassall, T. M. Phelps, M. R. Albrecht, C. A. Langsford, and W. Stillwell, *Biochim. Biophys. Acta* **939**, 393 (1988).
[13] H. DeBoeck and R. Zidovetzki, *Biochim. Biophys. Acta* **946**, 244 (1988).

phospholipid membranes is reviewed. The utility of the approach in membrane research is well documented.[14] This chapter focuses on recent studies of the effects of retinoids on order and fluidity within phospholipid model membranes, which demonstrated a distinction between the effects of all-*trans*-retinoic acid and those of all-*trans*-retinol and retinal.[10-12] Preliminary ESR work on the influence of retinoids on membrane permeability is also described, followed by a discussion of additional ESR spin label experiments that should prove informative in the future.

General Experimental Detail

Materials. Phospholipids are available from Avanti Polar Lipids (Birmingham, AL); all-*trans*-retinoids may be purchased from Aldrich (Milwaukee, WI). Molecular Probes, Inc. (Eugene, OR) is a source of spin labels.

Preparation of Samples

Multilamellar liposomes and sonicated unilamellar vesicles of phospholipids in the absence and presence of retinoids have been employed in ESR studies. Lipid mixtures are initially codissolved in chloroform. The solvent is then removed under a stream of nitrogen followed by vacuum pumping overnight. The addition of buffer and vortex mixing at temperatures at least 10° above the gel to liquid crystalline phase transition produce multilamellar liposomes. Subsequent sonication results in unilamellar vesicles. A Tekmar VC 250 Ultrasonic Cell Disruptor is suitable for this purpose. A water-cooled jacket is necessary to prevent excessive sample heating during a typical 5-min period of sonication, which is performed under nitrogen to reduce oxidation. Titanium fragments from the sonicating probe are removed by briefly spinning (~5 min) in a benchtop centrifuge.

Electron Spin Resonance

An IBM/Bruker ER 200D X-band ESR spectrometer operating at 9.2 GHz is utilized in all our investigations. The spectrometer is interfaced and controlled by a Hewlett Packard 9816 computer system with graphics display and Winchester disk drive. The signals are detected as the first derivative of the absorption. Spectral parameters are as follows: microwave power, 12 mW; field strength, 3294 G; sweep width, 80–100 G;

---

[14] D. Marsh, *in* "Membrane Spectroscopy—Molecular Biology, Biochemistry and Biophysics (E. Grell, ed.), Vol. 31, p. 51. Springer-Verlag, Berlin, 1981.

sweep time, 160–200 sec; time constant, 500 msec; modulation amplitude, 1.0–2.0 G; and dataset, 1000–2000 points. Temperature is regulated within 0.1° by an Omega Engineering, Inc. (Stamford, CT) Model 149 Controller and monitored immediately adjacent to the sample by a copper–constantan thermocouple. Samples are contained in a capilliary tube and are approximately 25 μl in volume.

Electron Spin Resonance Studies of Retinoid–Phospholipid Interactions

Basic Concepts

Observation of an ESR spectrum requires the presence of an unpaired electron. In biological systems this is achieved by the introduction of a stable free radical, usually a nitroxide spin label attached to the molecule of interest. The spectrum for such a group contains three lines, the frequency (g value) and splitting (A value) of which depend on the orientation of the group with respect to the applied magnetic field.[15] At sufficiently low concentrations when label–label interactions are negligible, the shape of a spectrum, in practice, is consequently determined by the anisotropy and rate of motions undergone by the spin-labeled group on the ESR time scale (10^{-11}–10^{-7} sec).

Acyl Chain Order and Fluidity

Molecular ordering and dynamics within phospholipid membranes are monitored with 5-, 7-, 10-, 12-, and 16-doxylstearic acids [β-5-, 7-, 10-, 12-, and 16-(4′,4′-dimethyloxazolidinyl-N-oxyl) stearic acids] intercalated at low concentration (1 mol%) into the membrane. Although spectra simu-

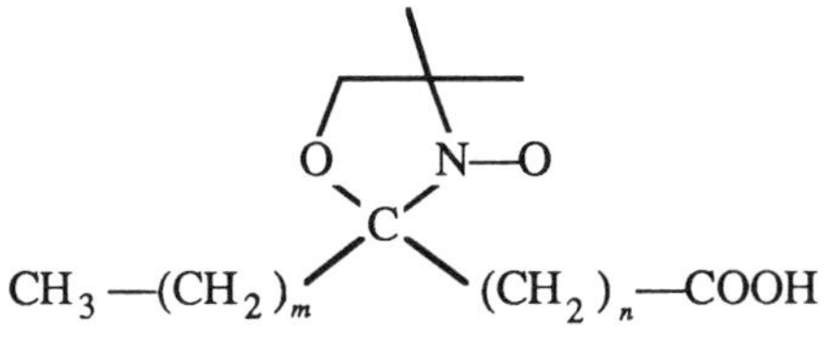

$$CH_3-(CH_2)_m \overset{\displaystyle O \quad N-O}{\underset{\displaystyle C}{\diagup\diagdown}} (CH_2)_n-COOH$$

Doxylstearic Acid

lations can provide more detailed analysis,[16] information is directly derived from the spectra in a relatively simple manner. Order parameters S

---

[15] I. D. Campbell and R. A. Dwek, in "Biological Spectroscopy," p. 179. Benjamin/Cummings, Menlo Park, California, 1984.

[16] M. Moser, D. Marsh, P. Meier, K-H. Wassmer, and G. Kothe, Biophys. J. 55, 111 (1989).

and correlation times τ_c are calculated according to the equations

$$S = \frac{A_\| - A_\perp - C}{A_\| + 2A_\perp + 2C} (1.66) \qquad (1)$$

where $A_\|$ and $A_\perp$ are the apparent parallel and perpendicular hyperfine splitting parameters, the constant $C = 1.4 - 0.053(A_\| - A_\perp)$ is an empirical correction for the difference between the true and apparent values of $A_\perp$, and the factor 1.66 is a solvent polarity correction factor[17]; and

$$\tau_c = 6.5 \times 10^{-10} W_0[(h_0/h_{-1})^{1/2} - 1] \qquad (2)$$

where W_0 is the peak to peak width of the central line and h_0/h_{-1} is the ratio of the heights of the central and high-field lines, respectively.[18]

Calculation of the order parameter is limited to the upper portion of the fatty acid chain (5, 7, and 10 positions), where molecular motion is sufficiently anisotropic to produce spectra for which outer and inner hyperfine extrema are discernible (Fig. 1). It is related to the amplitude of acyl chain motion at the position labeled and can take values in the range $0 \leq S \leq 1$, the respective limits representing isotropic motion and fast axial rotation with no flexing. This may be visualized if motion of the spin label is treated as a random wobbling that is restricted in amplitude to within a cone of angle γ about the normal to the membrane surface, in which case

$$S = \tfrac{1}{2} (\cos \gamma + \cos^2 \gamma) \qquad (3)$$

so that $S = 0$ for $\gamma = 90°$ and $S = 1$ for $\gamma = 0°$.[15] The correlation time equation assumes isotropic motion and is appropriate in the lower portion of the fatty acid chain (12 and 16 positions), where molecular motion is approximately isotropic and produces spectra characteristic of high disorder (Fig. 1). As the motion is not truly isotropic, it should be realized that the τ_c values obtained are not true correlation times. They are determined by the degree of anisotropy of motion as well as by the rate, and they can only be related crudely to microviscosity or fluidity within the membrane.

The effect of 0–20 mol% incorporation of all-*trans*-retinol, retinal, and retinoic acid on order parameters measured for 7- and 10-doxylstearic acid (1 mol%) intercalated into multilamellar liposomes of 1% (w/v) DPPC (dipalmitoylphosphatidylcholine) in 20 mM phosphate/1 mM EDTA buffer (pH 7.5) at 50°[12] is plotted in Fig. 2. The concentration dependence at the 7 position demonstrates that retinol and retinal slightly increase

[17] B. J. Gaffney, *in* "Spin Labeling: Theory and Applications" (L. J. Berliner, ed.), p. 567. Academic Press, New York, 1976.
[18] A. Keith, G. Bulfield, and W. Snipes, *Biophys. J.* **10**, 618 (1970).

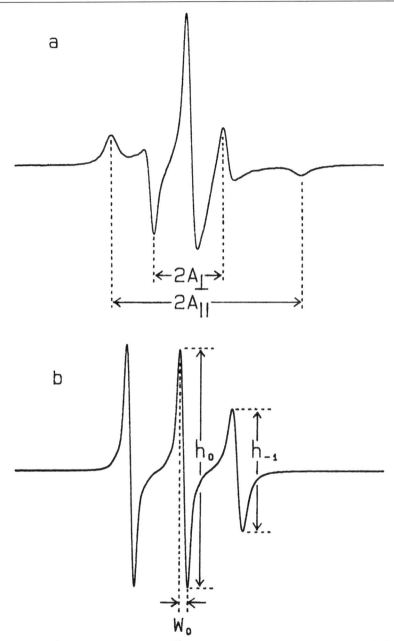

Fig. 1. Typical ESR spectra for spin-labeled stearic acids intercalated into phospholipid membranes. (a) Order parameter S calculation in the upper region of the acyl chain (5, 7, and 10 positions). (b) Correlation time τ_c calculation in the lower region of the acyl chain (12 and 16 positions).

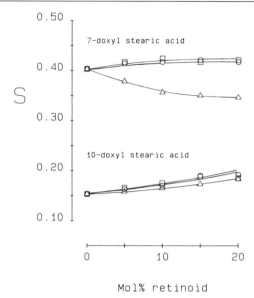

Mol% retinoid

Fig. 2. Dependence on all-*trans*-retinoid concentration of order parameters S for 7- and 10-doxylstearic acids (1 mol%) intercalated into 1% w/v multilamellar dispersions of DPPC in 20 mM phosphate/1 mM EDTA buffer (pH 7.5) at 50°: ○, retinol; □ retinal; and △, retinoic acid.

order within the membrane, whereas retinoic acid causes disordering by as much as 14% at 20 mol% concentration. Similar behavior is seen higher up the chain at the 5 position. In contrast, at the 10 position greater order is produced by all three retinoids. The increase is approximately 22% in each case when 20 mol% is present. Qualitatively, the same trend persists further down the chain at the 12 and 16 positions, where correlation times are increased by retinol, retinal, and retinoic acid.

The strongly hydrophilic nature of the carboxylic group was proposed to be responsible for the different profile of effect on acyl chain motion produced by retinoic acid.[12] It was argued that retinoic acid would be placed higher within the membrane than retinol or retinal. The consequent disturbance to molecular packing at the membrane surface would thus lead to increased disorder in the upper region of the chain. This is as opposed to the small perturbation near the aqueous interface anticipated owing to a lower location within the membrane for the less strongly hydrophilic alcohol group of retinol or aldehyde group of retinal. Experimental evidence in favor of the explanation is offered by ESR study of 5-doxylstearic acid in DPPC membranes, which shows that decanol has a negligible effect on order whereas decanoic acid decreases order.[12] The bulky hydrophobic cyclohexene group of all three retinoids, however, would be expected to locate and hinder molecular motion toward the center of the membrane.

ESR experiments on the influence of all-*trans*-retinoids on acyl chain order and fluidity within unsaturated DOPC (dioleoylphosphatidylcholine) membranes have also been performed.[11] They confirm that retinol and retinal restrict acyl chain motion in the lower portion of the chain and have little effect in the upper portion, whereas retinoic acid similarly restricts acyl chain motion in the lower portion of the chain but reduces order in the upper portion. Interestingly, the disordering produced by retinoic acid in DOPC membranes is much smaller at 35 than 45°. This is shown for the 5 position in Fig. 3. Perhaps the same trend continues at lower temperatures until retinoic acid decreases order to an even smaller or negligible extent. If so, it might be speculated that the disruption to acyl chain packing produced by the bend associated with the cis double bond (Δ^9) of the oleic fatty acid chain creates a space into which the cyclohexene group is constrained to fit. The result this could have on the depth of penetration of retinoic acid into the membrane is illustrated by Fig. 4, where the cyclohexene group is placed in the space beneath the

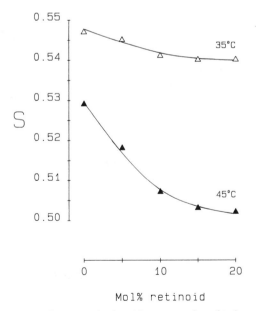

FIG. 3. Dependence on all-*trans*-retinoic acid concentration of order parameters S for 5-doxylstearic acid (1 mol%) intercalated into 1% (w/v) multilamellar dispersions of DOPC in 20 mM phosphate/1 mM EDTA buffer (pH 7.5) at 35 and 45°.

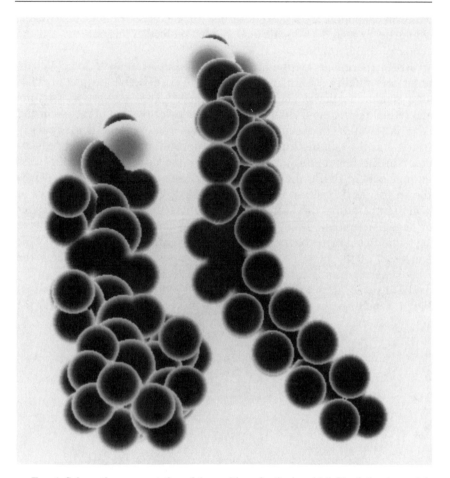

FIG. 4. Schematic representation of the position of retinoic acid (left) relative to an oleic acid acyl chain (right).

bend at the double bond in an otherwise all-*trans*-oleic acid chain. As can be seen, a lower location within the membrane would be imposed on the retinoic acid so that its carboxylic group would no longer protrude into the aqueous interface. Thus, the distinction in effect on molecular ordering of retinoic acid with respect to retinol and retinal would be lost.

A reservation concerning the use of spin-labeled fatty acids as probes of membrane order and fluidity should be mentioned here. The bulky nitroxide moiety constitutes a perturbation.[19] ESR data consequently

[19] M. G. Taylor and I. C. P. Smith, *Biochim. Biophys. Acta* **733**, 256 (1983).

must be considered qualitative, although the direction of effect (e.g., increased order) detected is usually correct. In view of this, although quantitative differences are expected, the discrepancy between the changes arising from retinoids seen by ESR and in a recent ^2H NMR (deuterium nuclear magnetic resonance) study of DPPC membranes[13] is surprising. The latter study concluded that addition of 33 mol% all-*trans*-retinoic acid causes a slight increase in order and that the same concentration of retinol decreases order throughout the membrane.

Permeability

Ascorbate reduction methods enable measurement of the permeability of membranes to water-soluble spin label molecules.[20] A typical protocol for observation of uptake of the cation TEMPOcholine (2,2,6,6-tetramethylpiperidinyl-1-oxycholine) into sonicated unilamellar vesicles is described here. Initially (time $t = 0$), equal volumes (1.5 ml) of vesicles [5% (w/v) phospholipid] and TEMPOcholine (6 mM) in 0.1 M KCl/10 mM Tris

TEMPOcholine Chloride

buffer (pH 7.5) are mixed as a master sample, which is subsequently maintained at the desired temperature. Aliquots (50 μl) are removed after various times intervals ($t = 0, 20, 40, \ldots, x$ min) and immediately mixed with 10 μl of 0.35 M ascorbate solution. The ascorbate reduces, hence eliminating the ESR signal from, TEMPOcholine remaining outside the vesicles. Measurement of the intensity (peak to peak height) of the low-field line in the ESR spectrum for each aliquot then reveals the time evolution of the spin label leaking into the vesicles. This consists of a signal intensity that increases with time according to the rate of permeability toward an asymptotic limiting value determined by the internal volume of the vesicles. To minimize ascorbate and TEMPOcholine leakage in a given aliquot, mixing with the ascorbate and collection of the spectra are performed at 0°. Control aliquots without ascorbate are run simultaneously to facilitate signal intensity normalization.

[20] D. Marsh, A. Watts, and P. F. Knowles, *Biochemistry* **15**, 3570 (1976).

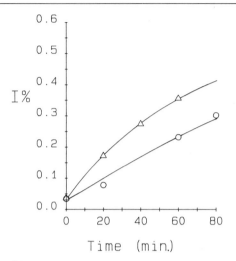

FIG. 5. Leakage of TEMPOcholine into vesicles (fractional intensity $I\%$ of ESR signal from TEMPOcholine inside vesicles) versus time for DMPC (O) and DMPC/10 mol% retinoic acid (△) vesicles at 37°.

The results of a preliminary investigation of membrane permeability to TEMPOcholine are shown in Fig. 5 for DMPC (dimyristoylphosphatidylcholine) vesicles pepared with and without 10 mol% all-*trans*-retinoic acid.[21] The graph follows uptake of the cation into the vesicles at 37°, as represented by the variation with time of the fractional intensity ($I\%$) of the spin label signal remaining after ascorbate reduction. The slope of the curve is clearly greater when retinoic acid is present, which supports previous reports of retinoid associated enhancement of membrane permeability.

Future Directions

There is plenty of scope to extend current ESR spin label work on molecular ordering and dynamics within the interior of model membranes containing retinoids. Natural membranes are comprised of a vast array of phospholipid and acyl chain constituents. Mixed membranes with PE (phosphatidylethanolamine) are of particular interest owing to Schiff base formation by retinal,[22] while the apparent health benefits of fish oils[23]

[21] C. A. Langsford, W. Stillwell, and S. R. Wassall, unpublished results (1987).
[22] H. Shichi and R. L. Somers, *J. Biol. Chem.* **249**, 6570 (1974).
[23] A. P. Simopoulos, R. R. Kifer, and R. E. Martin (eds.), "Health Effects of Polyunsaturated Fatty Acids in Seafoods." Academic Press, New York, 1986.

suggest that polyunsaturated membranes are worthy of attention. Our preliminary measurement of inc. ased permeability to TEMPOcholine in the presence of retinoic acid furthermore encourages studies with all three retinoids.

Other aspects of the interaction of retinoids with phospholipid membranes are accessible to future investigation by ESR. The application of ESR spin label methods to study membrane fusion, a process that has been linked to vitamin A,[24,25] has been reviewed.[26] Importantly, mxing of both aqueous and lipid environments may be monitored. The water-soluble spin-label TEMPO (2,2,6,6-tetramethylpiperidinyl-1-oxy) manifests a marked increase in membrane solubility at the gel to liquid crystalline phase transition[27] and may be exploited to determine the influence of retinoids on phase behavior in aqueous phospholipid dispersions. Specifically, the dependence on temperture is measured for the intensities of the two components, corresponding to TEMPO dissolved in water and membrane, in the high-field line of the ESR spectrum. Rates of lateral diffusion and flip-flop within phospholipid membranes may also be sensitive to the presence of retinoids and are cited as final examples of potential ESR work. Estimation of the former quantity relies on the dominant contribution exchange broadening makes to ESR spectra for membranes containing over 1 mol% spin-labeled fatty acid or phospholipid.[28] Analysis of the spectral broadening as a function of spin label concentration yields the diffusion coefficient.[29] Measurement of the flip-flop of head group spin-labeled   PC   (1,2-dipalmitoyl-sn-glycero-3-phosphorylTEMPOcholine) across the membrane of unilamellar vesicles entails an ascorbate reduction procedure.[30] The transfer of spin-labeled molecules from the inner to outer vesicle layer is observed following initial abolition of the ESR signal from the outside layer.

Conclusion

The intention of this chapter is to summarize previous ESR spin label work on phospholipid–retinoid interactions and to indicate possible aver-

[24] P. Dunham, P. Babiarz, A. Israel, A. Zerial, and G. Weissmann, *Proc. Natl. Acad. Sci. U.S.A.* **74,** 1580 (1977).

[25] A. H. Goodhall, D. Fisher, and J. A. Lucy, *Biochim. Biophys. Acta* **595,** 1 (1980).

[26] H. M. McConnell, *in* "Spin Labeling: Theory and Applications" (L. J. Berliner, ed), p. 525. Academic Press, New York, 1976.

[27] E. J. Shimshick and H. M. McConnell, *Biochemistry* **12,** 2351 (1973).

[28] P. Devaux, C. J. Scandella, and H. M. McConnell, *J. Magn. Reson.* **9,** 474 (1973).

[29] E. Sackmann, H. Traüble, H.-J. Galla, and P. Overath, *Biochemistry* **12,** 5360 (1973).

[30] R. Kornberg and H. M. McConnell, *Biochemistry* **10,** 1111 (1971).

nues for further research. The aim of these studies is to contribute toward an eventual definition of a relationship between the molecular structure of vitamin A and its physiological role. The current description that arises from ESR studies establishes a distinction between the response of phospholipid membranes to all-*trans*-retinoic acid and to all-*trans*-retinol and retinal. All three retinoids restrict acyl chain motion to a similar extent approaching the center of the membrane, but toward the surface of the membrane retinol and retinal are a negligible perturbation to order whereas retinoic acid causes disordering.[10–12] Consistent with this difference in behavior, greater enhancement of membrane permeability is measured with retinoic acid than retinol or retinal.[12] In the context of retinoid toxicity, it is thus noted that retinoic acid induces hypervitaminosis A more effectively than retinol.[31]

Acknowledgments

It is a pleasure to thank Marvin D. Kemple for use of the ESR spectrometer, Fritz W. Kleinhans and Timothy M. Phelps for development of spectral collection and analysis routines, Carol A. Langsford for performing experiments, Margo Page for typing the manuscript, and Ellen Chernoff for photographic assistance.

[31] G. A. J. Pitt, *Proc. Nutr. Soc.* **42,** 43 (1983).

[42] Transfer of Retinol from Retinol-Binding Protein Complex to Liposomes and across Liposomal Membranes

By GÖRAN FEX and GUNVOR JOHANNESSON

Introduction

The mechanism(s) of cellular uptake of retinol from its complex with plasma retinol-binding protein (RBP) and the subsequent intracellular distribution of retinol are largely unknown. According to the current view, retinol is taken up into cells after the retinol–RBP complex has interacted with a cell surface receptor for retinol-binding protein. The nature of this receptor is unknown. Using indirect methods the receptor has been demonstrated on bovine pigment epithelial cells,[1] monkey intestinal cells,[2]

[1] J. Heller, *J. Biol. Chem.* **250,** 3613 (1975).
[2] L. Rask and P. A. Peterson, *J. Biol. Chem.* **251,** 6360 (1976).

and in testis.[3] There is one account of binding of RBP to a component derived from pigment epithelium membranes,[4] but further information is lacking. Retinol can, however, be taken up and metabolized in an apparently normal way by cells even if not presented as a complex with retinol-binding protein, for example, in complex with albumin or simply added to the cells dissolved in organic solvent.[5-7] This newly taken up retinol is esterified within minutes, indicating that it rapidly traverses the plasma membrane and distributes to intracellular sites where esterification of retinol occurs. Thus, it seems that there is no obligatory pathway from the receptor on the cell surface to intracellular sites which has to be followed for normal metabolism of retinol. Also, the results of Green et al.,[8] which show extensive recirculation of retinol between plasma and tissues, imply equilibration of retinol between tissues and plasma RBP. This led us to consider that there may be other ways for retinol to enter cells beside via the RBP receptor and that there had to be mechanisms for the rapid transport of retinol across and between membranes. The experimental systems described below were used to test this idea.[9,10]

Materials and Methods

all-*trans*-[³H]Retinol (specific radioactivity 23.5 Ci/mmol) and [¹⁴C]triolein (specific radioactivity 120 Ci/mmol) were from New England Nuclear (Dreieich, FRG). DEAE-Sepharose and Sephadex G-25 were from Pharmacia (Uppsala, Sweden), all-*trans*-retinol, all-*trans*-retinal, all-*trans*-retinyl acetate, α-tocopherol, egg lecithin, cholesterol, butylated hydroxytoluene (BHT), octyl-β-D-glucopyranoside, horse liver alcohol dehydrogenase (ADH, lyophilized, 2 U/mg), and pyruvate were all from Sigma (St. Louis, MO). Hog muscle lactate dehydrogenase [LDH, suspension in 3.2 *M* ammonium sulfate, 10,000 U (10 mg)/ml] and NAD (lithium salt) were from Boehringer Mannheim (Mannheim, FRG). 1-Butanol, ethyl acetate, acetonitrile, 2-mercaptoethanol were from BDH (Poole, UK), Spectrapor 3 dialysis tubing from Spectrum Industries (Los Angeles, CA), and Bio-Beads from Bio-Rad (Richmond, CA).

[3] M. Krishna Bhat and H. R. Cama, *Biochim. Biophys. Acta* **587**, 273 (1979).
[4] B. C. Laurent, C. O. Båvik, U. Eriksson, H. Melhus, M. H. L. Nilsson, J. Sundelin, L. Rask, and P. A. Peterson, *Chem. Scr.* **27**, 185 (1987).
[5] J. E. Smith, C. Borek, and D. S. Goodman, Cell (*Cambridge, Mass.*) **15**, 865 (1978).
[6] J. C. Saari, L. Bredberg, and S. Futterman, *Invest. Ophthalmol. Visual Sci.* **19**, 1301 (1980).
[7] P. D. Bishop and M. D. Griswold, *Biochemistry* **26**, 7511 (1987).
[8] M. H. Green, L. Uhl, and J. B. Green, *J. Lipid. Res.* **26**, 806 (1985).
[9] G. Fex and G. Johannesson, *Biochim. Biophys. Acta* **901**, 255 (1987).
[10] G. Fex and G. Johannesson, *Biochim. Biophys. Acta* **944**, 249 (1988).

Proteins

Human retinol-binding protein (RBP)[11] and transthyretin (TTR)[12] are isolated from plasma/urine as described. RBP is saturated to over 80% with retinol by incubation in the dark under N_2 at room temperature for 1 hr with a 3- to 5-fold molar excess of unlabeled retinol in ethanolic solution. The resulting retinol–RBP complex is purified by chromatography on TTR linked to CNBr–Sepharose 4B.[11] The concentration of holo- and apo-RBP in the preparations is determined spectrophotometrically (ε_{330} = 46,000 M^{-1} cm^{-1} and ε_{280} = 47,760 M^{-1} cm^{-1} for holo-RBP and ε_{280} = 40,400 M^{-1} cm^{-1} for apo-RBP).[13] The TTR concentration is determined based on an ε_{280} value of 76,140 M^{-1} cm^{-1}.[14] Proteins are iodinated as described by Greenwood *et al.*[15]

[³H]Retinol–RBP Complex

Two hundred microliters of an approximately 1.5 mg/ml (76 nmol/ml) solution of RBP in 20 mM Tris-HCl, pH 7.4, containing 0.2 mM Na₂EDTA (buffer A) is incubated in the dark under N_2 with 5 μl of an ethanolic solution of [³H]retinol (5 μCi) for 60 min. The mixture is then applied on a 0.7 × 1 cm DEAE-Sepharose CL-4B column equilibrated in buffer A. The column is washed with 8 volumes buffer A and the retinol–RBP complex eluted with 0.3 M NaCl in buffer A. Fractions (0.5 ml) are collected and the ³H radioactivity in the fractions determined. Recovery of ³H radioactivity in the eluate is usually around 70% and of RBP about 95%.[9] Assume that 100% of the RBP elutes in the ³H radioactivity peak, and use this to calculate the concentration of RBP in the three fractions with the highest ³H radioactivity. The retinol saturation of RBP is assumed not to be changed by the addition of the small amount of [³H]retinol or by the chromatographic procedure. The concentration of RBP in the pooled fractions can be determined afterward by electroimmunoassay[16] or from the absorbance at 280 and 330 nm (see above).

Liposomes

Single-wall liposomes are prepared essentially as described by Mimms *et al.*[17] For the experiments with retinol transfer from RBP to liposomes,

[11] A. Vahlquist, S. F. Nilsson, and P. A. Peterson, *Eur. J. Biochem.* **20**, 160 (1971).
[12] G. Fex, E. Thulin, and C.-B. Laurell, *Eur. J. Biochem.* **75**, 181 (1977).
[13] J. Horwitz and J. Heller, *J. Biol. Chem.* **248**, 6317 (1973).
[14] J. Heller and J. Horwitz, *J. Biol. Chem.* **248**, 6308 (1973).
[15] F. C. Greenwood, W. M. Hunter, and J. S. Glover, *Biochem. J.* **89**, 114 (1963).
[16] C-B. Laurell, *Scand. J. Clin. Lab. Invest.* **29** (Suppl. 124), 21 (1972).
[17] L. Mimms, G. Zampighi, Y. Nozaki, C. Tanford, and J. Reynolds, *Biochemistry* **20**, 833 (1981).

the latter are prepared with a high cholesterol content to mimic the lipid composition of the plasma membrane, and the detergent is removed by dialysis overnight. For the studies of retinol transfer across membranes, pure lecithin membranes containing retinol are used, and detergent is removed by gel filtration, which is quicker, to minimize oxidation of the retinol. α-Tocopherol is included in all liposomes as an antioxidant. The concentrations of retinol, retinyl acetate, and α-tocopherol in organic solvents are determined using their respective absorbance coefficients (ε_{325} = 46,000 M^{-1} cm^{-1},[13] ε_{325} = 51,500 M^{-1} cm^{-1},[13] ε_{292} = 3190 M^{-1} cm^{-1} [18]).

Lecithin–Cholesterol Liposomes. Mix the following in a round-bottomed Pyrex glass tube: 4 mg egg lecithin (5.3 μmol), in chloroform/methanol, 9/1 (v/v), 1.8 mg cholesterol (4.7 μmol), in chloroform, 23 μg α-tocopherol (53 nmol), in ethanol, and 1 μCi[^{14}C]triolein, in toluene. Evaporate to dryness as a thin film with N_2 and add 44 mg octyl-β-D-glucopyranoside and 240 μl of 20 mM Tris-HCl buffer, pH 7.4, containing 0.2 mM Na$_2$EDTA and 50 mM NaCl (buffer B). Mix gently to dissolve the octyl-β-D-glucopyranoside and to disperse the lipids. The resulting solution should be clear.

Dialyze the mixture in Spectrapor 3 dialysis tubes overnight at room temperature, away from light, against 2 liters of buffer B containing 2 g/liter Bio-Beads SM 2 to trap the octyl-β-D-glucopyranoside.[19] Spin at 12,000 g to remove particles. During the dialysis the volume increases to about 0.7 ml, which means that the phospholipid concentration is about 5.7 mg/ml (7.6 μmol/ml). Phospholipid phosphorus[20] and cholesterol[21] are determined after extraction of the lipids according to Folch *et al.*[22] The liposomes obtained with this protocol are single walled according to electron microscopy, with a mean diameter of 40 nm. The cholesterol to lecithin molar ratio is close to 1 : 1.[9]

Lecithin–Retinol Liposomes. Mix the following in a round-bottomed Pyrex glass tube: 4 mg (5.3 μmol) egg lecithin in chloroform/methanol, 9/1 (v/v), 23 μg (53 nmol) α-tocopherol in acetonitrile (1 mol% relative to lecithin), 1.5 μg (5.3 nmol) retinol in acetonitrile (0.1 mol% relative to lecithin), and 0.1 μCi [^{14}C]triolein in toluene. Evaporate to dryness as a thin film under N_2 and add 23 mg octyl-β-D-glucopyranoside and 150 μl of 0.14 M Tris-HCl, pH 8.5 (buffer C), and mix until the solution is clear.

Apply the mixture on a Sephadex G-25 column (0.9 $\times$ 30 cm) and elute with buffer C. Collect 1-ml fractions. Liposomes elute in the void volume.

[18] D. W. Nierenberg and D. C. Lester, *J. Chromatogr.* **345,** 275 (1985).
[19] J. Phillipot, S. Mutaftschiev, and J. P. Liautard, *Biochim. Biophys. Acta* **734,** 137 (1983).
[20] P. S. Chen, T. Y. Toribara, and H. Warner, *Anal. Chem.* **28,** 1756 (1956).
[21] D. Webster, *Clin. Chim. Acta* **7,** 277 (1962).
[22] J. Folch, M. Lees, and G. H. Sloane-Stanley, *J. Biol. Chem.* **226,** 497 (1957).

Determine the ^{14}C radioactivity in the fractions. The concentration of phospholipid in the fractions can be approximated from the ^{14}C radioactivity, assuming 100% recovery of phospholipid and radioactivity. A pool of peak fractions usually contains 2–3 mg phospholipid/ml. Phospholipid phosphorus concentration can be checked afterward.[20] Lecithin–retinol liposomes containing encapsulated ^{125}I-labeled synthetic peptide are used to control that the liposomes remained tight during incubation.[10] The size of the liposomes can be determined by Sephacryl S-200 chromatography as described[23] using latex particles of known diameter as calibrators. Usually, liposomes prepared in this way are about 80 nm in diameter.[10]

DEAE Columns for Separation of [^{3}H]Retinol–RBP and ^{14}C-Labeled Liposomes

DEAE-Sepharose is equilibrated in buffer B and packed in small disposable columns to a bed size of 0.7×0.5 cm. They are then prerun with sonicated phospholipids as follows: 7.5 mg (10 μmol) lecithin is taken to dryness under N_2, 2 ml buffer B is added, and the mixture is sonicated 3 times for 5 min each on ice. One hundred microliters of the resulting phospholipid suspension (0.35 mg) is applied on each DEAE column and the columns washed with 5 column volumes of buffer B.

Retinol Transfer between RBP and Lecithin–Cholesterol Liposomes

Mix the following on ice to obtain the indicated final concentrations per milliliter buffer B: 104 μl of the pooled [^{3}H]retinol–RBP fractions (1 nmol[^{3}H]retinol–RBP), 81 μl of liposome solution (0.65 μmol liposomal phospholipid), 815 μl buffer A, and buffer B to obtain a final concentration of 50 mM NaCl. Start incubation at 37° immediately. At intervals (0–60 min), 50-μl aliquots are removed and applied to the 0.7×0.5 cm DEAE-Sepharose columns which have been prerun with sonicated lecithin in buffer B (above). The columns are eluted at room temperature with 1.5 ml of buffer B directly into scintillation vials. The zero-time point is obtained by mixing the components on ice and separating them immediately on the small DEAE columns. Control experiments are run without liposomes (i.e., [^{3}H]retinol–RBP only).

Calculation of Retinol Transfer to Liposomes

[14]C/^{3}H in the incubation mixture multiplied by ^{3}H/^{14}C in the eluate and retinol concentration (nmol/ml) in the mixture equals nanomoles retinol

[23] J. A. Reynolds, Y. Nozaki, and C. Tanford, *Anal. Biochem.* **130,** 471 (1983).

transferred to liposomes. This figure is corrected by subtracting the blank (experiment without liposomes), which is calculated as follows: 1/total ^{3}H applied on DEAE column $\times$ ^{3}H in the eluate $\times$ retinol concentration (nmol/ml) in the mixture = nanomoles retinol "transferred" in the blank.

Retinol Transfer across the Liposomal Membrane

To study transfer of retinol across liposomal membranes, we use the ability of horse liver alcohol dehydrogenase (ADH) to oxidize retinol.[24,25] Because retinol is not a very good substrate for ADH, it is important to ensure that there are no competing substrates present. Therefore, ADH without ethanol and lactate dehydrogenase without glycerol should be used.

Mix the following solutions (prewarmed to 37°) to obtain the following concentrations per milliliter: 400 μl lecithin–retinol liposomes corresponding to 1 μmol lecithin and 1 nmol retinol, 375 μl (15 U) ADH as a 20 mg/ml solution in buffer C, 160 μl (1.6 μmol) NAD as a 10 mM solution in buffer C, 20 μl (2 μmol) sodium pyruvate as a 0.1 M aqueous solution, 2.3 μl (8.5 U) LDH as a suspension in 3.2 M ammonium sulfate, and 43 μl buffer C. The reaction is started with ADH. At intervals (0–2 min) 200-μl samples are withdrawn by pipette and added to 1.8-ml polypropylene tubes on dry ice containing[18] 100 μl butanol/ethyl acetate, 1/1 (v/v), containing 100 mg/liter BHT, 13 μl of 1.25 mM α-tocopherol (freshly made in ethanol), 3 μl of 14 mM 2-mercaptoethanol (freshly made in water), 3 μl of 60 μM retinyl acetate internal standard (freshly made in acetonitrile), and 40 μl acetonitrile. The tubes are either processed immediately or stored at $-70°$ overnight before processing. Zero-time incubations are made by mixing all the above components (buffer instead of ADH solution) in the same proportions on dry ice. Controls are without added ADH and with encapsulated ^{125}I-labeled synthetic peptide to check for possible leakage of the liposomes during incubation.

The samples are thawed at room temperature while shaking on a manifold shaker at high speed for 5 min. Sixty microliters of K$_2$HPO$_4$ (1.2 g/ml)[18] is then added followed by mixing for another 30 sec. After centrifugation for 1 min in an Eppendorf centrifuge (12,000 g), the supernatant is transferred to a new polypropylene tube. Ten microliters is injected on an HPLC column, equipped with a 15-cm ODS 5 μm column and with acetonitrile/0.1 M ammonium acetate, 85/15 (v/v), at a flow rate of 1.5 ml/min as the mobile phase.[10]

[24] S. Futterman and J. Heller, *J. Biol. Chem.* **247**, 5168 (1972).
[25] D. Ong, B. Kakkad, and P. McDonald, *J. Biol. Chem.* **262**, 2729 (1987).

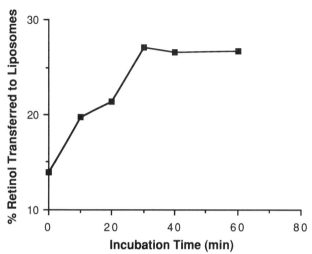

FIG. 1. Time course for the transfer of retinol from retinol-binding protein to liposomes. Concentrations and conditions as in the text.

Calculation of Rate of Inside–Outside Retinol Transfer

The retinol remaining after different incubation times is calculated as the percentage of the concentration in the zero-time samples. The retinol/retinyl acetate peak area (or peak height) ratio is used for these calculations. The logarithm of the percent retinol remaining after various incubation times is plotted against incubation time. The slope of the "final" part of the curve (i.e., after about 50% of the retinol has been oxidized), which corresponds to the rate of transfer of retinol from inside to outside, is calculated.

Results and Discussion

Retinol transfer from the retinol-binding protein complex to liposomes is a relatively rapid process which is essentially completed in 30 min (Fig. 1). This suggests a half-life for the process of the order of 10 min or less. The time course and extent of transfer were similar to previous results.[9] It is obvious that transfer occurred also on ice or during the separation of retinol–RBP and liposomes. It is known[26] that the affinity of lipophiles for membranes is higher at temperatures above the transition point of the

[26] P. Brecher, R. Saouaf, M. Sugarman, D. Eisenberg, and K. LaRosa, *J. Biol. Chem.* **259**, 13395 (1984).

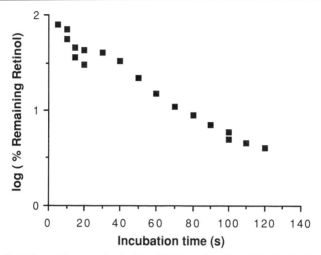

FIG. 2. Oxidation of liposomal retinol by ADH as a function of incubation time. Concentrations and conditions as in the text.

membrane lipids. Egg lecithin if fluid at all temperatures above $-10°$, and some transfer should, therefore, be expected even on ice.

Oxidation of retinol by ADH is very rapid (Fig. 2). After 2 min, less than 5% of the retinol remains. The first 50% of the retinol is oxidized with a half-life of about 12 sec; the second 50% is oxidized with a half-life of about 26 sec. If it is assumed that, at the start, retinol is equally distributed between the inner and the outer leaflets of the membrane, the retinol oxidized first should be that present on the outside and the last retinol to be oxidized should be that present on the inside. The rate of oxidation of that last retinol should, thus, give a measure of the rate of transfer of retinol from inside to outside. The half-life for the inside–outside transfer, therefore, should be about 26 sec, which is close to our previous results in a similar system.[10]

The fact that retinol transfers rapidly between RBP and phospholipid membranes and across such membranes does not mean that this is necessarily so *in vivo*. However, as all biological membranes contain areas with phospholipid bilayer structure,[27] it seems probable that these mechanisms are also operative *in vivo*. The extent and the rate at which these transfers occur may, of course, be different *in vivo,* but not necessarily less or slower. There seems to be some dependence of the affinity of a lipophilic

[27] B. DeKruyff, A. Rietveld, and P. R. Cullis, *Biochim. Biophys. Acta* **600,** 343 (1980).

compound for a membrane on the lipid composition of the membrane.[27,28] The protein content of the membrane also seems to promote partitioning into the membrane.[29] Thus, retinol transfer processes may, in fact, be more pronounced and more rapid *in vivo* than *in vitro*. Until more has been learned about the postulated RBP receptor and its importance in cellular retinol uptake, other ways for retinol to enter cells, although less specific, should not be ruled out.

Acknowledgments

This work was supported by grants from the Swedish Medical Research Council (03x-03364), the Albert Påhlsson Foundation, and the Lunds Sjukvårdsdist donation fund.

[28] M. Leonard, N. Noy, and D. Zakim, *J. Biol. Chem.* **264**, 5648 (1989).
[29] S. Lund-Katz, H. M. Laboda, L. R. McLean, and M. C. Phillips, *Biochemistry* **27**, 3416 (1988).

[43] Exchange of Retinoids between Lipid Vesicles and Rod Outer Segment Membranes

By WILLEM J. DE GRIP and FRANS J. M. DAEMEN

Introduction

For transport of lipid compounds through intracellular or extracellular aqueous compartments in the body, nature has designed special carrier systems. When bound to a water-permeable carrier a hydrophobic molecule can not only easily cross the aqueous phase, but it may also be protected from degradative processes like oxidation and finally can be properly guided to target tissue(s) and/or to further metabolic processing. Such carrier systems may either have a broad spectrum of "substrates" [like bovine serum albumin (BSA), chylomicrons, lipoproteins] or possess a high degree of specificity (like the fatty acid-binding proteins, retinoid-binding proteins, phospholipid transfer proteins). In the latter category particular protein families with common structural motifs have been identified.[1,2]

Intracellular as well as intercellular traffic of retinoids is an essential element of a physiologically very important process occurring in the ver-

[1] J. Godovac-Zimmermann, *Trends Biochem. Sci.* **13**, 64 (1988).
[2] I. Bernier and P. Jolles, *Biochimie* **69**, 1127 (1987).

tebrate retina, the visual cycle. This cycle functions to reactivate visual pigment, which has become inactivated following absorption of a photon. On light absorption, the chromophore of the visual pigment, 11-Z-retinal, is isomerized to the all-E form and eventually released. In order to regenerate, the inactive protein has to recombine with 11-Z-retinal, which must be generated by reisomerization of the all-E isomer to the 11-Z isomer (all-trans and 11-cis, respectively, in the older nomenclature). The latter, rather complex, pathway is called the visual cycle and involves enzymatic reduction to all-E-retinol (vitamin A) in the photoreceptor cell and transfer of the latter compound to the retinal pigment epithelium, where it is acylated and stored as a long-chain fatty acid retinyl ester and/or isomerized to the 11-Z geometry and eventually recycled as 11-Z-retinal back to the photoreceptor cell to regenerate active visual pigment (see elsewhere, this volume). Since several aqueous compartments have to be crossed in the cycle during this very active retinoid metabolism, one would expect the presence of specific retinoid-binding or transfer proteins in this part of the retina.

Indeed, in the retinal pigment epithelium the presence of the ubiquitous cellular retinol-binding protein (CRBP) has been demonstrated, as well as a photoreceptor tissue-specific 11-Z-retinoid-binding protein (CRALBP).[3,4] In addition, a less specific binding protein for retinol, retinal, and several other lipids, the interstitial retinol-binding protein (IRBP), has been detected in the extracellular space of the interphotoreceptor matrix (see Refs. 5–7 and elsewhere, this volume). On the other hand, no retinoid-binding proteins have so far been detected in the photoreceptor cell.[4,6,7] This apparent discrepancy can be explained by the subsequent observation that rapid exchange of retinol and retinal can occur between lipid vesicles and cells or microsomes through the aqueous space.[8–11] Hence, the close opposition of the membranes in the photoreceptor outer segment probably allows sufficiently rapid transfer of these retinoids without the need for a binding protein. As a matter of fact, several arguments can be presented that retinoid exchange between the

[3] J. C. Saari, this series, Vol. 81, p. 819.

[4] J. C. Saari and D. L. Bredberg, *Exp. Eye Res.* **46,** 569 (1988).

[5] Y.-L. Lai, B. Wiggert, Y. Liu, and G. Chader, *Nature (London)* **298,** 848 (1982).

[6] G. I. Liou, C. D. B. Bridges, S.-L. Fong, and R. A. Alvarez, *Vision Res.* **22,** 1457 (1982).

[7] A. Bunt-Milam and J. Saari, *J. Cell. Biol.* **97,** 703 (1983).

[8] R. R. Rando and F. W. Bangerter, *Biochem. Biophys. Res. Commun.* **104,** 430 (1982).

[9] G. W. T. Groenendijk, W. J. de Grip, and F. J. M. Daemen, *Vision Res.* **24,** 1623 (1984).

[10] M.-T. P. Ho, J. B. Massey, H. J. Pownall, R. E. Anderson, and J. G. Hollyfield, *J. Biol. Chem.* **264,** 928 (1989).

[11] G. Fex and G. Johanneson, *Biochim. Biophys. Acta* **944,** 249 (1988).

photoreceptor and pigment epithelium microvilli also might involve direct transfer: (1) No evidence has yet been presented that the photoreceptor or pigment epithelium contain receptors for IRBP. (2) The presence of IRBP does not accelerate exchange of retinol between lipid vesicles and actually reduces mass transfer.[10] Hence, IRBP might function as a temporary water-soluble store or trap for retinoids rather than as an exchange carrier. (3) The outer segment plasma membrane and the pigment epithelial microvillar membranes are sufficiently closely juxtapositioned to allow rapid transfer through the aqueous phase.

In vitro studies on retinoid uptake and metabolism in, for example, isolated retinas or in isolated photoreceptor or pigment epithelium cells do require a suitable carrier to administer the retinoid. Mere injection, for instance, of a concentrated solution of retinoid in an organic solvent, into an aqueous medium will cause severe aggregation and loss owing to adherence to whatever surfaces are present,[12] preventing reliable kinetic or dose-dependent studies. As an alternative, lipid vesicles as retinoid carriers have been successfully used to regenerate visual pigment in isolated retinas,[13,14] to study retinoid transfer between lipid systems,[8-11] and to supply retinol to isolated retinal pigment epithelium cells.[15,16] In these studies, lipid vesicles proved to function as a stable and reliable carrier system, which, in view of the arguments listed above, probably even mimics the physiological situation.

In the following sections we discuss the preparation of lipid vesicles for retinoid transfer studies, describe methods to measure exchange between vesicles and rod outer segment membranes, and present some characteristic features of the latter system.

Preparation of Lipid Vesicles as Retinoid Carrier

Principles. Most studies have employed small unilamellar vesicles (SUV, 30–50 nm in diameter) prepared by sonication, as these can easily be separated from the large rod outer segment membranes by centrifugation.[9,10] However, if this separation is not required, larger multilamellar liposomes can be used almost equally well.[17] The latter allow higher molar ratios of retinoid to lipid (e.g., up to 50 mol% for retinal).

[12] A. M. M. Timmers, W. A. H. M. van Groningen-Luyben, F. J. M. Daemen, and W. J. de Grip, *J. Lipid Res.* **27**, 979 (1986).
[13] S. Yoshikami and G. N. Nöll, this series, Vol. 81, p. 447.
[14] G. N. Nöll, *Vision Res.* **24**, 1615 (1984).
[15] A. M. M. Timmers, Ph.D. Thesis, University of Nijmegen (1987).
[16] A. M. M. Timmers and W. J. de Grip, this series, Vol. 190 [1].
[17] W. J. de Grip and J. van Oostrum, unpublished results (1985).

The lipid composition of the vesicles is not critical. Pure egg *sn*-1,2-di-acylglycerophosphocholine (PC),[8,10] dioleoyl-PC,[13,14] soy PC,[15-17] or combinations of PC with *sn*-1,2-diacylglycerophosphatidic acid (PA),[9] *sn*-1,2-diacylglycerophosphoethanolamine (PE),[9,10] or cholesterol[8] all perform satisfactorily. However, when the geometric configuration of the donor retinoid is critical (e.g., if a pure all-*E* or 11-*Z* isomer should be administered), no aminophospholipids, such as PE or *sn*-1,2-diacylglycerophosphoserine (PS), should be included since these efficiently catalyze isomerization, in particular of the retinals via Schiff base intermediates.[18,19] A simple system like egg PC or soy PC is therefore generally applicable and recommended. As a check on exchange arising from vesicle fusion rather than transfer, donor or acceptor vesicles can be charged with minor amounts of a nonexchangeable label like cholesteryl oleate[8,9,15,16] or triolein.[10] Antioxidants such as vitamin E[13,14] or butylated hydroxytoluene (BHT)[10] need not be included for short incubation times (2–4 hr) when the carrier vesicles have been prepared and stored under an inert atmosphere (N_2 or, preferably, argon).

Charging of the vesicles with retinoid can be done prior to liposome formation and sonication[8,13,14] or afterward by rapid mixing of a vesicle dispersion with a concentrated solution of retinoid in dimethylformamide (DMF).[9,10,16] The latter method is preferred since it presents less risk of retinoid degradation, and it is the method of choice to charge the rod outer segment membranes. Molar ratios of retinoid to lipid of 5×10^{-4} to 5×10^{-2} have been applied. Higher ratios are not recommended, since the vesicles then tend to become fusogenic. In addition, one should be aware that retinol is lytic to cells, and high doses of this retinoid should be avoided.

The stability of the carrier vesicles is excellent. During a 2-hr incubation at 37° no significant degradation of retinoid is observed and isomerization of all-*E* to 13-*Z* is usually less than 4%.[15,17] When frozen rapidly, the carrier vesicles can be stored under argon for at least several days at −80°. Repeated freezing and thawing should, however, be avoided as this promotes vesicle fusion.

Procedure. A typical batch requires from 5 to 20 mg of egg or soy PC (~6.5–26 μmol). Both are commercially available (e.g., from Lipid Products, South Nutfield, Surrey, UK) and supplied in chloroform/methanol or benzene/ethanol solution. To the lipid solution 0.02 mol% of [^{14}C]cholesteryl oleate is added, the solvent is evaporated by a stream of nitrogen,

[18] G. W. T. Groenendijk, C. W. M. Jacobs, S. L. Bonting, and F. J. M. Daemen, *Eur. J. Biochem.* **106,** 119 (1980).

[19] P. S. Bernstein, B. S. Fulton, and R. R. Rando, *Biochemistry* **25,** 3370 (1986).

and the lipid residue is thoroughly dried to remove all traces of solvent (30 min, vacuum pump). The residue is then suspended in 1 ml of the required buffer or tissue culture medium (see next section) by vigorous shaking on a mechanical shaker (15 min) with argon or N_2 as the top gas. This yields a creamy liposome suspension, which is converted to a weakly opalescent dispersion of small vesicles by sonication on ice [e.g., Branson B12 (Branson, Danbury, CT) sonifier with microtip, half-maximal intensity; 15–20 min]. The resulting vesicle suspension is cleared of titanium particles and remaining liposomal structures by centrifugation (100,000 g, 10 min, 4°).

When the geometric configuration of the retinoid is of importance, all subsequent manipulations should be performed in dim red light (e.g., Kodak safety light, Eastman Kodak, Rochester, NY) to prevent photoisomerization. ^{3}H-Labeled retinoid (10–20 μCi for exchange studies, 50–75 μCi for metabolic studies) is mixed with the desired amount of unlabeled material. For metabolic studies, purification of the retinoid by preparative HPLC is strongly recommended (e.g., Ref. 16 or this series, Vol. 190). Finally, the required amount of retinoid is carefully dissolved under argon in a small volume of DMF. This concentrated solution is rapidly diluted with a volume of vesicle suspension such that the final DMF concentration is below 2.0% (v/v) and the molar ratio retinoid/lipid does not exceed 5 mol%. The resulting mixture is thoroughly mixed (hand or vortex mixer) for several minutes with argon as the top gas. This allows incorporation of 95–100% of the retinoid into the vesicle membrane, which can be checked by centrifugation of a vesicle aliquot (100,000 g, 4°, 60 min) and liquid scintillation analysis of the supernatant and vesicle pellet. The carrier vesicles thus obtained are ready for use, or they may be stored under argon after rapid freezing for at least several days at −80° in small batches. Repeated freezing/thawing should be avoided. The same procedure can be applied to charge isolated rod outer segment membranes, except that, in the case of intact outer segments, mixing with the concentrated retinoid solution should be performed very carefully, preferably in the presence of a high sucrose concentration (250–600 mM), which stabilizes the outer segment structure.[20]

Incubation of Vesicles with Rod Outer Segment Membranes and Analysis of Exchange

Principles. Several media and conditions have been satisfactorily used: Krebs–Ringer or an equivalent ionic buffer, either oxygenated or nonoxygenated,[8,10,13,14] tissue culture media such as RPMI 1640 DM,[15,16]

---

[20] P. P. M. Schnetkamp and F. J. M. Daemen, this series, Vol. 81, p. 110.

and buffered 0.25 M sucrose.[9] Addition of EDTA (0.5–1 mM) is recommended to trap heavy metal ions, which catalyze oxidative degradation of both phospholipids and retinoids. It should be noted that higher ionic strength reduces the transfer rate proportionally,[10] probably owing to lowering of the water/lipid phase distribution coefficient. The rate of transfer of retinol increases about 7-fold going from 10 to 35°,[10] so measurements can easily be performed over the relevant range of 0–37°.

A range in concentration of donor or acceptor vesicle has been used: 50 μM–5 mM in phospholipid.[8,9,10,13,16] Between 50 μM and 1 mM, transfer rates are not dependent on vesicle concentration.[10] When rod outer segment membranes are used as the acceptor system, they should be present in at least 3- to 5-fold molar excess on a phospholipid basis with respect to the donor vesicles, in view of their relative low affinity for retinoids.[9]

In order to measure net exchange between donor and acceptor several approaches can be taken. First, net transfer of retinoid is determined most accurately and with highest sensitivity by measuring the retinoid distribution in donor and acceptor over time using radioactively labeled retinoid. For this, donor and acceptor have to be physically separated. For vesicles and intact rod outer segments or rod outer segment membranes, this can be easily accomplished by short centrifugation (30,000 g, 4°, 5 min) which precipitates over 95% of the membranes without significant contamination with vesicles.[9,10] This method is not applicable for liposomes; in this case, however, rod outer segment membranes can be selectively precipitated by means of a more complicated technique, such as immunoprecipitation using specific anti(rhod)opsin antibodies. Transfer can also be analyzed with unlabeled retinoids, but even with fluorometric analysis (retinol, retinyl ester) detection limits are at least 10- to 100-fold higher than for labeled material.

In the second approach, when an acceptor-dependent secondary reaction is to be measured, separation of donor and acceptor systems is not required or recommended.[13,14,17] This is the case, for example, when illuminated retina or rod outer segments are incubated with 11-Z-retinal in vesicles or liposomes to produce rhodopsin, which can be measured spectroscopically or electrophysiologically,[13,21] and when production of retinyl ester from retinol in the acceptor system is followed by TLC or HPLC.[15] Under these conditions, however, the kinetics are complex, and it will be difficult to obtain kinetic parameters for individual reactions (e.g., the rate of transfer), unless it can be unequivocally demonstrated that a specific reaction is rate-limiting.

[21] D. R. Pepperberg, this series, Vol. 81, p. 452.

Finally, when transfer is so rapid that accurate kinetic data cannot be obtained with the first approach, one should resort to more elaborate analytical techniques, such as (de)quenching of fluorescence using a rapid kinetic stopped-flow setup, which has a time resolution of several milliseconds.[10] This third approach is applicable both to the fluorescent retinols and retinyl esters (using a quencher molecule in the donor or acceptor system[10]) and to the nonfluorescent retinals (using a nonexchangeable quenchable fluorescent probe in the acceptor system). For most applications the first procedure, radiolabeling, suffices, and this is described in some detail.

Procedure. Because the interphotoreceptor matrix is densely populated with mucopolysaccharides,[22] an exchange medium containing sucrose will resemble the *in vivo* situation more closely than a purely ionic medium. In addition, sucrose will stabilize intact outer segments and retard their sedimentation.[20] Therefore, as exchange medium we routinely apply 0.25 M sucrose, 1 mM EDTA in 10 mM Tris-HCl, pH 7.4. Higher concentrations of sucrose (up to 0.6 M), however, have successfully been used.

Equal volumes of suspensions of retinoid carrier vesicles (100–200 μM phospholipid/ml) and rod outer segment membranes (0.4–1.0 mM phospholipid/ml) in exchange medium are combined and gently mixed ($t = 0$). The lipid content is determined by a phosphate assay (e.g., Ref. 23). The mixture is incubated at 20° in a polyethylene or polypropylene reaction tube or centrifuge tube in a thermostatted waterbath. At several time intervals, 0.1-ml aliquots are withdrawn and centrifuged in a microcentrifuge (5 min, 30,000 g). Equilibrium is usually reached within 30–60 min. Care should be taken that homogeneous samples are taken, and, if required, the mixtures should be occasionally gently swirled to maintain a homogeneous distribution of acceptor and donor systems.

After centrifugation, 50 μl of the supernatant is thoroughly mixed with 4 ml of scintillation fluid (Opti-Fluor, Packard, Downers Grove, IL) and counted by scintillation spectrometry in the double-label mode ($^{14}C/^{3}H$). The pellet is washed once with exchange medium, resuspended in exchange medium to a final volume of 0.1 ml, and counted as described above. The total recovery is determined from the recovery of ^{14}C label and should be over 95%. Based on this figure, the recovery of retinoid ^{3}H label varies between 95 and 103%, indicating that no significant loss of retinoid occurs. Exchange of ^{14}C label between donor and acceptor should

[22] A. T. Hewitt, *in* "The Retina, Part II" (R. Adler and D. Farber, eds.), p. 169. Academic Press, New York, 1986.
[23] E. H. S. Drenthe and F. J. M. Daemen, this series, Vol. 81, p. 320.

be less than 5%, otherwise membrane fusion events will have interfered significantly. The percent exchange is calculated by taking the average of the amount of label in the acceptor system and the loss of label in the donor system, correcting this figure for the percent recovery, and dividing the outcome by the amount of label originally present in the donor vesicles. Of course all figures should be normalized to the same amount of donor lipid.

It is recommended, even for studies investigating only net exchange, that a final sample of incubation mixture be subjected to retinoid analysis[16,18] to ensure that no extensive degradation or change in class distribution (e.g., by metabolic processing) has occurred.

Present Status

General Patterns

For retinals and retinols, irrespective of geometric configuration, rapid equilibration between donor and acceptor lipid systems is observed, with half-times of the order of seconds[10] (vesicles versus vesicles,[8–11] vesicles versus erythrocytes,[8] or vesicles versus microsomes[9]). Available evidence indicates that transfer takes place through the aqueous phase[8–10] and is therefore dependent on the equilibrium distribution coefficient K_d of retinoid between the aqueous and the lipid phase. The K_d is of comparable magnitude for retinol and retinal[9] but is appreciably smaller for the ester retinyl palmitate.[9] This explains why retinyl palmitate does not exchange and why the less hydrophobic retinyl acetate represents an intermediate situation.[9,24] Exchange from or into rod outer segments or rod outer segment membranes partially deviates from this general pattern and is discussed for the three retinoid classes (retinal, retinol, retinyl ester) separately.

Rod Outer Segment(s) (Membranes)

Retinal. Retinal rapidly (within minutes) equilibrates between vesicles and rod outer segments.[9,17] Donor vesicles charged with 11-Z-retinal effectively promote regeneration of bleached rhodopsin.[13,14,17] This requires a large excess of 11-Z-retinal over opsin for isolated retinas (>1000-fold[14]), whereas for isolated rod outer segment membranes a 2- to 3-fold excess suffices.[17] Diffusion over the large area of retinal membranes and aminophospholipid-catalyzed isomerization to all-E and 13-Z[18] strongly reduce the regeneration efficiency in the retina. On regeneration of opsin

---

[24] M.-T. P. Ho, H. J. Pownall, and J. G. Hollyfield, *J. Biol. Chem.* **264,** 17759 (1989).

in rod outer segment membranes, the transfer of retinal appears to be rate-limiting, as under identical conditions in detergent solution the rate of regeneration is about 5-fold higher.[17] In addition, retinal does not equilibrate randomly between egg or soy PC vesicles and rod outer segment membranes. Even for a three-fold excess of membrane lipid over vesicle lipid, only 40 to 45% of retinal equilibrates with the rod membranes.[9] These observations suggest that highly unsaturated membranes such as the rod outer segment membrane have a somewhat lower affinity and/or binding capacity for retinal than the less unsaturated membranes produced from egg or soy PC. Another remarkable observation in this context is that the rate of exchange is 2- to 3-fold slower for rod outer segment membranes relative to intact rod outer segments.[9] Nevertheless, the equilibrium distribution of retinal is similar for intact rod outer segments and rod outer segment membranes.

Retinol. Relatively rapid equilibration of retinol between vesicles and rod outer segment membranes has been reported, although actual rates rather diverge, from less than 1 min for a hypotonic ionic medium[10] to at least 3 min for a buffered isotonic sucrose medium.[9] This may relate to effects of ion strength or solvent polarity on the rate of transfer.[10] Again no random distribution is achieved (35–40% retinol in 3- to 5-fold excess of rod membrane lipid), retinol, like retinal, favoring the less unsaturated lipid phase of the vesicle.[9,10] In the case of retinol, on the other hand, intact rod outer segments behave quite differently: they take up very little retinol and, when charged with retinol, rather slowly lose over 90% to added vesicles ($t_{1/2}$ ~3 min), even at very low molar ratios of retinol to lipid (1/1000).[9] This suggests that there is either a one-way diffusion barrier from rod outer segment plasma membrane to disk membrane or a "purge mechanism" which rapidly clears the disk membrane. Either mechanism would accelerate transfer of retinol, produced on illumination, from the rod outer segment to the pigment epithelium. This is particularly important at high illumination levels, in view of the lytic properties of retinol.

Retinyl Ester. Retinyl ester demonstrates puzzling behavior as well. As expected, it does not transfer from vesicle to rod outer segments.[9] This may explain the fact that vesicles charged with 11-Z-retinyl palmitate do not promote rhodopsin regeneration in the isolated frog retina.[14] However, rod outer segments charged with retinyl palmitate lose 90–95% of their label within 2 min when incubated with vesicles.[9] The exchanged label could be positively identified as retinyl palmitate by HPLC. This unidirectional exchange is probably mediated by direct membrane–membrane interaction as up to 10% of cholesteryl oleate did exchange as well.[9] The latter phenomenon was observed only in the system of retinyl palmitate-charged rod outer segments versus lipid vesicles. A possible explanation is that the binding capacity of the rod outer segment membrane for retinyl ester is so low that the retinoid is clustered in the membrane instead of randomly distributed. Such membrane-perturbing clusters may promote membrane interaction or membrane fusion events.

Conclusion

Lipid vesicles have proved to be a reliable, stable, and reproducible carrier system to introduce retinol or retinal into retinoid-consuming or retinoid-metabolizing cells. The special properties of the rod outer segment lipid matrix may, however, present special problems, which need to be investigated in more detail. Properties such as the affinity and binding capacity of intact rod outer segments or the entire rod outer segment membrane population for the various retinoid classes and their subclasses of isomers need to be accurately determined. When special characteristics of the membrane matrix would indeed favor extrusion or unidirectional flow of retinoids, this might add new elements to the physicochemical properties of membranes as well as provide important new clues into the functioning of the visual cycle.

[44] Intermembranous Transfer of Retinoids

By Robert R. Rando and Faan Wen Bangerter

Introduction

Hydrophobic compounds undergo intermembranous transfer at vastly different rates. If transfer were to require that the hydrophobic material leave one membrane and enter the aqueous phase before inserting into the second, then it would be expected that the more hydrophobic the compound, the slower the rate of exchange. The more important kinds of exchange processes probably involve this type of mechanism. As part of the visual cycle, retinol (vitamin A) and retinal (vitamin A aldehyde) undergo intermembranous transfer within the retina.[1] This transfer occurs mainly between the rod outer segment discs, the rod outer segment plasma membrane, and the pigment epithelium cell plasma membrane. It is not clear whether or not this transfer is mediated by carrier proteins,

[1] G. Wald, *Science* **162**, 230 (1968).

although several retinol- and retinal-binding proteins have been found in the retina.[2-5] The physiological functions of these proteins are currently unknown.

It is generally assumed that intermembranous hydrophobic ligand transfer proteins are reserved for compounds which undergo very slow transfer in the absence of a catalyst.[6] For example, cholesterol undergoes fairly rapid transfer from membrane to membrane by itself, whereas phosphatidylcholine (PC) derivatives do not.[7] Indeed, well-characterized phospholipid transfer proteins have been reported in the literature, whereas none have been reported for unmodified cholesterol.[8] It was of some interest, then, to determine the basal rate of intermembranous transfer of vitamin A and its derivatives in the absence of putative transfer proteins. It could be shown that all-*trans*-retinol, all-*trans*-retinal, and 11-*cis*-retinol undergo exceedingly facile intermembranous transfer, whereas the all-*trans*-retinyl palmitate does not.[9]

General Methods

Materials. all-*trans*-Retinal, all-*trans*-retinol, and all-*trans*-retinyl palmitate are products of Sigma (St. Louis, MO). 11-*cis*-Retinal was a generous gift of Dr. William Scott of Hoffman-LaRoche, Inc. all-*trans*-[15-^{3}H]Retinol was purchased from New England Nuclear, (Boston, MA). all-*trans*-[15-^{3}H]Retinyl palmitate is synthesized from the alcohol by the published procedure.[10] all-*trans*-[15-^{3}H]Retinal is prepared by oxidizing the alcohol with manganese dioxide by the published procedure.[11] 11-*cis*-[15-^{3}H]Retinol is prepared by the reduction of the aldehyde with sodium[^{3}H]borohydride, also by the published procedure.[10] The specific activities of the [^{3}H]retinoids are in the 2–10 mCi/mmol range. Sodium

[2] S. Futterman, J. C. Saari, and D. E. Swanson, *Exp. Eye Res.* **22,** 419 (1976).

[3] J. C. Saari, S. Futterman, and L. Bredberg, *J. Biol. Chem.* **253,** 6432 (1978); W. Stillwell and M. Ricketts, *Biochem. Biophys. Res. Commun.* **97,** 148 (1980).

[4] G. W. Stubbs, J. C. Saari, and S. Futterman, *J. Biol. Chem.* **254,** 8529 (1979).

[5] Y.-C. Lai, B. Wiggert, Y. Lui, and G. Chader, *Inst. Opthalmol. Visual Sci.* **20** (Suppl.), 210 (1981).

[6] J. M. Backer and E. A. Dawidowicz, *Biochemistry* **20,** 3805 (1981).

[7] J. M. Backer and E. A. Dawidowicz, *Biochim. Biophys. Acta* **551,** 260 (1979). J. M. Backer and E. A. Dawidowicz, *J. Biol. Chem.* **256,** 586 (1981).

[8] A. M. Kasper and G. M. Helmkamp, *Biochemistry* **18,** 3624 (1981).

[9] R. R. Rando and F. W. Bangerter, *Biochem. Biophys. Res. Commun.* **104,** 430 (1982).

[10] G. W. T. Groenendijk, C. W. M. Jacobs, S. L. Bonting, and F. S. M. Daemen, *Eur. J. Biochem,* **106,** 119 (1980).

[11] M. Akhtar, P. T. Blosse, and P. B. Dewhurst, *J. Biochem.* **110,** 693 (1968).

[³H]borohydride and [¹⁴C]cholesterol oleate are products of New England Nuclear. Lactosyl ceramide is a product of Miles Laboratories (Naperville, IL).

Ricin, the β-galactosyl-binding agglutinin from *Ricinus communis*, was obtained from Boehringer Mannheim, (Indianapolis, IN). Egg yolk lecithin (EYL) is prepared and purified according to the method of Litman.[12] Egg phosphatidylethanolamine (PE) was purchased from Sigma. The phospholipid concentration is determined as inorganic phosphate after ashing and acid hydrolysis.[13] Bovine erythrocytes are collected in heparin-containing NaCl/Tris or phosphate (140 mM NaCl, 50 mM Tris, pH 7.4) from a slaughterhouse on the day that they are to be used. The erythrocytes are repeatedly washed with NaCl/Tris (phosphate) (pH 7.4) and centrifuged (2000 g). The cells are washed continuously until no more white layer appears. At this point the cells are ready for the transfer measurements. The small unilamellar vesicles (SUV) used in these studies are prepared by both the sonication method and the ethanol injection technique.[14,15]

Transfer Procedures

Vesicle–Vesicle Transfer. The transfer technique used for vesicle–vesicle transfer is adapted from the work of Backer and Dawidowicz.[7] In the experiments, donor vesicles are made up containing the [³H]retinoid, and acceptor vesicles are made up containing approximately 10% lactosylceramide and trace quantities of [¹⁴C]cholesterol oleate. The donor to acceptor concentrations are varied from 1 : 1 to 1 : 10. The nontransferable lactosylceramide enables the acceptor vesicles to be precipitated by adding ricin agglutinin, and the nonexchangeable [¹⁴C]cholesterol oleate marker is added to determine the efficiency of precipitation. The donor vesicles (0.05–0.1 μmol/ml phospholipid) are incubated with the acceptor vesicles (0.1–1 μmol/ml phospholipid) in the NaCl/Tris buffer at various temperatures. At various times, aliquots of the sample are removed and added to 1 ml of 200 mg/ml ricin agglutinin in the standard buffer at 10°. The sample is allowed to aggregate for 10 min and then centrifuged in a clinical centrifuge. The pellet is washed with cold buffer dissolved in ethanol and counted after removal of the precipitated protein. The

12 B. J. Litman, *Biochemistry* **12**, 2545 (1973).
13 B. M. Ames and D. T. Dubin, *J. Biol. Chem.* **235**, 769 (1960).
14 Y. Barenholz, D. Gibbes, B. J. Litman, J. Goll, T. E. Thompson, and F. D. Carlson, *Biochemistry* **16**, 2806 (1977).
15 B. Kwok-Keung Fung and L. Stryer, *Biochemistry* **17**, 5241 (1978).

amount of ^3H transferred to the acceptor vesicles as a function of time should theoretically yield a transfer rate.[7]

Vesicle–Erythrocyte Transfer. For vesicle–erythrocyte transfer, the donor vesicles containing the $[^3H]$retinoid are mixed with freshly prepared bovine erythrocytes. With time, the erythrocytes are removed by centrifugation and counted. The vesicles, of course, remain in the supernatant. These experiments are carried out similarly to other transfer experiments in this laboratory.[16]

Two milliliters of freshly obtained bovine erythrocytes (1×10^9 cells) in phosphate-buffered saline and 5 mM lactose is treated with vesicles (0.5 mmol/ml phospholipid) containing 25 mol% cholesterol and 5 mol% $[^3H]$retinoid at 25° with gentle shaking. At various times aliquots are removed, centrifuged, and washed twice with cold buffer. After 2 hr, approximately 30% of the erythrocytes have lysed, a result not at odds with the known surface-active effects of the retinoids.[17] The pellet containing the bovine erythrocytes is dissolved in Protosol and counted in Aquasol (New England Nuclear) by standard techniques.

Results of Transfer Experiments

Vesicle–Vesicle Transfer. Initial experiments on the intermembranous transfer of the $[^3H]$retinoids are done using SUV. Donor vesicles are loaded with $[^3H]$retinoid and incubated with acceptor vesicles containing nontransferable lactosylceramide and $[^{14}C]$cholesterol oleate. With time, aliquots are removed and treated with the β-galactosyl-binding agglutinin ricin to precipitate the acceptor vesicles. Surprisingly, under the conditions of the experiment, the transfer is so rapid that rates cannot be measured by this technique (Table I). This technique is not very sensitive to rapid transfer because of the relatively long period of time (minutes) required to fully precipitate the acceptor vesicles with ricin. Nevertheless, several interesting conclusions can be drawn from these experiments. First, the rate of transfer of the *cis*- and *trans*-retinols and *trans*-retinal is exceedingly rapid. By way of comparison, the rate $t_{1/2}$ for $[^3H]$cholesterol transfer under the same conditions is approximately 15 min. The results with the free vitamin are to be contrasted with those obtained with retinyl palmitate, where the transfer rate is negligible (Table I). This is expected owing to the increased hydrophobicity of the ester. The small nonequilibrium transfer of ~5% of the ester radioactivity presumably is due to nonspecific processes occurring during analysis.

The rapid rate of transfer of the retinoids appears not to depend on the

[16] R. R. Rando, J. Slama, and F. W. Bangerter, *Proc. Natl. Acad. Sci. U.S.A.* **77**, 2510 (1978).

[17] J. T. Dingle and J. A. Lucy, *Biol. Rev.* **40**, 422 (1965).

TABLE I

INTERMEMBRANOUS TRANSFER OF [³H]RETINOIDS[a]

[³H]Retinoids in donor vesicle (mol%)	Relative concentration of acceptor to donor	Incubation time (min)	[³H]Retinoid transferred to acceptor (%)
5% all-*trans*-[³H]Retinol/PC	1	2, 5, 10, 30[b]	41 ± 3
5% all-*trans*-[³H]Retinol/PC	2	2, 5, 10, 30	41 ± 4
5% all-*trans*-[³H]Retinol/PC	5	2, 5, 10, 30	53 ± 4
5% all-*trans*-[³H]Retinol/PC	10	2, 5, 10, 30	83 ± 5
5% 11-*cis*-[³H]Retinol/PC	10	2, 5, 10, 30	87 ± 5
5% all-*trans*-[³H]Retinol/PC	10	2, 5, 10, 30	79 ± 5
5% all-*trans*-[³H]Retinol/PC/PE-10%[c]	10	2, 5, 10, 30	84 ± 6
5% all-*trans*-[³H]Retinyl palmitate/PC	10	2, 5, 10, 30	5.7 ± 4

[a] Small unilamellar donor vesicles are prepared in 50 mM Tris buffer (pH 7.5) plus 140 mM NaCl with [³H]retinoid and pure egg phosphatidylcholine (PC) plus 10 mol% pure egg phosphatidylethanolamine (PE). Acceptor vesicles are prepared from PC plus 10% lactosyl ceramide plus 10% [¹⁴C]cholesterol oleate by procedures identical to those noted above. The nonexchangeable lactosylceramide enables the acceptor vesicles to be precipitated by adding ricin agglutinin, and the nonexchangeable [¹⁴C]cholesterol oleate marker is added to determine the efficiency of precipitation. The donor vesicles (0.1 μmol/ml phospholipid) are incubated with the acceptor vesicles (0.1–1.0 μmol/ml phospholipid) in the NaCl/tris buffer at various temperatures. At various times, aliquots of this sample are removed and added to 1 ml of 200 mg/ml ricin agglutinin in the standard buffer at 10°. The sample is allowed to aggregate for 10 min then centrifuged in a clinical centrifuge. The pellet is washed with cold buffer, dissolved in ethanol, and counted after removal of the precipitated protein. The amount of ³H transferred to the vesicles with time theoretically gives a transfer rate. Under the conditions of these experiments a time course could not be measured.
[b] Experiments were performed at 5, 10, and 25° with identical results. Remaining experiments were performed at 25°.
[c] In this experiment the acceptor vesicles also contained 10% PE.

level of oxidation nor the stereochemistry about the 11–12 bond. Of special interest here are the experiments with all-*trans*-retinal and vesicles prepared from either phosphatidylcholine or phosphatidylcholine plus phosphatidylethanolamine. The Schiff base formed between the aldehyde and the amine does not appear to slow the transfer rate markedly, although, of course, the time scale of the assay employed exerts a leveling effect on the rates. In addition, there is little or no effect of temperature on the transfer rate. In all cases, the retinoids can be completely transferred out of the donors to the acceptors, assuming a large enough excess of acceptor phospholipid is made available. However, it appears that retinoid–lipid interactions are slightly more favorable in the donor vesicles as compared to the acceptor vesicles (Table I).

Vesicle–Erythrocyte Transfer. The results from the vesicle–vesicle transfer experiments indicate that the rates of free retinol(al) transfer are exceedingly rapid and much too rapid to quantitate by that method. A direct centrifugation assay is required here, without an intervening agglutination step. To this end the transfer of the [³H]retinoids into bovine erythrocytes, which could be rapidly removed by centrifugation, is measured. The transfer of *trans*-[³H]retinol from EYL-based SUV at two different phospholipid concentrations into bovine erythrocytes is measured (Fig. 1). Even here the rates of transfer are barely measurable, with half-times of 1–2 min.[9] The rates of cholesterol transfer under similar conditions are 10–20 fold slower. To compare the rates for 11-*cis*-retinol and all-*trans*-retinol, transfer vesicles are made up by the ethanol injection technique, as sonication tends to isomerize the 11-*cis*-retinol. These rates are again approximately measured as before and are very similar, if not identical.[9] This suggests that 11-*cis*-retinol is not bound especially to the *cis* fatty acyl moieties in membranes. Finally, experiments are conducted measuring the rate of all-*trans*-[³H]retinal transfer from vesicles prepared either from pure EYL or from mixtures of EYL and phosphatidylethanolamine into erythrocytes.[9] The rates are again very rapid and not quantifiable by this technique, but there appears to be a retardation of transfer rate in the presence of the phosphatidylethanolamine.[9] In addition, the amount of transfer appears to be less in the presence of phosphatidylethanolamine. Both phenomena could easily be ramifications of Schiff base formation between the membrane-bound retinal and the primary amino moiety of phosphatidylethanolamine.[10]

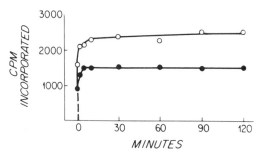

FIG. 1. Transfer of all-*trans*-[³H]retinol to bovine erythrocytes. SUV were prepared containing approximately 5% all-*trans*-[³H]retinol, 20% cholesterol, and 75% EYL. The vesicles were mixed with the erythrocytes, and the amount of ³H incorporation into these cells was measured with time according to the protocol described in the text. (●) refers to an incubation in which the final EYL concentration was 0.5 μmol/ml, and for (○) the final concentration was 1.25 μmol/ml. The extent of transfer for (●) was 19% and for (○) it was 14%. The zero time point in the experiment is actually in the neighborhood of 27 sec.

Conclusions

In conclusion, it can be shown that nonesterified retinoids undergo rapid intermembranous transfer in the absence of catalytic transfer or exchange proteins. The rates of transfer are at least an order of magnitude faster than that for cholesterol. These results suggest that carrier protein assistance is not essential for catalyzing bulk transfer of the retinols(als) in the retina and retinal pigment epithelium. The intermembrane transfer techniques discussed here are clearly of only qualitative use for rapid transfer processes such as those found in the case of retinol and retinal. This is because the standard separation techniques are useful only when the half-times for transfer occur on a time scale of minutes, whereas the actual transfer rates for the retinol(al)s occur within seconds. In order to measure more rapid rates of intermembranous transfer, spectroscopic techniques coupled with rapid kinetic measurements of the kind described by Storch and Kleinfeld are required.[18] These techniques have been recently applied to measure rates of intramembranous transfer of retinol, and first-order transfer rates of 0.85 sec^{-1} were determined.[19] However, the results of transfer experiments described here certainly demonstrate enormous differences in the rates of intermembranous transfer of retinol(al) compared to retinyl palmitate, for which appreciable intermembranous transfer did not occur. The considerable hydrophobicity of the retinyl esters renders their rates of transfer low. These results imply that carrier protein assistance need not be required for bulk transfer of retinol(al) in the retina/pigment epithelium. However, putative carrier proteins might function by targeting the retinoids to specific membranes and protecting them from possible chemical degradation. Since long-chain fatty acyl retinyl esters are not transferable at substantial rates, putative binding proteins for these derivatives would be expected to play a catalytic role in intermembranous transfer processes.

Acknowledgments

These studies were funded by National Institutes of Health Grant EY-03624.

[18] J. Storch and A. M. Kleinfeld, *Biochemistry* **25,** 1717 (1986).
[19] M.-T. Ho, M. B. Massey, H. J. Pownall, R. E. Anderson, and J. G. Hollyfield, *J. Biol. Chem.* **264,** 928 (1989).

[45] Incorporation of β-Carotene into Mixed Micelles

By LOUISE M. CANFIELD, THOMAS A. FRITZ, and THOMAS E. TARARA

Introduction

Dietary β-carotene is correlated with decreased incidence of a variety of cancers,[1] and it may have biological functions independent of its provitamin A activity.[2,3] These discoveries have led to renewed interest in the biological functions of carotene and the need to develop reliable *in vitro* assay systems. However, owing to its structure (Fig. 1), β-carotene is insoluble in water and has limited solubility in organic solvents.[4] In addition, β-carotene is easily oxidized both by light and by components in biological fluids, for example, metals and active oxygen species.[5] These factors have complicated the study of β-carotene metabolism; in fact, the differential solubility of β-carotene may explain the disparate results obtained by different laboratories.[6]

This chapter presents a method for the preparation of mixed micelles containing known quantities of β-carotene. The preparation is stable, reproducible, and simple to prepare. Although these micelles may have general application as a delivery vehicle in a variety of *in vitro* systems, the procedure was modified from an earlier preparation[7] designed to simulate micelles formed in the lumen of the human small intestine. Thus, the preparation is particularly well suited for the study of intestinal absorption and metabolism of β-carotene.

Reagents and Materials

Stock Solutions

Bile salts, 0.1 M: A stock solution containing 30 mM sodium glycocholate, 30 mM sodium glycochenodeoxycholate, 15 mM sodium glycodeoxycholate, 10 mM sodium taurocholate, 10 mM sodium taurochenodeoxycholate, and 5 mM sodium taurodeoxy-

[1] R. Peto, R. Doll, J. D. Buckley, and M. B. Sporn, *Nature (London)* **290**, 201 (1980).
[2] T. E. Edes, W. Thornton, and J. Shah, *J. Nutr.* **119**, 796 (1989).
[3] A. Bendich, *Clin. Nutr.* **7**, 113 (1988).
[4] E. DeRitter and A. E. Purcell, *in* "Carotenoids as Colorants and Vitamin A Precursors" (J. C. Bauernfeind, ed.), p. 815. Academic Press, New York, 1981.
[5] G. W. Burton and K. U. Ingold, *Science* **224**, 569 (1984).
[6] D. S. Goodman and J. A. Olson, this series, Vol. 15, p. 462.
[7] M. El-Gorab and B. A. Underwood, *Biochem. Biophys. Acta* **306**, 58 (1973).

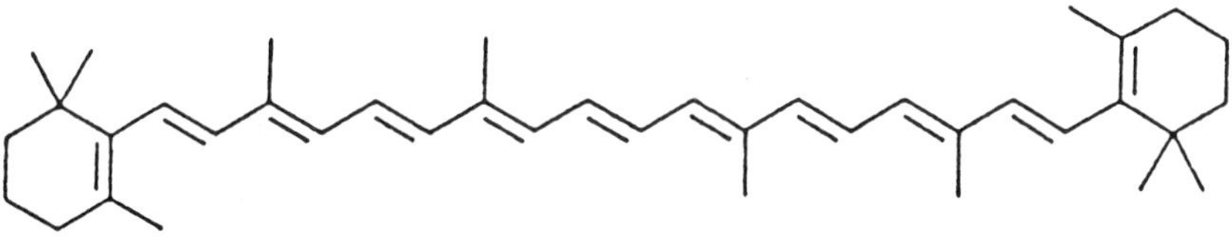

FIG. 1. Structure of β-carotene.

cholate is prepared as previously described.[7] The bile salts are mixed at 37° in doubly distilled, deionized water (MilliQ, Millipore, Bedford, MA) to provide a final bile salt concentration of 0.1 M. All bile salts are certified over 98% pure and are products of Sigma (St. Louis, MO). Stock solutions are stored at −20° until use.

β-Carotene: Stock solutions of β-carotene (Fluka, Buchs, Switzerland), typically 1.5 mM, in HPLC-grade tetrahydrofuran (THF) with 0.25 g/liter butylated hydroxytoluene (BHT) are prepared daily from crystalline β-carotene and maintained at −20° under argon until use. Concentrations and purity are verified spectroscopically.

Fatty acids: Fatty acids, certified over 98% pure are products of Sigma. Individual stock solutions, typically 100–150 mM, of triolein, monoolein, oleic acid, and phosphatidylcholine are prepared in HPLC-grade hexane and maintained in darkened glass vials at −20° under argon until use.

Krebs–Ringer bicarbonate buffer: Krebs–Ringer bicarbonate medium (without calcium chloride and sodium bicarbonate) is purchased from Sigma; 15 mM sodium bicarbonate and 1 mM calcium chloride is added, and the buffer is adjusted to pH 7.4 with NaOH. The final sodium ion concentration in the buffer is 0.14 M.

Preparation of Micellar Solutions

All manipulations involving carotenoids are performed under subdued light. The use of aluminum foil to cover solutions is strictly avoided as β-carotene is readily photooxidized by reflection of stray light.

Appropriate volumes of fatty acid stock solutions in hexane are added to a glass flask to provide final concentrations as follows: triolein (1.13 mM), monoolein (2.50 mM), oleic acid (7.5 mM), and phosphatidylcholine (0.68 mM). Appropriate amounts of β-carotene from stock solutions are added to the fatty acids to provide concentrations in the final micellar solutions from 20 to 200 μM (Fig. 2). The solvents are evaporated just

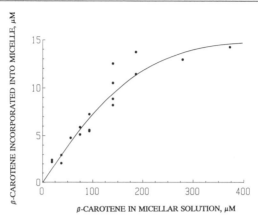

FIG. 2. Incorporation of β-carotene into a mixed micellar solution. β-Carotene was sonicated at 37° for 2 hr in the presence of bile salts and fatty acids as described in the text. The micellar phase was separated by centrifugation at 106,000 g for 18 hr at 4° and analyzed for β-carotene by the spectrophotometric assay.

until dryness under argon, and a volume of bile salt stock (37°) to provide a final bile salt concentration of 12 mM is added with shaking to emulsify the β-carotene. Krebs–Ringer buffer (37°) is added to volume, providing a final sodium concentration of 0.15 M. The resulting suspension is mixed by vigorous shaking until the solution is homogeneous. The preparation is sonicated in glass flasks for 2 hr at 37° in a bath sonicator (Bransonic 220) with occasional shaking. On completion of the sonication procedure, the turbid mixture is transferred to precoated Beckman Ultra-Clear ultracentrifuge tubes[8] and centrifuged at 4° for 2–24 hr at 106,000 g using a 50 Ti rotor (Beckman, Palo Alto, CA). Extent of incorporation of β-carotene did not vary significantly when mixtures were centrifuged for 2, 5, 18, or 24 hr. For convenience, we typically used 2 or 18 hr. Following centrifugation, the upper lipid layer is removed from the clear aqueous solution by aspiration with a Pasteur pipette. Solubilization of β-carotene in the lower aqueous phase is taken to be evidence of micellar formation.

Spectrophotometric Assay

Concentrations of β-carotene in micelles are determined by scanning from 250 to 600 nm in a Beckman DU-40 scanning spectrophotometer. The absorption maximum of β-carotene in the micellar phase is at 460 nm, as opposed to 450–452 nm in organic solutions.[4] β-Carotene concentra-

---

[8] L. Holmquist, *J. Lipid Res.* **23**, 1249 (1982).

tions are determined at the absorption maximum using an extinction coefficient of 2620 $(E^{1\%}_{1\,cm})^9$ after correction for the absorption of a blank micelle solution not containing carotene. To ensure that the absorption being measured in the micellar solutions is not quenched by interaction of lipids in the micelles, and that the extinction coefficient is not significantly altered, β-carotene is extracted from micelles and concentrations verified by HPLC.

Chromatographic Assay

Extraction of β-Carotene from Micelles. β-Carotene is extracted from micelles with 4 volumes of hexane/ethanol (3 : 1, v/v) containing 0.25 g/liter BHT and a known concentration of C_{45} β-carotene as internal standard.[10] Samples are vortexed 2 min and phases separated by centrifugation for 10 min at 1000 *g*. The upper hexane layer is aspirated with a pipette, dried under nitrogen, blanketed with argon, and stored at $-70°$ until assayed. The extraction efficiency in a typical procedure in our laboratory was 87.5 ± 4.4% (*n* = 6).

Assay of β-Carotene. Samples are resuspended in acetonitrile/THF (85 : 15, v/v) containing 0.25 g/liter BHT, injected onto a Waters (Bedford, MA) HPLC system using a Beckman Ultrasphere ODS 5-μm C_{18} column (4.6 mm × 25 cm), and eluted with acetonitrile/THF (85 : 15, v/v) containing 0.25 g/liter BHT at a flow rate of 2.5 ml/min.[11] Concentrations are determined from a standard curve run daily using C_{45} β-carotene as the internal standard.

β-Carotene in Micellar Solutions

As shown in Fig. 2, the maximum concentration of β-carotene incorporated into micelles using this method is 15 μM, well above physiological concentrations.[12] Thus, this preparation provides a method for delivering β-carotene within the physiological range for most biological systems. Incorporation of β-carotene into micelles is linear from 20 to 100 μM added carotene. When β-carotene is added in concentrations greater than 400 μM, solutions are turbid and crystals, possibly bile salts, are visible in the solution. Independent measurements of β-carotene extracted from micelles and injected onto the HPLC agree within 12 ± 5.8%

[9] O. Isler, H. Lindlar, M. Montavon, R. Rugge, and P. Zeller, *Helv. Chim. Acta* **39**, 249 (1956).
[10] Gift of Dr. Frederick Khachik, Nutrition Composition Laboratory, U.S. Department of Agriculture, Beltsville, Maryland.
[11] Y. M. Peng and J. Beaudry, *J. Chromatogr.* **273**, 410 (1983).
[12] J. Erdman, *Clin. Nutr.* **7**, 101 (1988).

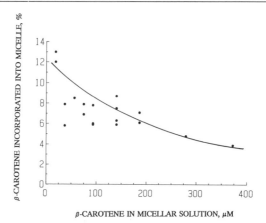

β-CAROTENE IN MICELLAR SOLUTION, μM

FIG. 3. Percent β-carotene incorporated into a mixed micellar solution as a function of β-carotene added. Experimental conditions and procedures were as indicated in Fig. 2.

($n = 6$) of the spectroscopic measurements. The percent β-carotene that is incorporated into micelles varies from about 4 to 13% of the initial concentration, and the percent incorporation decreases with increasing concentrations (Fig. 3). When stored at 4°, the micelles are stable for 24 hr as determined by spectrophotometric assay (<5% decrease in absorption). By 48 hr, absorption at 460 nm decreases by about 10%, apparently indicating degradation of β-carotene and/or micelles.

Acknowledgment

The work was supported in part by Contract N01 HD2992, grant #CA-27502, and grant #CA-23074 from the NIH, Bethesda, MD.

Section III

Enzymology and Metabolism

[46] Carotenoid Conversions

By JAMES ALLEN OLSON and M. RAJ LAKSHMAN

Introduction

β-Carotene is converted metabolically in nature to a wide variety of products, including retinal, β-apocarotenals of varying chain length, β-ionone, crocetin, 3- and 4-hydroxylated species, epoxy analogs, and acetylenic derivatives.[1-6] These transformations are almost invariably oxidative in nature, and in most cases involve molecular oxygen. Other carotenoids seem to be converted to analogous derivatives by similar reactions.

Microorganisms and plants, which make carotenoids from acetylcoenzyme A via geranyl geranyl pyrophosphate,[7] oxidize and cleave them in a variety of ways.[3,4] Both central and excentric cleavage occur in these organisms. Fish are somewhat unique in being able to reduce oxygenated carotenoids (xanthophylls) to β-carotene.[2-5] Mammals cleave carotenoids to vitamin A as well as oxidizing them in other, less well-defined ways.

The best studied reaction in mammals is the oxidative cleavage of all-trans-β-carotene to two molecules of all-trans-retinal, a reaction catalyzed by β-carotenoid 15,15'-dioxygenase.[8-22] The enzymatic cleavage of

[1] O. Isler (ed.), "Carotenoids." Birkhauser Verlag, Basel, 1971.
[2] J. A. Olson, in "Biosynthesis of Isoprenoid Compounds" (J. W. Porter and S. L. Spurgeon, eds.), Vol. 2, p. 371. Wiley, New York, 1983.
[3] K. L. Simpson and C. O. Chichester, Annu. Rev. Nutr. 1, 351 (1981).
[4] B. H. Davies, Pure Appl. Chem. 57, 679 (1985).
[5] J. A. Olson, J. Nutr. 119, 105 (1989).
[6] A. Bendich and J. A. Olson, FASEB J. 3, 1927 (1989).
[7] S. L. Spurgeon and J. W. Porter, in "Biosynthesis of Isoprenoid Compounds" (J. W. Porter and S. L. Spurgeon, eds.), Vol. 2, p. 1. Wiley, New York, 1983.
[8] D. S. Goodman and H. S. Huang, Science 149, 879 (1965).
[9] J. A. Olson and O. Hayaishi, Proc. Natl. Acad. Sci. U.S.A. 54, 1364 (1965).
[10] D. S. Goodman, H. S. Huang, and T. Shiratori, J. Biol. Chem. 242, 1929 (1966).
[11] D. S. Goodman, H. S. Huang, M. Kanai, and T. Shiratori, J. Biol. Chem. 242, 3543 (1967).
[12] D. S. Goodman, Am. J. Clin. Nutr. 22, 963 (1969).
[13] M. R. Lakshmanan, J. L. Pope, and J. A. Olson, Biochem. Biophys. Res. Commun. 33, 347 (1968).
[14] N. H. Fidge, F. R. Smith, and D. S. Goodman, Biochem. J. 114, 689 (1969).
[15] D. S. Goodman and J. A. Olson, this series, Vol. 15, p. 462.
[16] M. R. Lakshmanan, H. Chansang, and J. A. Olson, J. Lipid Res. 13, 477 (1972).
[17] A. M. Gawienowski, M. Stacewicz, and R. Longley, J. Lipid Res. 15, 375 (1974).
[18] H. Singh and H. R. Cama, Biochim. Biophys. Acta 370, 49 (1974).

carotenoids to β-apocarotenals has not been unambiguously demonstrated in mammalian tissues, although chemical reactions of this kind occur at a slow rate in the presence of oxygen.[5,23] This chapter focuses, therefore, on methods for measuring β-carotenoid 15,15'-dioxygenase, but we also include a brief treatment of excentric cleavage in microorganisms.

Assay Method

Principle. The carotenoid to be used as substrate must first be purified by chromatography. A small amount of the carotenoid substrate is appropriately solubilized in a detergent and incubated with the enzyme in a suitable medium in the presence of oxygen under yellow light. After inactivation of the enzyme, the substrate and its products are extracted with organic solvents, separated by chromatography, and quantitated by UV/VIS absorption and/or by radioactivity. Derivatives of the product can be made, followed by further purification.

Use of Radioactive β-Carotene. Early studies,[8,9] summarized in a jointly written methodological paper,[15] employed radioactive β-carotene. The radioactive β-carotene is synthesized from sodium [1-14C]acetate by *Phycomyces blakesleeanus,* purified on deactivated alumina, and crystallized to constant specific activity from benzene/methanol.[15] β-[15,15'-3H2]Carotene, synthesized chemically, is also used. The radioactive β-carotene is stored at −20° in the dark in hexane containing α-tocopherol. Benzene containing 5 mg% butylated hydroxytoluene (BHT) is the preferred solvent for the storage of β-carotene at −80°.

Rat Liver.[9] The excised liver is minced on a cold Petri plate and homogenized in a loose-fitting Potter–Elvehjem homogenizer in 5 volumes of cold 0.15 *M* Tris buffer, pH 8.0, that contains 0.1 *M* nicotinamide and 0.1 *M* cysteine. The homogenate is centrifuged first at 10,000 *g* for 10 min and then at 105,000 *g* for 60 min. The resultant supernatant solution, which contains the enzyme, is stored in the cold until used. The liver enzyme preparation is fairly stable when frozen and stored at −20°.

Just before each experiment, 1–6 nmol of labeled β-carotene per incubation flask is purified through a small column of 6% water-deactivated

[19] D. Sklan, *Int. J. Nutr. Res.* **53,** 23 (1983).
[20] D. Sklan and O. Havely, *Br. J. Nutr.* **52,** 107 (1984).
[21] F. J. Schweigert, M. Wierich, W. A. Rambeck, and H. Zucker, *Theriogenology* **30,** 923 (1988).
[22] M. R. Lakshman, I. Mychovsky, and M. Attlesey, *Proc. Natl. Acad. Sci. U.S.A.* **86,** 9124 (1989).
[23] S. Hansen and W. Maret, *Biochemistry* **27,** 200 (1988).

alumina, dried in the dark under nitrogen, dissolved in 50 μl of 20% Tween 40 in distilled acetone, and made up to a suitable volume (0.2 ml/ reaction flask) with 0.15 M Tris buffer, pH 8. The micellar solution of labeled β-carotene (1–6 nmol in 0.2 ml Tris buffer) is added to 5 ml of liver supernatant solution and gently rocked in the dark at 36° in air for 60 min. Thereafter, 20 ml distilled acetone, which contains approximately 60 nmol of nonradioactive β-carotene, 30 nmol each of retinal and retinol, and 120 nmol of α-tocopherol, is added, followed by equal volumes of diethyl ether and distilled water. The organic layer is removed, dried over anhydrous sodium sulfate, evaporated to dryness, and taken up in diethyl ether. The products are separated on thin-layer plates of silica gel, eluted, and counted.

Rat Intestine.[8] Mucosal scrapings are homogenized in a loose-fitting Potter–Elvehjem homogenizer in 0.1 M potassium phosphate buffer, pH 7.7, containing 30 mM nicotinamide and 4 mM MgCl$_2$. After the homogenate is centrifuged first at 2000 g for 20 min and then at 104,000 g for 60 min, an aliquot of the resultant supernatant solution (11 mg protein) is incubated at 37° in air in the dark for 1–2 hr with biosynthesized β-[^{14}C]carotene (1.3 nmol), washed cell particles (to provide phospholipids), 12 μmol sodium taurocholate, 2.5 μmol α-tocopherol, 10 μmol glutathione, 36 μmol nicotinamide, 4.8 μmol MgCl$_2$, and 200 μmol potassium phosphate buffer, pH 7.7, in a total volume of 2.0 ml. β-Carotene and α-tocopherol are added in 75 μl acetone. After incubation, lipids are extracted with 20 volumes of chloroform/methanol (2 : 1) in the presence of dilute H$_2$SO$_4$. The chloroform phase is evaporated, and the lipid extract is chromatographed on an alumina column (Sigma, St. Louis, MO, activity grade III). Radioactivity is measured in each of five chromatographic fractions (β-carotene, retinyl ester, retinal, retinol, and polar acids) by conventional means.

Use of Nonradioactive Substrates.[13] In assessing the activity of β-carotenoid 15,15′-dioxygenase toward various substrates, several nonradioactive analogs are used. The small intestines from 48-hr fasted rabbits served as the enzyme source. The excised intestine is washed free of its contents with ice-cold 0.9% NaCl, cut into segments 15–20 cm long, and slit along the mesenteric line. The mucosa is scraped off with a cold microscope slide and homogenized with 8 volumes cold 0.1 M potassium phosphate buffer, pH 7.8, in a Potter–Elvehjem homogenizer. After centrifugation at 43,500 g for 60 min, the supernatant solution is fractionated with (NH$_4$)$_2$SO$_4$. The fraction precipitating between 25 and 50% saturation is stored at $-20°$ and dissolved just before use in a small volume of 0.1 M potassium phosphate buffer, pH 7.8.

The assay solution contains 50 nmol substrate, 50 mg reduced glutathione (GSH), 25 μmol sodium lauryl sulfate, and 10–40 mg enzyme in 25 ml of 0.1 M potassium phosphate buffer, pH 7.8. After incubation for 1 hr at 37° in air in the dark, the reaction mixture is extracted 2 times with 10-ml batches of peroxide-free ether. The pooled ether extract is washed with water, dried over anhydrous Na_2SO_4, evaporated to dryness in the dark under reduced pressure, and dissolved in a small volume of petroleum ether (40–60°).

Substrate and products are separated by the use of a column of 5% water-deactivated alumina (10 g; 1 × 12 cm). Compounds are eluted with the following percentages of ether in hexane: 4% (β-carotene, epoxy derivatives of β-carotene), 6–8% (none), 10% (retinal), and 25% (β-apocarotenals). Retinal, characterized by its chromatographic behavior and by the formation of its thiobarbituric acid addition product, is the sole product formed. The most active substrates in this study are β-apo-10′- and β-apo-8′-carotenal.

Recent Modifications.[22] The procedure is similar to that described above, except as noted below. The proximal one-third of the intestine from anesthetized white rabbits (pentobarbital, 50 mg/kg body weight) is excised, washed, and slit. The scraped mucosa is homogenized with 5 volumes of 0.1 M potassium phosphate buffer, pH 7.8, containing 1 mM dithiothreitol. After centrifugation at 100,000 g for 1 hr at 0°, the supernatant solution is brought to 60% saturation with $(NH_4)_2SO_4$ and centrifuged at 16,000 g for 15 min at 0°. The pellet is dissolved in the homogenizing buffer (final protein concentration of 10 mg/ml) and fractionated with cold (−30°) acetone into 0–45% and 45–60% (v/v) fractions. Both pellets are redissolved in the homogenizing buffer containing 1 mM GSH and brought to 50% saturation with $(NH_4)_2SO_4$. Under these conditions, the enzyme activity is stable for a month at −80°.

β-Carotene (100 nmol in 0.1 ml benzene) is mixed with 0.18 ml of 10% aqueous Tween 20. Benzene is removed under nitrogen. The assay solution contains 0.18 ml of the substrate in Tween 20, 1 mM GSH, 1 mM $FeSO_4$, 15 mM nicotinamide, 0.1 M potassium phosphate, pH 7.8, and 7 mg of the enzyme preparation in a final volume of 2 ml. After incubation with shaking at 37° for 60 min in air under a safe yellow light, the reaction is stopped with 2 ml methanol. Radiolabeled β-[15,15′-^{14}C]carotene, diluted with nonradioactive β-carotene to a specific activity of 17,600 dpm/ nmol, can be substituted for nonradioactive β-carotene at the same concentration in the incubation mixture.

As indicated below, the incubated, methanol-inactivated mixture can be treated with O-ethylhydroxylamine, followed by extraction of the O-ethylretinaloxime. The O-ethylretinaloxime is then separated and quantified by high-performance liquid chromatography (HPLC).

Chromatography and Characterization of Retinal

Chromatographic Systems. In early studies, retinal, other products, and carotenoid substrates were separated by use of thin-layer chromatography (TLC) on silica[9] or on 5–10% water-deactivated alumina columns.[8,10–17] In more recent studies, HPLC has been employed. Some systems used are a reversed-phase 5-μm C_{18} 9% carbon-loaded column (Vydac TP) with a linear gradient from methanol/water (92:8) to methanol/water (98:2) in the presence of 0.5% NH_4OOCCH_3 and detection at 360 nm[22]; a reversed-phase 5-μm RP C_{18} column (Merck, Darmstadt, FRG, LiChrosorb) with aqueous 1% NH_4OOCCH_3/CH_3CN (28:72) and detection at 350 nm[23]; a reversed-phase 5-μm Si-60 (LiChrosorb) column with dioxane/*n*-hexane (16:100) and detection at 365 nm (retinal and retinol) and at 436 nm (β-carotene)[21]; and a normal-phase Du Pont Zorbax-Sil golden series column with various amounts of dichloroethane or acetone in hexane (±0.1% acetic acid) and detection at 313 nm (retinol) at 340 nm (retinoic acid), or at approximately 370 nm (retinal).[24]

Characterization. As indicated above, retinal was separated initially from carotenoid substrates by chromatography on various adsorbents. Retinal has been further characterized as the semicarbazone[8] or thiobarbituric acid conjugate,[13,18] as retinol after treatment with $NaBH_4$[9,17] or after reduction by alcohol dehydrogenase,[22] as anhydroretinol after acidic dehydration of the resultant retinol,[9] as retinyl palmitate after its enzymatic esterification,[22] and as *O*-ethylretinaloxime following treatment with *O*-ethylhydroxylamine.[22] Retinal semicarbazone[8] and *O*-ethylretinaloxime[22] derived from radiolabeled β-carotene have been crystallized repeatedly to constant specific activity. The oxime has also been characterized by mass spectrometry.[22]

Precautions

Although β-carotenoid 15,15'-dioxygenase activity has been demonstrated *in vitro* in many laboratories, others have been unable to detect the biological conversion of β-carotene to retinal, or indeed, to any biologically derived product. In general, negative results are not published. A recently published, carefully conducted, but unsuccessful study,[23] therefore, is well worth perusing in regard to possible problems that might arise in dealing with this enzyme. One must also bear in mind that from 1930, when Thomas Moore first unambiguously demonstrated the *in vivo* conversion of β-carotene to vitamin A,[25] until 1965, when central cleavage

[24] J. L. Napoli and K. R. Race, *J. Biol. Chem.* **263**, 17372 (1988).
[25] T. Moore, "Vitamin A" Elsevier, Amsterdam, 1957.

was shown in cell-free homogenates,[8,9] attempts by many workers were unsuccessful in showing this conversion by cell-free preparations.[25] Some possible reasons for such difficulties are as follows:

1. β-Carotene and other carotenoids are unstable in the presence of light, oxygen, and heavy metals.[1,26] Thus, a variety of oxidation products, incuding the β-apocarotenals and retinal, are formed chemically.[17,23] The substrates must therefore be stored with antioxidants and carefully purified just before use.

2. The intestinal enzyme, although sufficiently active maximally to yield 200 times the nutritional requirement for vitamin A in rabbits[2,5] and presumably in other species,[5] shows a relatively low specific activity in comparison with other dioxygenases.[23]

3. As a consequence, when a large amount of β-carotene is initially present, the percent conversion to retinal is small. Concomitantly, the chemical oxidation of β-carotene to artifactual products is proportional to the amount of β-carotene added. Under these circumstances, the "background" may exceed, and consequently mask, the biological formation of retinal.

4. The enzyme, which contains sulfhydryl groups, is sensitive to poisoning by heavy metals. Thus, purified reagents and deionized water free of such contaminants must be used. The enzyme must be stored in the presence of sulfhydryl-protecting agents such as reduced glutathione or dithiothreitol.

5. Solvents used for extraction must be free of peroxides and other oxidizing agents. Furthermore, the use of 5 mg% BHT in all the organic solvents is strongly recommended to serve as an antioxidant.

6. The preparation of micellar solutions of β-carotene and of other carotenoid substrates in a suitable detergent requires careful attention. Some detergents are inhibitory at higher concentrations, where others do not solubilize carotenoids well. In early studies, phospholipid was found to promote the enzyme activity.[8,15] Detergents, many of which are not pure compounds, can vary in properties from batch to batch. Commercial conjugated bile salt preparations often contain impurities that may be inhibitory. In essence, one must often test several detergents and detergent combinations to find one that is most suitable.

7. As the enzyme is purified, it becomes unstable. Frozen, stored preparations often lose activity quickly. Perhaps the most suitable storage form is as a 50% saturated $(NH_4)_2SO_4$ solution containing 1

---

[26] J. A. Olson, in "Handbook of Vitamins" (L. J. Machlin, ed.), p. 1. Dekker, New York, 1984.

mM reduced glutathione.[22] The lower the storage temperature, the better the stability ($-80°$ is preferred).[22]

Enzyme Properties

With few exceptions, the partially purified enzyme, regardless of the tissue or species studied, has very similar properties.[2] It is localized in the cytosolic fraction, requires molecular oxygen, and yields retinal as the sole identified enzymatic product. The K_m for β-carotene is 2–10 μM, and the V_{max} for β-carotene is 0.6–1.7 nmol cleaved/mg protein/hr at 37° by purified preparations. The pH optimum is 7.5–8.5. Various iron and copper chelators, such as α,α-dipyridyl, o-phenanthroline, and possibly EDTA, as well as sulfhydryl-binding agents, like N-ethylmaleimide, iodoacetamide, p-chloromercuribenzoate, and heavy metal ions, inhibit the enzyme. Reduced glutathione is protective. Ferrous, but not ferric, ions enhance its activity. Nicotinamide may be protective.

The substrate specificity is fairly broad.[2] Compounds cleaved at an appreciable rate, that is, at least 25% that of β-carotene, are β-carotene, 3,4,3′,4′-tetradehydro-β-carotene, 5,6-epoxy-β-carotene, β,ε-carotene, 5,6-epoxy-β,ε-carotene, 3,4-dehydro-β,ψ-caroten-16-al, 10′-β-apocarotenal, and 10′-β-apocarotenol.[2,13,18]

Excentric Cleavage of Carotenoids

Excentric cleavage clearly occurs in plants and microorganisms, as evidenced by the presence of β-apocarotenals and derivatives such as azafrin, bixin, and crocetin in nature. Whether excentric cleavage occurs in mammals is still debated. As yet, however, enzymes that carry out such reactions have not been isolated and characterized from any species. Nonetheless, in some cases, as described below, evidence for the presence of a given enzyme is compelling.

β-Carotenoid 7,8,7′,8′-Oxygenase of Microcystis.[27] The cyanobacterium *Microcystis* PCC 7806 is cultivated in 300-ml culturing tubes at 27° in a 0.27% (v/v) CO_2/air mixture under fluorescent light.[28] The medium contains 0.6 mM $CaCl_2$, 8 mM $NaNO_3$, 0.4 mM K_2HPO_4, 0.4 mM $MgSO_4$, 10 μM NaFeEDTA, 10 μM H_3BO_3, 2 μM Na_2MoO_4, 0.2 μM $ZnSO_4$, 0.2 μM $CuSO_4$, and 0.2 μM $CoSO_4$ in distilled water. Cultures are harvested by centrifugation at various times up to 16 days.[28]

Activation and Assay of β-Carotenoid 7,8,7′,8′-Oxygenase.[27] The enzyme is activated in 3 ml of cell suspension (10^8 cells/ml) by freezing at

[27] F. Juttner and B. Hoflacher, *Arch. Microbiol.* **141,** 337 (1985).
[28] F. Juttner, J. Leonhardt, and S. Mohren, *J. Gen. Microbiol.* **129,** 407 (1983).

−16° for at least 3 hr. Activation is also induced by adding 20% NaCl (w/v) or 20% ethanol (v/v) to the cells or by disrupting them in a French press or by sonification. To prevent the immediate oxidative cleavage of endogenous carotenoids in the cells upon thawing, cells are frozen under argon in the presence of glucose, glucose oxidase, and catalase.

Microcystis cells contain significant amounts of endogenous β-carotene and zeaxanthin as well as other carotenoids. Frozen cells (5×10^9) are thawed in the presence of oxygen and incubated at room temperature for 30 min in 50 ml water. β-Cyclocitral and 3-hydroxy-β-cyclocitral are collected on 150 mg of Tenax TA (60–80 mesh) by aeration in a closed-loop system or by liquid–solid sorption. Crocetindial is extracted from cells with 60 ml of acetone/MeOH (7 : 3).

Chromatographic Analysis.[27] β-Cyclocitral is assayed on a 25-m glass capillary column coated with UCON 50 NB 5100 at 100° with 2-decanone as the internal standard. 3-Hydroxy-β-cyclocitral is characterized by mass spectrometry (M^+, m/z 168). Crocetindial is applied in 0.5 ml acetone/methanol (7 : 3) to a C_8 Bond Elut column, eluted with methanol/acetone/water (20 : 4 : 3), and rechromatographed on a methanol-deactivated neutral alumina column (initially activity grade I, Merck No. 1077). Crocetindial is eluted with pentane.

β-Carotene, zeaxanthin, and other pigments of *Microcystis* are extracted twice from the cell pellet with acetone/methanol (7 : 3). After the evaporation of solvents, the residue is dissolved in chloroform/methanol (2 : 1) and chromatographed on reversed-phase TLC plates of silica gel G (Merck), impregnated with fat and containing 2,6-di-*tert*-butyl-4-methylphenol, by use of methanol/acetone/water (20 : 8 : 3). After development, separated pigments are eluted with ethanol and quantitated spectrophotometrically.[27]

Enzyme Properties.[27] Endogenous β-carotene is converted to 2.8 mol/mol of β-cyclocitral (theory 2.0), and the decrease in endogenous zeaxanthin and β-carotene is accompanied by a 0.88 mol/mol increase in crocetindial (theory 1.0). The enzyme, which acts both on β-carotene and on zeaxanthin, is membrane-bound. On release from cells by sonication, it becomes very sensitive to inactivation by oxygen. The enzyme is inhibited by N-ethylmaleimide and 4-hydroxymercuribenzoate, but is not protected in activated cells by sulfhydryl-containing reagents. Various antioxidants either prevent the activation of the enzyme or inhibit its activity. Ferrous, ferric, and possibly calcium ions somewhat enhance the activity, whereas zinc and copper ions are inhibitory. *o*-Phenanthroline also inhibits the enzyme. Thus, the *Microcystis* cleavage enzyme, apart from its substrate specificity, seems to possess properties that are both similar and dissimilar to the carotenoid 15,15'-dioxygenase of mammalian tissues.

[47] Carotenoid Cleavage: Alternative Pathways

By L. E. GERBER and K. L. SIMPSON

Introduction

Although it is well-established that carotene and other carotenoids require cleavage to shorter products for nutritional value as vitamin A, the nature of the cleavage remains in dispute. Some argue strongly for central cleavage by a putative 15,15'-dioxygenase (EC 1.13.11.21) with production of retinol directly.[1] Others argue strongly for excentric cleavage to 8'-, 10'-, and 12'-apo-β-carotenals followed by serial oxidation to retinal by mechanisms similar to β-oxidation.[2] The best evidence for excentric cleavage of carotenoids is the demonstration of the presence of 8'-, 10'-, and 12'-β-apocarotenals in small intestinal tissue taken from animals previously consuming carotenoids.[3-5] Other evidence is provided by observations of the oxidative fission of carotenoids in model systems devoid of cells or cellular extracts.[6-9] In addition, recent evidence suggests that excentric cleavage of carotenoids is enhanced by some factors present in cell-free small intestinal extracts.[10]

Oxidative fission of β-carotene using alkaline permanganate as a catalyst was demonstrated by Karrer and Solmssen[6] and revealed that the major products were apo-β-carotenals. Glover and Redfearn[7] and Hasani and Parrish[8] have also demonstrated the chemical oxidation of β-carotene to 8'-, 10'-, and 12'-apo-β-carotenals. A detailed study of the site of initial attack of β-carotene by molecular oxygen in a model system devoid of biological material indicates that an enzyme system would be needed in order to produce central fission.[9]

[1] J. A. Olson, *J. Nutr.* **119**, 105 (1989).
[2] J. Ganguly and P. S. Sastry, *World Rev. Nutr. Diet.* **45**, 198 (1985).
[3] G. N. Festenstein, Ph.D. Thesis, University of Liverpool, England, 1951.
[4] A. Winterstein and B. Hegedus, *Chimia* **14**, 18 (1960).
[5] R. V. Sharma, S. N. Mathur, A. A. Dmitrovskii, R. C. Das, and J. Ganguly, *Biochim. Biophys. Acta* **485**, 183 (1977).
[6] P. Karrer and L. Solmssen, *Helv. Chim. Acta* **20**, 682 (1937).
[7] J. Glover and E. R. Redfearn, *Biochem. J.* **58**, 15 (1954).
[8] S. M. A. Hasani and P. B. Parrish, *J. Agric. Food Chem.* **15**, 943 (1967).
[9] A. H. El-Tinay and C. O. Chichester, *J. Org. Chem.* **35**, 2290 (1970).
[10] S. Hansen and W. Maret, *Biochemistry* **27**, 200 (1988).

Isolation of 8'-, 10'-, and 12'-apo-β-carotenal from intestinal tissue of horses, chickens, and rats has been accomplished. Festenstein[3] isolated both 10'- and 12'-apo-β-carotenals from horse intestine. Both Winterstein and Hegedus[4] and Sharma et al.[5] have isolated various apo-β-carotenals from rat small intestines, whereas only Sharma et al.[5] have isolated apo-β-carotenals from chicken small intestine.

A recent report by Hansen and Maret[10] demonstrated the production of 8'-, 10'-, and 12'-apo-β-carotenals from β-carotene when the crude enzyme extract of rat small intestine was incubated with β-carotene. It is of interest that the production of both 8'- and 12'-apo-β-carotenals was approximately twice as high for small intestinal tissue taken from vitamin A-deficient rats compared to vitamin A sufficient rats. Their protein-free control incubations also showed production of 8'-, 10'-, and 12'-apo-β-carotenals, as found in investigations of model systems of carotenoid cleavage devoid of biological materials.

Assay Method: General

Purified β-carotene is cleaved in the presence of a cell-free homogenate fraction from the small intestines to form 8'-, 10'-, and/or 12'-apo-β-carotenals. These cleavage products are extracted from lyophilized samples with acetone, separated by HPLC, and detected by spectrophotometry at a wavelength of 450 nm.

Preparation of Cell-Free Homogenate Fraction

As identified previously in this series,[11] intestinal mucosa scrapings are homogenized in 0.1 M potassium phosphate buffer, pH 7.7, and centrifuged at 2000 g for 20 min at 4°. The resulting supernatant is then centrifuged at 104,000 g for 60 min at 4° to obtain a soluble fraction containing cleavage activity. Cleavage activity can be further purified by ammonium sulfate precipitation isolating the 20–45% ammonium sulfate precipitation fraction.

Purity of β-Carotene

The purity of commercially available β-carotene is assessed by high-performance liquid chromatography (HPLC),[12] and the material is recrystallized, if necessary, using a modification of the procedure of Britton and

[11] D. S. Goodman and J. A. Olson, this series, Vol. 15, p. 462.
[12] G. Arroyave, C. O. Chichester, H. Flores, J. Glover, L. A. Mejia, J. A. Olson, K. L. Simpson, and B. A. Underwood, "Biochemical Methodology for the Assessment of Vitamin A Status." The Nutrition Foundation, Washington, D.C., 1982.

Goodwin.[13] The purity of the recrystallized β-carotene is then assessed by HPLC before use. Procedures for the proper handling and analysis of carotenoids have been published.[12]

Incubation of β-Carotene with Cell-Free Ammonium Sulfate Fraction

Assay procedures are those outlined by Goodman and Olson for β-carotene conversion.[11] The cell-free homogenate fraction previously isolated is dissolved in 10 mM potassium phosphate buffer, pH 7.7. The resulting solution is dialyzed for several hours against a large solution of phosphate buffer to remove ammonium sulfate. A 1-ml aliquot of this solution is used for each incubation. An additional amount of 200 μmol potassium phosphate buffer, pH 7.7, is added as well as several chemicals which enhance β-carotene cleavage. These enhancing factors include 30 μmol nicotinamide, 10 μmol glutathione, 12 μmol sodium glycocholate, and 400 μg egg lecithin. In addition, a nonpolar antioxidant such as α-tocopherol should be added in small amounts. To the resulting incubation mixture is added a total of 1 μg β-carotene in 50 μl of acetone. Incubations are carried out either in amber Erlenmeyer flasks or in clear Erlenmeyer flasks under red light. After 1 hr of shaking at 37° in a water bath, the reaction is stopped by rapid freezing and the sample lyophilized.

Extraction and Detection of Carotenoids, Retinoids, and Apocarotenals

Lyophilized samples are extracted 3 times with acetone using pyrogallol as antioxidant. The pooled extracts are evaporated with a rotary evaporator at 37° and resolubilized in the appropriate carrier system prior to chromatography.[12] For retinoid analysis, extracts are redissolved in hexane/ethyl acetate/methanol (95 : 4 : 1). Retinoids are then detected after HPLC separation using a Spherisorb-S-5-CN column (5 μm, 0.46 × 25 cm) with a carrier solvent of hexane/ethyl acetate/methanol (95 : 4 : 1) at a flow rate of 1 ml/min. Detection of retinol, retinal, and retinoic acid is accomplished at 350 nm. For analysis of β-carotene and 8'-, 10'-, and 12'-apo-β-carotenal, extracts are redissolved in acetonitrile/dichloromethane (7 : 3). A Zorbax ODS column (10 μm, 0.46 × 25 cm) is used for separation with a carrier solvent system of acetonitrile/dichloromethane (7 : 3) at 1 ml/min. Detection is accomplished at 450 nm.

Biological Significance

It appears that evidence of excentric cleavage of β-carotene in mammalian systems is undeniable. Further experiments are required using the

---

[13] G. Britton and T. W. Goodwin, this series, Vol. 18C, p. 654.

described techniques to determine if the reactions are enhanced by enzymatic activity and what enzyme is responsible for the cleavage. Clearly, apo-β-carotenals are present in the small intestine, but enhanced production in small intestine tissue owing to enzymatic activity may not be a normal physiological event.

Acknowledgment

The above manuscript is contribution #2502 of the Rhode Island Agricultural Experiment Station, Kingston, RI.

[48] Isoenzymes of Alcohol Dehydrogenase in Retinoid Metabolism

By Xavier Parés and Pere Julià

Introduction[1]

$$\text{all-}trans\text{-Retinol} + NAD^+ \rightleftarrows \text{all-}trans\text{-retinal} + NADH + H^+$$

Alcohol dehydrogenase (alcohol : NAD^+ oxidoreductase, EC 1.1.1.1) (ADH) is an enzyme widely distributed in animals, plants, and microorganisms.[2] It catalyzes the reversible oxidation of a great variety of alcohols to the corresponding aldehydes and ketones.[3] ADH exhibits several isoenzymes in most species studied. In mammals the isoenzymes have been grouped in three classes.[4] Class I shows a wide substrate specificity, a low K_m for ethanol, and low K_i for pyrazole. It is abundant in liver. Class III is a glutathione-dependent formaldehyde dehydrogenase.[5] Its activity with alcohols is restricted to long-chain primary alcohols, is not inhibited by pyrazole, and is detected in most tissues. Class II shows intermediate properties. In general, ADH isoenzymes are very active toward long-chain hydrophobic alcohols.[3] Retinol and retinal are, therefore, sub-

[1] Supported by Grant 86/0156 of Dirección General de Investigación Científica y Técnica (Spain). We acknowledge Dr. B. L. Vallee and Dr. W. P. Dafeldecker for the gift of the CapGapp-Sepharose.
[2] H. Jörnvall, B. Persson, and J. Jeffery, *Eur. J. Biochem.* **167,** 195 (1987).
[3] R. Pietruszko, in "Biochemistry and Pharmacology of Ethanol" (E. Majchrowicz and E. P. Noble, eds.), Vol. 1, p. 87. Plenum, New York, 1979.
[4] B. L. Vallee and T. J. Bazzone, *Isozymes: Curr. Top. Biol. Med. Res.* **8,** 219 (1983).
[5] M. Koivusalo, M. Baumann, and L. Uotila, *FEBS Lett.* **257,** 105 (1989).

strates for most of them. Only the all-trans isomers have been used as substrates in the study reported here.

Assay Method

Principle. Activity with retinoids can be measured spectrophotometrically at 400 nm,[6] or at 410 nm,[7] where retinol does not absorb and retinal has a large extinction coefficient. A problem in determining ADH activity with retinoids is their low solubility in aqueous solutions. Three alternative solubilizing agents have been assayed: bovine serum albumin (BSA), acetone, and the detergent Tween 80. Addition of 0.02% Tween 80 to the assay buffers was chosen because of the reproducibility of the measurements,[8,9] although the presence of the detergent reduces the ADH activity with retinoids, in a concentration-dependent fashion. The 0.02% concentration employed should be regarded as a compromise between solubility and activity.

Procedure. A solution is first prepared with 6 mg of all-*trans*-retinol or all-*trans*-retinal (Sigma, St. Louis, MO) in 0.5 ml of acetone. A 350-μl sample of the acetone solution is mixed with 50 ml of the assay buffer which contains 0.02% Tween 80. The resulting substrate concentration is about 300 μM. The aqueous solutions are prepared daily and stored at 4° protected from light. Concentrations of retinol and retinal are determined spectrophotometrically using an ε_{328} value of 39,500 M^{-1} cm^{-1} for retinol and an ε_{400} of 29,500 M^{-1} cm^{-1} for retinal.

The retinol dehydrogenase activity of ADH is monitored spectrophotometrically by recording the increase in absorbance at 400 nm, at 25°. The reaction mixture contains 150 μM retinol in 0.1 M sodium phosphate, pH 7.5 (or 0.1 M glycine/NaOH, pH 10.0), with 40 mM NaCl, 0.02% Tween-80, and 4 mM NAD$^+$.

The retinal reductase activity of ADH is determined by measuring the decrease in absobance at 400 nm, at 25°, in a cuvette of 0.2-cm path length. The reaction mixture contains 100 μM retinal in 0.1 M sodium phosphate, pH 7.5, with 40 mM NaCl, 0.02% Tween 80, and 0.77 mM NADH. For both retinol oxidation and retinal reduction activities, the background change of absorbance is first measured in the presence of buffer, retinoid, and coenzyme in the cuvette, and activity is subsequently determined by adding the enzyme. An alternative method is based in the reaction of retinal and thiobarbituric acid.[8,10] HPLC methods

[6] M. Reynier, *Acta Chem. Scand.* **23**, 1119 (1969).
[7] E. Mezey and P. R. Holt, *Exp. Mol. Pathol.* **15**, 148 (1971).
[8] P. Julià, J. Farrés, and X. Parés, *Biochem. J.* **213**, 547 (1983).
[9] P. Julià, J. Farrés, and X. Parés, *Exp. Eye Res.* **42**, 305 (1986).
[10] S. Futterman and L. D. Saslaw, *J. Biol. Chem.* **236**, 1652 (1961).

have been recently used for the determination of retinol oxidation in tissue homogenates.[11,12]

Starch Gel Electrophoresis of Alcohol Dehydrogenase Isoenzymes

ADH isoenzymes are heterogeneously distributed in animal tissues. Prior to purification from a particular organ it is convenient to analyze the isoenzyme composition by starch gel electrophoresis.[13]

Procedure. Tissue samples are homogenized in 10 mM Tris-HCl, pH 7.9, 0.5 mM dithiothreitol (DTT) and centrifuged. About 10 μl of supernatant is used for analysis. Electrophoresis is carried out at 4° on horizontal gels containing 13% starch (Sigma), 0.74 mM NAD$^+$, 20 mM Tris-orthophosphoric acid, pH 7.6, for 6 hr at 720 V, using a Multiphor instrument (LKB, Bromma, Sweden). The gels are sliced in two and stained for activity by incubating at 37° in 50 mM Tris-HCl, pH 8.6, 11 mM sodium pyruvate, 0.6 mM NAD$^+$, 0.24 mM nitroblue tetrazolium, and 65 μM phenazine methosulfate. One of the gel slices is stained with 0.1 M pentanol (or crotyl alcohol) to reveal the three isoenzyme classes. The second slice is stained with retinol in the following way.[14] 125 mg of retinol is dissolved in 6.25 ml of acetone, and 6.25 ml of 50 mM Tris-orthophosphoric acid, 11 mM sodium pyruvate, pH 8.6, is then added. The starch gel slice is incubated in this solution for 5–10 min, and subsequently 125 ml of the staining solution is added. The incubation is continued for 1–2 hr at 37°. The whole process is performed in the dark.

Purification

Procedures for the purification of ADH isoenzymes from a variety of mammalian species have been reported (see Julià *et al.*[15] for references). In this chapter we describe the purification of the rat ADH isoenzymes.[15] Three ADH isoenzymes are detected in rat tissues, which have been named ADH-1, ADH-2, and ADH-3 according to their mobility on starch gel electrophoresis. ADH-1 is the most anodic form (pI 5.1), characteristic of the stomach, ocular tissues, and external epithelia.[16] ADH-1 shows K_m values of 5 M for ethanol and 20 μM for retinol (pH 7.5),[9] and it has been considered a class II isoenzyme although no clear correspondence

---

[11] M. J. Connor and M. H. Smit, *Biochem. J.* **244**, 489 (1987).
[12] J. L. Napoli and K. R. Race, *Arch. Biochem. Biophys.* **255**, 95 (1987).
[13] X. Parés, P. Julià, and J. Farrés, *Alcohol* **2**, 43 (1985).
[14] A. Koen and C. R. Shaw, *Biochim. Biophys. Acta* **128**, 48 (1966).
[15] P. Julià, J. Farrés, and X. Parés, *Eur. J. Biochem.* **162**, 179 (1987).
[16] M. D. Boleda, P. Julià, A. Moreno, and X. Parés, *Arch. Biochem. Biophys.* **274**, 74 (1989).

has been found with the human isoenzymes.[15] ADH-2 (pI 5.95–6.30) is distributed in all rat organs, is specific for long-chain primary alcohols [K_m (octanol) 0.5 mM] although it shows very low activity with retinol. ADH-2 has been identified as a class III isoenzyme.[15,17] ADH-3 (pI 8.25–8.40) is the characteristic liver enzyme (class I), of wide substrate specificity, with K_m values of 1.4 mM and 0.7 μM for ethanol[15] and retinol,[6] respectively.

Standard Assay for Rat Alcohol Dehydrogenase Isoenzymes. The activity of the rat isoenzymes during the purification procedure is measured using the 0.1 M glycine/NaOH, pH 10.0, buffer and the following specific conditions of substrate and coenzyme for each isoenzyme: 1 mM octanol and 4 mM NAD$^+$ for ADH-1; 1 mM octanol and 2.4 mM NAD$^+$ for ADH-2; and 33 mM ethanol and 2.4 mM NAD$^+$ for ADH-3. One unit of activity equals 1 μmol of NADH produced/min at 25° based on an absorption coefficient of 6220 M^{-1} cm^{-1} for NADH at 340 nm. Total protein is determined by the Coomassie blue method, using bovine serum albumin as standard.[18]

Purification Procedure

Sprague-Dawley rats (150–250 g) are used. The animals are sacrificed by decapitation, and the liver and stomach are quickly removed. The stomach is cut, cleaned, and washed in ice-cold distilled water. The organs are stored at −80°.

Purification of ADH-1 (Class II). In a typical experiment (Table I) 30 stomachs (44 g) are thawed, cut in small pieces, and homogenized in 10 mM Tris-HCl, pH 7.9 (1:1, w/v). The homogenate is centrifuged at 27,000 g for 90 min at 4°. The pH of the supernatant is adjusted to 7.9 by addition of NaOH. The supernatant is dialyzed against 10 mM Tris-HCl, pH 7.9, and applied to a column of DEAE-Sepharose CL-6B (2.5 × 20 cm) equilibrated with the same buffer. The column is eluted at a flow rate of 40 ml/hr with 450 ml of buffer followed by a 1-liter linear gradient of NaCl (0–150 mM) in the same buffer. Fractions of 7.5 ml are collected. Active fractions are pooled (50 ml) and concentrated to 15.5 ml. The preparation is dialyzed against 10 mM Tris-HCl, pH 7.6, and applied to an AMP-Sepharose 4B column (1.4 × 25 cm) equilibrated with the same buffer. The column is washed at a flow rate of 25 ml/hr with the buffer and fractions of 4 ml are collected. ADH-1 binds weakly to the column and elutes in the buffer wash, although it is retarded with regard to the major protein peak. Active fractions are pooled, concentrated to 16 ml, and

[17] P. Julià, X. Parés, and H. Jörnvall, *Eur. J. Biochem.* **172,** 73 (1988).
[18] M. Bradford, *Anal. Biochem.* **72,** 248 (1976).

TABLE I
PURIFICATION OF RAT STOMACH ADH-1 (CLASS II)[a]

Step	Total protein (mg)	Total activity (units)	Specific activity (units/mg protein)	Yield (%)
Extract	1564	81.4	0.05	100
DEAE-Sepharose	62	53.6	0.86	66
First AMP-Sepharose	14	37.0	2.7	46
Second AMP-Sepharose	0.3	9.5	32.5	12

[a] Data from Julia et al.[15] Activity is measured with 1 mM octanol and 4 mM NAD$^+$, in 0.1 M glycine/NaOH, pH 10.0, at 25°.

loaded on a second AMP-Sepharose column. Chromatography under the same conditions used for the first column generally results in homogeneous ADH-1, as judged by sodium dodecyl sulfate (SDS)–polyacrylamide gel electrophoresis. In the instances where some contaminating proteins are still present, the sample is further purified by gel filtration on a Sephacryl S-300 superfine column (2.5 × 89 cm). The column is eluted at 11 ml/hr with 0.1 M Tris-HCl, pH 8.3, and fractions of 2.2 ml are collected. ADH-1 isoenzyme is stored at 4°, in 0.1 M Tris-HCl, pH 8.5. ADH-1 is found to be unstable at pH values below 8.0.

Purification of ADH-2 (Class III). ADH-2 is isolated from rat liver homogenate by using chromatography on DEAE-Sepharose under the same conditions described for the ADH-3 purification (see below) followed by AMP-Sepharose chromatography.[15] We do not detail here the purification of this isoenzyme because it is practically inactive toward retinoids at neutral pH.

Purification of ADH-3 (Class I). Four rat livers (45 g) are thawed, cut, and added to 45 ml of 10 mM Tris-HCl/0.5 mM dithiothreitol, pH 8.3. The material is homogenized and centrifuged at 10,400 g for 1-hr at 4°. The supernatant is filtered through glass wool and centrifuged at 85,000 g for 1 hr at 4°. The supernatant is then dialyzed against 10 mM Tris-HCl/0.5 mM dithiothreitol, pH 8.3, and applied to a DEAE-Sepharose column (1.5 × 25 cm) equilibrated with the same buffer. The column is eluted at a flow rate of 25 ml/hr with 500 ml of buffer followed by a 500-ml linear gradient of 0–150 mM NaCl in the same buffer. Fractions of 8 ml are collected and assayed for ADH activity. ADH-3 appears in the buffer wash, whereas ADH-2 elutes in the NaCl gradient and is used for the subsequent ADH-2 purification (see above). Fractions with ADH-3 activity are pooled, concentrated to 2.5 ml, and dialyzed against 50 mM sodium phosphate/0.5

TABLE II
PURIFICATION OF RAT LIVER ADH-3 (CLASS I)[a]

Step	Total protein (mg)	Total activity (units)	Specific activity (units/mg protein)	Yield (%)
Extract	990	13.5	0.014	100
DEAE-Sepharose	538	8.8	0.016	65
CapGapp-Sepharose	4.7	6.0	1.3	44
Sephadex G-25	2.3	3.0	1.3	22

[a] Data from Julia et al.[15] Activity is measured with 33 mM ethanol and 2.4 mM NAD$^+$, in 0.1 M glycine/NaOH, pH 10.0, at 25°.

mM dithiothreitol, pH 7.5. NAD$^+$ is added to the dialyzed sample to reach a concentration of 2 mM. The enzyme preparation is then applied to a column (1.2 × 5 cm) of 4,3-N-(6-aminohexanoyl)aminopropylpyrazole-Sepharose (CapGapp-Sepharose),[19] equilibrated with 50 mM sodium phosphate/0.5 mM dithiothreitol, 2 mM NAD$^+$, pH 7.5. The column is washed exhaustively at 25 ml/hr with the buffer, and the enzyme is eluted with 0.5 M ethanol in 50 mM sodium phosphate/0.5 mM dithiothreitol, pH 7.5. Fractions of 4 ml are collected and assayed for ADH-3 activity. The enzyme is concentrated to 6 ml and dialyzed against 0.1 M Tris-HCl/0.5 mM dithiothreitol, pH 8.5. Purified ADH-3 is precipitated with ammonium sulfate, 0.6 g/ml, in an ice bath, and centrifuged at 9,800 g for 20 min at 4°. The precipitated enzyme is solubilized in 0.1 M Tris-HCl/0.5 mM dithiothreitol, pH 8.5, and gel filtered through a column of Sephadex G-25 (1.3 × 35 cm) equilibrated with the same buffer. ADH-3 is stored in this buffer at 4°. Results of purification are indicated in Table II.

[19] L. G. Lange and B. L. Valle, Biochemistry 15, 4681 (1976).

[49] Measurement of Acyl Coenzyme A-Dependent
Esterification of Retinol

By A. CATHARINE ROSS

Introduction

There is now evidence that retinol can be esterified with long-chain
fatty acids by microsomal enzymes from a variety of sources in a manner
that is stimulated by exogenous acyl-CoA.[1-5] Additionally, there exist
mechanisms of retinyl ester synthesis that do not depend directly on
activated fatty acid.[6-8] The exact relationship between these types of
activities is not yet clear and must await separation of the two enzymes, if
they are independent, and their characterization. In our experience with
rat liver microsomes or rat mammary gland microsomes, whether the
observed esterifying activity is stimulated by exogenous acyl-CoA is related to at least two factors: (1) the concentration of retinol substrate
provided and (2) the means of presentation of retinol, that is, whether
retinol is presented in dispersed form or is presented bound to one of the
intracellular retinol-binding proteins.[8,9] When dispersed retinol at concentrations exceeding approximately 5 μM is provided to either hepatic or
mammary gland microsomes, the addition of exogenous fatty acyl-CoA
leads to substantially greater rates of retinol esterification as compared to
the basal reaction in the absence of fatty acyl-CoA.[1,2,8,9] In comparison,
when retinol bound to cellular retinol-binding protein is presented to these
microsomes, the addition of fatty acyl-CoA to an incubation mixture has
only a small stimulatory effect, and membrane-associated phospholipid
acts as the immediate source of fatty acid for transfer to retinol.[6-10] In
comparison, Saari and Bredberg[11] have reported recently that microsomes prepared from bovine retinal pigment epithelium briskly esterify

[1] A. C. Ross, J. Lipid Res. **23**, 133 (1982).
[2] A. C. Ross, J. Biol. Chem. **257**, 2453 (1982).
[3] P. Helgerud, L. B. Petersen, and K. R. Norum, J. Lipid Res. **23**, 609 (1982).
[4] H. Torma and A. Vahlquist, J. Invest. Dermatol. **88**, 398 (1987).
[5] M. D. Ball, H. C. Furr, and J. A. Olson, Biochem. Biophys. Res. Commun. **128**, 7 (1985).
[6] D. E. Ong, P. N. MacDonald, and A. M. Gubitosi, J. Biol. Chem. **263**, 5789 (1988).
[7] J. C. Saari and D. L. Bredberg, J. Biol. Chem. **263**, 8084 (1988).
[8] R. W. Yost, E. H. Harrison, and A. C. Ross, J. Biol. Chem. **263**, 18693 (1988).
[9] R. K. Randolph and A. C. Ross, unpublished observations.
[10] P. N. MacDonald and D. E. Ong, J. Biol. Chem. **263**, 12478 (1987).
[11] J. C. Saari and D. L. Bredberg, J. Biol. Chem. **264**, 8636 (1989).

METHODS IN ENZYMOLOGY, VOL. 189

retinol added in dispersed form, even in the absence of acyl-CoA. This reaction has been characterized as a transesterification supported by endogenous microsomal phosphatidylcholine.[11]

This chapter presents the method we have used previously[1,2,8] to measure the reaction rate of the acyl-CoA-stimulated esterification of retinol presented to microsomes in dispersed form. This enzymatic reaction is referred to as acyl-CoA : retinol acyltransferase, or ARAT (EC 2.3.1.76, retinol fatty-acyltransferase).

Principle

The rate of retinol esterification is determined under conditions which optimize pH and in which product formation is linear with incubation time and amount of microsomal protein (enzyme). We have found it most convenient to utilize radiolabeled retinol of known specific radioactivity and to separate the labeled retinyl ester product from remaining labeled retinol on small columns of alumina oxide. It is also possible to separate product from reactants by high-performance liquid chromatography (HPLC)[2,5] and thus to utilize either radiolabeled or unlabeled substrate or, additionally, to characterize the species of esters formed.[2,8,10] Generally, both the retinol and fatty acyl-CoA substrates are present at or near saturating concentrations. We have routinely used palmitoyl-CoA as the activated fatty acid because retinyl palmitate predominates among most tissue retinyl esters; however, other activated fatty acids have also been studied and are stimulatory to varying extents.[1,2]

Method

Enzyme. Analytical subcellular fractionation of rat liver has revealed that the microsomal fraction is the locus of nearly all ARAT activity.[12] Accordingly, microsomes are prepared from rat liver (or other tissue) using standard methods of differential centrifugation which we have adapted[1,2] from the method of Amar-Costesec *et al.*[13] The microsomal pellet is washed once by resuspension and recentrifugation.[2] This washed microsomal fraction is suspended in 0.15 M potassium phosphate buffer, pH 7.4, containing 1 mM dithiothreitol. Aliquots are dispensed into small vials and stored at $-70°$. Good stability of the enzymatic activity has been observed for at least 2 months.

[12] E. H. Harrison, W. S. Blaner, D. S. Goodman, and A. C. Ross, *J. Lipid Res.* **28,** 973 (1987).
[13] A. Amar-Costesec, H. Beaufay, M. Wibo, D. Thines-Sempoux, E. Feytmans, M. Robbi, and J. Berthet, *J. Cell Biol.* **61,** 201 (1974).

Substrate. Tritiated retinol for substrate is prepared as a stock and is kept in ethanol at $-20°$. The stock substrate is prepared by mixing unlabeled retinol [prepared by saponification and extraction of retinyl acetate (Sigma, St. Louis, MO)] with [^{3}H]retinol ([1-^{3}H(N)]vitamin A_1, New England Nuclear, Boston, MA) such that the specific radioactivity is approximately 6–20 $\mu Ci/\mu mol$. The mixture is purified by chromatography in columns containing 1.5 g of deactivated (with 5% water) neutral alumina oxide eluted sequentially with 6 ml of hexane, 12 ml of 3% diethyl ether in hexane, 12 ml of 8% diethyl ether in hexane, and 20 ml of 50% diethyl ether in hexane. The latter fraction contains retinol. The solvent is evaporated under nitrogen, and the purified, labeled, substrate is reconstituted to 0.5–3 mM in ethanol. A small portion is diluted in ethanol so that the concentration may be determined by UV spectrophotometry at 325 nm using the molar extinction coefficient of 52,275 cm^{-1} M^{-1}.[14] The final specific radioactivity is determined by counting an aliquot of this diluted material. Stock substrate is stored under nitrogen or argon in a foil-wrapped container at $-20°$. For each experiment, a portion is transferred into a small vial, and the ethanol is evaporated under nitrogen. Dimethyl sulfoxide (5 μl per incubation) is immediately added and mixed with the [^{3}H]retinol. All procedures with retinol are conducted under dim light.

Incubations. For each experiment, freshly thawed microsomes are diluted in 0.15 M potassium phosphate buffer, pH 7.4. Our typical reaction volume is 0.25 ml; thus, the desired amount of protein in 0.15 ml of buffer is dispensed into 13 × 100 mm tubes on ice. For rat liver, 0.05–0.1 mg of microsomal protein per incubation is appropriate;[2,8] however, the linear range for protein should be determined for each new type of sample. We routinely set up duplicates for each condition, with a third tube that is subjected to 1–2 min in a boiling water bath serving as a blank. The desired reaction additives are prepared as a 2.5-fold concentrated solution. Generally, this includes fatty acyl-CoA (final concentration in additive mixture is 250 μM such that the final concentration in the reaction tube will be 100 μM), bovine serum albumin (final reaction concentration 20 μM), and dithiothreitol (final reaction concentration 5 mM) in potassium phosphate buffer. This mixture is kept on ice until a few minutes before the reactions are initiated, when it is equilibrated at 37°. Enzyme incubations are timed such that each microsomal sample is warmed at 37° for 50 sec prior to addition of 100 μl of the reaction additives mixture. Ten seconds later, 5 μl of the [^{3}H]retinol substrate in dimethyl sulfoxide is added rapidly, and the sample is immediately vortexed and returned to the incubation bath for the desired time interval. Reactions are stopped by

---

[14] A. C. Ross, *Anal. Biochem.* **115**, 324 (1981).

addition of 1 ml of ethanol containing 0.1% butylated hydroxytoluene (BHT).

Analysis. Neutral lipids are extracted by sequential addition, with mixing, of 4 ml hexane (also containing BHT) and 1 ml water.[1] Tubes are capped and centrifuged briefly to produce a clear hexane upper phase. An aliquot (3 ml) of the upper phase is pipetted into a small tube, and the hexane is evaporated under nitrogen to remove traces of dissolved ethanol or water. Hexane is added, and separation of retinyl esters is performed by alumina chromatography as previously described.[1,2,13] The fraction eluted with 3% diethyl ether in hexane contains [³H]retinyl esters. Solvent is evaporated under an air stream, and the retinyl ester product is quantified by liquid scintillation counting. An aliquot of the working substrate is also counted directly to determine the counts and, hence, to calculate the mass of substrate added. Counting efficiencies are established with reference to calibrated [³H]toluene (New England Nuclear) disintegrations per minute (dpm) standards. Data are corrected for the boiled blanks and are then expressed as picomoles of retinyl ester product formed per milligram of microsomal protein per minute.

Comments. We have found that the acyl-CoA-dependent esterifying activity of microsomes is inhibited by mercurial reagents (such as *p*-chloromercuribenzoate) but not by the serine esterase inhibitor, phenylmethylsulfonyl fluoride (PMSF).[1,2] In contrast, PMSF inhibits esterification of retinol presented to microsomes by cellular retinol-binding protein (type II).[6,10] In our experience, PMSF at concentrations above 1.5 mM inhibits the acyl-CoA-independent activity (presumably lecithin-retinol acyltransferase, or LRAT) but does so only in the absence of thiol-reducing agents.[9] The activity of ARAT is also sensitive to inhibition by several detergents and by the addition of progesterone, a property shared with acyl-CoA : cholesterol acyltransferase.[1,2] Again in contrast, the esterification of retinol bound to cellular retinol-binding protein, which is not significantly stimulated by exogenous fatty acyl-CoA, is increased slightly, rather than inhibited, by progesterone.[8] Thus, there are significant differences in the enzymatic properties of the activities that esterify dispersed retinol in an acyl-CoA-stimulated manner as compared to the esterification of protein-associated retinol in a non-CoA-dependent reaction.

Acknowledgments

This work was supported by National Institutes of Health Grants HL-22633 and HD-16484 and by funds from the Howard Heinz Endowment. Dr. Ross is recipient of an NIH Research Career Development Award (HD-00691).

[50] Acyl Coenzyme A-Dependent Retinol Esterification

By MARK D. BALL

Introduction

One of the reactions whereby retinol is esterified for storage in the liver is acyl-CoA-dependent transacylation, catalyzed by the microsomal enzyme acyl-CoA : retinol *O*-acyltransferase (ARAT; EC 2.3.1.76, retinol fatty-acyltransferase). ARAT transfers the acyl group directly from the acyl-CoA derivative to retinol,[1] thereby forming the corresponding retinyl ester. ARAT from the liver[1-3] and small intestine[4,5] has been the most thoroughly studied, although ARAT-like activity has been identified in other tissues as well. Described below is an *in vitro* assay method for ARAT from rat liver.

Assay of Acyl-CoA : Retinol *O*-Acyltransferase

Principle of Assay. The basic approach to measurement of ARAT activity is that of net retinyl ester synthesis. After incubation of microsomes with substrates, newly formed retinyl ester is extracted from the reaction mixture and then quantified by high-performance liquid chromatography (HPLC). The amount of product formed during incubation is simply the difference between the amount present after incubation and the amount endogenous to the microsomes. Under normal circumstances, the use of radiolabeled substrates is unnecessary.

Stock Solutions. Required stock solutions are listed in Table I. Convenient final volumes can be calculated by consulting Table I in conjunction with Table II, which lists the volume of each stock solution needed for the reaction mixtures.

Preparation of Stock Solution 5. Palmitoyl-CoA (Sigma, St. Louis, MO) is dissolved in stock solution 1, at a ratio of 3 mg of anhydrous solid per milliliter. The final concentration, which should be 3 m*M*, is checked spectrophotometrically. An aliquot of the solution is diluted 100-fold into

[1] A. C. Ross, *J. Biol. Chem.* **257**, 2453 (1982).
[2] M. Rasmussen, P. Helgerud, L. B. Petersen, and K. R. Norum, *Acta Med. Scand.* **216**, 403 (1984).
[3] D. Sklan and O. Halevy, *Br. J. Nutr.* **52**, 107 (1984).
[4] P. Helgerud, L. B. Petersen, and K. R. Norum, *J. Lipid Res.* **23**, 609 (1982).
[5] P. Helgerud, L. B. Petersen, and K. R. Norum, *J. Clin. Invest.* **71**, 747 (1983).

TABLE I
REQUIRED STOCK SOLUTIONS

Solution	Composition	Remarks
1	150 mM KH$_2$PO$_4$, pH 7.4	For reaction mixtures and for preparing stock solution 6
2	100 mM KH$_2$PO$_4$, pH 7.0	10–20 ml total needed; used for determining concentration of stock solution 6
3	250 mM sucrose in 15 mM KH$_2$PO$_4$, pH 7.4	Volume needed: about 5 times the volume of liver to be used
4	1 mM bovine serum albumin in 150 mM KH$_2$PO$_4$, pH 7.4	Albumin should be essentially globulin- and fatty acid-free; the molecular weight of the anhydrous powder is assumed to be 66,300
5	250 mM dithiothreitol in 150 mM KH$_2$PO$_4$, pH 7.4	—
6	3 mM palmitoyl-CoA in 150 mM KH$_2$PO$_4$, pH 7.4	See text
7	30 mM all-*trans*-retinol in dimethyl sulfoxide	See text

100 mM KH$_2$PO$_4$, pH 7.0 (stock solution 2), the absorbance at 259.5 nm is read, and the concentration is then calculated from an absorption coefficient of 15,400 M^{-1} cm^{-1}. The solution can be stored in capped microcentrifuge tubes at $-70°$ for several weeks.

TABLE II
COMPOSITION OF REACTION MIXTURES

	Volume of component added to tube (μl)	
Component	Test mixture	Control mixture
Microsomes	Equivalent to 150 μg of microsomal protein	Equivalent to 150 μg of microsomal protein (heat-inactivated)
Stock solution 1	469 $-$ (μl microsomes)	469 $-$ (μl microsomes)
Stock solution 4	10	10
Stock solution 5	10	10
Stock solution 6	10	10
Stock solution 7	1	0
Dimethyl sulfoxide	0	1

Preparation of Stock Solution 6. Retinol should be handled only under subdued laboratory light or under lights that do not emit wavelengths under about 500 nm. all-*trans*-Retinol is dissolved in dimethyl sulfoxide at a ratio of 8.6 mg/ml. The final concentration, which should be 30 mM, is checked spectrophotometrically. An aliquot of this solution is diluted 3000-fold into methanol or absolute ethanol, the absorbance at 325 nm is read, and the concentration is then calculated from an absorption coefficient of 52,480 M^{-1} cm^{-1}. For storage, the retinol solution is sealed under argon in capped microcentrifuge tubes, which are then kept at $-70°$ in darkness.

Preparation of Rat Liver Microsomes. For anesthesia, Innovar-Vet (Pitman-Moore, Washington Crossing, NJ), which is injected intraperitoneally at 60 μl/100 g body weight, is recommended. Livers are excised from anesthetized rats, then rinsed and minced in 250 mM sucrose buffered with 15 mM KH$_2$PO$_4$, pH 7.4 (stock solution 3). The tissue is homogenized in 3 volumes of fresh stock solution 3 by means of a tight-fitting, motor-driven Teflon pestle. Microsomes are then prepared by differential centrifugation at 4°. The homogenate is first centrifuged for 20 min at 13,000 g, then the resulting supernatant solution at 100,000 g for 70 min, at the same temperature. The pellet is resuspended in stock solution 3, and the centrifugations are repeated. Microsomes are suspended in a volume (typically 1–4 ml) of 150 mM KH$_2$PO$_4$, pH 7.4 (stock solution 1), such that the final concentration of protein is 10–15 mg/ml. Microsomal protein can be quantified by any of the standard methods. For storage, the suspension is distributed into capped microcentrifuge tubes (100-μl aliquots are convenient), quick-frozen in liquid nitrogen, and thereafter kept at $-70°$ until needed for the assay. Normally, the enzyme is stable at this temperature for several weeks. Making the microsome suspension 1 mM in dithiothreitol helps to stabilize activity.

Preparation and Incubation of Reaction Mixtures. The most convenient tubes for incubation are 13 × 100 mm, because the final volume of the mixture is 0.5 ml, and because the reaction mixture will be extracted in the tube itself after incubation. For a simple measurement of ARAT activity in a sample of rat-liver microsomes, the basic assay requires two reaction mixtures (which, of course, should be duplicated): the test mixture, with all components, and the control mixture, with the same components but with microsomes that have been heat-inactivated. For such inactivation the thawed microsome suspension (100 μl or less) is immersed in boiling water for at least 30 sec. All components of the reaction mixture, *except* stock solution 7 (retinol), are pipetted into the incubation tube according to Table II. The tube is then placed in a shaking water bath at 37°, where it is left for 2 min, at the end of which time 1 μl of the retinol solution is added by means of a microliter syringe (dimethyl sulfoxide is added likewise to the control tube); this initiates the reaction, and the mixture is allowed to incubate in the water bath for 10 min more.

Termination of Reaction and Extraction of Product. At the end of the 10-min incubation, 1 ml cold absolute ethanol is added to the tube to stop the reaction. Significant precipitation occurs at this point. The resulting mixture is then extracted in the tube 2 times with 1 ml hexane each; this is accomplished easily by inserting a tight-fitting cap into the open end of the tube, then inverting or vortexing. The extracts are combined, and the solvent is removed under a stream of argon (evaporation is hastened by gentle heating). The residue (not visible to the naked eye) is dissolved in 100 μl n-butyl acetate; if, however, residual water is visible in the tube after removal of the hexane, the butyl acetate solution should be dried with a standard drying agent. The final solution is then analyzed by HPLC.

Chromatographic Analysis. The retinyl palmitate formed during incubation can be quantified by reversed-phase HPLC with the system described earlier in this volume.[6] However, even when palmitoyl-CoA is the only exogenous thioester used in the assay, small amounts of other retinyl esters are formed, including retinyl oleate, stearate, and linoleate, all of which are likely to appear on the chromatogram. These arise presumably from the corresponding acyl-CoA thioesters endogenous to the microsomes.[1] For a given retinyl ester, the amount formed is the amount in the test mixture minus the amount in the control mixture. Data are usually expressed as picomoles retinyl ester formed per minute per milligram microsomal protein.

Alternate Acyl-CoA Substrates. Many other acyl-CoA thioesters are effective substrates for ARAT *in vitro,* including derivatives of all saturated fatty acids from C_{12} to C_{20}, as well as oleoyl-CoA. Polyunsaturated derivatives, however, are poor substrates.[7] These various other thioesters can be used in lieu of palmitoyl-CoA in this assay, and their stock solutions can be prepared similarly (unsaturated derivatives should be stored under argon).

[6] A. C. Ross, this volume [7].
[7] M. D. Ball and J. A. Olson, *FASEB J.* **2** Abstr. 3183 (1988).

[51] Assay of Lecithin–Retinol Acyltransferase

By Paul N. MacDonald and David E. Ong

Introduction

Retinol (vitamin A alcohol) is esterified with long-chain fatty acyl groups during the absorptive process in the small intestine. Retinyl esters are also the predominant form of vitamin A in the liver, potentially serving as an active reservoir of vitamin A for the animal. New evidence suggests that the enzyme responsible for retinol esterification in the intestine and liver is a recently described microsomal activity termed lecithin:retinol acyltransferase, or LRAT.[1,2] LRAT catalyzes the transesterification of fatty acyl moieties from phosphatidylcholine to retinol to produce retinyl esters. This activity is specific for phosphatidylcholine. Phosphatidylethanolamine, phosphatidic acid, acyl-CoA, or free fatty acids are not effective acyl substrates. LRAT also demonstrates positional selectivity in that only the fatty acid at the *sn*-1 position of phosphatidylcholine is transferred. The preponderance of palmitate/stearate groups at the *sn*-1 position of phosphatidylcholines *in vivo* and in the majority of retinyl esters observed *in vivo* implicates LRAT in the physiological mechanism of retinol esterification.

Cellular retinol-binding protein (CRBP) and cellular retinol-binding protein, type II (CRBP II) are two of several binding proteins that carry retinol as an endogenous ligand *in vivo*.[3] These proteins may be involved in the transport of retinol to intracellular sites of action and metabolism. Consequently, retinol complexed to CRBP and CRBP II are used as substrates for the *in vitro* esterifications described here. Both complexes are effective substrates for LRAT. However, retinol–CRBP and retinol–CRBP II are not esterified by the distinct, acyl-CoA-dependent activity called acyl-CoA:retinol acyltransferase (ARAT, EC 2.3.1.76, retinol fatty-acyltransferase), which is also present in the microsomal preparations. The use of retinol–CRBP and retinol–CRBP II as substrates appears to minimize interference from other esterifying activities in these crude preparations. The procedures described here detail the methods used in our laboratory to examine retinol esterification by LRAT.

[1] P. N. MacDonald and D. E. Ong, *J. Biol. Chem.* **263**, 12478 (1988).
[2] P. N. MacDonald and D. E. Ong, *Biochem. Biophys. Res. Commun.* **156**, 157 (1988).
[3] D. E. Ong, *Nutr. Rev.* **43**, 225 (1985).

Materials

Preparation of [³H]Retinol–CRBP II and
[³H]Retinol–CRBP Complexes

Generation of Apoproteins. CRBP II and CRBP are purified as holo complexes from rat intestine[4] and rat liver,[5] respectively. The apoproteins are generated by treating the purified holo preparations with ultraviolet light to destroy the retinol ligand. Generally, 1 ml of a 1 mg/ml preparation of retinol–CRBP or retinol–CRBP II is placed in a quartz cuvette and stoppered with Parafilm. The cuvette is positioned on its side on ice and covered with a glass plate onto which is placed an XX-15 Black-Ray ultraviolet lamp (Ultra-Violet Products, San Gabriel, CA). The protein solution is illuminated for 15–30 min at a distance of approximately 5 cm. Destruction of the retinol is monitored by measuring the loss of absorbance at 350 nm.

Chromatographic Purification of [³H]retinol. [³H]Retinol is purified after reduction of retinaldehyde by $NaB[^3H]_4$ as previously described.[6] Generally, 300–400 nmol of [³H]retinol is dried under N_2, redissolved in 100 μl of acetonitrile, and injected onto a 5-μm, C_{18} Econosphere reversed-phase high-performance liquid chromatography (HPLC) column (4.5 × 25 mm) equilibrated in acetonitrile/water (70 : 30, v/v). The flow rate is 2.0 ml/min, and the elution is monitored with an absorbance detector at 325 nm. all-*trans*-Retinol elutes in approximately 15 min. The [³H]retinol peak, contained in approximately a 2-ml volume, is collected in a 15-ml conical glass extraction tube. This preparation is extracted 3 times with 4 ml of hexane. The combined hexane extracts are dried under N_2, and the oily residue is redissolved in 25–50 μl of dimethyl sulfoxide.

Reconstitution of [³H]Retinol and Apo-CRBP II or Apo-CRBP. The purified [³H]retinol is added dropwise with stirring directly to the cuvette of UV-treated CRBP II or CRBP using a 25-μl Hamilton syringe. Formation of the complex is observed by monitoring the characteristic absorbance spectrum of the complex between 240 and 400 nm.[5,7] Retinol is added until the ratio OD_{350}/OD_{280} is approximately 1.0–1.3 : 1, and the reconstituted complex is subjected to gel filtration on a PD-10 column (Sephadex G-25, Pharmacia, Piscataway, NJ) equilibrated in 0.2 M KH_2PO_4, pH 7.2. The protein eluting in the void volume is collected,

---

[4] W. H. Schaefer, B. Kakkad, J. A. Crow, I. A. Blair, and D. E. Ong, *J. Biol. Chem.* **264,** 4212 (1989).

[5] D. E. Ong and F. Chytil, *J. Biol. Chem.* **253,** 828 (1978).

[6] G. Liau, D. E. Ong, and F. Chytil, *J. Cell Biol.* **91,** 63 (1981).

[7] D. E. Ong, *J. Biol. Chem.* **259,** 1476 (1984).

pooled, and the concentration of bound ligand determined by measuring the absorbance at 350 nm with an extinction coefficient of 55,000 M^{-1} cm^{-1}. Specific activities of the final preparation range from 3 to 4.5 Ci/ mmol. Approximately 6 hr is required for the entire procedure, and enough [³H]retinol–protein complex is obtained to carry out 300–400 assays. The complex is stable for several weeks at 4° when stored in the dark.

Preparation of Positionally Defined Phosphatidylcholines

Phosphatidylcholines with chemically defined fatty acids at the sn-1 and sn-2 positions are prepared by the base-catalyzed condensation of commercially available 1-acyl-2-lysophosphatidylcholine and an appropriate fatty acid anhydride.[8] Lysophosphatidylcholines (25 mg) are dissolved in a 1.5-ml glass Reactivial (Pierce, Rockford, IL) with 300 μl of chloroform freshly distilled over P_2O_5. A 5-fold molar excess of the acyl anhydride is added followed by 1 mol equiv (based on lysophosphatidylcholine) of 4-pyrrolidinopyridine. An additional 600 μl of dry, freshly distilled chloroform is added, the vial is sealed under N_2, and the reaction is stirred with a Teflon triangular stir bar for 2–3 hr in a 35° dry oven.

The reaction mixture is quantitatively transferred to a 5-ml conical extraction tube with 300 μl of additional chloroform. The reaction mixture is acid-extracted following the addition of 600 μl of methanol and 440 μl of 0.1 N HCl. The lower chloroform phase is isolated after vigorous mixing and centrifugation to separate the phases. The upper phase is back-extracted with 800 μl of dry chloroform, and the combined lower phases are taken to dryness by rotary evaporation. The residue is dissolved in 600 μl of dry chloroform and applied as a single 200-μl streak across three separate IB2 (20 × 20 cm) flexible silica gel thin-layer chromatography (TLC) plates (J. T. Baker, Phillipsburg, NJ). The sheets are developed in chloroform/methanol/28% ammonium hydroxide (65 : 35 : 5). Thin strips are cut from the edges of each sheet, and the lipids on these strips are visualized with I_2 vapors.

The diacylphosphatidylcholine products, identified by comigration with authentic dimyristoylphosphatidylcholine on a separate TLC sheet, are extracted from the scraped silica with two 15-ml washes with chloroform/methanol (1 : 2) and one 15-ml wash with methanol. The silica is mixed and warmed to 50° with each elution, then subjected to centrifugation at 10,000 g for 15 min to pellet the silica. The solvent is filtered through glass wool to remove residual silica. The combined extracts are taken to dryness by rotary evaporation, the residue is redissolved with warming in 5 ml of acetone/chloroform (95 : 5), and the phosphatidylcho-

[8] J. T. Mason, A. V. Broccol, and C. Huang, *Anal. Biochem.* **113**, 96 (1981).

line product is allowed to crystallize on ice in 15-ml tube. Following centrifugation at 10,000 g for 15 min, the solvent is carefully removed, and the pelleted crystals are dried under a very gentle stream of N_2 and by warming for several hours under a heat lamp. A typical yield for the preparation is 15 mg of phosphatidylcholine.

Preparation of Phospholipid Substrates

Potential phospholipid substrates are added to the reaction mixture as sonicated aqueous dispersions. Concentrated stock solutions of phospholipids are prepared in chloroform. Generally, 400–500 nmol of phospholipid is taken to dryness under N_2 and residual solvent is removed under vacuum for 30 min on a lyophilizing apparatus. One milliliter of 0.2 M KH_2PO_4, pH 7.2, is added to the oily film on the bottom of the tube. The phospholipid is dispersed into this buffer with 60 1-sec pulses of a Branson Sonifier fitted with a microtip on a power setting of 2. The phospholipid preparations are made fresh daily and stored on ice until used in the assay.

Isolation of Microsomal Preparations

Normal, chow-fed Sprague-Dawley rats (200–300 g body weight) are sacrificed by decapitation, and the organs of interest are removed, weighed, and processed as rapidly as possible. When obtaining intestinal preparations, the upper half of the small intestine is removed, and the intestinal contents are flushed with ice-cold isotonic saline. The intestine is slit longitudinally, placed on an ice-cold glass plate, and the mucosa is scraped from the muscularis with two glass microscope slides. The scraped mucosa is weighed, resuspended in 4 volumes of isotonic saline, and subjected to centrifugation at 200 g for 10 min at 4°. The washed mucosa (or other tissues of interest) is disrupted in 4 volumes of 0.2 M potassium phosphate, pH 7.2, with six or seven passes of a Teflon pestle in a Potter–Elvehjem homogenizer. The homogenate is subjected to centrifugation at 20,000 g for 15 min. Floating fat is carefully removed with a Pasteur pipette, and the supernatant solution is subjected to centrifugation at 106,000 g for 1 hr at 4°. The microsomal pellet is resuspended in 2 volumes of 0.2 M potassium phosphate, pH 7.2, and again subjected to centrifugation at 106,000 g for 1 hr. The washed microsomal pellet is resuspended in 0.2 M KH_2PO_4, pH 7.2, that contains 1.0 mM dithiothreitol (DTT) at a protein concentration of 5–15 mg/ml. The preparations are divided into 100- to 200-μl aliquots, quick-frozen in dry ice/ acetone, and stored at −70°. Little, if any, loss in the ability of these preparations to esterify retinol has been observed for periods up to 1 year under these conditions.

Assay Methods

Microsomal preparations from rat liver and small intestine catalyze retinyl ester synthesis from exogenous sources of retinol. The esterification proceeds when retinol and microsomal preparations are combined in neutral phosphate buffer at 37°. The microsomal ester synthase does not require additional components or cofactors. The source of acyl groups is presumably a lipid component of the microsomal vesicles. Potential exogenous acyl donors are examined by including them in the assay and analyzing the products of the reaction for the appropriate retinyl ester by HPLC. Retinol is added as a complex with its physiological carrier proteins, CRBP II and CRBP. Unbound retinol dispersed from dimethyl sulfoxide (DMSO) into the reaction buffer is an effective substrate as well. However, unbound retinol may also be esterified by microsomal ARAT. Therefore, if unbound retinol is to be used as a substrate, it should be used in conjunction with chemical inhibitors that can distinguish between the LRAT and ARAT activities (see section on inhibitor studies, below).

Two *in vitro* assays are routinely used by this laboratory to characterize LRAT. The procedures differ mainly in the method of analysis. When the individual retinyl esters synthesized in the assay are of interest, the products are analyzed by an HPLC system that separates the retinyl esters based on their acyl chain length and degree of saturation. If the total amount of retinyl ester synthesis is the only concern, then [3H]retinol–CRBP II or [3H]retinol–CRBP are used as substrates and the total [3H]retinyl esters synthesized are determined after batch elution from columns of aluminum oxide.

Method 1

Assay Conditions. All manipulations are carried out under dim yellow light. Of primary importance is that all assays are carried out in new borosilicate test tubes. We have observed the gradual accumulation of inhibiting substances in reusable, detergent-washed tubes that requires a chromic acid wash for removal. To examine retinol esterification with the endogenous acyl donor of the microsomal preparations, 30–60 μg of microsomal protein is added to 0.2 M KH_2PO_4, pH 7.2, in a 12 × 100 mm borosilicate glass tube. Retinol–CRBP or retinol–CRBP II (generally 1.5 nmol) is added to a final volume of 0.5 ml to initiate the reaction. The tubes are incubated for 10 min in a 37° shaking water bath. When exogenous phospholipids are examined, they are added as sonicated aqueous dispersions (described above) to a final concentration of 40–80 μM. When chemical inhibitors are utilized, 30–60 μg of microsomal protein is preincubated with appropriate concentrations of inhibitor in approximately 450 μl of 0.2 M KH$_2$PO$_4$, pH 7.2, for 10 min at 37°. The reactions are initiated with the addition of substrates to a final volume of 0.5 ml and incubated at 37° for an additional 10 min.

Extraction of Lipids. Following the 10-min incubation, the total reaction volume is removed to a 15-ml conical extraction tube containing 2.0 ml of ice-cold ethanol with 100 μg of butylated hydroxytoluene (BHT) per milliliter. After the addition of 8.0 ml of hexane and 2.0 ml of distilled water, the tubes are mixed with a vortex apparatus, then shaken vigorously by hand. Although the phases rapidly separate, centrifugation at 1000 rpm for 5 min is necessary to obtain a clear upper phase. Retinoids are extracted with greater than 95% efficiency. A measured volume (generally 7.5 ml) of the upper hexane phase is removed to a 15-ml conical tube and dried under a gentle stream of N$_2$.

Analysis of Hexane Extract by HPLC. The lipid residue is dissolved in 100 μl of methanol and injected onto a 5-μm Waters C$_{18}$ Nova-pak HPLC column (3.9 × 150 mm) fitted with a 5-μm C$_{18}$ guard column (2 cm). The mobile phase is 100% methanol at a flow rate of 2 ml/min. The retention times of authentic retinol, retinyl laurate, retinyl myristate, retinyl linoleate, retinyl palmitate/oleate, and retinyl stearate are 2, 5, 7.5, 8, 11, and 15 min, respectively. Retinyl palmitate and retinyl oleate coelute under these conditions. However, these esters are easily resolved by including 23 mM AgNO$_3$ in the mobile phase when the resolution of retinyl palmitate and retinyl oleate is important.

Method 2

The alternative method uses [³H]retinol–CRBP or [³H]retinol–CRBP II as a substrate, and the [³H]retinyl esters synthesized by the microsomal preparations are analyzed as a class by batch elution from aluminum oxide columns. This method is more rapid and more sensitive than the HPLC method described above. Up to 20 individual assays can be completed within a 3-hr period.

Assay Conditions. Incubation conditions are similar to the previous method with the exceptions that the total reaction volume is reduced to 120 μl, with similar reductions in substrate and protein, and the assay is carried out in 10 × 75 mm borosilicate tubes.

Extraction of Lipids. The reactions are stopped by removing 100 μl of the 120-μl reaction volume into 0.4 ml of ice-cold ethanol (100 μg/ml in BHT) in a 15-ml conical extraction tube. The reactions are extracted following the addition of 1.6 ml of hexane (100 μg/ml BHT) and 0.4 ml of distilled water.

Aluminum Oxide Chromatography. Generally, 500 µl of the hexane upper phase is added directly to an 8 × 180 mm glass column (Isolab, Inc., Akron, OH) containing 1.2 g of aluminum oxide (activated, neutral, Brockmann I) that has been deactivated with 10% (w/v) distilled water. The sample is washed into the column bed with two 0.5-ml aliquots of hexane. [^{3}H]Retinyl esters are then eluted with 5 ml of hexane containing 2% diethyl ether (v/v), collecting directly into a scintillation vial. [^{3}H]Retinol is not eluted under these conditions. The solvent is removed with a stream of N_2 prior to determination of retinyl esters by scintillation counting. Control reactions use microsomal preparations that have been heat-inactivated at 100° for 5 min. Control reactions generally yield 100–200 cpm of radiation in the retinyl ester fraction, which is approximately 0.01% of the total radioactivity added to the assay.

Properties

Kinetic Parameters. The microsomal esterification of retinol–CRBP and retinol–CRBP II increases linearly with increasing amounts of liver and intestinal microsomal protein up to 150 µg/ml. The activities exhibit a rather broad pH optimum, with neutral pH being most effective. The reactions with both liver and intestinal preparations display typical Michaelis–Menten kinetics. The K_m and V_{max} values for retinol–CRBP II and the intestinal enzyme are 0.2 µM and 200 pmol/min/mg protein, respectively. Similar results are obtained for liver preparations with retinol–CRBP as substrate.

Substrate Specificity. The selectivity of intestinal and hepatic LRAT for various exogenous lipids has been examined.[1,2] The preferred exogenous acyl donor is phosphatidylcholine. Retinyl myristate is generated only if dimyristoylphosphatidylcholine is included in the assay. Little if any formation of retinyl myristate is observed if dimyristoylphosphatidylethanolamine, dimyristoylphosphatidic acid, myristoyl-CoA, or myristic acid are provided as potential acyl donors. When intestinal or liver microsomal preparations are incubated with retinol–CRBP II or retinol–CRBP and 1-myristoyl-2-lauroylphosphatidylcholine, only retinyl myristate is produced from the exogenous phospholipid (Fig. 1). Conversely, when 1-lauroyl-2-myristoylphosphatidylcholine is included, only the production of retinyl laurate is observed. Thus, LRAT demonstrates a positional selectivity for acyl transfer from the *sn*-1 position of phosphatidylcholine.

Inhibitor Studies. The effect on LRAT activity of several reagents that preferentially modify serine hydroxyl or cysteine sulfhydryl residues have been examined (Fig. 2).[1,2] The esterification of retinol–CRBP and retinol–CRBP II is sensitive to the sulfhydryl-modifying reagent *N*-ethylma-

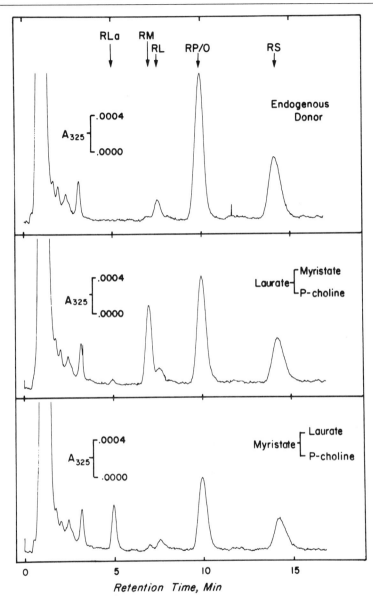

FIG. 1. Positional specificity in the transfer of acyl groups from phosphatidylcholine to retinol–CRBP II. 1-Myristoyl-2-lauroylphosphatidylcholine (middle) and 1-lauroyl-2-myristoylphosphatidylcholine (bottom) were assayed for acyl transfer to retinol–CRBP II. Esterification without exogenous phosphatidylcholine (top) is also illustrated (Endogenous Donor). (From MacDonald and Ong.[1])

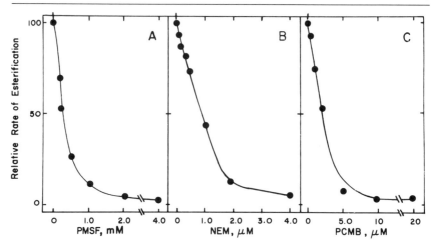

FIG. 2. Effect of various inhibitors on the esterification of retinol–CRBP II with endogeneous acyl donor. Intestinal microscomes (12 μg microsomal protein) were preincubated for 10 min at 37° in the presence of increasing concentrations of phenylmethylsulfonyl fluoride (A), N-ethylmaleimide (B), or p-chloromercuribenzoate (C) in 0.2 M KH$_2$PO$_4$, pH 7.2. [^{3}H]Retinol–CRBP II (0.3 nmol, 4 Ci/mmol) was added to a total volume of 0.1 ml. The reaction was incubated for 10 min at 37°, extracted into hexane, and the extract analyzed for [^{3}H]retinyl esters by batch analysis on columns of alumina. (From MacDonald and Ong.[1])

leimide (NEM). Microsomal preparations preincubated with as little as 1 μM Nem are no longer able to catalyze retinyl ester synthesis from retinol–CRBP II or retinol–CRBP. The LRAT reaction is also sensitive to p-chloromercuribenzoate, another reagent that modifies sulfhydryl residues. These results suggest an essential sulfhydryl residue(s) in the catalytic mechanism of LRAT. Another critical residue was suggested when the antiesterase phenylmethylsulfonyl fluoride (PMSF) was included in the LRAT assay. PMSF modifies serine hydroxyl residues, and LRAT is completely inhibited when microsomal preparations are pretreated with 1.0 mM PMSF. The effect of PMSF on LRAT can be exploited to distinguish between LRAT and the CoA-dependent activity ARAT since ARAT is essentially insensitive to PMSF.[1,9,10] Thus, if unbound retinol is to be used as substrate, one could include PMSF in a control reaction to be certain that the activity under study is not ARAT. Retinol bound to its physiological carrier proteins circumvents this problem as these substrates are unavailable to the esterfication catalyzed by ARAT. However, we have observed esterification of protein-bound retinol by micro-

[9] A. C. Ross, J. Lipid Res. 23, 133 (1982).
[10] H. Torma and A. Vahlquist, J. Invest. Dermatol. 88, 398 (1987).

somes from pancreas that is not PMSF-sensitive and consequently is not LRAT or ARAT. It might be a reverse esterase reaction. At this time, sensitivity to both PMSF and NEM appears to be diagnostic for LRAT activity.

[52] Bile Salt-Independent Retinyl Ester Hydrolase Activities Associated with Membranes of Rat Tissues

By EARL H. HARRISON and JOSEPH L. NAPOLI

Introduction

Hydrolysis of retinyl esters plays a major role in the overall metabolism of vitamin A in the body. In liver, retinyl ester hydrolysis occurs both during the uptake of chylomicron remnants and prior to the mobilization of retinyl esters stored in cellular lipid droplets.[1] Although the majority of the endogenous retinoids in mammals consists of retinyl esters stored in the liver, liver is not the only site of retinyl ester storage. In kidney, lung, and testes esters can represent as much as 70% of the total neutral retinoids.[2-4] Being the enzymes required to release retinol accumulated and stored as esters, retinyl ester hydrolases are likely to play a key role in retinoid metabolism.

Most studies of the hydrolysis of long-chain retinyl esters in liver and other tissues have focused on a neutral retinyl ester hydrolase activity that is markedly stimulated by millimolar concentrations of bile salts or their analogs.[3,5-10] This activity has an unusual distribution among subcellular fractions of rat liver, with most of the activity being distributed in the "nuclear" fraction and high-speed supernatant fraction of homogenates;

[1] D. S. Goodman and W. S. Blaner, *in* "The Retinoids" (M. B. Sporn, A. B. Roberts, and D. S. Goodman, eds.), Vol. 2, p. 1. Academic Press, New York, 1984.

[2] J. B. Williams, B. C. Pramanik, and J. L. Napoli, *J. Lipid Res.* **25**, 638 (1984).

[3] J. L. Napoli, A. M. McCormick, B. O'Meara, and E. A. Dratz, *Arch. Biochem. Biophys.* **230**, 194 (1984).

[4] D. S. Goodman, H. S. Huang, and T. Shiratori, *J. Lipid Res.* **6**, 390 (1965).

[5] E. H. Harrison, J. E. Smith, and D. S. Goodman, *J. Lipid Res.* **20**, 760 (1979).

[6] J. H. Prystowsky, J. E. Smith, and D. S. Goodman, *J. Biol. Chem.* **256**, 4498 (1981).

[7] W. S. Blaner, J. H. Prystowsky, J. E. Smith, and D. S. Goodman, *Biochim. Biophys. Acta* **794**, 419 (1984).

[8] W. S. Blaner, G. Halperin, O. Stein, Y. Stein, and D. S. Goodman, *Biochim. Biophys. Acta* **794**, 428 (1984).

[9] D. A. Cooper and J. A. Olson, *Biochim. Biophys. Acta* **884**, 251 (1986).

[10] D. A. Cooper, H. C. Furr, and J. A. Olson, *J. Nutr.* **117**, 2066 (1987).

very little activity is associated with the microsomal fraction. In addition, the absolute activity varies markedly among liver homogenates from individual rats. Also, when assayed in the presence of millimolar concentrations of bile salts (e.g., cholate) most preparations show higher apparent specific activity toward cholesteryl oleate and/or triolein than toward retinyl palmitate.

We have recently detected bile salt-independent retinyl ester hydrolases in rat liver, kidney, lung, and testes.[11,12] These enzymes are clearly distinct from the previously described bile salt-dependent enzyme(s). This chapter describes the properties of these bile salt-independent retinyl ester hydrolases in liver and kidney, the two most studied tissues.

Assay

Retinyl ester hydrolase activity is determined using the sensitive, radiometric assay of Prystowsky et al.,[6] as described in detail elsewhere in this volume.[13] Briefly, reactions are carried out in a final volume of 0.2 ml by adding buffer and a source of enzyme in a final volume of 0.19 ml. Reactions are initiated by adding radioactive retinyl palmitate in 10 μl of ethanol and incubated at 37°. Reactions are terminated by adding 3.25 ml of methanol/chloroform/heptane (1.4 : 1.25 : 1) and 1 ml of potassium carbonate/borate buffer, pH 10, and the product, radioactive palmitic acid, is extracted into the alkaline aqueous upper phase. An aliquot of this phase is mixed with scintillation solvent and counted in a liquid scintillation counter. The amount of palmitic acid released is determined from the partition coefficient (~75%) of palmitic acid and the specific radioactivity of the substrate.

We have used both [1-^{14}C]palmitic acid and [9,10-^{3}H]palmitic acid for the preparation of labeled retinyl palmitate. Retinyl [1-^{14}C]palmitate is used at final specific activities of 2.5–10.0 μCi/μmol. Retinyl [9,10-^{3}H]palmitate is used at final specific activities of 17.0–23.0 μCi/μmol.

For work with liver homogenates we routinely assay with 50 mM Tris–maleate, pH 8.0 (the pH optimum), and 100 μM retinyl [1-^{14}C]palmitate. Under these conditions the reaction is linear for 30 min and with up to approximately 15 μg of liver homogenate protein. For work with kidney homogenates routine assays are carried out using 10 mM HEPES, pH 8.0, and 100 μM retinyl [9,10-^{3}H]palmitate. Under these conditions the reaction is linear for 45 min and with up to 10 μg protein.

[11] E. H. Harrison and M. Z. Gad, J. Biol. Chem. 264, 17142 (1989).
[12] J. L. Napoli, E. B. Pacia, and G. J. Salerno, Arch. Biochem. Biophys. 274, 192 (1989).
[13] D. A. Cooper, this volume [55].

Heptatic Retinyl Ester Hydrolases

We present below information on the properties of the bile salt-independent retinyl ester hydrolase (REH) activity of rat liver.[11] In particular, three areas are discussed which serve to distinguish the bile salt-independent hydrolase(s) from the bile salt-dependent enzyme(s).

Variability in Activity

When homogenates of the livers of individual rats are assayed for retinyl palmitate hydrolase activity in the presence of 20 mM cholate, there is marked variation in absolute activity.[5,6] Table I shows the neutral retinyl palmitate hydrolase (RPH) activity for eight individual animals assayed in the presence or the absence of cholate. As expected the activities in the presence of cholate varied extensively. In marked contrast, the activities in the absence of bile salt showed little variation. Indeed, in this series of animals the addition of cholate inhibited the neutral RPH activity in seven of the homogenates. These results are consistent with there being a neutral, bile salt-independent RPH activity that is inhibited by bile salt and a separate bile salt-dependent activity present to various extents in the livers of individual animals.

TABLE I

BILE SALT-DEPENDENT AND BILE
SALT-INDEPENDENT RETINYL ESTER HYDROLASE
ACTIVITIES IN INDIVIDUAL RAT LIVERS[a]

Rat	Retinyl ester hydrolase activity (nmol/h/g liver)	
	No cholate	20 mM cholate
1	1244	0
2	1143	0
3	1044	0
4	881	135
5	1268	278
6	1567	350
7	1224	876
8	1151	2339

[a] Whole homogenates of rat livers were assayed in 30-min incubations containing 50 mM Tris–maleate buffer (pH 8.0) and 100 μM retinyl palmitate, with and without 20 mM cholate. Reproduced with permission.[11]

Subcellular Distribution

Table II provides information on the distribution of the neutral, bile salt-independent RPH activity among subcellular fractions prepared by differential centrifugation of rat liver homogenates. Most of the enzyme activity is recovered in the microsomal fraction of the homogenate, and its distribution is very similar to those of marker enzymes for the endoplasmic reticulum (glucose-6-phosphatase and o-nitrophenyl acetate esterase) and plasma membrane (alkaline phosphodiesterase). Thus, in terms of subcellular distribution, the bile salt-independent RPH activity is entirely unlike the previously studied, cholate-stimulated RPH that is nearly absent from microsomes and, rather, is found associated with the "nuclear" and high-speed supernatant fractions of the liver homogenate.

TABLE II
DISTRIBUTIONS OF BILE SALT-INDEPENDENT RETINYL ESTER HYDROLASE AND MARKER
ENZYMES IN SUBCELLULAR FRACTIONS OF RAT LIVER[a]

Constituent	Percentage of recovered amount[b]				Recovery (%)[c]
	N	ML	P	S	
Protein (4)	24 ± 4	20 ± 4	19 ± 2	37 ± 6	97 ± 5
N-Acetyl-β glucosaminidase (lysosomes) (4)	19 ± 7	48 ± 6	28 ± 4	4 ± 4	94 ± 10
Glucose-6-phosphatase (ER) (3)	26 ± 2	11 ± 3	60 ± 5	3 ± 0	105 ± 8
Esterase[d] (ER) (1)	29	7	60	4	91
Alkaline phosphodiesterase (plasmalemma) (4)	23 ± 5	17 ± 9	58 ± 13	2 ± 1	110 ± 13
Retinyl ester hydrolase (4)	23 ± 5	8 ± 5	56 ± 9	13 ± 6	99 ± 11

[a] Reproduced with permission.[11]
[b] Results are given as means ± 1 S.D. The number of experiments is given in parentheses. Relative values are presented for the distribution of each constituent among the four fractions: nuclear (N), mitochondrial–lysosomal (ML), microsomal (P), and supernatant (S). Values are percentages of the constituent recovered in each fraction relative to the amounts recovered in all four fractions (taken as 100%).
[c] Recovery represents the total amount recovered in all four fractions relative to the amount in the whole homogenate.
[d] "Nonspecific" esterase was assayed using o-nitrophenyl acetate as the substrate.

TABLE III

RELATIVE SPECIFIC ACTIVITIES OF BILE SALT-INDEPENDENT RETINYL ESTER
HYDROLASE AND MARKER ENZYMES IN PLASMA MEMBRANE FRACTIONS
OF RAT LIVER HOMOGENATES[a,b]

	Exp. 1		Exp. 2	
Constituent	RSA[c]	Recovery[d]	RSA[c]	Recovery[d]
Protein	1	96	1	89
N-Acetyl-β-glucosaminidase	1	82	2	91
Glucose-6-phosphatase	2	103	1	96
Esterase[e]	—	—	1	89
Alkaline phosphodiesterase	29	89	15	101
Retinyl ester hydrolase	35	115	19	124

[a] Reproduced with permission.[11]
[b] Plasma membrane-rich fractions were isolated from the microsomal fraction by the method of O. Touster, N. N. Aronson, J. T. Dulaney, and H. Hendrickson, *J. Cell Biol.* **47**, 604 (1970).
[c] Relative specific activity (RSA) is the percentage of activity recovered in the plasma membrane fraction divided by the percentage of total homogenate protein recovered in the fraction. The isolated plasma membrane fractions contained 0.5–1.0% of the homogenate protein.
[d] Recoveries represent the sum of the constituent in the plasma membrane-rich fraction and all other fractions relative to the amount observed in the unfractionated homogenate.
[e] "Nonspecific" esterase was assayed using o-nitrophenyl acetate as the substrate.

In order to further resolve the membrane components of the microsomal fraction, fractions rich in plasma membranes are prepared from it. The analysis of these preparations is presented in Table III. The results agree well with those presented in the original description of the method. Thus, the isolated fraction is highly enriched in plasma membrane fragments as assessed by the enrichment of alkaline phosphodiesterase activity. Significantly, the preparation shows no enrichment of endoplasmic reticulum markers such as glucose-6-phosphatase and o-nitrophenyl acetate esterase. The bile salt-independent REH activity is enriched in the isolated plasma membrane fraction to the same extent as alkaline phosphodiesterase. This result suggests that most or all of the bile salt-independent REH activity is localized in the plasma membrane. In this regard it is entirely different from the cholate-dependent activity, which has been shown to be absent in highly enriched plasma membrane fractions of rat liver.

Differential Responses to Pancreatic Cholesteryl Ester Hydrolase Antibodies

A final demonstration that the bile salt-independent RPH activity is due to an enzyme(s) different than that (those) responsible for the bile salt-stimulated hydrolysis is indicated by the differential inhibition of the activities by antibodies to pancreatic cholesteryl ester hydrolase (EC 3.1.1.13). In order to make this comparison, whole homogenates of rat liver are prescreened for retinyl palmitate hydrolase activity in the presence and absence of 20 mM cholate. A homogenate is then chosen that shows substantial bile salt-dependent REH activity. As shown in Fig. 1, incubation of this homogenate with anti-pancreatic cholesteryl ester hydrolase IgG led to marked inhibition (70%) of the bile salt-dependent RPH activity without affecting the bile salt-independent activity. In a separate experiment (data not shown), 15 μg of IgG led to 80% inhibition of the bile salt-dependent activity without affecting the bile salt-independent activity. In both experiments, the remaining activity may be due to the hydrolysis of the substrate by the bile salt-independent hydrolase(s) also present in these whole homogenates.

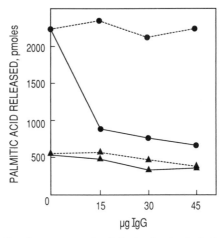

FIG. 1. Effects of anti-rat pancreatic cholesteryl ester hydrolase on the retinyl ester hydrolase activities of rat liver homogenates. Aliquots of rat liver homogenate were incubated for 16 hr at 4° with normal rabbit IgG (dashed lines) or with rabbit anti-rat pancreatic cholesteryl ester hydrolase IgG (solid lines). They were then assayed in incubations containing 100 μM retinyl palmitate and 50 mM Tris–maleate buffer (pH 8.0). (●) Assays for the bile salt-dependent activity contained 20 mM cholate; (▲) assays for the bile salt-independent activity did not contain cholate. Reproduced with permission.[11]

Extrahepatic Retinyl Ester Hydrolases

Detergent Dependence

Retinyl ester hydrolysis in rat kidney, testes, and lung homogenates,[12] assayed at a protein concentration in the range of linear rates, is inhibited 33–66% by 18 mM cholate, a concentration that produces maximum stimulation of the cholate-dependent "pancreatic" esterase (Table IV). At a protein concentration beyond the linear range, cholate inhibits hydrolysis in lung and testes homogenates, but it provides marginal stimulation in kidney homogenates. There are reports that up to 300 mM cholate or {3-[(3-cholamidopropyl)dimethylammonio]propane sulfonate} (CHAPS), concentrations that are about 2 orders of magnitude greater than the critical micelle concentration (CMC), in combination with Triton X-100, produce at least 10-fold stimulation of retinyl ester hydrolase activity.[9,10] No stimulation is observed, however, when assays are done in the linear range of rat kidney homogenate protein (Table V). These detergents do rescue hydrolysis activity from inhibition by Triton X-100. Modest stimulation of hydrolysis by CHAPS or cholate, alone or in combination with Triton X-100, occurs only when substrate and/or protein concentrations are high, suggesting that the detergents aid in substrate solubilization and/ or disruption of protein aggregates. The previous work showing that CHAPS in the presence of Triton X-100 stimulated retinyl ester hydroly-

TABLE IV

EFFECTS OF CHOLATE ON RETINYL PALMITATE HYDROLASE ACTIVITY IN EXTRAHEPATIC RAT TISSUES[a]

Tissue	Cholate	Activity (nmol product)	
		7.5 μg protein/assay	750 μg protein/assay
Kidney	−	0.3 ± 0.02	1.0 ± 0.06
	+	0.2 ± 0.02[b]	1.3 ± 0.07[b]
Lung	−	0.5 ± 0.07	1.5 ± 0.06
	+	0.2 ± 0.02[b]	0.7 ± 0.06[b]
Testes	−	0.6 ± 0.10	1.7 ± 0.11
	+	0.2 ± 0.01[b]	1.0 ± 0.07[b]

[a] Homogenates from the tissues of seven rats were assayed in the absence or presence of 18 mM cholate. Incubations were done for 45 min with 100 μM all-*trans*-retinyl palmitate. Data are the means ± 1 S.D. of four replicates.

[b] Significantly different from the assay done in the absence of cholate ($p < .05$).

TABLE V
DETERGENT EFFECTS ON RETINYL PALMITATE HYDROLASE
ACTIVITY IN HOMOGENATES OF RAT TISSUE[a]

Detergent	Activity (nmol product) substrate (μM)/protein (μg)		
	100/7.5	500/7.5	500/750
None	0.3 ± 0.03	0.7 ± 0.04	3.7 ± 0.1
Cholate	0.2 ± 0.03^b	0.7 ± 0.04	4.5 ± 0.3^b
CHAPS	0.2 ± 0.04^b	1.2 ± 0.05^b	5.0 ± 0.5^b
Triton X-100	0	0	3.6 ± 0.1
Cholate plus Triton X-100	0.2 ± 0.03^b	0.5 ± 0.04^b	4.8 ± 0.6^b
CHAPS plus Triton X-100	0.3 ± 0.05	1.0 ± 0.10^b	7.7 ± 0.9^b

[a] Assays were done for 30 min with a homogenate prepared from the kidneys of seven rats. Cholate and CHAPS were used at 200 mM, and Triton X-100 was used at 2 mg/ml. Data are the means $\pm$ 1 S.D. of triplicates.
[b] Significantly different from the assay done in the absence of detergent ($p < .05$).

sis was done with high amounts of protein (0.35–0.7 mg) and high substrate concentrations (0.5 mM). Even though stimulation was observed in these reports, as is observed here under similar conditions, the rates of hydrolysis were no greater than those obtained here with low protein concentrations in the absence of detergent. The idiosyncratic conditions of high detergent/high protein assays, therefore, provide only a higher yield of product, not an enhanced rate. Retinyl ester hydrolase activity is also inhibited in kidney microsomes by cholate, CHAPS, and Triton X-100 when linear assay conditions are used (see below, Subcellular Distribution).

Subcellular Distribution

Eighty-four percent of the kidney retinyl ester hydrolase activities is associated with the membrane fractions. The cytosolic fraction is the only one with decreased specific activity relative to the homogenate (Table VI). The "microsomal" fraction is the only one with increased specific activity. Activity in the "microsomal" fraction is inhibited 50 to 100% by cholate (10–61 mM), CHAPS (3–400 mM), Triton X-100 (1–5 mg/ml), and CHAPS (100 mM) in the presence of Triton X-100 (2 mg/ml) when assayed under linear conditions.

TABLE VI
DISTRIBUTION OF RAT KIDNEY RETINYL PALMITATE
HYDROLASE ACTIVITY[a]

Fraction	Specific activity (nmol/min/mg protein)	Activity (%)	Protein (%)
Kidney homogenate	1.5 ± 0.5	—	—
P_1 (580 g pellet)	1.7 ± 0.3	44 ± 4	34 ± 7
P_2 (8140 g pellet)	1.7 ± 0.6	21 ± 5	16 ± 2
P_3 (125,000 g pellet)	2.5 ± 0.4[b]	19 ± 2	10 ± 1
S (supernatant)	0.5 ± 0.1[b]	16 ± 4	39 ± 8

[a] Data are the means ± 1 S.D. of five experiments. Each experiment was done with kidneys from 5 to 20 rats. Specific activity was determined with 100 μM substrate in assays done for 30 min with 2 to 5 μg of protein.
[b] Significantly different from the homogenate ($p < .01$).

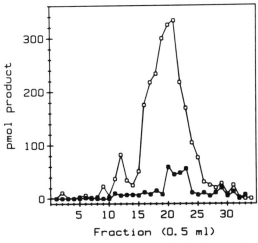

FIG. 2. Gel-permeation chromatography of solubilized microsomal retinyl ester hydrolase activity. Retinyl ester hydrolase (○) and triolein hydrolase (●) were determined in separate 20-μl aliquots of each fraction. Chromatography was done with a TSK G3000SW column (0.5 × 5.0 cm, Toyo Soda Manufacturing Co., Tokyo, Japan) eluted with 10 mM HEPES and 100 mM NaCl, pH 7.5, at 0.5 ml/ml. Standards eluted in fractions 12 (blue dextran, 2000 kDa), 15 (α-amylase, 200 kDa), 16.5 (alcohol dehydrogenase, 150 kDa), 17 (albumin, 68 kDa), 20 (carbonic anhydrase, 29 kDa), and 24.5 (cytochrome c, 12.4 kDa). A plot of log(kDa) of standards versus fraction was linear ($r = -0.996$), with a slope of -0.128 and a y-intercept of 7.16.

TABLE VII
RETINYL PALMITATE, ACYLGLYCEROL, AND CHOLESTERYL
OLEATE HYDROLASE ACTIVITIES IN RAT KIDNEY FRACTIONS

	Relative rate		
Fraction	Retinyl palmitate	Acylglycerol	Cholesteryl oleate
Homogenate	1.0[a]	0.2	0.06
Microsomes	1.7	0.7	0.2
Solubilized[b]	3.8	0.8	0.03
F16-19[c]	4.1	0.4	ND[d]
F20-24[c]	6.8	1.9	ND[d]

[a] Data are the rates relative to a rate of 1.7 nmol/min/mg protein measured in a 30-min assay with 2–5 μg of protein and 100 μM substrate.
[b] Data represent the activities solubilized with OSG.
[c] F16-19 and F20-24 refer to pools of fractions 16–19 and 20–24, respectively, from the gel-permeation column (Fig. 2).
[d] ND, not detectable, less than 1 pmol of product generated.

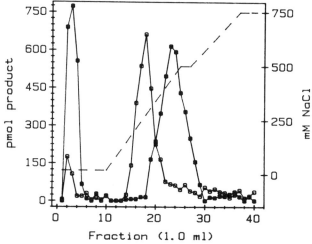

FIG. 3. Anion-exchange chromatography of high and low molecular weight retinyl ester hydrolase pools. Each sample was applied to a Mono Q column (0.5 × 5.0 cm, Pharmacia, Piscataway, NJ) in a buffer of 20 mM Tris-HCl, pH 7.5, containing 20 mM NaCl. In each analysis, the column was eluted at a flow rate of 1 ml/min with the same buffer containing increasing NaCl as shown (dashed line): (○) retinyl ester hydrolase activity from analysis of the high molecular weight pooled fractions (fractions 16–19); (●) retinyl ester hydrolase activity from analysis of the low molecular weight pooled fractions (fractions 10–24). Approximately equivalent amounts of total protein were loaded, and equivalent aliquots of each fraction were assayed in each analysis.

Solubilization

Membrane-bound (P_1 = 580 g pellet, P_2 = 8140 g pellet, and P_3 = 125,000 g pellet) enzymes with retinyl ester hydrolase activity can be solubilized efficiently with 1-S-octyl-β-D-thioglucopyranoside (OSG). The membrane fraction (5 mg of protein/ml of buffer) is stirred gently with a magnetic stirrer at 4° for 1 hr in solubilization buffer (10 mM HEPES, 150 mM KCl, 0.4% OSG). The mixture is centrifuged at 125,000 g for 1 hr. The solubilization is routine; on the average 45% of the protein and 64% of the units are solubilized from P_1 and 45% of the protein and 82% of the units are solubilized from P_3 (assays with purified fractions are done in the presence of the 18 mM cholate to aid in solubilization of substrate). The activity not released into the supernatant can be accounted for in the pellet.

Multiplicity and Substrate Specificity

The cholate-dependent retinyl ester hydrolase activity in rat liver coelutes with cholesterol esterase and triacylglycerol lipase activities through at least three columns that separate by charge or hydrophobicity.[7] The partially purified lipase(s) has higher specific activities assayed with cholesterol oleate (2-fold) and triolein (10-fold) than when assayed with retinyl palmitate. In contrast, the cholate-independent retinyl ester hydrolase has higher specific activity with retinyl palmitate than with cholesteryl oleate or triolein (Table VII).

Fast protein liquid chromatography (FPLC) is useful for purification and resolution of retinyl ester hydrolase activity from other lipases (Figs. 2 and 3). The retinyl esterase activity can be distinguished from cholesteryl esterase and triolein hydrolase (Fig. 2 and Table V). A combination of gel-permeation chromatography (Fig. 2) and anion-exchange chromatography indicates that there are at least four lipases with retinyl ester hydrolase activity.

Acknowledgments

We are grateful for the technical assistance of Gregory Salerno, Emmanuel Pacia, Nancy Kelso, Mohamed Gad, and Donald Lewis. This work was supported in part by grants from the National Institutes of Health (AM 32096, CA42094, and HL22633) and the Howard Heinz Endowment.

[53] Quantification and Characteristics of Retinoid Synthesis
from Retinol and β-Carotene in Tissue Fractions and
Established Cell Lines

By Joseph L. Napoli

Introduction

Retinoic acid, the most potent physiologically occurring retinoid known, is an activated metabolite of retinol that is synthesized in many retinoid target tissues.[1-3] Sensitive and specific assays for the quantification of retinoic acid are essential for the study of its biogenesis and to evaluate its role in the etiology of retinoid-responsive diseases and in morphogenesis. A gas chromatography–mass spectrometry (GC/MS) assay has been developed which is sensitive to 0.25 pmol of retinoic acid,[4,5] and high-performance liquid chromatography (HPLC) assays for retinol and retinoic acid have been described.[5,6] This chapter relates modifications in HPLC assays made since the previous reports and describes conditions to assay the metabolism of retinol into retinal and retinoic acid, retinal into retinol and retinoic acid, and β-carotene into retinol, retinal, and retinoic acid. Homogenates, cytosol, and microsomes convert retinol and retinal to retinoic acid.[3,7] Cytosol converts β-carotene to retinol and retinoic acid.[8] Cultured mammalian cells convert retinol and retinal to retinoic acid[9] and β-carotene to retinoic acid.[10] Incubation, extraction, and HPLC conditions are tailored to each of these situations.

Purification and Handling of Substrates

The laboratory is kept under gold lighting to minimize isomerization of retinoids. The retinoids and β-carotene available require purification be-

[1] J. B. Williams and J. L. Napoli, *Proc. Natl. Acad. Sci. U.S.A.* **82**, 4658 (1985).
[2] J. B. Williams, C. O. Shields, L. M. Brettel, and J. L. Napoli, *Anal. Biochem.* **160**, 267 (1987).
[3] J. L. Napoli and K. R. Race, *Arch. Biochem. Biophys.* **255**, 95 (1987).
[4] J. L. Napoli, B. C. Pramanik, J. B. Williams, M. I. Dawson, and P. D. Hobbs, *J. Lipid Res.* **26**, 387 (1985).
[5] J. L. Napoli, this series, Vol. 123, p. 112.
[6] A. P. De Leenheer, W. E. Lambert, and I. Claeys, *J. Lipid Res.* **23**, 1362 (1982).
[7] K. C. Posch, W. J. Enright, and J. L. Napoli, *Arch. Biochem. Biophys.* in press (1989).
[8] J. L. Napoli and K. R. Race, *J. Biol. Chem.* **263**, 17372 (1988).
[9] J. L. Napoli, *J. Biol. Chem.* **261**, 13592 (1986).
[10] J. L. Napoli, unpublished results.

fore use, especially if detection of picomole amounts of products is to be meaningful. Retinol, for example, contains up to 4% of a peak that co-migrates with retinal on HPLC, as well as other polar contaminants. β-Carotene can also contain several percent of a peak that comigrates with retinal on HPLC. Substrates are purified every 1 to 2 weeks by HPLC. Retinol is eluted from a Waters (Milford, MA) μPorasil column (1 $\times$ 30 cm) in 60 ml with a mobile phase of 0.01% methanol in acetone/hexane (1 : 19). Retinal is eluted from a Du Pont Zorbax-Sil column (0.6 $\times$ 8 cm) in 18 to 20 ml with a mobile phase of acetone/hexane (1 : 99), or from the μPorasil column. β-Carotene is eluted from a Whatman ODS-2 column (1 $\times$ 25 cm) in 48 ml with a mobile phase of tetrahydrofuran/methanol (1 : 3). Substrates are monitored at 340 nm with a UV detector. Purified substrates are stored under an inert gas (nitrogen or argon) at $-20°$ in brown vials. Retinol and retinal are stored in 2-propanol/hexane (1 : 5). β-Carotene is stored in acetone/2-propanol/hexane (3 : 2 : 5, v/v). Shortly before use, the storage solvent is evaporated under a gentle stream of nitrogen, and the substrates are dissolved in dimethyl sulfoxide (DMSO). The DMSO solutions of substrates should be prepared no sooner than 1 to 2 hr before use.

Internal Standards

A host of synthetic retinoids are available as internal standards. The ones used most frequently in our laboratory are all-*trans*-7-(1,1,3,3-tetra-methyl-5-indanyl)-3-methylocta-2,4,6-trienoic acid (TIMOTA) as an internal standard for retinoic acid and all-*trans*-9-(4-methoxy-2,3,6-trimeth-ylphenyl)-3,7-dimethyl-2,4,6,8-nonatetraen-1-ol (TMMP-retinol) as an internal standard for retinol. These are available from Hoffmann-La Roche (Nutley, NJ).

Incubations and Extractions

Homogenates and Subcellular Fractions. Incubations with homogenates or subcellular fractions are done in disposable glass tubes (13 $\times$ 100 mm) at 37° in a buffer containing 150 mM KCl, 2 mM dithiothreitol (DTT), 2 to 4 mM NAD (see Table I), and 20 mM HEPES. The pH optimum ranges between 6.0 and 8.5, depending on the source of activity (species, tissue, or subcellular fraction) and the substrate (retinol, retinal or β-carotene), and should be determined for each specific case.[3,7,10] To determine the rate of retinol synthesis from retinal, 2 mM NADH and 20 mM [2-(N-morpholino)ethanesulfonic acid] MES buffer are used, and the pH is adjusted between 5.5 and 6.5 (experimentally determined). Linearity of

TABLE I
EXTRACTION VOLUMES AND NAD CONCENTRATIONS
USED WITH VARIOUS INCUBATION VOLUMES FOR
DETECTING RETINOIC ACID SYNTHESIS *in Vitro*

Component	Incubation volume (ml)[a]		
	0.25	0.5	1.0
NAD (mM)	4	2	2
Hexane (ml)[b]	2.5	2.5	5
0.025 N KOH/ethanol (ml)	0.25	0.5	1
4 N HCl (ml)	0.025	0.05	0.25

[a] The amounts of base and acid given are for an incubation pH of 8.5. These must be adjusted according to the incubation pH as detailed in Table II.
[b] Volume of hexane to be used for each extraction (hexane-1 and hexane-2).

rate with protein and time also depends on the substrate and the source of activity, and also should be determined for each case. Generally, 0.25 mg of protein or less, and incubations of 30 min or less, are within the linear ranges for homogenates and cytosol. In contrast, with microsomes, 0.3 to 0.5 mg of protein or greater is used, as are shorter incubation times (10 min).

NAD stimulates retinoic acid production from retinol or retinal *in vitro,* but the degree of the effect is pH dependent: the relative enhancement increases as the pH of the incubation buffer increases. NADP also stimulates retinoic acid production by rat intestinal cytosol. NADH is a potent inhibitor of the conversion of retinol to retinoic acid. The reaction supported by 2 mM NAD is inhibited 70% by 0.5 mM NADH.[3] Even in the presence of NADH, however, retinoic acid, not retinol, is the major metabolite of retinal. Pyridine nucleotide cofactors are usually not required to observe retinoid production from β-carotene (at pH 7.5).[8] In some tissues, however, NAD does provide 1.5- to 2.2-fold stimulation of retinoic acid synthesis from β-carotene, but it has a marginal and inconsistent effect on retinol synthesis. NADH does not seem to stimulate retinol synthesis from β-carotene, and in some cases may inhibit retinoic acid synthesis 40 to 60%, compared to the NAD-supported rate. Consequently, it is prudent to routinely add NAD to incubations meant to study retinol and retinoic acid synthesis from β-carotene, at least during preliminary trials.

To initiate the reaction, substrate is added in 2.5–10 μl of dimethyl sulfoxide, depending on the incubation volume. Detergents are not required to detect the synthesis of retinoic acid from retinoids or from β-carotene. In fact, detergents seem to have an adverse effect, especially with low concentrations (20 μM or less) of retinal or β-carotene as substrate.[10] Controls consist of incubations done in the presence of buffer without an added source of enyzmatic activity and/or with boiled protein. The incubations are quenched with 0.025 N KOH in ethanol, and the internal standards are added in ethanol (10–100 μl). Neutral retinoids and β-carotene are extracted from the alkaline aqueous phase with hexane (hexane-1) by vortexing 5 min with a multitube vortexer. The tubes are centrifuged for 2 to 5 min to effect clean separation of the phases. The alkaline aqueous phase is then acidified with 4 N HCl. Retinoic acid and the other acidic products are extracted by vigorous vortexing with a second volume of hexane (hexane-2). The exact amount of each solvent depends on the pH and the volume of the incubation buffer (Tables I and II). The extraction volumes (and the concentration of NAD in the incubation) are critical to observing rates of retinoic acid synthesis with decreasing incubation volumes. It is also important to have a pH greater than 12 for the alkaline phase and less than 2 for the acidic phase to effect maximum partitioning of retinoic acid into the hexane-2 phase. The concentration of base must be kept low, because a final concentration of 0.2 N KOH or NaOH, for example, causes the conversion of all-*trans*-retinoic acid to geometric isomers, including 13-*cis*-retinoic acid.[5,6]

Cultured Cells. Incubations with established mammalian cell lines are done with confluent 100-mm plates of cells containing 6 ml of medium.[9,10] Both defined medium without serum and medium with as much as 10% fetal calf serum have been used. Controls consist of incubating the substrate in medium without cells and, especially with a new cell line or medium, incubating cells and medium without substrate to determine

TABLE II

AMOUNTS OF BASE AND ACID FOR EXTRACTION OF
RETINOIDS FROM INCUBATIONS AS FUNCTION OF
INCUBATION pH[a]

pH	0.025 N KOH/ethanol (ml)	4 N HCl (μl)
6.0	1.5	100
7.0	1.5	75
7.5	1.0	75
8.0	0.5	50
8.5	0.5	50

[a] These amounts are to be used with an incubation volume of 0.5 ml.

background. The reactions are initiated by the addition of substrate in 5 to 20 μl of dimethyl sulfoxide, or by adding medium containing a detergent emulsion of substrate. No detergent is required with retinol as substrate; however, with β-carotene as substrate, emulsions made by sonicating the substrate in medium with 6 mM CHAPS increases the rate of retinoic acid synthesis in HT-29 cells about 12-fold over the rate observed in the absence of detergent.

At the end of the incubation period, the cells and the medium are transferred to a 15-ml screw-capped glass tube. Internal standards, 0.1 N KOH in ethanol (3 ml), and hexane (4 ml) are added, and the neutral retinoids and β-carotene are removed into the hexane phase by rotation on an extraction wheel for 10 min. The tubes are centrifuged for 2 to 5 min to effect clean separation of the layers. To ensure clean chromatograms for the analysis of the hexane-2 phase, the alkaline phase is reextracted with a second 4-ml portion of hexane. These hexane phases are combined (hexane-1). The aqueous phase is acidified with 4 N HCl (0.5 ml) and extracted with 4 ml of hexane (hexane-2) to recover the retinoic acid. Table III is a compilation of established mammalian cell lines that convert retinol to retinoic acid.

High-Performance Liquid Chromatography

The products are analyzed with an automated HPLC system consisting of an automatic sample injector capable of handling 90 samples (Waters 712 WISP), a tunable UV absorbance detector (Waters 484), a microprocessor-controlled pump with gradient capability (Waters Model 600E), and an integrator (Spectra Physics, San Jose, CA, SP4270). Retinoic acid, retinol, and retinal are detected at 340, 325, and 370 nm, respectively. For normal-phase HPLC, Du Pont Zorbax-Sil Reliance-3 cartridges are used (3-μm beads, 0.6 $\times$ 4 cm). For reversed-phase HPLC, Du Pont Zorbax-ODS is used, either in a Reliance-3 cartridge or in a Golden Series column (3-μm beads, 0.62 $\times$ 8 cm). Formerly, normal-phase analyses are conducted with a Golden Series column and a Model 440 fixed-wavelength UV detector.[5] The change of the UV detector reduced the noise in the signal, and the change of the column reduced analysis time, elution volume, and peak width.

Immediately prior to HPLC analysis, the solvents are evaporated from the hexane phases with a gentle stream of nitrogen, and the residue is dissolved in 0.1 ml of mobile phase. Samples are not stored in mobile phases containing either acid or chlorinated hydrocarbons.

The retinol or retinal in the hexane-1 phase is analyzed by one of three systems. A reversed-phase system can be used to detect retinal synthesis

TABLE III
RETINOIC ACID SYNTHESIS FROM RETINOL BY ESTABLISHED
MAMMALIAN CELL LINES

Cell line	Species	Type	Relative rate[a]
MCDK	Dog	Kidney epithelia	0.8[b]
A-431	Human	Epidermal carcinoma	0.5
BeWo		Choriocarcinoma	ND[c]
HepG2		Hepatoma	0.4
HT-29		Colon adenocarcinoma	1.2[b]
JEG-3		Choriocarcinoma	ND
SCC-9		Tongue carcinoma	0.3
SCC-15		Tongue carcinoma	0.3
T-24		Bladder carcinoma	1.0
U-937		Monoblast-like	ND
F9	Mouse	Embryonal carcinoma	ND[d]
LLC-PK$_1$	Pig	Kidney epithelia	1.0
H4-II-E-C3	Rat	Hepatoma	0.04[e]
IEC-6		Small intestinal epithelia	ND
IEC-18		Ileum epithelia	0.1
Rice		Leydig tumor	ND
RLC		Hepatoma	0.4
ROS 17/2		Osterosarcoma	0.2

[a] A relative rate of 1 is approximately equivalent to 150 pmol/hr/100-ml plate of confluent cells obtained at a retinol concentration of 10 μM.
[b] Done with serum-less medium. All other cell lines were screened with medium containing 10% fetal calf serum.
[c] ND, Not detected with an HPLC assay sensitive to 10 pmol;[5] less than 10 pmol/hr/plate.
[d] Not detected with an HPLC assay but detected by using 60 Ci/mmol [³H]retinol.
[e] Detected by combining two plates of cells before HPLC analysis.

from retinol. The mobile phase is acetonitrile/water (1 : 4) with an elution rate of 2 ml/min. Retinol elutes in 18 ml and retinal elutes in 25 ml (Fig. 1). Normal-phase systems can be used to measure retinol synthesis from retinal or β-carotene. With a mobile phase of acetone/hexane (1 : 19, v/v) at a flow rate of 2 ml/min, retinol elutes in 9.5 ml and retinal elutes in 3.6 ml (Fig. 2). β-Carotene elutes in less than 2 ml and interferes with the retinal peak. To further separate retinal from β-carotene, a gradient system can be used. The initial mobile phase is acetone/hexane (1 : 49, v/v) at 2 ml/min for 5 min. At 5 min a linear gradient is started to reach a final mobile phase of acetone/hexane (1 : 24, v/v) in 1 min. β-Carotene elutes in less than 3 ml. Retinal elutes in 5.6 ml and retinol elutes, after final conditions are established, in 21 ml. In experiments with β-carotene sub-

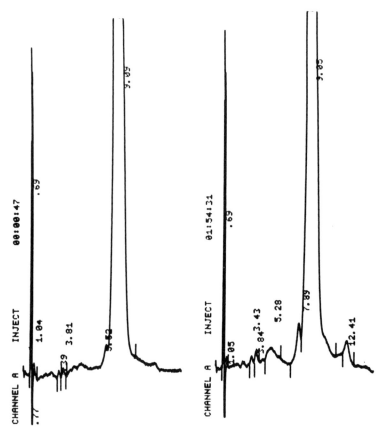

Fig. 1. Reversed-phase HPLC for the detection of retinal synthesized from retinol. The mobile phase is acetonitrile/water (1 : 4) at 2 ml/min. The figure shows a no protein control (left) and an incubation (pH 7.5) with ADH+ deermouse kidney microsomes (0.25 mg protein) (right) of 50 μM retinol for 10 min. The large peak eluting in 9 min is retinol. Retinal (25 pmol) elutes in 12.4 min. The flow rate was 2 ml/min.

strate concentrations greater than 20 μM, excess β-carotene is removed before HPLC analysis to reduce interference with retinoid peaks by concentrating the hexane-1 phase to 2 ml and applying the concentrate to a small silica column (Waters Sep-Pak). The carotenoid(s) are eluted with 2 ml of acetone/hexane (7.5 : 92.5, v/v). Retinol is eluted with 2 ml of acetone.[8]

The amount of retinoic acid in the hexane-2 phase is analyzed by normal-phase HPLC with a mobile phase of 0.35% acetic acid in 1,2-dichloroethane/hexane (1 : 9, v/v), usually at a flow rate of 2 ml/min. The

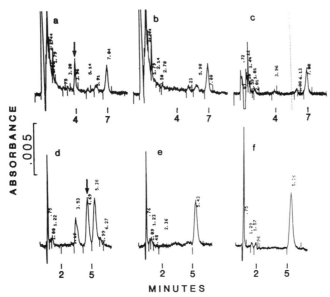

FIG. 2. Normal-phase HPLC analysis of retinol and retinoic acid production form β-carotene. Incubations (pH 7.5) were done with rat testes cytosol (0.25 mg protein) for 30 min with (a, d) 10 μM β-carotene; (b, e) 10 μM β-carotene (boiled cytosol): (c, f) no β-carotene. (a, b, c) Analyses of retinol by the normal-phase isocratic system (see text for HPLC details); (d, e, f) analyses of retinoic acid. The arrows denote retinol (a) (24 pmol) and retinoic acid (d) (72 pmol). The internal standards are the peaks eluting about 7 min (a, b, c) and 5.4 min (d, e, f). The baseline is noisier than in the other figures because the Waters Model 440 rather than the Model 484 UV detector was used. The flow rate was 2 ml/min.

exact proportions of solvents in this mobile phase should not be considered as absolute; the constituents of the mobile phase need to be adjusted periodically to respond to differences in solvent batches, relative humidity, and specific columns. The following guidelines are helpful in such situations: acetic acid suppresses ionization of retinoic acid, that is, makes it act as a nonionic species, and also is the predominant determinant of k' (elution volume); 1,2-dichloroethane changes the α characteristics (selectivity) of the mobile phase. Thus, increasing the acetic acid decreases the elution volume, and increasing the 1,2-dichloroethane increases the resolution of all-*trans*-retinoic acid from its isomers, such as 13-*cis*-retinoic acid, and decreases the volume between all-*trans*-retinoic acid and the internal standard. Examples are shown of the sensitivity of the retinoic acid assay (Fig. 3) and of the analyses of retinoic acid generated from retinol by cytosol (Fig. 4) and microsomes (Fig. 5), and from β-carotene by cytosol (Fig. 2) and cultured human cells (Fig. 6).

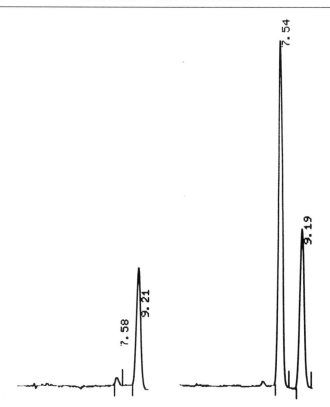

FIG. 3. Sensitivity of the normal-phase HPLC system used for the analysis of retinoic acid. The chromatogram on the left-hand side shows detection of 2.5 pmol of retinoic acid (7.58 min) generated by incubating retinol with rat testes cytosol. The chromatogram on the right-hand side shows 100 pmol of the retinoic acid standard injected to generate the standard curve. The peak at 9.2 min is the internal standard. The flow rate was 1 ml/min.

Cytosolic Alcohol Dehydrogenases and Retinol Dehydrogenase

Two types of evidence demonstrate that cytosolic retinol dehydrogenases exist that are not cospecific with the alcohol dehydrogenases which metabolize ethanol. First, cytosol from the alcohol dehydrogenase-negative deermouse mutant converts retinol to retinoic acid (Table IV and Fig. 4).[7] Second, ethanol and inhibitors of ethanol metabolism inhibit retinol conversion only partially in some tissues and are totally ineffective in others.[3,7,10] 4-Methylpyrazole in concentrations less than 0.1 mM is a potent inhibitor of ethanol metabolism. It cannot completely inhibit reti-

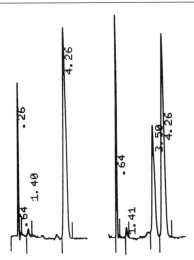

FIG. 4. Normal-phase HPLC analysis of retinoic acid generated from retinol with cytosol from the ADH− deermouse. Retinol (7.5 μM) was incubated (pH 7.0) with liver cytosol from the deermouse mutant lacking cytosolic alcohol dehydrogenase for 20 min: no enzyme (left); 0.15 mg of cytosolic protein (right). The internal standard eluted at 4.26 min, and retinoic acid eluted in 3.5 min (55 pmol, right chromatogram). The flow rate was 2 ml/min.

TABLE IV
RELATIONSHIP BETWEEN RETINAL (RCHO) AND RETINOIC ACID
(RA) PRODUCTION FROM RETINOL BY CYTOSOL[a]

	ADH+		ADH−	
Tissue	RA	RCHO	RA	RCHO
Kidney	400 ± 72	84 ± 21	55 ± 6	<5
Liver	409 ± 17	262 ± 34	53 ± 3	<5
Lung	23 ± 2	<5	18 ± 2	<5
Testes	40 ± 7	<5	24 ± 5	<5
	In presence of 10 mM 4-methylpyrazole			
Kidney	23 ± 2	<5	28 ± 8	<5
Liver	34 ± 2	<5	37 ± 2	<5
Lung	17 ± 3	<5	16 ± 2	<5
Testes	28 ± 7	<5	29 ± 12	<5

[a] Assays were conducted with 7.5 μM retinol and 0.1 mg of cytosolic protein from the deermouse at pH 7.0: ADH+, alcohol dehydrogenase positive; ADH−, alcohol dehydrogenase negative. The assays were repeated in the presence of 10 mM 4-methylpyrazole. Data are pmol of product ± SD ($n = 4$).

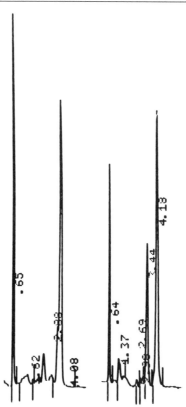

FIG. 5. Normal-phase HPLC analysis of retinoic acid produced from retinol by rat testes microsomes. Retinol (20 μM) was incubated (pH 8.5) for 20 min with no enzyme (left) or 0.5 mg of microsomal protein (right). The internal standard eluted in 4.18 min, and retinoic acid eluted in 3.44 min (36 pmol, right). The flow rate was 2 ml/min.

noic acid synthesis in the deermouse (Table IV). Most notably, in the presence of 10 mM 4-methylpyrazole, cytosols prepared from the same tissues had the same rates of retinoic acid synthesis regardless of the tissue source (ADH+ or ADH− deermouse). Retinoic acid synthesis in intact LLC-PI$_1$ cells is not affected by 10 mM 4-methylpyrazole, nor by 220 mM ethanol.[9] Retinoic acid synthesis from 20 μM retinol by cytosol from MDCK cells is not inhibited by 10 mM 4-methylpyrazole and is stimulated by ethanol concentrations up to 1.43 M (perhaps by enhancing the solubility of substrate)![10] These data are consistent with multiple cytosolic dehydrogenases recognizing retinol, including ones that do not recognize ethanol.

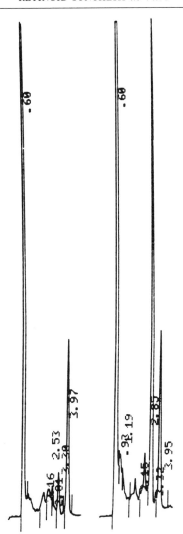

FIG. 6. Normal-phase HPLC analysis of β-carotene production by an established human intestinal cell line (HT-29). Serumless medium (6 ml) containing 20 μM β-carotene and 6 mM CHAPS was sonicated and added to 100-mm plates of HT-29 cells or plates without cells. After 1 hr the samples were extracted and analyzed for retinoic acid. The chromatogram at left shows the control, and the chromatogram at right shows the plate with the cells. In the control 5.5 pmol of retinoic acid was detected, and in the experimental 108 pmol of retinoic acid was detected (2.85 min). The peak at 3.95 min is the internal standard. The flow rate was 2 ml/min.

Retinal Detection and Retinoic Acid Synthesis

Other than as a cofactor of opsin, retinal has no known direct function. Retinal has been presumed to be an intermediate in β-carotene metabolism and in the converison of retinol to retinoic acid, but steady-state levels of retinal have not been detected in nonocular tissues. This may be the result of rapid metabolism; retinal, added to cells in culture or to subcellular fractions in lower micromolar concentrations, is converted rapidly to retinol and retinoic acid.[9,10] In fact, retinal is undetectable (<5 pmol) as a product of retinol metabolism in most tissues when physiological concentrations of retinol are used. Moreover, when retinal is detected, there is no apparent quantitative relationship between it and the amount of retinoic acid synthesized (Table IV). These data indicate that retinal production is not a reliable qualitative or quantitative measure of the ability of a sample to synthesize retinoic acid.

Microsomes and Retinol Metabolism

Microsomes convert retinol to retinal and retinoic acid (Fig. 5), but their specific activity is much lower than cytosol. Typically, 400 to 500 μg of microsomes and higher substrate concentrations are needed to observe measurable retinal and retinoic acid production. Low amounts of microsomes (25–100 μg of microsomal protein), added to cytosol, inhibit the cytosolic conversion of retinol to retinal and retinoic acid by 60–100%. As little as 10 μg can cause 55% inhibition. Contamination of cytosol by membrane fractions, therefore, should be avoided, or retinoic acid synthesis may be depressed or undetectable.

Acknowledgments

The technical assistance of Kevin Race and William Enright is gratefully acknowledged. This work was supported by National Institutes of Health Grants DK36870 and DK38885 and U.S. Department of Agriculture Grant 87-CRCR-1-2332.

[54] Retinol 4-Hydroxylase

By M. A. Leo and C. S. Lieber

Introduction

Drugs that induce liver microsomes have been shown to result in a depletion of hepatic vitamin A.[1] A similar effect was also observed after administration of ethanol[2,3] and other xenobiotics that are known to interact with liver microsomes, including carcinogens.[4] These observations suggested that microsomal breakdown of retinol might play a role in the hepatic depletion. Liver microsomes have been shown to be capable of esterifying retinol to retinyl esters[5]; however, no oxidation of retinol in liver microsomes has been reported, and only retinoic acid has been shown to be degraded in microsomes of either hamsters[6] or rats.[7,8] In both species, the reported activity was very low compared to the degree of hepatic vitamin A depletion. These observations prompted the search for alternate pathways of retinol metabolism in liver microsomes. Subsequently, a new pathway of retinol metabolism was described: rat liver microsomes, when fortified with NADPH, converted retinol to polar metabolites, including 4-hydroxyretinol.[9] This activity was also demonstrated in a reconstituted monooxygenase system containing purified forms of rat cytochromes P-450,[9] including P-450IIB1 (a phenobarbital-inducible isozyme). Finally metabolism of retinol and retinoic acid was also demonstrated with human liver microsomes and purified cytochrome P-450IIC8.[10]

Microsomal Preparation

Rat microsomes are prepared as follows. Liver is homogenized in 4 volumes of 50 mM Tris-HCl buffer, pH 7.4, at 4° containing 1.15% KCl

[1] M. A. Leo, N. Lowe, and C. S. Lieber, *Am. J. Clin. Nutr.* **40**, 1131 (1984).
[2] M. Sato and C. S. Lieber, *J. Nutr.* **11**, 2015 (1981).
[3] M. A. Leo and C. S. Lieber, *N. Engl. J. Med.* **307**, 597 (1982).
[4] T. V. Reddy and E. K. Weisburger, *Cancer Lett.* **10**, 39 (1980).
[5] A. C. Ross, *J. Biol. Chem.* **257**, 2453 (1982).
[6] A. B. Roberts, L. C. Lamb, and M. B. Sporn, *Arch. Biochem. Biophys.* **199**, 374 (1980).
[7] M. Sato and C. S. Lieber, *Arch. Biochem. Biophys.* **213**, 557 (1982).
[8] M. A. Leo, S. Iida, and C. S. Lieber, *Arch. Biochem. Biophys.* **234**, 305 (1984).
[9] M. A. Leo and C. S. Lieber, *J. Biol. Chem.* **260**, 5228 (1985).
[10] M. A. Leo, J. M. Lasker, J. L. Raucy, C. Kim, M. Black, and C. S. Lieber, *Arch. Biochem. Biophys.* **269**, 305 (1989).

and 1 mM EDTA, and centrifuged at 10,000 g for 30 min. The supernatant is spun at 140,000 g for 70 min. The fluffy part of the pellet (without the sticky so-called glycogen portion) is resuspended in 100 mM sodium pyrophosphate, pH 7.4, at 4° with 0.1 mM EDTA and centrifuged again as described above. The final washed microsomal pellet is resuspended (1 g liver/ml) in 10 mM potassium phosphate, pH 7.4, at 4° with 0.25 M sucrose. Human microsomes are prepared with a similar procedure except that the washed microsomes are resuspended in 100 mM Tris-HCl buffer, pH 7.4, containing 100 mM KCl, 1 mM EDTA, 1 mM phenylmethylsulfonyl fluoride (PMSF), and 20% glycerol and frozen at $-80°$ until use.

Substrate Preparation

[11,12-(n)-^{3}H]Retinol (specific activity 40–60 Ci/mmol) is obtained from Amersham Corp. (Arlington Heights, IL), and its purity is verified by high-performance liquid chromatography (HPLC). To prepare the substrate, a fresh stock solution of all-*trans*-retinol (Ro-1-4955), kindly provided by Dr. P. F. Sorter, Hoffman-LaRoche, Inc. (Nutley, NJ) [~10 mg/ml absolute ethanol containing 1 mg butylated hydroxytoluene (BHT)/ml] is made. Then an aliquot of labeled retinol, together with another aliquot of unlabeled retinol, is evaporated to near dryness but never completely, and bovine serum albumin (fraction V), 3.2 mg/ml, is added in a ratio of 6.5 μg of albumin per nanomole of retinol. The substrate (10–20 μl) is counted, and the specific activity is calculated as distintegration per minute (dpm) per nanomole of retinol.

Incubation Procedures

Microsomes. Reaction mixtures (0.4 ml) contain 0.2 mg liver microsomes, 0.1–0.15 mM retinol including 2 μCi labeled retinol, 0.1 M Tris-HCl, 0.15 M KCl buffer (pH 7.4 at 37°), 20 mM potassium phosphate buffer, pH 7.4, 5 mM MgCl$_2$, and 1 mM EDTA. When antibodies to P 450 isozymes are used, microsomes are preincubated with anti-P-450 IgG or an equivalent amount of control preimmune IgG, for 3 min at 37° prior to the addition of the other components. The incubation is carried out at 37° in a shaking water bath (120 cycles/min) under air. After a 3-min preincubation, the reaction is started with the addition of NADPH (1 mM). The different blanks used are (1) the complete system but without cofactor; (2) the complete system but without microsomes; and (3) the complete system with boiled microsomes. Reactions are usually negligible with all three systems. For routine purposes, blank 1 can be adopted. After 5 or 10 min of incubation, the reaction is stopped by adding a mixture of

L-ascorbic acid sodium salt and EDTA tetrasodium salt (0.25 mg each) and rapid freezing in dry ice–acetone. The samples are then lyophilized, extracted overnight with 0.5 or 1.0 ml of methanol (containing 20 μg of BHT/ml), and centrifuged at 2000 g for 10 min. The methanol extract is transfered to HPLC vials, and an aliquot (50 μl) is counted for radioactivity. Recovery of radioactive substrates in the extraction procedure is 90.7 ± 1.3%.

Reconstituted Systems. A reconstituted system contains (per milliliter) 0.05–0.1 nmol purified P-450, 0.5–1.0 units cytochrome-P-450 reductase (Fp), and 15–30 μg dilauroylphosphatidylcholine. The enzymes and lipid are first mixed together in a small volume (30–60 μl), preincubated for 3 min at 37°, and placed on ice. In a total incubation volume of 0.4 ml the remaining reaction components (as described above for microsomes) are then added. In immunoinhibition assays, anti-cytochrome P-450 IgG or control IgG (5 mg protein/nmol P-450) are added to the enzyme–lipid mixture. After a 3-min preincubation at 37°, the reaction is started with or without 1 mM NADPH. A retinoic acid-metabolizing system can also be reconstituted with or without P-450IIC8, with or without antibodies to P-450IIC8, and the same reaction mixture is indicated above for the retinol system. The final retinoic acid concentration used is 1 μM. Labeled retinoic acid (kindly provided by Dr. P. F. Sorter, Hoffmann LaRoche) must be purified by HPLC.[8]

Chromatographic Procedure

HPLC analyses are carried out with either a Hewlett Packard Model HP-1084B liquid chromatograph equipped with a variable wavelength detector and a fraction collector (for labeled metabolites) or a Model HP-1090 liquid chromatograph equipped with a diode-array detector (for spectral identification), a Chemstation, and an HP-7470A plotter (Hewlett-Packard, Avondale, PA). A Zorbax ODS column (0.46 × 15 cm) (Mac-Mod Analytical Inc., Chadds Ford, PA) is used, for which the mobile phase is programmed with a gradient from 25% mobile phase B (75% A) to 90% mobile phase B (10% A) in 10 min and continued for 8 min more at a flow rate of 1.5 ml/min. The mobile phase A is acetonitrile/water/acetic acid (49.75 : 49.75 : 0.5, v/v/v), and the mobile phase B is acetonitrile/water/acetic acid (90 : 10 : 0.04, v/v/v) both containing 10 mM ammonium acetate. All HPLC analysis are done at 30–35°. Fractions are collected every 30 sec, except for the 4-hydroxyretinol peak which is collected every 7.5–15 sec. After the addition of 5 ml of Aquasol (New England Nuclear, Boston, MA), fractions are counted in a Beckmann LS 7800 liquid scintillation counter (Beckman Scientific Instruments, Palo

Alto, CA). Handling of chemicals and solvents has been described before.[9]

Calculation

The enzyme activity is calculated from the specific activity of the prepared substrate [11,12-(n)-³H retinol] as follows:

Nanomole 4-hydroxyretinol
produced/min/mg
protein or nanomole P-450

$$= \frac{\text{dpm in 4-hydroxyretinol peak from HPLC}}{\text{specific activity of substrate (dpm/nmol)}}$$

$$\times \frac{0.5 \text{ or } 1.0}{0.1} \times \frac{1}{5 \text{ or } 10} \times \frac{1}{0.2 \text{ or } [0.02-0.08]}$$

where 0.5 or 1.0 is the volume of methanol added for extraction (ml); 0.1, the volume subjected to HPLC (ml); 5 or 10, the incubation time (min); 0.2, the microsomal protein in incubation (mg); and [0.02–0.08], the P-450 isoenzymes in incubation (nmol).

When 4-hydroxyretinol standard is available, activity can be calculated through peak area comparison. Recovery of 4-hydroxyretinol added to microsomes is 91.6 ± 2.2% (n = 6). As previously reported,[10] values obtained by the radiometric (³H) and spectrophotometric (UV) methods are not significantly different. Purified 4-hydroxyretinol was kindly donated by Drs. A. B. Barua and J. A. Olson (Iowa State University, Ames, IA) and Dr. J. L. Napoli (SUNY, Buffalo, NY).

Retinol 4-Hydroxylase Activity in Microsomes and Reconstituted Systems

A typical HPLC profile of metabolites produced from all-*trans*-[³H]retinol by microsomes of rats given phenobarbital is shown in Fig. 1. The K_m of the microsomal retinol 4-hydroxylase varies depending on drug treatment of the animal (Table I). Ethanol (100 mM) and dimethylnitrosamine (10 mM) had no significant effect on retinol 4-hydroxylation in microsomes of either phenobarbital- or ethanol-treated rats. By contrast, benzphetamine exerted a striking competitive inhibition of the retinol 4-hydroxylation in microsomes of phenobarbital-treated rats.[11] A similar inhibition (K_i of 1.3 mM) was observed with aminopyrine in microsomes obtained from ethanol-fed animals.

[11] M. A. Leo, N. Lowe, and C. S. Lieber, *Biochem. Pharmacol.* **35**, 3949 (1986).

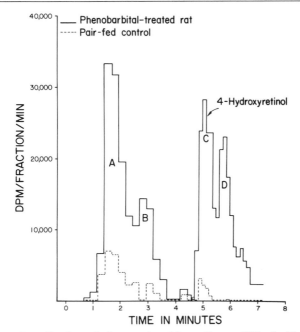

TIME IN MINUTES

FIG. 1. HPLC profile of metabolites produced from all-*trans*-[³H]retinol by microsomes of rats given phenobarbital or pair-fed controls. This representative tracing (obtained with an HP-1084B chromatograph) shows that microsomes of animals fed phenobarbital, when incubated with all-*trans*-[³H]retinol, produced an increased amount of 4-hydroxyretinol and other polar metabolites. [From M. A. Leo and C. S. Lieber, *J. Biol. Chem.* **260**, 5228 (1985).]

TABLE I

EFFECT OF ETHANOL OR PHENOBARBITAL TREATMENT ON
4-HYDROXYLATION OF all-*trans*-[³H]RETINOL BY RAT
LIVER MICROSOMES[a,b]

Treatment	K_m (mM)	V_{max} (nmol/min/mg protein)
Control ($n = 4$)	0.028 ± 0.006	0.11 ± 0.02
Ethanol ($n = 5$)	0.040 ± 0.009	0.23 ± 0.04^d
Phenobarbital ($n = 4$)	0.080 ± 0.005^c	0.98 ± 0.01^e

[a] Mean ± SEM.
[b] [From M. A. Leo, N. Lowe, and C. S. Lieber, *Biochem. Pharmacol.* **35**, 3949 (1986).]
[c] $p < .01$ significantly different compared to control.
[d] $p < .05$ significantly different compared to control.
[e] $p < .001$ significantly different compared to control.

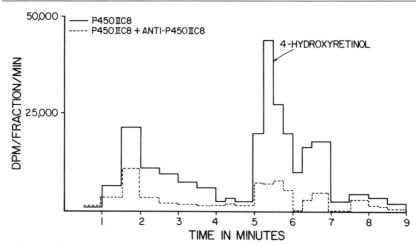

FIG. 2. HPLC profile of metabolites produced from all-*trans*-[3H]retinol in a system reconstituted with purified human cytochrome *P*-450IIC8, cytochrome-*P*-450 reductase, dilauroylphosphatidylcholine, and NADPH. In such a system, active conversion of retinol to polar metabolites, including 4-hydroxyretinol, was observed. This activity was inhibited by antibodies against cytochrome *P*-450IIC8. The metabolites were separated with an HP-1084B liquid chromatograph. [From M. A. Leo, J. M. Lasker, J. L. Raucy, C. Kim, M. Black, and C. S. Lieber, *Arch. Biochem. Biophys.* **269,** 305 (1989).]

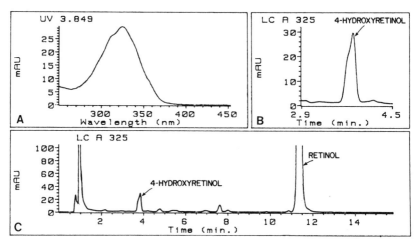

FIG. 3. HPLC profile (UV absorbance) of 4-hydroxyretinol (C) produced by the cytochrome *P*-450IIC8-reconstituted system. 4-Hydroxyretinol was identified by its typical spectrum (A), virtually identical to that of retinol. (B) Magnification of 4-hydroxyretinol peak. The metabolites were separated with an HP-1090 liquid chromatograph (see text). [From M. A. Leo, J. M. Lasker, J. L. Raucy, C. Kim, M. Black, and C. S. Lieber, *Arch. Biochem. Biophys.* **269,** 305 (1989).]

In reconstituted systems, active retinol metabolism was shown with rat cytochrome *P*-450b and *P*-450f.[9] Among the human cytochromes *P*-450 tested, *P*-450IIC8 had high activity.[10] The HPLC profile of metabolites produced from [³H]retinol in a system reconstituted with cytochrome *P*-450IIC8 is shown in Fig. 2 for the radioactive metabolites produced and in Fig. 3 for the products measured by UV absorbance. This activity was inhibited by antibodies against cytochrome *P*-450IIC8. In such a reconstituted monooxygenase system containing human liver cytochrome *P*-450IIC8, retinol was converted to 4-hydroxyretinol and other polar metabolites, with a K_m of 0.071 mM and a V_{max} of 1.73 nmol/min/nmol cytochrome *P*-450.[10]

Conclusions

By revealing that microsomes, including human ones, are capable of actively metabolizing retinol and by elucidating that cytochrome *P*-450 isozymes are involved, one has acquired a better understanding of the physiological role of a family of microsomal cytochrome *P*-450 isozymes that appear to be poorly inducible by xenobiotics. It is now increasingly apparent that microsomal cytochromes *P*-450 play a role not only in the detoxification of foreign compounds but also in important physiological processes, including vitamin A metabolism and maintenance of vitamin A homeostasis.[12] Retinol and retinoic acid have key functions not only in terms of cellular differentiation and maintenance of the normal integrity of mucosal tissues, but also for the prevention of carcinogenesis through various mechanisms, including the inhibition of the microsomal activation of chemical carcinogens.[11] Retinoids, however, may also acquire cellular toxicity in the liver and other tissues.[13] Enzymes such as cytochrome *P*-450IIC8, shown to be involved in vitamin A metabolism,[10] may thereby participate in maintaining the delicate balance between those retinol concentrations that promote cellular integrity and oppose the development of cancer and those that cause cellular toxicity.

Acknowledgments

Original studies referred to in this chapter were supported, in part, by U.S. Department of Health and Human Services Grants DK 32810, AA 03508, and AA 05934 and by the Veterans' Administration. The authors are grateful to Dr. C. Kim for her contribution to the preparation of the manuscript, to Ms. N. Lowe for her expert technical assistance, and Ms. R. Cabell and P. Walker for skillful typing of the paper.

[12] M. A. Leo, C. Kim, and C. S. Lieber, *J. Nutr.* **119**, 993 (1989).
[13] M. A. Leo and C. S. Lieber, *Hepatology* **8**, 412 (1988).

[55] Assay of Liver Retinyl Ester Hydrolase

By DALE A. COOPER

Introduction

The hydrolysis of long-chain fatty acyl esters of retinol is an important event in the release of stored hepatic vitamin A. Two widely used methods for the assay of retinyl ester hydrolase are based on the observation that liver preparations catalyze this reaction in the presence of added trihydroxy bile salts.[1] One of these methods (Method 1), developed by Prystowsky *et al.*,[2] is based on stimulation of activity with sodium cholate and the measurement of the release of [14C]palmitic acid from retinyl [14C]palmitate. A second method (Method 2), developed by Cooper *et al.*,[3,4] utilizes a 3-[(3-cholamidopropyl)dimethylammonio]propane sulfonate (CHAPS)/Triton X-100 detergent system and high-performance liquid chromatography (HPLC) to measure the production of unlabeled retinol. The protocols for these assays and some of the enzymatic properties of retinyl ester hydrolase are presented in this chapter.

Method 1

Tissue and Substrate Preparation. Originally designed to measure rat liver retinyl ester hydrolase activity, this procedure has also been used with rat kidney, intestine, testes, and lung,[5] and chick liver.[6] Homogenates are prepared by mincing rinsed liver in 3 or more volumes (w/v) of ice-cold 50 mM Tris–maleate (pH 8.0), followed by homogenization with several strokes of a motor-driven mortar and pestle (Thomas Scientific, Philadelphia, PA) in an ice-water bath. Liver homogenates can be frozen and stored at $-20°$ for at least 1 year without loss of activity.

Liver homogenates that contain high concentrations of retinyl esters must be diluted so that the amount of hepatic retinyl esters added to the

[1] S. Mahadevan, N. I. Ayyoub, and O. A. Roels, *J. Biol. Chem.* **241**, 57 (1966).
[2] J. H. Prystowsky, J. E. Smith, and D. S. Goodman, *J. Biol. Chem.* **256**, 4498 (1981).
[3] D. A. Cooper and J. A. Olson, *Biochim. Biophys. Acta* **884**, 251 (1986).
[4] D. A. Cooper, H. C. Furr, and J. A. Olson, *J. Nutr.* **117**, 2066 (1987).
[5] J. L. Napoli, A. M. McCormick, B. O'Meara, and E. A. Dratz, *Arch. Biochem. Biophys.* **230**, 194 (1984).
[6] D. Sklan and S. Donoghue, *Biochim. Biophys. Acta* **711**, 532 (1982).

assay mix does not affect the measured enzyme activity by changing the concentration and specific activity of the substrate. This condition is met if the homogenate contributes less than a few percent of the retinyl esters in the assay mix. Also, dilution may be necessary to ensure that the amount of protein added (microgram quantities) is linearly proportional to the reaction rate.

Retinyl [1-^{14}C]palmitate is prepared by the reaction of the symmetric anhydride of [1-^{14}C]palmitic acid (New England Nuclear, Boston, MA) with all-*trans*-retinol (Eastman Kodak, Rochester, NY).[2] The procedures for the preparation of the anhydride[7] and of retinyl palmitate are too lengthy to be presented here, but they are identical to published procedures used in the synthesis of cholesteryl esters.[8]

Assay Procedure and Activity Calculations. The reaction mixture is prepared in a screw-top glass test tube (16 × 100 mm) and consists of 100 μl of 0.1 M Tris–maleate, pH 8.0, 40 μl of sodium cholate (1.3 mg, Sigma, St. Louis, MO), 50 μl of enzyme-containing sample, and 10 μl of retinyl [1-^{14}C]palmitate in ethanol (2 nmol and ~0.05 μCi or 100,000 cpm). Tubes are incubated in a 37° water bath for 15–30 min. The reaction is quenched by the addition of 3.25 ml of methanol/chloroform/heptane (1.41 : 1.25 : 1.00, v : v : v) containing 0.1 mM palmitic acid (Sigma) as a carrier (organic solvents used in Methods 1 and 2 are purchased from Fisher, Minneapolis, MN). The liquid phases are separated by centrifugation at 1000 rpm in a clinical centrifuge at room temperature for 15 min. Radioactivity of free fatty acids is determined by liquid scintillation spectrometry of 1 ml of the upper phase after addition to 10 ml of Scintiverse I (Fisher).

To calculate retinyl ester hydrolase activity, the radioactivity in test assays given in counts per minute (cpm) is corrected for radioactivity in an assay mixture from which the enzyme has been omitted. The corrected counts are multiplied by 2.45 to account for the total volume of the upper phase, and the product is divided by the partition coefficient of palmitic acid. The partition coefficient is that portion of palmitic acid extracted from a tube containing an enzyme-free reaction mixture to which 10 μl of 200 μM [1-^{14}C]palmitic acid in ethanol has been added. The partition coefficient for palmitic acid in this solvent system is approximately 0.70.

Level of Activity and Enzymatic Properties. Rat liver retinyl ester hydrolase activity ranges from 5 to 250 pmol/min/mg; this range is larger when portions of liver are assayed.[2,9] In spite of this variability, it is well

[7] Z. Selinger and Y. Lapidot, *J. Lipid Res.* **7**, 174 (1966).
[8] B. R. Lentz, Y. Barenholz, and T. E. Thompson, *Chem. Phys. Lipids* **15**, 216 (1975).
[9] W. S. Blaner, J. E. Smith, R. B. Dell, and D. S. Goodman, *J. Nutr.* **115**, 856 (1985).

established that the assay gives consistent activity for a given enzyme preparation.[2,10]

The apparent K_m and V_{max} values of an approximately 200-fold purified preparation of retinyl ester hydrolase are 11.2 μM and 141 nmol/min/mg, respectively.[11] The most potent inhibitors of hydrolase activity, with concentrations giving 50% inhibition, include the following: ether analogs of cholesterol esters [especially the linoleyl ether (0.16 μM)],[12] ether analogs of triglycerides (>10 μM),[12] phylloquinone (20 μM),[5] and α-tocopherol (100 μM).[5]

Method 2

Tissue and Substrate Preparation. This procedure has been used with pig liver and kidney[3] and rat liver[4]; however, pig and rat liver hydrolases differ with regard to the substrate and CHAPS concentrations and pH giving maximal activity. The differences are noted below. Tissue homogenates are prepared as described for Method 1. Homogenates are diluted so that the amount of homogenate protein added to the reaction mixture is less than 1 mg for pig liver or 750 μg for rat liver; usually 200–300 μg is sufficient. Under these conditions, the amount of liver-derived vitamin A added to the assay mix is negligible when the liver reserves are less than 200 μg/g (higher levels have not been tested).

For each 1 ml of substrate solution, 2.63 mg (for pig assay) or 26.3 mg (for rat assay) of retinyl palmitate (Sigma) is dissolved in hexanes, and the concentration is verified by measuring the absorption at 325 nm of a known dilution. An appropriate volume of this solution is added to a screw-top test tube, and the hexanes are evaporated under a stream of argon. The residue is dissolved in 20 mg Triton X-100 (Sigma) per milliliter of assay buffer, a small stir bar is added, and the solution is stirred vigorously. The substrate solution can be used up to 2 days later when stored in the dark at 4° in an argon-flushed container. A surface layer of retinyl ester may form during storage but the solution will be homogeneous if it is warmed to room temperature and stirred prior to use.

Assay Procedure and Activity Calculations. Assays are carried out in 12 × 75 mm glass tubes with caps. For the pig enzyme, 250 μl of 200 mM CHAPS (Sigma) in 50 mM Tris–maleate, pH 8.0, is combined with 50 μl

[10] E. H. Harrison, J. E. Smith, and D. S. Goodman, *J. Lipid Res.* **20**, 760 (1979).
[11] W. S. Blaner, J. H. Prystowsky, J. E. Smith, and D. S. Goodman, *Biochim. Biophys. Acta* **794**, 419 (1984).
[12] W. S. Blaner, G. Halperin, O. Stein, Y. Stein, and D. S. Goodman, *Biochim. Biophys. Acta* **794**, 428 (1984).

of 5 mM retinyl palmitate suspension. After a 5-min preincubation at 37°, 50 μl of homogenate is added to initiate the reaction. The mixture is vortexed gently and incubated for 30–60 min in a Dubnoff shaking incubator at 25 strokes/min.

The assay mix for the rat enzyme is composed of 350 μl of 393 mM CHAPS in 100 mM Tris–maleate, pH 7.0, 50 μl of 50 mM retinyl palmitate suspension, and 50 μl of liver homogenate. Incubation is for 30–45 min. Reaction mixtures containing over 900 μM retinyl palmitate may be turbid; however, this does not affect the assay. To stop the hydrolase reaction, 500 μl absolute ethanol is added and the solution is vortexed. Next, the assay mix is extracted twice with 2 ml of hexanes containing 0.1% butylated hydroxytoluene (BHT) (Sigma). The combined hexane extracts are dried under a gentle stream of argon, and the residue is redissolved in 500 μl 2-propanol and stored in 1.5-ml Eppendorf tubes at $-20°$.

Retinol can be analyzed by using any HPLC method that separates retinol from contaminating peaks. A proven method is based on chromatography with the following equipment: a 4.6 × 250 mm Partisil 10/25 ODS-2 column (Whatman, Clifton, NJ), an Uptight guard column (Upchurch Scientific, Oak Harbor, WA) filled with pellicular octadecylsilane (Vydac, The Separation Group, Hesperia, CA), an LDC Constametric III pump (Riviera Beach, FL), an ISCO variable absorbance detector (Lincoln, NE), and a Shimadzu C-R3A integrator (Kyoto, Japan). A Waters Associates (Milford, MA) WISP Model 710 HPLC autosampler can be used to process multiple samples. After every second or third injection of 25 μl, the column should be flushed with 1:1 methanol/tetrahydrofuran (v/v) to elute retinyl esters. all-*trans*-retinol is eluted in 7.3 min with 9:1 (v/v) methanol/water pumped at a flow rate of 2 ml/min. The ultraviolet absorbance (325 nm) of the column effluent is monitored, and retinol is quantitated by comparison of peak area to that of a standard concentration curve.

Retinyl ester hydrolase activity is calculated from the amount of retinol produced during the assay corrected for retinol formed in control incubations containing all assay components except the homogenate. If liver homogenates contain high levels of retinol, correction should be made for retinol extracted from unincubated assay mixtures (minus retinyl palmitate). This assay will detect the production of as little as 4 pmol retinol. The interassay variability is 4.7%.

Level of Activity and Enzymatic Properties. Liver retinyl ester hydrolase activity in 24 pigs[3] was 13–34 pmol/min/mg. The enzyme has an apparent K_m of 140 μM.[3] Other excellent substrates and their rates of hydrolysis relative to all-*trans*-retinyl palmitate are as follows: the linolenic (2.1), myristic (2.1), oleic (0.8), linoleic (0.7), and stearic (0.6) esters of retinol as well as the 9-cis (2.4), 13-cis (5.7), and 9,13-cis (6.8) isomers of retinyl palmitate.[13]

Rat liver retinyl ester hydrolase activity in 25 rats was 100–500 pmol/ min/mg and had an apparent K_m of 1.3 mM.[4] In comparison to equimolar sodium cholate, CHAPS gives 9.3 times higher activity on average and reduced interanimal variability.[4] Ethanol (0.01 to 0.5 M) stimulates hydrolase activity 20 to 86%.[14]

Conclusions

Although the methods described in this chapter have not been compared side by side, they have similar ease of use, sensitivity, and reproducibility. The assays differ largely with respect to composition, methodology, and the interanimal variability in activity they give.

---

[13] D. A. Cooper and J. A. Olson, *Arch. Biochem. Biophys.* **260**, 705 (1988).
[14] H. Friedman, S. Mobarhan, J. Hupert, D. Lucchesi, C. Henderson, P. Langenberg, and T. J. Layden, *Arch. Biochem. Biophys.* **269**, 69 (1989).

---

[56] Assay of the Retinoid Isomerase System of the Eye

By PAUL S. BERNSTEIN and ROBERT R. RANDO

Introduction

The all-*trans*- to 11-*cis*-retinoid isomerase is the key enzyme of the vertebrate visual cycle. It is responsible for the endergonic isomerization of vitamin A compounds from the all-trans isomeric form to the higher energy 11-cis configuration essential for rhodopsin regeneration in the eye. The existence of an enzyme of this type was postulated in the early 1950s by Wald and Hubbard when the vertebrate visual cycle was first discussed.[1] The reproducible demonstration of the existence of an isomerase was not achieved until recently, however, requiring the use of high specific-activity radioactive retinoid substrates and high-performance liquid chromatographic (HPLC) analysis.[2]

Although the isomerase has not yet been completely purified, a number of its properties are well established: (1) the isomerase is localized to

---

[1] R. Hubbard and G. Wald, *J. Gen. Physiol.* **36**, 269 (1952).
[2] P. S. Bernstein, W. C. Law, and R. R. Rando, *Proc. Natl. Acad. Sci. U.S.A.* **84**, 1849 (1987).

the pigment epithelium of the eye, as opposed to the retina;[2] (2) the
activity of the isomerase has been detected in a variety of vertebrates
ranging from frogs to cows to humans,[2-4] (3) the isomerase is membrane-
associated, and its activity is sensitive to inactivation by many detergents
and by the presence of relatively low concentrations of organic solvents
such as ethanol,[3,5] (4) despite its sensitivity to many detergents, the isomerase has recently been solubilized in Zwittergent-3,14[6]; (5) retinoid isomerase activity is eliminated by boiling, by proteinase K, and by phospholi-
pase C[5]; (6) the all-*trans*- to 11-*cis*-retinoid isomerization reaction
catalyzed by the isomerase proceeds with stereochemical inversion at the
prochiral C-15 and with cleavage of the carbon–oxygen bond at C-15.[7,8]

Further characterization of the retinoid isomerase has been hampered
by the instability of the enzyme and by the persistent presence in enzyme
preparations of large amounts of retinyl ester synthetase activity, an enzyme that competes for and utilizes the same all-*trans*-retinol precursor
routinely used in a standard isomerase assay. Although it has been clearly
established in both *in vivo* and *in vitro* double-label radioisotope experiments that all-*trans*-retinoid to 11-*cis*-retinoid isomerization occurs at the
alcohol oxidation state,[5,9] there has been a question as to whether it is all-
trans-retinol or a long-chain fatty-acid ester of all-*trans*-retinol that is the
immediate substrate for 11-*cis*-retinol production mediated by the retinoid
isomerase. Initial studies of the retinoid isomerase showed that added
retinyl esters were almost completely resistant to isomerization, whereas
all-*trans*-retinol was effectively converted to the 11-cis configuration.[2,5] A
direct all-*trans*-retinol to 11-*cis*-retinol pathway, however, leaves the obligate energy source for this process unidentified. More recently, in membrane preparations in which remaining endogenous retinoids had been
removed by preirradiation with ultraviolet light and repeated washing, 11-
cis-retinoid production has been detected with an all-*trans*-retinyl palmitate substrate.[8] The hypothesis of an all-*trans*-retinyl palmitate to 11-*cis*-
retinol pathway is quite attractive because the energy released by
ester-bond cleavage would be sufficient to drive the endergonic conversion of an all-*trans*-retinoid to an 11-*cis*-retinoid.[8]

The discovery and characterization of the retinoid isomerase have
significant implications. It is an enzyme that completes a fundamental

[3] B. S. Fulton and R. R. Rando, *Biochemistry* **26,** 7938 (1987).
[4] C. D. B. Bridges and R. A. Alvarez, *Science* **236,** 1678 (1987).
[5] P. S. Bernstein, W. C. Law, and R. R. Rando, *J. Biol. Chem.* **262,** 16848 (1987).
[6] R. J. Barry, F. J. Cañada, and R. R. Rando, *J. Biol. Chem.* **264,** 9231 (1989).
[7] W. C. Law and R. R. Rando, *Biochemistry* **27,** 4147 (1988).
[8] P. S. Deigner, W. C. Law, F. J. Cañada, and R. R. Rando, *Science* **244,** 968 (1989).
[9] P. S. Bernstein and R. R. Rando, *Biochemistry* **25,** 6473 (1986).

biochemical cycle that is apparently common to all vertebrate visual systems. It is also the first known isomerase whose physiological task is to drive the enzymatic conversion of a trans double bond to a higher energy cis configuration. Further study of this enzyme is certain to yield invaluable information on the role of this enzyme in the physiology and pathophysiology of vision and would also be expected to provide novel insights into the biochemical mechanisms of cis to trans isomerization by enzymes. The enzyme has been detergent solubilized and substantially purified,[6] and its properties are described in a review.[10] The objective of this chapter is to describe in detail the methods used to assay this important enzyme.

Retinoid Substrates

All retinoids, and all 11-*cis*-retinoids in particular, are inherently unstable compounds prone to oxidation, thermal isomerization, and photoisomerization. To prevent oxidation and thermal isomerization, retinoid solutions should be stored at as low a temperature as possible (4° during routine handling and −70° for storage). If storage is planned for a significant period of time, the solution should have an added antioxidant such as 1 mg/ml butylated hydroxytoluene (BHT) and an inert nitrogen or argon atmosphere. Although strict darkness is not required when handling retinoids since most retinoids do not have significant absorbance above 400 nm, it is advisable to keep room lights off and to use red-filtered safelights. All tubes containing retinoids should be wrapped in opaque materials whenever possible.

The assays described below utilize radioactively labeled retinoids. This permits maximal sensitivity of retinoid detection and minimizes interference from any endogenous retinoids or other chromophoric compounds in the tissue preparations that may coelute with substrates or products during the HPLC separations. The most convenient commercially available radioactive retinoid is all-*trans*-[11,12-³H]retinol (40–60 Ci/mmol), which can be purchased from Amersham (Arlington Heights, IL) or New England Nuclear (Boston, MA). Its tritium label is not exchangeable with solvent during storage or during biological reactions such as trans to cis isomerization. Commercial lots of this compound have wide variations in isomeric purity and stability, and it is thus wise to perform frequent purity checks by HPLC and to repurify by HPLC if necessary. If other radioactive retinoid substrates such as retinal or retinyl esters are desired, they can be prepared from

---

[10] R. R. Rando, P. S. Bernstein, and R. J. Barry, *Prog. Retinal Res.* **10**, in press (1991).

all-*trans*-[11,12-^{3}H]retinol by standard methods.[11] For specialized double-label and stereochemical experiments all-*trans*-[15-^{3}H]retinol and all-*trans*-[15-^{14}C]retinol are also available commercially.

Tissue Extract Preparation

Retinoid isomerase activity is found in the pigment epithelium of the vertebrate eye.[2] The northern leopard frog (*Rana pipiens*) is an excellent source of fresh pigment epithelium. These animals have large eyes, are easily cared for, and are available year-round. By using a living source of pigment epithelium it is possible to control the degree of dark-adaptation of the animal and to perform pharmacological manipulations before sacrifice. In a standard preparation procedure, the eyes of a light-adapted frog are dissected out and immersed in a frog Ringer's solution (80 mM NaCl, 2 mM KCl, 0.1 mM CaCl$_2$, 0.1 mM MgCl$_2$, in a 15 mM sodium phosphate buffer, pH 7.8). The cornea, iris, lens, and vitreous are then dissected away and discarded. The retina and pigment epithelium can then be carefully removed from the scleral cup. If desired, the retina and pigment epithelium may be separated with forceps under a dissecting microscope.

The tissue from one or two frog eyes is placed in a 2-ml centrifuge tube, and 0.45 ml of 50 mM sodium phosphate buffer (pH 7.2–8.0) is added. For tissue homogenization, 10 sec of sonication with a microultrasonic cell disrupter (Kontes, Bell River, NJ) has proved to be the method that maximally releases isomerase activity. After homogenization, two successive 10-min centrifugations at 600 g are performed to sediment the pigment granules, nuclei, and unbroken cells. At this point, the 600 g supernatant may be used immediately for isomerase assays, or it can be stored at 4° for several days. Alternatively, further purifications can be performed, such as (1) repeated washing and sedimentation of the membrane material at 50,000 g to remove soluble nicotinamide cofactors utilized by competing redox enzymes[2]; (2) fractionation of the membrane material on Percoll gradients[5]; (3) preparation of a 100,000 g supernatant fraction relatively depleted of competing retinyl ester synthetase activity[5]; (4) irradiation of the tissue extract with ultraviolet light to destroy endogenous retinoids that may be present[8]; (5) detergent solubilization of the retinoid isomerase in Zwittergent-3,14 and purification by column chromatography.[6]

When large amounts of retinoid isomerase material are needed, it has proved expedient to forgo using freshly sacrificed animals as a source of tissue. Bulk quantities of frozen bovine eye cups with the retinas already

---

[11] C. D. B. Bridges and R. A. Alvarez, this series, Vol. 81, p. 463.

removed are available commercially from companies that harvest the retinas to supply scientific researchers. Large quantities of retinoid isomerase activity can be obtained by gently brushing the pigment epithelium from the thawed eye cups with a camel hair brush in the presence of phosphate buffer.[3] The tissue extracts can then be prepared as described above. Retinoid isomerase activity has also been detected in extracts of homogenized pigment epithelium cells grown in tissue culture.[12]

Isomerase Assay

In a standard isomerase assay, 100 μl of pigment epithelium extract (typically 1–5 mg of protein/ml) is placed in a 2-ml centrifuge tube along with 100 μl of 50 mM phosphate buffer (pH 7.2–8.0). Ten microliters of 10% (w/v) bovine serum albumin (fatty acid-free) is then added to facilitate solubilization of the hydrophobic radioactive retinoid substrate. Finally, 1 μCi of all-$trans$-[11,12-^{3}H]retinol (40–60 Ci/mmol) is added in 1 μl of ethanol. Alternatively, the radioactive retinoid substrate in ethanol can be added to the incubation tube first, and the ethanol can then be evaporated under a gentle stream of nitrogen. Tubes are wrapped in foil and incubated at room temperature for 2 hr on a Clay-Adams (Parsippany, NJ) Nutator for gentle orbital mixing. If mammalian pigment epithelium is used instead of amphibian pigment epithelium, the incubation is done at 37° in a shaking water bath. A control tube containing no eye tissue, but otherwise identical, is prepared for each assay.

The final ethanol concentration of the incubation mixture is never more than 0.5%, and care must be taken not to exceed this value, as the activity of the retinoid isomerase is severely inhibited at ethanol concentrations greater than 1.0%.[3,5] Excessive concentrations of ethanol in assay mixtures account for previous failures to detect isomerase activity in homogenates of vertebrate pigment epithelium.[13,14] It should be noted that the concentration of all-$trans$-retinol in a standard assay is approximately one-fifth of the experimentally determined apparent K_m of 0.49 μM.[5] This low concentration of labeled substrate is used whenever possible because operating near the K_m, in the concentration range where the reaction kinetics are still first-order, maximizes the percent conversion of the labeled substrate. This substantially improves the signal-to-noise ratio of the product on the analytical radiochromatograms, where nonspecific ra-

[12] S. R. Das and P. Gouras, Biochem. J. 250, 459 (1988).
[13] S.-L. Fong, C. D. B. Bridges, and R. A. Alvarez, Vision Res. 23, 47 (1983).
[14] M. T. Flood, C. D. B. Bridges, R. A. Alvarez, W. S. Blaner, and P. Gouras, Invest. Ophthalmol. Visual Sci. 24, 1227 (1983).

dioactive background can be quite high relative to the specific retinoid counts recovered, and where baseline separation of the isomeric retinoids may be quite difficult.[2]

Radioactive Retinol Analysis Following Isomerase Assay

At the end of an incubation, the reaction is terminated by the addition of 400 μl of methanol and 200 μl of water, followed by 600 μl of hexane containing 1 mg/ml butylated hydroxytoluene. After vigorous shaking at room temperature, the material is centrifuged briefly at 13,000 g at 4° to separate the hexane layer from the aqueous layer.

Portions (120 μl) of the hexane extract are mixed with 10 μl of a standard mixture of carrier retinol isomers in hexane prepared by photoisomerization of retinal in methanol, followed by reduction with NaBH$_4$.[11] This mixture is then injected onto a Waters (Milford, MA) HPLC system with a 5-μm Merck LiChrosorb RT Si 60 silica column (250 × 4.0 mm). Detection is by absorbance at 320 nm, and the eluant is 7% dioxane in hexane at 2 ml/min to provide optimum separation of 11-cis-retinol from 13-cis-retinol.[15] With this chromatographic system, however, 9-cis-retinol and all-trans-retinol coelute.

The chemical and isomeric identity of each of the radioactive retinol products has been confirmed by iodine isomerization and by chemical derivatization to their corresponding retinyl palmitate esters followed by HPLC analysis.[2] Since 9-cis-retinol is not found in the eye physiologically,[1] is an unfavored product kinetically,[16] and was never detected in the ester derivatization experiments,[2] the peak consisting of all-trans-retinol and 9-cis-retinol is referred to as simply all-trans-retinol.

The elution of each isomeric retinol peak is marked by the elution of the carrier retinoid standards. A Gilson 201 microprocessor-controlled fraction collector in the peak-detection mode collects each of these retinoid peaks with appropriate delay for the dead volume between the detector and the dropping needle. Samples are counted in 4.5 ml of Econofluor (Fisher, Fairlawn, NJ) on a Beckman (Palo Alto, CA) LS 1800 scintillation counter interfaced with an Apple II-Plus microcomputer for data analysis. All counts of radioactivity are corrected for quench and background. More recently, a Berthold in-line HPLC scintillation counter linked to an IBM PC-XT has been successfully employed in our laboratory to streamline the data collection and analysis process.

---

[15] G. M. Landers and J. A. Olson, *J. Chromatogr.* **291**, 51 (1984).
[16] R. R. Rando and A. Chang, *J. Am. Chem. Soc.* **105**, 2879 (1983).

Analysis of Other Radioactive Retinoids Following Isomerase Assay

The hexane extraction method described above was compared to a more complicated extraction method using NH_2OH and CH_2Cl_2, known to extract all retinoids from eye tissue quantitatively.[17] The simpler hexane extraction method described above was found to extract retinols quantitatively, but retinals and retinyl esters were not extracted well, although isomeric distributions were unaltered.[2] Briefly, the NH_2OH/CH_2Cl_2 method entails the addition of 300 μl of methanol and 100 μl of 1 M NH_2OH (pH 6.5) at the end of an incubation in order to form the oximes of retinal required for the quantitative extraction of retinoids without isomerization.[17] After brief centrifugation, the emulsion is then extracted twice with 300 μl of CH_2Cl_2. This organic extract is then dried under a stream of nitrogen and resuspended in hexane in preparation for HPLC analysis. Since *syn*-oximes of retinal isomers often coelute with retinol isomers in a dioxane/hexane chromatographic system, 8% ether in hexane is used; however, 11-*cis*-retinol and 13-*cis*-retinol coelute in this system,[17] and parallel standard hexane extractions are performed to distinguish between these two isomers. Retinyl palmitate esters are analyzed with 0.5% ether in hexane.

Typical Results of Isomerase Assay

In a 600 g frog retina/pigment epithelium supernatant, the least purified of the isomerase preparations, 11-*cis*-retinol production from added radioactive all-*trans*-retinol is linear for at least 3 hr, and, at the end of an assay, 20–40% of the total remaining retinols are in the 11-cis configuration.[2] This percentage is in marked contrast to the 0.1% 11-cis isomer present when retinoids are at thermal equilibrium.[16] Interpretation of the kinetics of retinoid isomerization in these unpurified tissue extracts is hampered, however, by several other enzymatic activities, retinyl ester synthetase and retinol dehydrogenase (EC 1.1.1.105), that also utilize the radioactive all-*trans*-retinol precursor. At all time points after the first 10 min of a standard assay of a 600 g supernatant, only 10–15% of originally added radioactivity is recovered as retinol, whereas 70–80% is converted to retinyl palmitate esters and 5–10% is isolated as retinals.[2] It is possible to suppress much of the retinol dehydrogenase activity by removing the nicotinamide redox cofactors through repeated sedimentation and washing of the membrane fraction or by dialysis.[2,5] Retinyl ester synthetase activity, on the other hand, has continually copurified in even the most highly purified isomerase preparations,[3,5,6] and it is a distinct possibility

[17] P. S. Bernstein, J. R. Lichtman, and R. R. Rando, *Biochemistry* **24**, 487 (1985).

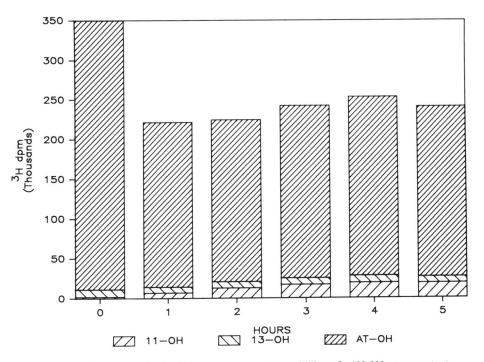

FIG. 1. Time course of retinoid isomerase assay. One milliliter of a 100,000 g supernatant of a frog (*Rana pipiens*) retina/pigment epithelium homogenate (4.47 mg protein/ml) was incubated in the dark with 50 µl of 10% (w/v) bovine serum albumin and 10 µCi of all-*trans*-[11,12-^{3}H]retinol. A control tube with no eye tissue was also prepared. Portions (200 µl) were removed hourly, and retinoids were extracted by the hexane method before HPLC analysis. 11-OH, 13-OH, and AT-OH denote 11-*cis*-retinol, 13-*cis*-retinol, and all-*trans*-retinol, respectively. There was no detectable formation of 11-*cis*-retinol in the control incubation in the absence of eye tissue.

that the retinoid isomerase and the retinyl ester synthetase may be part of a multienzyme complex.[3,6,8]

Detergent solubilization coupled with column chromatography gives the best purification of isomerase activity, but the isomerase activity is quite unstable.[6] For kinetic characterization of the retinoid isomerase, a more stable but less purified 100,000 g dialyzed supernatant preparation from frog eyes has been used.[5] As shown in Fig. 1, over 65% of the initially added all-*trans*-[^{3}H]retinol radioactivity is recovered as retinol at all time points after the reaction starts. Production of 11-*cis*-retinol is linear for at least 3 hr, and there is no significant formation of nonphysiological 13-*cis*-retinol. No formation of 11-*cis*-retinol was detected in control experiments in the absence of eye tissue. Figure 2 shows a substrate saturation curve of isomerase activity of a 100,000 g supernatant prepara-

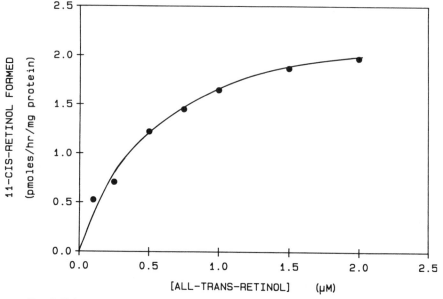

FIG. 2. Substrate saturation of retinoid isomerase activity. Portions (100 μl) of a 100,000 g supernatant of frog (*Rana pipiens*) retina/pigment epithelium homogenate (4.17 mg protein/ml) were incubated with 100 μl of phosphate buffer (pH 8.0), 10 μl of 10% (w/v) bovine serum albumin, 1 μCi of all-*trans*-[11,12-³H]retinol, and sufficient unlabeled retinol to achieve the final concentrations indicated. Incubation in the dark was for 2 hr, and retinoid extraction with hexane was used to prepare the samples for HPLC analysis. The apparent K_m of 11-*cis*-retinol production was 0.49 μM, and the apparent V_{max} was 2.5 pmol/hr/mg protein.

tion at its pH optimum of 8.0 with an all-*trans*-retinol substrate. An apparent K_m of 0.49 μM and an apparent V_{max} of 2.5 pmol/hr/mg protein were determined from this curve. The relatively low V_{max} found here is a consequence of the depletion of the retinyl ester synthetase (lecithin retinol acyl transferase) from the membranes. In crude bovine pigment epithelial membranes, the V_{max} is ~1–2 nmol/hr/mg protein. More recently, by using membrane preparations depleted of endogenous retinoids by preirradiation with ultraviolet light and repeated washing, 11-*cis*-retinol production has been assayed with all-*trans*-retinyl palmitate as the substrate.[8] Kinetic parameters have not yet been determined with retinyl esters as the isomerase substrate.

Conclusion

It is clear that the retinoid isomerase, an enzyme system that completes the visual cycle of bleaching and dark adaptation, will have an important role in the physiology and pathophysiology of vision, as well as providing insights into the study of isomerase enzymes in general. Future purification and characterization of this novel class of enzyme should clarify its relationship to other visual cycle enzymes, its regulation, and its mechanism of action.

Acknowledgments

This work was supported by U.S. Public Health Service Research Grant EY 04096 from the National Institutes of Health.

[57] Assay of all-*trans*→11-*cis*-Retinoid Isomerase Activity in Bovine Retinal Pigment Epithelium

By MARIA A. LIVREA and LUISA TESORIERE

Introduction

The use of radiolabeled substrate and HPLC analysis have recently contributed to the discovery of a membrane-bound retinoid isomerase concentrated to the retinal pigment epithelium (RPE) cell layer of vertebrate eyes.[1-5] Broadly distributed among all subcellular organelles,[3,5] the enzyme appeared remarkably active in the nuclear fraction.[3,6]

Here we report the assay of the isomerase activity in the membrane fraction obtained from isolated nuclei of bovine RPE, under experimental conditions where the use of tritiated all-*trans*-retinol is not required.

Treatment with CHAPS of the nuclear membrane produces a 200,000 *g* supernatant retaining 80% of the total isomerase activity, and results in partial purification of the enzyme. The use of fresh bovine eyes to carry out the membrane preparation, followed by the immediate extraction with CHAPS and by the enzyme assay, is the best way to assure the highest recovery of the enzyme activity.

[1] P. S. Bernstein, W. C. Law, and R. R. Rando, *Proc. Natl. Acad. Sci. U.S.A.* **48,** 1849 (1987).
[2] C. D. B. Bridges and R. A. Alvarez, *Science* **236,** 1678 (1987).
[3] B. S. Fulton and R. R. Rando, *Biochemistry* **26,** 7938 (1987).
[4] P. S. Bernstein, W. C. Law, and R. R. Rando, *J. Biol. Chem.* **262,** 16848 (1987).
[5] C. D. B. Bridges, *Vision Res.* **29,** 1711 (1989).
[6] M. A. Livrea, A. Bongiorno, L. Tesoriere, and C. Nicotra, *Ital. J. Biochem.* **38/1,** 46 (1989).

Another assay of retinoid isomerase activity is reported for the frog eye in this volume.[7]

Preparation of Enzyme Source

Freshly excised calf eyes (40–50) should be used within a short time from enucleation (1–2 hr). Transport must be in ice in a light-tight box, and all other operations should be carried out under dim red light, at 4°. Retinal pigment epithelium is carefully brushed from the eye cups with the aid of a few milliliters of 0.25 M sucrose in 50 mM Tris-HCl buffer, pH 7.5, containing 5 mM MgCl$_2$ and 2.5 mM KCl (TKM buffer). The cellular suspension is filtered through two layers of gauze to exclude fragments from injured retinal cells, and then the RPE cells are collected at 1000 g. Hand homogenization of RPE cells is carried out in 0.25 M sucrose–50 mM TKM buffer (0.25 ml/eye cup) with a glass–glass homogenizer, using a pestle with a mean clearance of 0.1 mm. Approximately 20 up-and-down strokes are satisfactory to break the cellular envelope without damage to the nuclei. Nuclei can be prepared by precipitation in hypertonic sucrose solutions of RPE homogenate. Sonic disruption of the nuclei and precipitation at 20,000 g is used to obtain nuclear membrane. The procedure described by Kasper,[8] has been verified as suitable for the purpose of isolating the material for the enzyme assay.

Endogenous retinoids occurring in the preparations to be used for the isomerase assay should be monitored. In the authors' laboratory it has been observed that substantial amounts of both all-*trans*- and 11-*cis*-retinoids (retinol and retinyl esters) may be stored in membrane fractions of RPE.

Nuclear membrane lacking of endogenous retinoids should be used, as the method reported herein for the isomerase assay is feasible without using labeled substrate. Under these circumstances, the quantitation of the reaction product is better performed in the absence of any initial retinoid in the membrane, in that small variations can be more easily appreciated. Furthermore, as all-*trans*-retinyl esters are obligate intermediates in the all-*trans*→11-*cis* isomerization of retinol,[9] amounts of endogenous retinyl esters can substantially affect the relationship between the substrate and product. UV irradiation of the enzyme source, as reported by Deigner *et al.*,[10] can be adopted to remove endogenous retinoids.

[7] P. S. Bernstein and R. R. Rando, this volume [56].
[8] C. B. Kasper, this series, Vol. 31, p. 279.
[9] A. Trehan, F. J. Canada, and R. R. Rando, *Biochemistry* **29**, 309 (1990).
[10] P. S. Deigner, W. C. Law, F. J. Canada, and R. R. Rando, *Science* **644**, 968 (1989).

Suspensions of nuclear membrane (2 mg protein/ml 50 mM Tris-HCl buffer pH 7.8) are mixed with CHAPS to give a final detergent concentration of 0.5% (w/v). The solution is allowed to sit at 0° in the dark for 10 min. Then centrifugation at 200,000 g for 50 min at 4° is performed and the supernatant is utilized for the isomerase assay. The recovery of the total protein by this extraction procedure is about 40%.

Isomerase Activity Assay and Analysis Procedures

Five nanomoles of all-*trans*-retinol in ethanol are dried under nitrogen on the bottom of a glass extraction tube and then resuspended in 100 μl of a 6% solution in 50 mM Tris-HCl, pH 7.8, of defatted bovine serum albumin. After a 10-min incubation at room temperature (20°), 0.3 mg protein of enzyme preparation in a volume of 1 ml is added, the tubes are sealed, and the samples are incubated for 10 min at 37° in a shaking water bath. The final concentrations of all-*trans*-retinol and albumin are 4.54 and 81 μM, respectively. Retinol blanks are carried out by incubating all-*trans*-retinol in the absence of tissue extract and under all other assay conditions. The reaction is stopped by mixing with 2 ml of absolute ethanol, and the sample is left at room temperature for 10 min. The reaction products are then extracted by two successive extractions with 8 and 2 ml of petroleum ether, and the extracts are pooled. One-half milliliter of absolute ethanol is added to the mixture to assure a sharp phase formation after the second extraction. With the use of all-*trans*-[³H]retinol as a probe, this procedure has proved capable of extracting 80% of the retinoids in the samples.

Extracted retinoids are dried under a gentle nitrogen stream, and then HPLC analysis is carried out using a μPorasil column (0.4 × 30 cm) in conjunction with a HPLC apparatus adjusted so that it can measure 1 ng quantities of every eluted retinoid. Retinol isomers are eluted with 7% dioxane in hexane at 1.8 ml/min and detected at a wavelength of 320 nm. With this procedure all major isomers are sufficiently separated to allow their quantitative determination. Calibration of the elution system is performed by injecting 1 to 100 ng of the reference compounds, and the samples can be quantitated by reference to a standard curve relating the amount of the retinoid injected to the peak area.

As a result of nonspecific isomerization owing to the manipulation and extraction procedure, limited amounts of 13-*cis*-retinol (up to 15% of the substrate added in the assay) can be found among the reaction products at the end of the incubation. A saturation curve can be made relating the all-*trans*→11-*cis*-retinoid isomerase activity extracted from the RPE nuclear membrane to the all-*trans*-retinol concentration. With the assay system

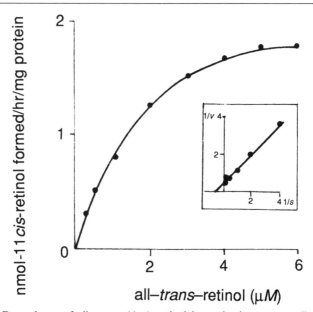

FIG. 1. Dependence of all-*trans*→11-*cis*-retinol isomerization rate on all-*trans*-retinol concentration. The assays were carried out by incubating 0.30 mg of CHAPS-solubilized protein from nuclear membrane in the presence of 81 μM BSA and variable concentrations of all-*trans*-retinol in a total volume of 1.1 ml. Data from double reciprocal plot (inset) were used to calculate the apparent V_{max} and K_m values for the reaction.

described, the rate of 11-*cis*-retinol formation approaches saturation at about 5 μM all-*trans*-retinol. The linear transformation of the saturation curve gives an apparent V_{max} of 2.5 nmol/hr/mg protein and a K_m of 1.6 μM (Fig. 1).

[58] Structure–Function Analyses of Mammalian Cellular Retinol-Binding Proteins by Expression in *Escherichia coli*

By MARC S. LEVIN, ELLEN LI, and JEFFREY I. GORDON

Rationale for Prokaryotic Expression of Cellular Retinoid-Binding Proteins

Several small (~15 kDa) intracellular retinoid-binding proteins have been purified from animal tissues and their primary structures determined. These include cellular retinol-binding protein (CRBP),[1-3] cellular retinol-binding protein II (CRBP II),[4,5] and cellular retinoic acid-binding protein.[6] They belong to a family of homologous, cytoplasmic, hydrophobic ligand-binding proteins which currently contains 10 known members, several of which bind long-chain fatty acids (reviewed by Sweetser *et al.*[7]).

Attempts to rapidly and easily purify these cellular retinoid binding proteins from their cells of origin have been limited by a number of factors. First, the yields of purified protein have been poor: 15 μg of CRBP per gram wet weight of rat liver and 170 μg of CRBP II per gram wet weight of rat small intestine.[8,9] Second, copurification of comparably sized, homologous cellular retinoid-binding proteins may occur when several are expressed in a given tissue. Last, the purified proteins contain bound ligand. To analyze apoprotein–ligand interactions, UV irradiation or organic extraction is required to remove endogenous retinoids, methods which may perhaps lead to structural modifications that could complicate interpretation of subsequent functional studies. By contrast, purification of these mammalian proteins after their expression in *Escherichia coli* offers several advantages: (1) large quantities of protein are easily obtained; (2) only one retinoid-binding protein is expressed because *E. coli* does not possess any endogenous retinoid-binding proteins; (3) because *E. coli* does not require retinoids for growth and endogenous retinoids are not present, the retinoid-binding proteins purified from this prokaryote can be obtained as apoproteins; (4) site-directed mutagenesis of cloned cDNAs encoding retinoid-binding proteins and their subsequent expression in *E. coli* represent a powerful method for analyzing protein structure–activity relationships; and (5) amino acid derivatives can be introduced into the proteins by growing suitable *E. coli* auxotrophs containing a prokaryotic expression vector in the presence of the analog. Analog-substituted retinoid-binding proteins can, in turn, be used for nu-

[1] M. M. Bashor, D. O. Toft, and F. Chytil, *Proc. Natl. Acad. Sci. U.S.A.* **70**, 3483 (1973).

[2] D. E. Ong and F. Chytil, *J. Biol. Chem.* **253**, 828 (1978).

[3] J. Sundelin, H. Anundi, L. Tragardh, U. Erikson, P. Lind, H. Ronne, P. A. Peterson, and L. Rusk, *J. Biol. Chem.* **260**, 6488 (1985).

[4] D. E. Ong, *J. Biol. Chem.* **259**, 1476 (1984).

[5] E. Li, L. A. Demmer, D. A. Sweetser, D. E. Ong, and J. I. Gordon, *Proc. Natl. Acad. Sci. U.S.A.* **83**, 5779 (1986).

[6] J. Sundelin, S. R. Das, U. Eriksson, L. Rask, and P. A. Peterson, *J. Biol. Chem.* **260**, 6495 (1985).

[7] D. A. Sweetser, R. O. Heuckeroth, and J. I. Gordon, *Annu. Rev. Nutr.* **7**, 337 (1987).

[8] D. E. Ong, A. J. Crow, and F. Chytil, *J. Biol. Chem.* **257**, 13385 (1982).

[9] W. H. Schaefer, B. Kakkad, J. A. Crow, I. A. Blair, and D. E. Ong, *J. Biol. Chem.* **264**, 4212 (1989).

clear magnetic resonance (NMR) studies of ligand–protein interactions (see below).

Expression of Cellular Retinol-Binding Proteins in *Escherichia coli*

General Principles

For a general discussion of useful strategies for expressing, identifying, and characterizing recombinant gene products in *E. coli*, the reader is referred to the review by Shatzman and Rosenberg in this series.[10] Two desirable components of prokaryotic expression vectors are (1) inducible promoters which can direct efficient transcription of foreign cDNAs and (2) translational control elements that allow efficient initiation of translation of the foreign mRNA transcript to occur. We have used pMON vectors[11] designed and provided by Monsanto, but a large number of commercially available expression vectors are also suitable.[12] The pMON series of vectors are derived from pBR327[13] and incorporate the *E. coli recA* promoter which can be induced by nalidixic acid and a ribosome binding site from the T7 bacteriophage gene 10 leader (G10L) sequence.[11] The pMON vectors that we have used contain a unique *Nde*I restriction site downstream from the *recA* promoter and the translational control element. By genetically engineering an *Nde*I site at its initiator ATG codon (if necessary), the cDNA of interest can be placed under the control of the *recA* promoter. Moreover, the distance between the vector-derived G10L ribosome-binding site and its translation start site is thereby maintained. One of the advantages of such a vector is that it directs the synthesis of a full-length "recombinant" polypeptide rather than one which represents a fusion between bacterial and eukaryotic protein sequences.

The bacterial strain used for expression of the recombinant protein must be selected with several caveats in mind. First, the induction system used must be compatible with the bacterial phenotype. For example, a *recA⁻* strain cannot be used if the *E. coli recA* promoter is utilized. Second, *E. coli* proteases such as the product of the *lon* gene (protease La) may cause proteolysis of some foreign proteins. Expression of these proteins may be improved by using strains that are protease deficient (e.g., Lon⁻ and Htpr⁻). Third, variables that can affect the efficiency of

---

[10] A. R. Shatzman and M. Rosenberg, this series, Vol. 152, p. 661.

[11] P. O. Olins, C. S. Devine, S. H. Rangwala, and K. S. Kavka, *Gene* **73**, 227 (1988).

[12] Examples of currently available vectors that are suitable for production of "nonfusion" proteins in *E. coli* include pBTac2 and pBTrp2 (Boehringer Mannheim Biochemicals); pNH18A (Stratagene); and pKK233-2 and pPL Lambda (Pharmacia LKB Biotechnology).

[13] X. Soberón, L. Covarrubias, and F. Bolivar, *Gene* **9**, 287 (1980).

production of recombinant proteins include incubation temperature, growth medium, cell density, timing of induction, and the length of fermentation after induction of foreign protein synthesis. Finally, it should also be noted that *E. coli* cannot support many critical posttranslational modifications of mammalian proteins (e.g., glycosylation).

Growth of Escherichia coli Transformed with pMON–CRBP or pMON–CRBP II

We have not found it necessary to utilize protease-deficient hosts for the expression of CRBP and CRBP II. For most purposes, we use *E. coli* strain JM101 transformed with pMON-based vectors (Monsanto). A fresh overnight culture of *E. coli* JM101 containing the recombinant pMON vector is diluted 1:50 in fresh Luria broth plus ampicillin (100 μg/ml) (pMON contains an *Amp*^{r} locus). After achieving a cell density equivalent to an OD_{600} of approximately 0.2 nm, nalidixic acid is added (1:200 dilution of a 10 mg/ml stock prepared in 0.1 *N* NaOH) to activate the *recA* promoter of the plasmid. Fermentation is continued with agitation for an additional 3 hr at 30° (at which time the OD_{600} is ~1.0). Cells are then harvested by centrifugation at 5,000 *g* or by filtration through a Millipore (Bedford, MA) Pellicon cassette (the Pellicon cassette is useful for harvesting cells from large-scale cultures, i.e., 20–100 liters). The cell paste is stored at −70°.

Purification of Escherichia coli Expressed Rat CRBP and CRBP II

Although foreign proteins that are expressed in *E. coli* may be soluble, many are present as insoluble aggregates known as inclusion bodies (reviewed by Marston[14]). Recovery of active proteins from these aggregates requires isolation of the inclusion bodies following cell lysis, subsequent denaturation to solubilize the protein, and finally a refolding reaction. Because soluble proteins can be purified directly, they are more likely to retain their native conformation. At the conclusion of the fermentation described in the preceding section, rat CRBP or CRBP II represents 10–15% of the total *soluble* proteins present in *E. coli* homogenates.

The purification of rat apo CRBP and apo CRBP II from *E. coli* is remarkably straightforward and relatively rapid, requiring two or three steps.[15,16] Our protocol is summarized in Table I. An aliquot of the *E. coli*

---

[14] A. Marston, *Biochem. J.* **240**, 1 (1986).
[15] E. Li, B. Locke, N. C. Yang, D. E. Ong, and J. I. Gordon, *J. Biol. Chem.* **262**, 13773 (1987).
[16] M. S. Levin, B. Locke, N. C. Yang, E. Li, and J. I. Gordon, *J. Biol. Chem.* **263**, 17715 (1988).

TABLE I
PURIFICATION OF RAT APO-CRBP AND APO-CRBP II FROM *Escherichia coli* LYSATES

1. Thaw cell pellet and resuspend in 2–3 volumes cold lysis buffer (see text).
2. Disrupt bacteria with a French press (~2000 psi).
3. Clear lysate by centrifugation at 5000 g for 10 min (4°).
4. Ammonium sulfate fractionation: Drip in an equal volume of cold filtered 100% saturated ammonium sulfate (Whatman) containing 15 mM dithiothreitol (pH adjusted to 7.1 with KOH) over 4 hr (keep on ice). Centrifuge as above to obtain a 50% ammonium sulfate supernatant. For CRBP II only, bring the 50% supernatant to 70% saturation by the addition of solid ammonium sulfate. Centrifuge as above to obtain the 50–70% pellet. Suspend this pellet in lysis buffer.
5. Dialyze the ammonium sulfate fractions (~0.1–0.2 liters) overnight at 4° against ~20 liters potassium phosphate (20 mM, pH 7.1), EDTA (1 mM), 2-mercaptoethanol (10 mM), sodium azide (0.05%), phenylmethylsulfonyl fluoride (0.02%, pH 7.1).
6. Concentrate the postdialysis preparations, if necessary, to a final protein concentration of ~5 mg/ml with an Amicon YM10 membrane.
7. Perform gel-filtration column chromatography with Sephadex G-50. Screen fractions by absorbance at 280 nm and by SDS–PAGE. Pool fractions containing CRBP or CRBP II.
8. If the acyl carrier protein of *E. coli* is present (detected by SDS–PAGE and amino-terminal amino acid sequencing) or if separation of Met$^+$ from Met$^-$ CRBP is desired, then proceed to FPLC, Step 9.
9. Perform FPLC with a Mono Q (Pharmacia) column equilibrated with imidazole (20 mM, pH 6.8) and eluted with a NaCl gradient (see Fig. 1B for details).

cell paste is first thawed to room temperature and resuspended in 2–3 volumes of lysis buffer [Tris (final concentration 50 mM, pH 7.9), sucrose (10%), and phenylmethanesulfonyl fluoride (0.5 mM)]. Bacteria are disrupted by passage through a French Press (~2000 psi). Cellular debris is removed by centrifugation at 5000 g for 10 min at 4°. Ammonium sulfate fractionation of the supernatant fraction is a very useful first step in the purification: most bacterial proteins are insoluble in solutions of NH$_4$SO$_4$ at 50% saturation, in contrast to the cellular retinol-binding proteins which remain soluble. Thus, the supernatant is adjusted slowly to 50% saturation with NH$_4$SO$_4$ (with constant stirring at 0–4°). After gentle stirring for an additional hour, the suspension is subjected to centrifugation at 41,000 g for 30 min at 4°. An additional 70% NH$_4$SO$_4$ "cut" is also beneficial when purifying CRBP II.

The over 50% NH$_4$SO$_4$ supernatant containing CRBP or the 50–70% NH$_4$SO$_4$ precipitate containing CRBP II is then dialyzed overnight at 4° against buffer A [potassium phosphate (20 mM, pH 7.1), EDTA (1 mM), glycerol (15%), sodium azide (0.05%), phenylmethanesulfonyl fluoride (0.5 mM), and 2-mercaptoethanol (10 mM)]. The dialyzed protein solution is then fractionated through a Sephadex G-50-80 column equilibrated with the same buffer. Column fractions are assayed for protein by measuring the absorbance at 280 nm. Selected fractions are then surveyed by electrophoresis through a 15% polyacrylamide gel containing 0.1% sodium dodecyl sulfate (SDS–PAGE). Typically, CRBP and CRBP II elute from the gel-filtration column at a relative retention volume (V_e/V_o) of 1.5 (see Fig. 1A).

When purifying CRBP (but not CRBP II) an additional purification step is desirable to avoid contamination with the acyl carrier protein of *E. coli*. The acyl carrier protein is a soluble 8.8-kDa protein which migrates on SDS–PAGE as an approximately 20-kDa protein and elutes like a 12-kDa protein from Sephadex G-50 (reviewed by Rock and Cronan[17]). Those fractions which appear to be enriched for CRBP by SDS–PAGE are pooled and further purified by fast protein liquid chromatography (FPLC). The pooled fractions are applied to a Pharmacia (Piscataway, NJ) Mono Q column equilibrated with imidazole (20 mM, pH 6.8) and sodium azide (0.05%). CRBP is eluted with a continuous NaCl gradient (15.5–25 mM) in the same buffer. Two CRBP peaks elute at approximately 19 and 20 mM NaCl (see Fig. 1B). These peaks correspond to CRBP without and with its initiator methionine residue, respectively. The proportion of Met$^+$ CRBP and Met$^-$ CRBP varies from protein preparation to preparation. The initiator methionine of foreign proteins expressed in *E. coli* may be removed coincidently, or shortly after synthesis, by a nonspecific aminopeptidase.[18,19] The efficiency of methionine removal from recombinant proteins is variable and affected by the physicochemical properties of the penultimate amino acid.[18,19] The spectrofluorimetric and ligand binding properties of Met$^+$ and Met$^-$ CRBP appear to be identical.[16] For some applications, such as protein crystallization and NMR studies, it may be desirable to separate the two species (see below).

Protein integrity and purity can be assessed by SDS–PAGE, isoelectric focusing, and automated sequential Edman degradation. Unlike the forms recovered from mammalian cells or cell-free translation systems (wheat germ or reticulocyte lysates), *E. coli*-derived rat CRBP and CRBP II do not have an acetyl group linked to their amino-terminal amino acid.[15,16] Therefore, the intact *E. coli*-derived proteins can be sequenced directly.

Purified *E. coli*-derived rat apo-CRBP and apo-CRBP II are stored in plastic tubes at 4° in phosphate buffer (20 mM, pH 7.4) supplemented with 2-mercaptoethanol (1 mM), EDTA (1 mM), and NaN$_3$ (0.05%). We have not found aggregation to be a problem even at protein concentrations as high as 15 mg/ml (~1 mM).

[17] C. O. Rock and J. E. Cronan, Jr., this series, Vol. 71, p. 341.
[18] J. L. Brown, *Biochim. Biophys. Acta* **221**, 480 (1970).
[19] J. L. Brown and J. F. Krall, *Biochem. Biophys. Res. Commun.* **42**, 390 (1971).

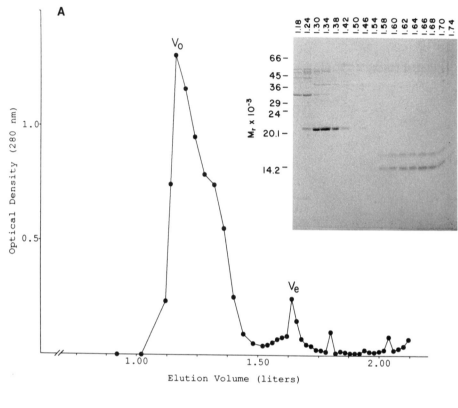

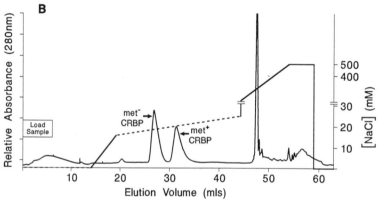

FIG. 1. Purification of rat CRBP from *E. coli* lysates. (A) Fractionation of ammonium sulfate fractions by Sephadex G-50 chromatography. A 5 × 150 cm Sephadex G-50 column was pretreated with 11 mg of bovine serum albumin (dissolved in 25 ml of buffer A, described in the text) and then equilibrated with several column volumes of buffer A. Approxi-

Quantitation

A reliable measure of CRBP and CRBP II concentration is required for many experiments (e.g., determination of ligand binding constants and stoichiometries). We have determined that the molar extinction coefficients for *E. coli*-derived CRBP and CRBP II are 28,080 and 25,506 M^{-1} cm^{-1}, respectively.[15,16] The concentrations of solutions of purified CRBP and CRBP II determined using these values are within 5% of those derived using a protein assay kit (Bio-Rad, Richmond, CA) based on the procedure of Bradford[20] with bovine serum albumin as a standard.

Conformational Analysis

The spectrofluorimetric properties of rat CRBP and CRBP II purified from liver/intestine have been well characterized.[2,4] Our observation that purified *E. coli*-derived CRBP and CRBP II do not absorb above 310 nm confirms that they do not contain any endogenous (bound) retinol.[15,16] Moreover, when excitation and emission spectra are collected from *E. coli*-derived CRBP and CRBP II complexed with a molar excess of all-*trans*-retinol, the results are virtually identical to those obtained from the respective rat liver and intestinal holoproteins (see Fig. 2). These latter observations suggest that purified *E. coli*-derived rat holo-CRBP and holo-CRBP II have folded into a (overall) conformation which is similar to that of the "authentic" tissue proteins.

[20] M. M. Bradford, *Anal. Biochem.* **72**, 248 (1976).

mately 100 ml of the postdialysis 50% NH$_4$SO$_4$ fraction containing around 360 mg protein was subsequently applied. Fractions (20 ml) were collected at 4° (flow rate 20 ml/hr). The OD$_{280}$ readings of selected fractions are shown plotted against the total elution volume. Twenty-five microliters of selected Sephadex G-50 fractions were reduced, denatured, and then fractionated by electrophoresis through a 15% polyacrylamide gel containing sodium dodecyl sulfate (0.1%). The results obtained after Coomassie blue staining of this gel are shown in the inset. The elution volume of each of the fractions is labeled. The position of migration of authentic rat liver CRBP is noted by the arrow. M_r markers are indicated at left. (B) Separation of Met$^+$ and Met$^-$ CRBP by fast protein liquid anion-exchange chromatography. Sephadex G-50 fractions containing CRBP were pooled. They were then equilibrated with imidazole buffer (20 mM, pH 6.8, plus 0.05% sodium azide) and concentrated using a Centriprep 10 concentrator (Amicon). Six milliliters of this protein solution (~5 mg) was loaded onto a 5 mm × 5 cm Mono Q HR 5/5 column (Pharmacia). The protein solution was fractionated using the NaCl gradient shown (flow rate 1 ml/min; temperature 22–25°). The OD$_{280}$ was monitored continuously.

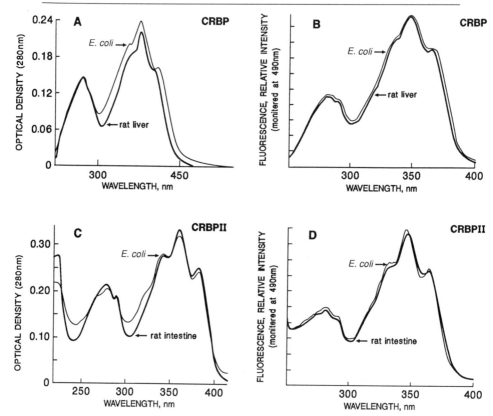

FIG. 2. Comparison of the spectral properties of *E. coli*-derived rat CRBP and CRBP II complexed with all-*trans*-retinol with the corresponding authentic tissue-derived proteins. Approximately 1.5 to 1.8 μM solutions of CRBP or CRBP II were used, and all measurements were made at 25°. *Escherichia coli*-derived CRBP or CRBP II are represented by thin lines, whereas rat liver CRBP or rat intestinal CRBP II are denoted by thick lines. (A, C) Absorption spectra of CRBP and CRBP II, respectively, complexed with all-*trans*-retinol. These spectra were taken on a Perkin-Elmer Lambda 5 UV-Vis spectrophotometer (slit width 2 nm). (B, D) Corrected fluorescence excitation and emission spectra of CRBP and CRBP II, respectively, complexed with all-*trans*-retinol. These spectra were obtained using a Perkin-Elmer MPF-66 spectrofluorimeter equipped with corrected spectra units and constant temperature cell holders.

Ligand–Protein Interactions

Spectrofluorimetric methods described by Cogan *et al.*[21] can be used to compare the affinities and binding stoichiometries of *E. coli*-derived apo-CRBP and apo-CRBP II for a variety of ligands (see Refs. 15, 16, and

[21] U. Cogan, M. Kopelman, S. Mokady, and M. Shinitzky, *Eur. J. Biochem.* **65,** 71 (1976).

TABLE II
PHYSICOCHEMICAL PROPERTIES OF *Escherichia*
coli-DERIVED RAT CRBP AND CRBP II

Property	CRBP[a]	CRBP II[a,b]
pI	4.73	5.61
	4.78	5.66
E_{280}^M (M^{-1} cm^{-1})	28,080	25,506
Fluorescence (maxima)		
Excitation (nm)[c]	282	282
Emission (nm)[d]	338	336
Stoichiometry of ligand binding	1 : 1	1 : 1
all-*trans*-Retinol binding (K_d', nM)	20	10
all-*trans*-Retinal binding (K_d', nM)	50	90
all-*trans*-Retinal competes with	No	Yes
all-*trans*-retinol for binding		
all-*trans*-Retinoic acid binding	No	No
Methylretinoate binding	No	No

[a] M. S. Levin, B. Locke, N. C. Yang, E. Li, and J. I. Gordon, *J. Biol. Chem.* **263**, 17715 (1988).
[b] E. Li, B. Locke, N. C. Yang, D. E. Ong, and J. I. Gordon, *J. Biol. Chem.* **262**, 13773 (1987).
[c] Emission: λ = 340 nm.
[d] Excitation: λ = 290 nm.

Table II). When all-*trans*-retinol is bound to CRBP or CRBP II, it has a distinct absorption spectra and exhibits enhanced fluorescence emission compared with that of unbound all-*trans*-retinol.[2,4] Quantitative binding studies can be done by exploiting these differences. Apo-CRBP and apo-CRBP II both contain four tryptophan residues in comparable positions (9, 89, 107, and 110). When the apoproteins are excited at around 290 nm, they demonstrate fluorescence with an emission maxima near 340 nm. However, when saturated with all-*trans*-retinol, 90% of the intrinsic protein (tryptophan) fluorescence is quenched. Because the efficiency of energy transfer is dependent on the distance between the tryptophan residue(s) and the bound retinol (αr^{-6}), this would suggest that at least one tryptophan residue in each of these proteins is located near the binding site. The capacity of nonfluorescent substances to bind to CRBP and CRBP II can also be assessed by monitoring their ability to quench the native fluorescence of these proteins.

Ligand binding studies have shown that both apo-CRBP and apo-CRBP II bind all-*trans*-retinal as well as all-*trans*-retinol with high affinity (Table II).[16] Neither can bind all-*trans*-retinoic acid nor methyl retinoate, an uncharged analog of retinoic acid.[16] The latter observation suggests that the inability to bind all-*trans*-retinoic acid is not simply due to the negative charge of the C-15 carboxylate group. Although both proteins can bind all-*trans*-retinal with high affinity in direct binding assays, monitoring of retinol fluorescence revealed that all-*trans*-retinal could displace all-*trans*-retinol bound to CRBP II but not to CRBP.[16,22] These observations raised the possibility that CRBP and CRBP II do not complex to all-*trans*-retinal in the same way, despite the fact that their apparent K_d values are essentially identical (Ref. 16 and below).

Isotopic Labeling of Retinoid-Binding Proteins Expressed in *Escherichia coli* for Nuclear Magnetic Resonance Analysis

NMR spectroscopy provides information about the structural environment, chemical properties, and motional characteristics of defined functional groups in protein molecules. Detection of site-specific changes in chemical shifts in the NMR spectra can be used to follow protein conformational changes during processes such as ligand binding, protein–lipid interactions, and denaturation–renaturation.

Although considerable information about protein structure and dynamics has been obtained by the use of ^{1}H NMR, assignment of resonances is a formidable task in the size range of the intracellular retinoid-binding proteins (~15 kDa). Isotopic labeling of a particular amino acid with a variety of magnetic nuclei (^{2}H, ^{13}C, ^{15}N, and ^{19}F) has been used to simplify the NMR spectrum ("spectral editing") of a number of proteins.[23] Incorporation of isotopically labeled amino acid analogs can be facilitated by selecting strains of *E. coli* which are auxotrophs for the amino acid of interest and using them to express foreign proteins.

[19]F nuclei are very useful as NMR probes. Their advantages include 100% natural abundance, a sensitivity close to that of protons, a large chemical shift range, and the absence of background signals. ^{19}F NMR spectroscopy represents an alternative to fluorescence spectroscopy for studying CRBP/CRBP II–retinoid interactions.

Escherichia coli provide a straightforward way of obtaining large quantities of the ^{19}F-labeled cellular retinoid-binding proteins. To do so, a tryptophan auxotroph (*E. coli* strain *w3110 trp A33*)[24] is transformed with the recombinant pMON expression vector containing CRBP or CRBP II cDNA. An overnight culture of transformed cells is diluted 1:15 in M9 medium[25] supplemented with 0.25% glucose, 1% casamino acids, 0.1%

[22] P. N. MacDonald and D. E. Ong, *J. Biol. Chem.* **262**, 10550 (1987).
[23] J. L. Markey and E. L. Ulrich, *Annu. Rev. Biophys. Bioeng.* **13**, 493 (1984).
[24] G. R. Drapeau, W. J. Brammer, and C. Yanofsky, *J. Mol. Biol.* **35**, 357 (1968).

vitamin B_1, 5 mg/liter $FeCl_3$, 0.4 mg/liter $ZnSO_4$, 0.7 mg/liter $CoCl_2$, 0.7 mg/liter Na_2MoO_4, and 0.1 mM L-tryptophan and incubated at 30°. After achieving an OD_{600} of 1.0, cells are harvested by centrifugation at 5000 *g* and resuspended in the above medium except that 6-fluorotryptophan (final concentration 0.1 mM) is substituted for L-tryptophan. The cells are incubated at 37° for 30 min, and expression of the cellular retinol-binding protein is induced by adding nalidixic acid to a final concentration of 50 μg/ml.

Following a 3-hr fermentation, cells are harvested, and the 6-fluoro-tryptophan-substituted rat cellular retinol-binding protein purified as detailed above. The purified protein solution is adjusted to 1 mM by ultra-filtration through YM10 filters (Amicon, Lexington, MA) and then equilibrated with D_2O by four cycles of buffer exchange using Centriprep 10 concentrators (Amicon). The efficiency of incorporation of fluorotryptophan analogs into CRBP or CRBP II is high, greater than 90%. This value was obtained by comparing the NMR signal intensities of known amounts of the fluorotryptophan-labeled protein and trifluoroacetic acid (the latter is included in the sample as an internal reference standard. The presence of analog does not alter the ligand binding properties of the protein as measured by fluorescence spectroscopic methods. The choice of amino acid analog (for example 4-fluoro-, 5-fluoro-, or 6-fluorotryptophan) is empirical and based on the resolution of spectral signals obtained from individual groups with a given analog. With CRBP II, a 1.5-liter culture yields sufficient amounts of purified 6-fluoro-tryptophan-substituted protein (5–10 mg) to obtain high-quality spectra (0.8 ml of a 1 mM solution). At this protein concentration, using a Varian VXR500 spectrometer, as few as 200–300 transients can be collected to obtain spectra with an excellent signal-to-noise ratio. Spectra can be obtained (with a larger number of transients) using solutions that are 10-fold more dilute.

Figure 3 shows uncoupled 470.3-MHz ^{19}F NMR spectra of a 1 mM solution of 6-fluorotryptophan-substituted *E. coli*-derived rat CRBP II in the absence of ligand and fully saturated with all-*trans*-retinol. Resonances (labeled A and B) corresponding to two of the tyrptophan residues (W_A and W_B), undergo large downfield changes in chemical shifts (2.0 and 0.5 ppm, respectively) associated with ligand binding. In contrast, the resonances labeled C and D, corresponding to two other tryptophan residues (W_C and W_D), undergo only minor perturbations in chemical shifts. The fact that only two of the four resonances arising from the four

[25] T. Maniatis, E. F. Fritsch, and J. Sambrook, "Molecular Cloning: A Laboratory Manual." Cold Spring Harbor Laboratory, Cold Spring Harbor, New York, 1982.

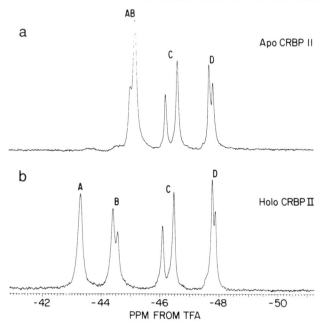

FIG. 3. ^{19}F NMR spectra (recorded at 470.3 MHz) of 6-fluorotryptophan-labeled *E. coli*-derived rat CRBP II with and without bound all-*trans*-retinol. (a) Spectrum collected on a Varian VXR 500 spectrometer using a 1 mM solution of the apoprotein at 22° (1344 transients). (b) Spectrum obtained at 22° with a 1 mM solution of protein after incubation with 1.1 mM all-*trans*-retinol (896 transients). A line broadening of 4 Hz was applied. The chemical shifts are referenced to the ^{19}F signal for trifluoroacetic acid.

tryptophan residues are sensitive to the binding of all-*trans*-retinol provides a functional assay for (1) examining the conformational effects of the binding of a series of retinoids with different structures to a given cellular retinoid-binding protein and (2) comparing the effects of binding a given retinoid on the structures of several retinoid-binding proteins. For example, comparison of the ^{19}F NMR spectra of 6-fluorotryptophan-substituted CRBP and CRBP II with and without bound all-*trans*-retinol and all-*trans*-retinaldehyde can provide an independent evaluation of the fluorescence spectroscopic data alluded to above, which suggested that all-*trans*-retinaldehyde interacts with each protein in a unique way.

It is important to note that the ^{19}F resonances can be extremely sensitive to amino-terminal heterogeneity in the protein sequence. The relative intensities of the two signals comprising resonances B, C, D in Fig. 3 correspond to the relative amounts of two protein species: Met$^+$ CRBP II, which has the initiator methionine, and Met$^-$ CRBP II, with an amino-terminal Thr residue (the second amino acid residue in the primary translation product). Thus, because of the sensitivity of ^{19}F NMR, care must be taken to purify Met$^+$ from Met$^-$ CRBP II.

The ^{19}F NMR spectra will be much easier to interpret once each resonance (W_A–W_D) can be definitively assigned to each of the four tryptophan residues in the cellular retinol-binding protein and when the tertiary structure of these proteins are known. The assignments may be facilitated by systematic site-directed mutagenesis of CRBP/CRBP II cDNA so that conservative substitutions are made for each Trp residue (e.g., replacement with another aromatic amino acid such as Phe). A comparison of the ^{19}F NMR spectra of the 6-fluorotryptophan-labeled wild-type and mutant proteins could then be undertaken. Resonances corresponding to the substituted tryptophan residue would be predicted to be absent in the ^{19}F NMR spectra of the mutant protein.

Crystallization of Cellular Retinol-Binding Proteins

Both bovine liver CRBP[26] and *E. coli*-derived CRBP II[27] have been crystallized, although the tertiary structures have not yet been solved. The conditions used to crystallize the *E. coli*-derived protein without bound retinol are remarkably similar to those used to crystallize the homologous CRBP. A 10 mg/ml solution of protein is prepared in a buffer containing 30 mM Tris, 75 mM NaCl, 2 mM EDTA, 2 mM dithiothreitol, and 0.05% sodium azide (final pH 8.0). Six microliters of this CRBP II preparation is mixed with an equal volume of a buffer containing 37% polyethylene glycol 4000 (Baker), 2 mM cadmium acetate, 2 mM EDTA, 2 mM dithiothreitol, and 0.05% sodium azide (final pH 7.9). Mixing takes place on a dimethyldichlorosilane-treated glass coverslip. The drop is then inverted over a diffusion chamber containing the polyethylene-containing buffer. Crystals usually appear within 1–2 weeks when using the hanging drop vapor diffusion method. The CRBP II crystals are triclinic with a P_1 space group. The unit cell contains two monomeric copies of the 134-residue protein. Its dimensions are as follows: $a = 36.8$ Å, $b = 64.0$ Å, $c = 30.4$ Å, $\alpha = 92.8°$, $\beta = 113.5°$, $\gamma = 90.1°$. Solution of the structures of crystalline CRBP and CRBP II should be aided by the recent determination of the structures of two homologous intracellular hydrophobic ligand binding proteins, namely, rat intestinal fatty acid-binding

[26] M. E. Newcomer, A. Liljas, U. Erikkson, L. Rask, and P. A. Peterson, *J. Biol. Chem.* **256**, 8162 (1981).
[27] J. C. Sacchettini, D. Stockhausen, E. Li, L. J. Banaszak, and J. I. Gordon, *J. Biol. Chem.* **262**, 15756 (1987).

protein[28] and the P2 protein from bovine peripheral nerve.[29] Comparison of the structures of apo-CRBP II/CRBP with that of the corresponding holoproteins should provide important insights about the mechanisms of ligand binding and the effects of the bound retinoid on protein structure.

Acknowledgments

Work from our laboratories cited in this chapter was supported by grants from National Institutes of Health (DK 30292), the Lucille P. Markey Charitable Trust Foundation, and the Monsanto Company. E.L. is a Lucille P. Markey Scholar. J.I.G. is an Established Investigator of the American Heart Association. We would like to acknowledge the contributions of Nien-chu C. Yang and Bruce Locke (University of Chicago), David Ong (Vanderbilt University), James C. Sacchettini, Andre d'Avignon, and Shi-jun Qian (Washington University), plus Leonard J. Banaszak (University of Minnesota) and Peter Olins (Monsanto) to various aspects of these studies.

[28] J. C. Sacchettini, J. I. Gordon, and L. J. Banaszak, *J. Mol. Biol.* **208**, 327 (1989).
[29] T. A. Jones, T. Bergfors, J. E. Sedzik, and T. Unge, *EMBO J.* **7**, 1597 (1988).

[59] NAD+-Dependent Retinol Dehydrogenase in Liver Microsomes

By M. A. Leo and C. S. Lieber

Introduction

The classic pathway for the conversion of retinol to retinal in the liver involves a cytosolic NAD+-dependent retinol dehydrogenase (CRD), believed to be similar, if not identical, to liver alcohol dehydrogenase (ADH, alcohol : NAD+ oxidoreductase, EC 1.1.1.1).[1,2] Our previous observation that a strain of deermice lacks this enzyme without apparent adverse effects[3] prompted a search for an alternate pathway for the production of retinal, the precursor of retinoic acid. Evidence was obtained in favor of the existence of an NAD+-dependent microsomal retinol dehydrogenase (MRD)[4] which can convert retinol to retinal using NAD+ as a cofactor. It is distinct from the cytochrome P-450 microsomal system on the one

[1] R. D. Zachman and J. A. Olson, *J. Biol. Chem.* **236**, 2309 (1961).
[2] E. Mezey and P. R. Holt, *Exp. Mol. Pathol.* **15**, 148 (1971).
[3] M. A. Leo and C. S. Lieber, *J. Clin. Invest.* **73**, 593 (1984).
[4] M. A. Leo, C. I. Kim, and C. S. Lieber, *Arch. Biochem. Biophys.* **259**, 241 (1987).

hand[5] and the cytosolic NAD$^+$-dependent retinol dehydrogenase (CRD) on the other hand.

Incubation Procedures

Microsomes are obtained with standard procedures.[5] The substrates are prepared as follows: for the MRD reaction, all-*trans*-retinol RO-1-4955, kindly provided by Dr. P. F. Sorter, Hoffman-LaRoche, Inc. (Nutley, NJ) or obtained from Sigma (St. Louis, MO), is dissolved in absolute ethanol with butylated hydroxytoluene (BHT) (1 mg/ml) at a concentration of 10–15 mg/ml. An aliquot is concentrated under nitrogen to a volume of about 15 μl. Then bovine serum albumin (BSA) (fraction V), 3.2 mg/ml, is added to the retinol, mixed vigorously, and subjected again to a stream of nitrogen for an additional 20 min. When retinal is used as a substrate, it is prepared as indicated above for retinol.

Microsomes (0.5 mg protein/ml) are incubated at 37° with retinol (0.15–0.4 mM) under air. Substrate inhibition is observed with higher concentrations. The reaction is performed in duplicate or triplicate in disposable borosilicate glass culture tubes (13 × 100 mm). A convenient total incubation volume is 0.4 ml. The reaction mixture contains 5 mM MgCl$_2$, 1 mM EDTA, in 0.1 M Tris-HCl, 0.15 M KCl buffer, pH 7.4 (at 37°). After a 3-min preincubation, the reaction is started with the addition of the cofactor, 3 mM NAD$^+$. In the absence of cofactor or with boiled microsomes, no significant reaction is detected; a small reaction is observed at 0 time, and this is used routinely as a blank. After a 10-min incubation, the reaction is stopped by adding 0.5 ml of cold absolute ethanol followed by rapid freezing in dry ice–acetone. Samples are extracted 3 times with 2 ml hexane containing BHT (1 mg/ml), vortexed well, and centrifuged for 10 min at 2000 g. After freezing in dry ice–acetone, the supernatant is rapidly decanted into a second set of disposable tubes, and the three combined supernatants are completely evaporated under nitrogen, resuspended in 200–500 μl of 2-propanol/water (9 : 1), and 50–100 μl is injected for high-performance liquid chromatographic (HPLC) analysis.

Chromatography Procedures

The HPLC analyses can be carried out with a Hewlett Packard HP-1090 liquid chromatograph equipped with a diode-array spectrophotometric detector, and a HPLC Chemstation or with a HP-1090L, equipped

[5] M. A. Leo and C. S. Lieber, this volume [54].

with a diode-array detector, a HP-3392A integrator, a Think-jet printer, and a HP-7440A Plotter (Hewlett-Packard, Avondale, PA). A Zorbax ODS column (0.46 × 15 cm) (Mac-Mod Analytical Inc., Chadds Ford, PA) is used for which the mobile phase is programmed with a gradient from 25% mobile phase B (75% A) to 90% mobile phase B (10% A) in 10 min and continued for another 8 min at a flow rate of 1.5 ml/min. The mobile phase A is acetonitrile/water/acetic acid (49.75 : 49.75 : 0.5, v/v/v) and the mobile phase B is acetonitrile/water/acetic acid (90 : 10 : 0.04, v/v/v), both containing 10 mM ammonium acetate as reported else-

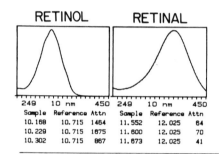

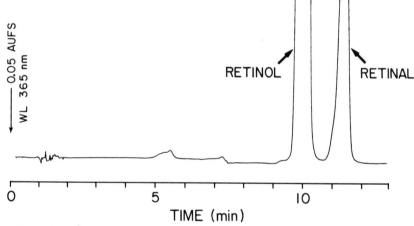

FIG. 1. Representative HPLC chromatogram of retinal production in liver microsomes. Microsomes were incubated with retinol (0.15 mM) and NAD$^+$ (3 mM) as described in the text. Retinal was formed and clearly separated from retinol as shown in the chromatogram. Absorption spectral analyses (inset), obtained from upslope, apex, and downslope of the two compounds, show perfect superposition of the three readings, suggesting purity. The absorption spectra maxima were at 325 and 379 nm, as for the pure standards of retinol and retinal, respectively. [From M. A. Leo, C. I. Kim, and C. S. Lieber, *Arch. Biochem. Biophys.* **259**, 241 (1987).]

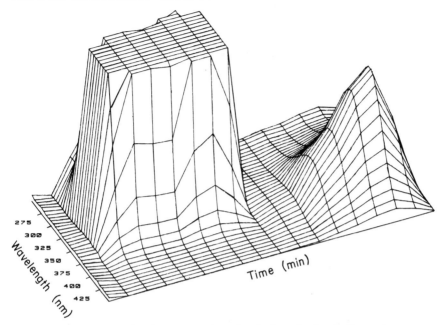

FIG. 2. Three-dimensional representation of retinal production in liver microsomes. Retinal (right-hand side) shows a typical spectral maximum at 379 nm, whereas the substrate retinol (left-hand side) has a maximum at 325 nm.

where.[6] Simultaneous multiwavelength detection at 325, 350, and 365 nm is usually obtained for all samples.

Enzyme Activity

Rat liver microsomes incubated with NAD$^+$ at pH 7.4 actively convert retinol to retinal, identified by HPLC (Figs. 1 and 2). Enzyme activity is measured by the amount of total retinal produced per minute per milligram of protein. No retinoic acid is detected under the conditions used. Optimal activity is observed with NAD$^+$ (3 mM): with retinol as a substrate, the enzyme shows saturation kinetics with a K_m of 0.12 ± 0.01 mM and a V_{max} of 3.3 ± 0.3 nmol/min/mg protein ($n = 7$) in chow-fed animals. With NADP$^+$, the K_m value remains the same but V_{max} is lower (1.3 ± 0.1 nmol/min/mg protein, $n = 4$). The activity is negligible with NADPH. This MRD activity is maximal at physiological pH. The reaction has no

[6] M. A. Leo and C. S. Lieber, *J. Biol. Chem.* **260**, 5228 (1985).

dependence on oxygen, and the activity is the same under air and under nitrogen. It is not inhibited by CO. Tested under optimal conditions, the reaction is linear up to 10 min, and it is proportional to the protein concentration up to 0.5 mg/ml. With retinal (0.2 mM) as substrate, retinol is produced by liver microsomes; with 3 mM NADH, the activity is 2.9 ± 0.1 nmol/min/mg protein ($n = 9$). In the presence of chloral hydrate (10 mM) this activity is modestly decreased to 2.4 ± 0.1 (17 ± 1% inhibition; $p < .001$), and with 3 mM NADPH as cofactor the activity is somewhat lower than with NADH.

Factors Affecting MRD Activity

Ethanol (50 and 100 mM) has no effect on MRD activity. Similarly, 4-methylpyrazole (1 mM) is not inhibitory. This lack of effect of ethanol and 4-methylpyrazole is strikingly in contrast with the marked inhibition of the cytosolic retinol dehydrogenase (CRD) by these compounds.[4] MRD activity is inhibited 61.8 ± 1.8% by Triton X-100 (0.1%, v/v).[4] Similarly, sodium deoxycholate (0.1%, v/v) reduces MRD activity by 91.1 ± 0.4% and Lubrol PX (0.1%, v/v) by 57.7 ± 0.7%. Other detergents such as sodium cholate (0.1%, v/v), CHAPS {3-[(3-cholamidopropyl)dimethylammonio]-1-propane sulfonate} (0.1%, v/v), and BigCHAP [N,N-bis(3-D-gluconamidopropyl)cholamide] (0.1%, v/v) are not inhibitory; in fact, BigCHAP may even exert moderate stimulatory effect (10–20%). Interassay variability does not exceed 2–3%. Repeated freezing and thawing strikingly affect the enzyme activity.

Conclusion

A microsomal NAD$^+$-dependent retinol dehydrogenase (MRD) has been described with optimal activity at physiological pH. The enzyme is present in liver microsomes of rats and also in a strain of deermice which lacks the cytosolic retinol dehydrogenase[4] (CRD). Unlike the latter enzyme, MRD is not inhibited by either ethanol or 4-methylpyrazole; its activity is insensitive to CO and is not oxygen dependent, in contradistinction with that of the microsomal cytochrome P-450 and NADPH-dependent retinol oxidase.[5]

Acknowledgments

Original studies referred to in this chapter were supported, in part, by U.S. Department of Health and Human Services Grants DK 32810, AA 03508, and AA 05934 and the Veterans' Administration. The authors are grateful to Dr. C. Kim for her contribution to the preparation of the manuscript, to Ms. N. Lowe for her expert technical assistance, and to Ms. R. Cabell and P. Walker for skillful typing of the paper.

[60] Metabolism of Retinoic Acid and Retinol by Intact Cells and Cell Extracts

By MARY LOU GUBLER and MICHAEL I. SHERMAN

Introduction

Retinoic acid (RA) is a potent promoter of differentiation of embryonal carcinoma (EC) cells. It is extensively metabolized by EC cells, and this metabolism is often inducible.[1-3] Studies have suggested that RA itself rather than a metabolite actively promotes differentiation; thus, the role and nature of the RA-metabolizing enzyme(s) remain to be elucidated. Unlike RA, retinol is relatively poorly metabolized, at least by cultured murine EC and fibroblast cells.[2-4] The major products appear to be fatty acid esters,[1,2] although it has been reported that there are low levels of conversion of retinol to RA in F9 cells.[1] Because of the increasing use of retinoids to modulate cell behavior (including the promotion of differentiation and/or the inhibition of proliferation of many cell types), there is a growing utility for assays designed to evaluate retinoid metabolism. We have used the following procedures to investigate the metabolism of retinoids in EC cells,[2,3,5] and to a lesser extent fibroblasts,[4] both *in vivo* and *in vitro*.

Analysis of Retinoid Metabolism by Intact Cells

Retinoic Acid

The assay for RA metabolism by intact cells can be carried out with radiolabeled or unlabeled RA. In our experience unlabeled RA is metabolized to products which are undetectable by absorbance. Unlabeled RA is, therefore, less useful as a substrate in such assays than is radioactive RA, and the latter is preferred particularly for quantitative determinations. We have found tritiated RA (all-*trans*-[11,12-³H]RA) to be particularly useful in our studies. This compound has been synthesized at a specific activity

[1] J. B. Williams and J. L. Napoli, *Proc. Natl. Acad. Sci. U.S.A.* **82,** 4658 (1985).
[2] M. L. Gubler and M. I. Sherman, *J. Biol. Chem.* **260,** 9552 (1985).
[3] M. I. Sherman, M. L. Gubler, U. Barkai, M. I. Harper, G. Coppola, and J. Yuan, *Ciba Found. Symp.* **113,** 42 (1985).
[4] J. Rundhaug, M. L. Gubler, M. I. Sherman, W. S. Blaner, and J. S. Bertram, *Cancer Res.* **47,** 5637 (1987).
[5] M. L. Gubler, P. Thomas, W. Levin, and M. I. Sherman, manuscript in preparation.

of 20–36 Ci/mmol at Hoffmann-La Roche (Nutley, NJ); [³H]RA is also available commercially. The purity of labeled retinoids should be routinely analyzed by HPLC prior to use. Preparations less than 90% pure should be repurified or discarded.

Assay. Since retinoids are in general light-sensitive, all operations should be carried out under amber lights. RA is added to cell cultures in ethanol so that the final concentration of vehicle is less than 0.5%. Radioactive RA is added at a final concentration of 10^{-7} M and 2 μCi/ml. Unlabeled RA can be added at concentrations as high as 10 μM; higher concentrations can be toxic and, in our experience, exceed the metabolic capacities of the cultured cells, making metabolism difficult to detect by disappearance of substrate. Cells (0.5×10^6–2×10^7) are incubated at 37° in 5 ml of fetal calf serum-containing medium plus RA for various time periods. If the cells are to be tested for induction of RA metabolism (see Induction of Retinoic Acid Metabolism), the incubation period should be 12–24 hr in duration to allow for adequate levels of enzyme induction and activity. Control samples, consisting of RA in medium without cells, should always be included during the incubation period. After incubation, the medium from the cells is removed and saved for extraction (since RA metabolites appear to be rapidly excreted). The monolayer is washed once with phosphate-buffered saline (PBS) and then dislodged with trypsin–EDTA or scraping followed immediately by extraction as described below.

If one wishes to follow the fate of RA at various times after it enters cells, pulse experiments can be performed by minor modification of the method described above: cells are incubated for a relatively short time (e.g., 5 hr) with labeled RA in order to load the cells, and the retinoid-containing culture medium is subsequently removed. The cell monolayer is washed twice with PBS, and fresh culture medium is added for further incubation periods followed by extraction of cells and medium as described above. If desired, after incubation with RA, cells can be fractionated by conventional means and analyzed for subcellular localization of RA and its metabolites.

Extraction of Retinoids. We have routinely used the extraction procedure of McClean *et al.*[6] for recovery of retinoids from tissue culture medium and cells, and we have found the recoveries of labeled retinoids to be essentially complete. The method is as follows: to a glass 12 × 75 mm tube is added 0.5 ml of cells or medium, 25 μl acetonitrile containing an internal standard (unlabeled if radioactive retinoids are used in the

[6] S. W. McClean, M. E. Rudel, E. G. Gross, J. J. DeGiovanna, and G. L. Peck, *Clin. Chem.* **28**, 693 (1982).

assay or vice versa, in order to monitor recovery and to provide a marker for the HPLC profile), and 350 μl acetonitrile/butanol (50:50, v/v). The mixture is vortexed thoroughly for 30 sec. After addition of 300 μl of a saturated (1 g/liter) K_2HPO_4 solution and thorough mixing, the sample is centrifuged for 10 min at 3000 g. The upper organic layer is transferred to an injector vial for HPLC analysis (see below). Samples can be stored at $-20°$ for up to 1 week before HPLC analysis.

Metabolism of Retinol and Other Retinoids by Intact Cells

Determination of retinol metabolism by intact cells is carried out essentially as described for RA. [H^3]Retinol has also been prepared at Hoffmann-La Roche at a specific activity in the range of 20–36 Ci/mmol. The conditions of the assay, the extraction, and the analysis are as for RA, except that different HPLC conditions must be used to elute retinol esters from the column (see below). It is important to note that nonenzymatic oxidative products of retinol are often formed during the incubation period, and some of these are similar to RA in terms of chromatographic behavior even though they are not acidic retinoids.[2] Therefore, if the objective of the experiment is to determine whether a particular cell type can convert retinol to RA, it is essential to confirm the identity of any putative RA by other means, for example, by derivatization (see below).

It should be possible to evaluate the metabolism of any retinoid by cultured cells in the same way as has been described above. We have, for example, studied the metabolism of, and induction of metabolic activity by, a number of retinoids, including the active arotinoid Ro 13-7410.[3,5] It is obviously necessary to define the chromatographic behavior of each retinoid prior to initiation of the study. Also, if the retinoid in question is unavailable in radioactive form, the compound must have an extinction coefficient high enough so that the retinoid can be followed chromatographically after extraction from cells and culture medium.

Retinoid Metabolism in Cell Extracts

Preparation of Cell Extracts

After investigating features of retinoid metabolic activity *in vivo*, one may wish to investigate further the regulation of retinoid metabolism or the nature of the enzyme(s) involved. We have adopted a procedure, first described by Roberts *et al.*[7] for RA metabolism in hamster tissues, to

---

[7] A. B. Roberts, C. A. Frolik, M. D. Nichols, and M. B. Sporn, *J. Biol. Chem.* **254**, 6306 (1979).

investigate cell culture extracts. Monolayers are washed twice with PBS and removed from the flask with trypsin–EDTA. The trypsin is inactivated with fetal calf serum, and the cell suspension is centrifuge at 500 g for 5 min. The cell pellet is washed twice with PBS, resuspended in 10 mM Tris, pH 7.4, 2 mM MgCl$_2$, and homogenized with a Dounce homogenizer containing a B Pestle. The homogenates are frozen in an ethanol–dry ice bath and stored at $-70°$.

Assay Conditions

A standard reaction mixture contains, in a total volume of 0.5 ml, 50 mM Tris (pH 7.4), 1 mM NADPH, 5 μg bovine serum albumin, 0.15 M KCl, 5 mM MgCl$_2$, 0.25 ml homogenate (containing 0.3–0.5 mg protein), and 25 pM of [11,12-^{3}H]RA added in 5 μl ethanol just prior to incubation. NADPH can be replaced with NADP plus a NADPH-regenerating system (10 mM glucose 6-phosphate and 2.5 units glucose-6-phosphate dehydrogenase). The reaction should be carried out in duplicate at 37° in 12 × 75 mm glass tubes for 30 min in a shaking water bath. Appropriate control samples (e.g., omission of homogenate or NADPH) should be included. Incubation is terminated by the addition of 1.5 μg unlabeled RA in 25 μl acetonitrile. The mixture is immediately extracted as described above.

It should be possible to determine the metabolism of RA analogs by extracts from induced cells in the same way as described for RA, provided that the same enzymes are involved in the metabolism. We have evidence to suggest that some retinoids might be metabolized by different enzyme systems,[5] but we have not sought conditions in cell extracts in which this metabolism can be followed. Similarly, we have not attempted to optimize conditions for conversion of retinol to RA by cell extracts as we have been unable to detect such a conversion in those intact cells which we have studied.

Induction of Retinoic Acid Metabolism

Retinoic acid and other acidic retinoids are metabolized extensively by cultured EC cells and murine fibroblasts.[1-5] However, kinetic studies demonstrate that in most instances the activity of the enzyme(s) involved can be increased by prior exposure to RA or other retinoids: for example, there is little RA metabolism after 5 hr of exposure to the retinoid, whereas after 24 hr RA metabolism is very extensive or complete.[2] In cell extracts we have been unable to detect RA metabolism unless the intact cells are pretreated for at least 4 hr with RA.[5] The kinetics of induction of RA metabolism may vary among cell lines and under differing culture conditions. Finally, since retinoids which are capable of inducing the metabolic enzyme system do not necessarily serve as substrates for the induced enzymes and vice versa,[5] it is possible to assay for each of these properties independently by preincubating cells with the retinoid of choice, preparing cell extracts, and then testing for metabolic activity with [³H]RA or another labeled retinoid.

Chromatographic Analysis of Retinoids

Chromatography Conditions

The HPLC system used routinely in our analyses of retinoids consists of a Model 110A solvent metering pump (Altex Scientific Inc., Berkeley, CA), a WISP autoinjector (Waters, Milford, MA), a Holochrome detector (Gilson) set at the optimal wavelength for the particular retinoid substrate (e.g., 360 nm for RA), and a 3390 A reporting integrator (Hewlett Packard, Avondale, PA). Injected volumes are 40 μl or less. Separations are carried out under isocratic conditions on a 4.6 mm i.d. × 15 cm reversed-phase Zorbax ODS column (DuPont). In our standard conditions, the mobile phase is acetonitrile/water (80:20, v/v) containing 0.1% ammonium acetate (final pH 6.7). The flow rate is 1.5 ml/min. Under these conditions all-*trans*-RA elutes at about 9.5 min and retinol at about 11.5 min. To resolve more polar compounds such as 4-hydroxy-RA or 4-keto-RA, which elute near the front in this system, the acetonitrile proportion can be decreased to 60%. When metabolites of retinol are being analyzed, the percentage of acetonitrile can be increased to 98% to elute retinoids esterified with fatty acids. In quantitate radioactive retinoids, eluted fractions of 0.5- or 1.0-min duration are collected and measured in 10 ml of Formula 989 with a liquid scintillation spectrometer. Calculations of metabolic activity should take into account nonenzymatic alteration of substrate (determined from control samples incubated without cell homogenate).

Derivatization

When it is important to further characterize the products formed after incubation of retinoids with whole cells or cell extracts, derivatization can be performed on experimental samples to confirm the presence of carboxyl, hydroxyl, or carbonyl residues. Methyl acid esters can be prepared with ethereal diazomethane as described by Vane et al.[8] (Note that diazomethane is potentially explosive. Operations with the reagent should be

[8] F. M. Vane, C. J. L. Bugge, and T. H. Williams, *Drug. Metab. Dispos.* **10**, 212 (1982).

carried out in a chemical hood. Ethereal diazomethane should be stored at $-20°$, and it should be discarded when the color of the solution becomes deep yellow.) To 100 μl extracted sample is added 150 μl methanol plus 1 ml ethereal diazomethane. After incubation at room temperature for 5 min, the sample is evaporated under a stream of nitrogen and the residue is dissolved in acetonitrile. Hydroxylated retinoids can be acetylated by treating 50 μl of extracted sample with 100 μl acetonitrile containing 10 μg dimethylaminopyridine plus 50 μl acetic anhydride. After 60 min on ice, the solution is evaporated to dryness with nitrogen, and the residue is dissolved in acetonitrile for HPLC analysis. In order to derivatize retinoids with keto or aldehyde residues, the methoxyamine-HCl procedure described by Vane et al.[8] is used as described. After derivatization, the samples are dried under a stream of nitrogen and redissolved in acetonitrile in preparation for HPLC analysis.

When samples containing labeled retinoids are being derivatized, appropriate unlabeled internal standards (including the substrate) should be added to the samples prior to derivatization so that the extent of derivatization, and position of migration, of the substrate can be monitored by absorbance in the HPLC profile. Under the standard HPLC conditions described above, methylated retinoic acids will be strongly retained by the column and will not be eluted until approximately 30 min unless the acetonitrile/water ratio is increased. A retinoic acid with derivatized hydroxyl or carbonyl groups should elute later than the unmodified retinoid but earlier than free RA.

[61] Retinoic Acid Formation from Retinol and Retinal Metabolism in Epidermal Cells

By Georges Siegenthaler

Introduction

Cultured epidermal cells (keratinocytes) offer an attractive model for studying the metabolism of natural retinoids in a target tissue since they contain the enzymatic system that transforms retinol into retinoic acid.[1] This system was found in the cytosol of differentiating human cultured keratinocytes[1] and in psoriatic plaques,[2] whereas normal human skin[2] and

[1] G. Siegenthaler, J.-H. Saurat, and M. M. Ponec, Biochem. J. 268, 371 (1990).
[2] G. Siegenthaler, D. Gumowski-Sunek, and J.-H. Saurat, J. Invest. Dermatol., in press (1990).

nondifferentiated cultured keratinocytes[1] showed only trace amounts of activity. This metabolic pathway seems to be different from other alcohol and aldehyde dehydrogenases as the formation of retinoic acid was not significantly affected by specific inhibitors of alcohol metabolism.[1] In contrast to other studies performed in murine tissues[3,4] in which nonphysiological amounts of substrate (retinol 10 μM–10 mM) were used, the rate of retinoic acid formation from retinol in epidermal cells was determined using physiological levels of retinol (0.6 μM), which excludes nonspecific reactions through alcohol dehydrogenases.

We demonstrated that when retinol is oxidized to retinoic acid, retinal is an intermediate metabolite which is released into the medium.[1] Retinal used as substrate can be reduced to retinol in the presence of NADH and oxidized to retinoic acid in the presence of NAD. A schematic representation of retinoids metabolism in epidermal cells is shown. Thus, retinal (β-carotene cleavage metabolite) could play a central role in the formation of either retinol or retinal in retinoid epidermal cells. Cultured human keratinocytes are thus a useful model for the measurement of retinoid metabolism for researchers interested either in modulation of keratinocyte differentiation, screening of new drugs, or studying the fate of synthetic retinoids in epidermal cells.

$$\text{ROL} \underset{\text{NADH}}{\overset{\text{NAD}}{\rightleftharpoons}} \text{RAL} \xrightarrow{\text{NAD}} \text{RA}$$

Methods

To study the formation of retinoic acid from retinol, the procedure consists of incubating the protein extract with tritiated retinol in the presence of NAD and subsequently extracting the radioactive material with hexane before analyzing by reversed-phase HPLC. Retinal metabolism is studied in the same way using tritiated retinal instead of retinol. Application of the oxidation route is performed in the presence of NAD, whereas reduction of retinal requires the reduced cofactor NADH. The sensitivity of the method is about 100 fmol per assay when using a specific activity of approximately 50 Ci/mmol for the retinoid. All retinoids solutions are handled under yellow light.

Human Cultured Keratinocytes

Keratinocytes, isolated from human foreskin, are cultured in the presence of lethally irradiated 3T3 fibroblasts using the Rheinwald–Green feeder technique.[4] The culture medium is a mixture of Dulbecco–Vogt

[3] J. L. Napoli and K. R. Race, *Arch. Biochem. Biophys.* **255**, 95 (1987).
[4] M. J. Connor and M. H. Smit, *Biochem. J.* **244**, 489 (1987).

and Ham's F12 (DVH) medium (3:1) supplemented with 5% fetal calf serum (FCS), 0.4 μg/ml hydrocortisone, 1 μM isoproterenol,[5] and 10 ng/ml epidermal growth factor (EGF).[6] The enzymatic activity can be slightly increased by separating the differentiated keratinocytes from the nondifferentiated cells lacking the enzymatic system for transforming retinol into retinoic acid. This separation can be performed when the cells grown to confluency at normal calcium (see above) are switched to low-calcium medium/calcium-free DVH (3:1) supplemented with 5% Chelex-treated FCS for 2–3 days.[7] The final calcium concentration should be approximately 60 μM, as determined by flame photometry. The differentiating cells are detached from the attached (nondifferentiated) cells by shaking the flask strongly and horizontally. The detached cells (differentiating) are collected by centrifugation and washed 3 times with ice-cold phosphate-buffered saline (PBS). When the whole population is used (not separated) they are washed 3 times with PBS before being scraped off in a minimum of distilled water. The cells are stored at −70° and lyophilized. The contamination of cell culture by 3T3 fibroblasts is negligible.

Preparation of Protein Extracts

Approximately 35 mg of lyophilized cells or tissues is homogenized in 600 μl of 50 mM Tris-HCl buffer containing 25 mM NaCl, 2.5 mM EDTA, 1 mM dithiothreitol (DTT), corrected to pH 7.5. A maximum activity is obtained with a pH of 8.5, but this is unphysiological. The homogenization is carried out with Polytron PT7 by three strokes at full speed for 30 sec, at 0°. Debris is removed by ultracentrifugation at 100,000 g for 1 hr at 4°. The supernatant is immediately used or frozen until use.

Preparation of Tritiated Retinal from Tritiated Retinol

The alcoholic solution of 0.25 ml of [11,12-^{3}H(N)]retinol (0.25 mCi, 42.5 Ci/mmol; Du Pont de Nemours, Paris, France) containing 1 mg/ml α-tocopherol is dried under a stream of N_2. The residue is dissolved in hexane and the solvent transferred to a small screw-capped glass tube. The solvent is again totally evaporated, and 80 μl of a saturated solution of hexane with retinoic acid (this solution is used as carrier), 240 μl hexane, and 2.4 mg of MnO_2[8] (Merck, Darmstadt, FRG) are added. The tubes are flushed with N_2 to remove air, closed with the cap, and gently agitated for 8 hr at room temperature. The solution is then filtered onto a

[5] J. G. Rheinwald and H. Green, *Cell (Cambridge, Mass.)* **6**, 331 (1975).
[6] J. G. Rheinwald and H. Green, *Nature (London)* **265**, 421 (1977).
[7] J. K. Brennan, J. Monsky, L. G. Roberts, and M. A. Lichtman, *In Vitro* **11**, 354 (1975).
[8] S. Ball, T. W. Goodwin, and R. A. Norton, *Biochem. J.* **42**, 516 (1948).

glass pipette with a glass wool filter (Whatman GF/C glass microfiber filter), and the filtered solution is evaporated down to 20 μl before being purified by HPLC (see below).

The elution profile is monitored at 313 nm, and the peak corresponding to retinal is collected in a tube containing 50 μl of a solution of butylated hydroxytoluene (BHT) (5 μg/ml). The organic solvent is evaporated and [³H]retinal is extracted from the aqueous phase with hexane (3 times, 2 ml each) and placed in a small brown glass tube. Hexane is again evaporated, and [³H]retinal is dissolved in absolute ethanol containing 50 μg/ml of BHT to give a final concentration of [³H]retinal of approximately 5 μM.

Enzyme Assays

All retinoids (labeled or unlabeled) are dissolved in absolute ethanol containing 50 μg/ml of BHT as antioxidant and stored at $-20°$. The required volume of alcoholic solutions of [³H]retinol or [³H]retinal ($\sim$5 μl) for a final concentration of 600 nM is deposited in a glass microtube and the solvent is evaporated under N_2. The microtubes are then placed on ice; subsequently, 80 μl of protein extract (0.15–0.4 mg protein) and 20 μl of a solution of NAD (to give a final concentration of 2 mM NAD) are added, and the mixture is incubated for 1 hr at 37° in a shaking water bath. The tubes are stoppered with a sheet of laboratory film (Parafilm). When [³H]retinal reduction is studied, NAD is replaced by a 20 μl solution of NADH at a final concentration of 5 mM.

The compounds to be tested (e.g., inhibitors or stimulators of the enzymatic system) are dissolved in the incubation buffer or in ethanol and are added and mixed before the addition of the cofactor. A blank containing only the solvent should be tested, and care should be taken not to produce micelles in the reaction mixture when the alcoholic solution of very hydrophobic compounds is added to the reaction medium (a cloudy solution may allow precipitation of the enzymes).

Analyses of Retinoid Metabolites by Chromatography

The enzymatic reaction is terminated by adjusting the pH of the medium to 6 (this must be checked) by addition of 8 μl of 0.1 M HCl and 100 μl of ethanol containing BHT. Subsequently, 10 μl of each of the 0.1 mM standard solutions (4-oxoretinoic acid, all-*trans*-retinoic acid, 13-*cis*-retinoic acid, retinol, and retinal) is added, and the retinoids are extracted 3 times with 1 ml of hexane. Diethyl ether, chloroform, and dichloromethane as extraction solvents are not advised, because they dramatically increase the polar metabolites of retinoids. The organic phase is collected in a conical glass tube and evaporated to dryness under N_2. The residue is dissolved in 50 μl of acetonitrile and applied to an HPLC column. Our HPLC system includes a Varian 9050 solvent delivery system, a variable-wavelength UV–visible detector, and an integrator (SP 4290 from Spectra-Physics, San Jose, CA). The radioactivity of the eluant is measured after collecting 600-μl fractions with a fraction collector. Separation takes place on a 10-μm reversed-phase Ultrasyl-ODS column (octadecasilyl-substituted column, 25 × 0.46 cm from Beckman, Geneva, Switzerland) with a guard column of LC_{18} (2 cm long, Superguard, Supelco, Inc., Bellafonte, PA). Retinol, retinal, retinoic acid, 13-*cis*-retinoic acid, and 4-oxoretinoic acid separation is achieved using an isocratic elution system composed of 60% of solvent A [50 mM ammonium acetate (pH 7), 10% (v/v) tetrahydrofuran, and 52% acetonitrile (v/v)] and 40% of solvent B (100% acetonitrile). The separation of peaks can be adjusted by slightly modifying the percentage of solvents A and B. The flow rate is adjusted to 2.4 ml/min, and nonlabeled retinoids are detected at 340 nm. The eluate of the column is collected in 600-μl fractions in counting vials, and the radioactivity is determined after addition of 3 ml of Pico-Fluor 15 (Packard).

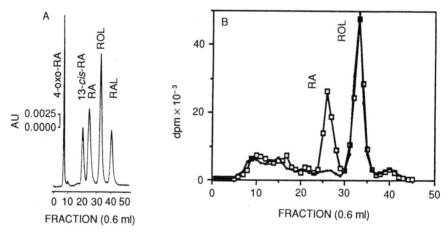

FIG. 1. Metabolism of [^{3}H]retinol in differentiating keratinocytes analyzed by HPLC. Aliquots of cytosol (180 μg protein in 100 μl buffer) were incubated with 600 nM [^{3}H]retinol in the presence of 2 mM NAD for 1 hr at 37°. Internal standards were added to the sample before the extraction of retinoids, and the extracts were chromatographed as described in the text. Absorbance was monitored at 340 nm; fractions (0.6 ml) were collected and the radioactivity determined. (A) Profile of the internal standards 4-oxo-*trans*-RA, 13-*cis*-RA, RA, retinol, and retinal; (B) radioactivity profile of retinoids in the cytosolic fraction, boiled (■) and untreated (□). (Reproduced with permission from Ref. 1.)

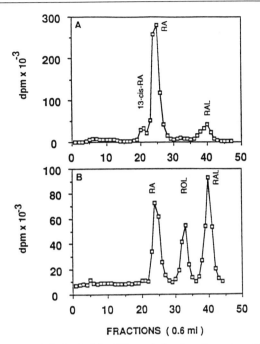

FIG. 2. Metabolism of [³H]retinal in differentiating keratinocytes. Aliquots of cytosol (100 μl containing 150 μg of protein) were incubated with 600 n*M* [³H]retinal for 1 hr at 37° in the presence of (A) 2 m*M* NAD or (B) 5 m*M* NADH. Samples were extracted and subjected to HPLC analysis, as described in the text. (Reproduced with permission from Ref. 1.)

Results

A typical chromatogram of the separation of internal standards measured at 340 nm is shown in Fig. 1A, and the radioactive elution profile when [³H]retinol and NAD are used as substrates is shown in Fig. 1B. When [³H]retinal and NAD are used as substrates the elution profile of the oxidation metabolites is as shown in Fig. 2A. The radioactive peak of 13-*cis*-retinoic acid at fraction 21 corresponds to partial isomerization during the extraction procedure. [³H]retinal reduction to retinol occurs when NAD is replaced by NADH (Fig. 2B). During retinal reduction, NADH is oxidized to NAD, which in turn can be utilized for the oxidation of retinal to retinoic acid. This explains why, in the presence of NADH, retinoic acid was also detected. The enzyme activities in the cytosol of differentiating keratinocytes are rather low when compared to other murine tissues but appear to have greater specificity for retinoids. Thus, the rate of conversion of retinol to retinoic acid is found to be 4.49 ± 0.17 pmol/hr/

mg protein.[1] Retinal is oxidized to retinoic acid at a rate of 51.6 pmol/hr/ mg protein and reduced at 8.2 pmol/hr/mg protein.[1]

Conclusion

The enzymatic system presented here is useful for the study of natural retinoid metabolism in normal and psoriatic plaques.[2] Moreover, we can demonstrate that retinoic acid and synthetic analogs such as nonretinoid compounds downregulate the transformation of retinol to retinoic acid (G. Siegenthaler, unpublished observations). This may be pertinent to the understanding of some pharmacological actions of synthetic retinoids and to the development of new agents designed to alter endogenous levels of retinoids in a tissue.

Acknowledgments

This work was supported in part by the Swiss National Science Foundation (Grant 3.874.0.88). I gratefully acknowledge R. Hotz and E. Leemans for excellent technical assistance, Dr. E. Johnson for rereading and S. Deschamps for typing the manuscript.

Author Index

Numbers in parentheses are footnote reference numbers and indicate that an author's work is referred to although the name is not cited in the text.

Subject Index

A

U

V

X

Z